ZABABAKHIN SCIENTIFIC TALKS - 2005

To learn more about the AIP Conference Proceedings, including the Conference Proceedings Series, please visit the webpage **http://proceedings.aip.org/proceedings**

ZABABAKHIN SCIENTIFIC TALKS - 2005

International Conference on
High Energy Density Physics

Snezhinsk, Russia   5 – 10 September 2005

EDITORS
Evgeniy N. Avrorin
Vadim A. Simonenko
*Russian Federal Nuclear Center -
Zababakhin All-Russia Research
Institute of Technical Physics
Snezhinsk, Russia*

SPONSORING ORGANIZATIONS
Federal Agency for Atomic Energy of the Russian Federation
Russian Federal Nuclear Center - Zababakhin All-Russia Research Institute
  of Technical Physics (Snezhinsk)
Russian Federal Nuclear Center - All-Russia Research Institute of Experimental Physics (Sarov)
Lebedev Institute of Physics of the Russian Academy of Sciences (Moscow)
Ural Branch of the Russian Academy of Sciences (Yekaterinburg)

75 Years of Service

Melville, New York, 2006
AIP CONFERENCE PROCEEDINGS ■ VOLUME 849

Editors:
Evgeniy N. Avrorin
Vadim A. Simonenko

Federal State Unitary Enterprise
Russian Federal Nuclear Center - Zababakhin All-Russia
Research Institute of Technical Physics
13 Vasiliev Street
Snezhinsk, 456770
Chelyabinsk Region
Russian Federation

E-mail: e.n.avrorin@vniitf.ru
        v.a.simonenko@vniitf.ru

L.C. Catalog Card No. 2006929840
ISBN 0-7354-0345-7
ISSN 0094-243X

Printed in the United States of America

CONTENTS

PLENARY

SECTION 1

CUMULATIVE PHENOMENA AND HIGH-INTENSIVE PROCESSES

SECTION 2

EXPLOSION AND DETONATION PHENOMENA

SECTION 3

DENSE PLASMA PHENOMENA AND INTENSIVE
ELECTROMAGNETIC PROCESSES

SECTION 5

PROPERTIES OF MATTER AT HIGH-INTENSIVE PROCESSES

PREFACE

Zababakhin Scientific Talks is a traditional International Conference on High Energy Density Physics (HEDP). The Conference is named in memory of E.I. Zababakhin (1917—1985), Member of the USSR Academy of Sciences, who was a Scientific Director of the Russian Federal Nuclear Center — Zababakhin All–Russian Research Institute of Technical Physics (RFNC — VNIITF, Snezhinsk, Urals, Russia) for 25 years. E.I. Zababakhin had a wide range of scientific interests, which covered many fields of HEDP and hydrodynamics, including evolution of high intensive processes, shock and detonation wave physics, material properties under extreme conditions and phase transitions in dynamic processes, hydrodynamic instabilities and turbulent mixing, and inertial thermonuclear fusion. Of his special interest were so called cumulative processes with sharpening of the profiles and unlimited enhancing of energy density. E.I. Zababakhin made a great contribution into nuclear explosive technology development, creation of nuclear explosive devices for peaceful purposes, and peaceful applications of nuclear explosions.

The first Conference of this series (14—16.01.87) was initiated by RFNC — VNIITF as a memorial and supported by Russian Federal Nuclear Center — All-Russian Research Institute of Experimental Physics (RFNC — VNIIEF at Sarov). Mainly scientists of these Centers attended it. The addressed problems embraced many fields of high energy density physics, which deserved wide discussions, so the next Conference (16—19.01.90) was organized with participation of scientists from various scientific centers of the USSR. All the subsequent Conferences (13—18.01.92; 14—20.10.95; 21—25.10.98; 24—28.09.01; 08—12.09.03) were international.

The objective of the Conferences is to cover traditional and new fields of research in high energy density physics. The directions usually discussed include methods for experimental generation of high intensive processes and high energy density states and their potential applications in science and technology; displaying of these processes and states in laboratory experiments, technological applications, and natural phenomena including astrophysics. The list of topics scrutinized at the Conferences includes, but is not limited to, the following:

- Shock waves and high intensive processes, flows with sharpened profiles as convergent shock waves and collapsing cavities, various options to create implosive flows;
- High explosive properties and compositions; detonation and explosion phenomena both chemical and thermonuclear;
- Thermodynamic and transport properties of matter in high energy density states and high intensive processes, including experimental technology and theoretical model development and research applications;
- Dense plasma physics and plasma properties; intensive electromagnetic phenomena; strong pulsed magnetic fields;
- High power and high intensive laser beams and their interaction with matter;

- Pulsed high power electric systems and electric explosion of conducting sets, resultant magnetic fields and their interaction with matter;
- Hydrodynamic instability and turbulent mixing, turbulent matter and energy transfer under various conditions including presence of magnetic fields; and
- Mathematical models, algorithms and codes to simulate matter properties, processes and phenomena, their development and applications.

The 8th Zababakhin Scientific Talks took place in Snezhinsk on September 05—10, 2005. The Conference was organized by Federal Agency for Atomic Energy of the Russian Federation, RFNC — VNIITF as a part of 50th Anniversary of the Center and supported by RFNC — VNIIEF, Lebedev Institute of Physics (Moscow) and Ural Branch of Russian Academy of Sciences. More than 200 scientists from five countries presented more than 160 papers.

The Conference activity proceeded in plenary and six topical sessions, including
1. Cumulative phenomena and high intensive processes;
2. Explosions and detonation phenomena;
3. Dense plasma and intensive electromagnetic phenomena;
4. Hydrodynamic instability and turbulence;
5. Matter properties in high intensive processes;
6. Numerical simulations, algorithms, codes and exact solutions.

After the Conference the American Institute of Physics made a proposal to publish the materials in the series of proceedings of APS conferences. The ZST Organizing Committee is grateful for this proposal. We asked the authors to send their papers in format required by AIP. Most of the authors responded to this call and we have assembled a representative set of presented paper, which would enable a reader to estimate the scope and professional level of the Conference. We are grateful to the participants, who despite their busy schedules dedicated time to the additional preparatory work. We hope to be better prepared for such an opportunity at the next ZST Conference, which is scheduled on September 2007, the year of 90th jubilee of E.I. Zababakhin.

The papers in the Proceedings are organized according to the sections where they were presented, and in the sections — according to the scientific program of the Conference.

E.N. Avrorin,
Member of Russian Academy of Sciences,
ZST Organizing Committee Chair.

V.A. Simonenko,
Professor,
ZST Program Committee Chair.

PLENARY

High Intensive Processes and Extreme States of Matter: Achievements and Problems

V.A. Simonenko

Federal State Unitary Enterprise "Russian Federal Nuclear Center — Zababakhin All–Russia Research Institute of Technical Physics", 456770, Snezhinsk, Chelyabinsk region, Russia

Abstract. The paper briefly presents some main highlights of High Energy Density Physics (HEDP) achievements starting from its origin in the 1940s to the current time. A decisive role of high explosives (HE) is emphasized in studying high intensive processes and high energy density states of matter. Mechanisms of detonation and kinetics of energy release still remain acute in the HE studying. Research and scientific applications of nuclear explosions opened a new stage in HEDP development. They provided a million-fold increase of energy density if compared to that of high explosives. High intensive heat waves and strong shock waves were studied and used to measure dense plasma opacities and matter properties under extreme conditions. This data remains important for the development of theoretical models of matter. Powerful pulsed facilities (lasers, electric explosion installations, and charged particle accelerators) were constructed to extend opportunities for the HEDP research. One of their main goals is to study inertial confinement fusion. HEDP technologies and results are very useful in space and astrophysical research, and on the contrary, astrophysical studies enrich HEDP with new models, problems and solutions.

INTRODUCTION

One of the main directions of modern physics is to study processes and phenomena with high intensive flows and high energy density. They are of great basic and applied importance. On one hand, they open new opportunities for laboratory study and deepen the understanding of a wide range of natural phenomena; while on the other hand, they stimulate the development of new technologies. When combined, they are often referred to as a new branch of physics – high energy density physics (HEDP).

This direction emerged in the middle of the last century when experiments were started on dynamic loading of dense materials with the help of high explosives (HE). However, its roots go back to the earlier stages of the science evolution. The primary purpose of these experiments was the fast assembling and compression of components made of fissile materials (plutonium and/or uranium) to reach the condition enabling development of a chain reaction. To meet this goal, it was necessary to understand a process of HE detonation, to know equations of state (EOS) of detonation products (DP) and structural materials, and to understand development of flows with sharpening profiles (implosive or cumulative flows) and to use them. A wide scope of theoretical models, experimental methods and registration equipment were developed, a great number of experiments and numerical simulations were conducted. This stage resulted in the creation of nuclear explosive devices (NED). Research and scientific

CP849, *Zababakhin Scientific Talks - 2005,*
edited by E. N. Avrorin and V. A. Simonenko
© 2006 American Institute of Physics 0-7354-0345-7/06/$23.00

applications of the nuclear explosions opened a new stage in HEDP research, dealing with a millions-fold increase of energy density if compared to that of high explosives.

However, nuclear explosive devices usually have a large energy yield. Therefore, even their scientific applications result in large-scale destructions, e.g. in the surrounding rocks. For this reason an important step in HEDP development was the creation of laboratory facilities, which can provide a higher energy density than that of HE. Among them, the most advanced are high power laser facilities, installations for sharpened electric explosions of wire arrays, and high intensive charged particle beams. They provide new research opportunities and state new problems. One of the most popular is inertial confinement fusion (ICF), which has the goal to burn thermonuclear fuel (at first, a mixture of heavy isotopes of hydrogen: D, T) in a small pellet to avoid strong destructions as in nuclear explosions.

Scientific technology and the results of high energy density physics are widely used in space physics and astrophysics, and on the contrary, astrophysical studies enrich the HEDP with new models, problems, and solutions. For example, the EOS for matter of Earth mantel and core, for matter of other planets were constructed on the basis of shock wave data for corresponding materials. Physical models and mathematical codes developed for traditional HEDP applications, being extended and adapted to astrophysical problems, are very useful in exploration of evolutionary and explosive phenomena of ordinary stars, dwarfs, and neutron stars. Theoretical models and codes developed for underground explosions were used for the problems of crater formation caused by collisions of space objects in the Solar System. Moreover, the future inevitable collisions of small space objects (asteroids and comets) with the Earth can be prevented in the frame of current technology with the help of nuclear explosions.

In parallel with basic and applied scientific applications, high intensive processes are applied in various technologies. The leading role is played by high explosives. Their applications have a one-and-a-half century history. A list of well known civilian and military applications can be extended with some new ones, such as explosive welding and punching, cutting of hard materials, and perforating. There was a successful program for peaceful nuclear explosions in our country, which was terminated after the signing of the Comprehensive Test Ban Treaty. There is an extension of powerful laser applications for the cutting and welding of materials, and the development of pulsed sources of electromagnetic radiation and charged particles. Electric explosions of conductors are being used in testing facilities for dynamic testing. High-intensive charged-particles beams can be used to provide volumetric distribution of released energy and material loading. Success of the ICF programs will open a promising perspective for energy generation.

1. HIGH EXPLOSIVES AND DETONATION

At the beginning of the 1940s, when there was a technological need for fast assembly and compression of fissile materials to reach above-critical states, high explosives attracted the attention of the scientists (e.g., [i]). They are usually metastable. However, when impacted by high pressure pulse or intensive energy flux, they start to release energy intensively, and the energy releasing region spreads as a detonation wave.

To use detonation, it was necessary to know the characteristics of this process, and the properties of detonation products. The important results were obtained on the eve of these events. A.A. Grib [ii] and Ya.B. Zeldovich [iii] (in our country), J. Von Neiman [iv] (in the USA) and W. Döring [v] (in Germany) independently proposed a phenomenological theory of detonation. According to this theory, matter is compressed at the lead front of the detonation wave primarily by the shock jump, which is followed by chemical decomposition in the subsequent layers. This process terminates at the Jouget point, where velocity of the detonation front is $D = u + c$, (u and c are local mass and sound velocities).

Interaction of the detonation wave with the adjacent matter is determined by pressure at the detonation front P_d and unloading process through release of detonation products (DP). At that time there was no experimental data for these characteristics. Estimates made by A. Schmidt [vi] and based on gas-like EOS for DP gave rather low values, e.g., for trinitritoluol $P_d = 12$ GPa.   L.D. Landau and K.P. Stanukovich [vii] proposed the use of liquid-like EOS for DP at high pressures, which gave more favorable $P_d = 19$ GPa for trinitrotoluol and power law $P \sim \rho^{n_{ef}}$ for DP release with exponent index $n_{ef} \simeq 3$.

So, a primary task was to experimentally estimate these parameters. It was done at the Institute of Experimental Physics in Sarov in 1947-48 with three independent methods (e.g., [viii]). The obtained data confirmed Landau's and Stanukovich's model, which was a very valuable result. The model stimulated the development of a more physically correct EOS for DP. It enabled the use of a simple form of EOS in labor-intensive numerical simulations, and stimulated the development of an experimental technique for accurate measurements of DP properties. The next step was to introduce a new generation of EOS for detonation products based on precise experimental data [ix, x, xi, xii]. This became possible in the 1960s-70s with the advent of a new generation of computers and numerical codes.

Subsequent use of HE for scientific research and applications stimulated studies of initial ignition of chemical reactions in hot spots, their kinetic and spreading. It is necessary to provide precise ignition and formation of the detonation front, to account action of DP on thin layers, to control sensitivity of new HE compositions, etc.

Two approaches can be identified in the development of kinetic models though there is no well defined boundary between them. The first one is a phenomenological approach, based on coarse simple models and using extended sets of experimental data to define functions and constants included in these models. The examples can be found in [xiii, xiv, xv]. One such model was proposed by specialists of our Center [xvi]. Another approach is based on the first principles. Development of such models was based essentially on the new theoretical models, like molecular dynamics, accurate quantum-mechanical description of molecular interactions and evolution of molecular structures, and ensured by new computational capabilities. One such model was proposed and worked out by K.F. Grebionkin and his colleagues [xvii]. They paid attention to the dielectric-to-conducting phase transition in material during the shock-induced compression. The conducting electrons strongly increased thermal conductivity, providing an extension of the reacting front.

So, high explosives remain important for high energy density physics. They are still the objects to be studied and are used as an essential component in the wide class of experimental assemblies to create high-intensive flows, and to study high energy density states.

2. STUDY OF DYNAMIC COMPRESSIBILITY OF DENSE MATTER

One of the most efficient scientific applications of high explosives is to study matter properties under dynamic processes. For this purpose, many different assemblies were developed, often called as explosive gasdynamic generators (EGG). Their design includes layers of high explosives and passive materials configured to ensure a necessary shape of the detonation wave, formation and movement of shock wave in passive materials, and acceleration of some passive layers when it is needed. EGGs enable creation of pressures up to ten times higher than that of the detonation wave front, acceleration of matter to the velocities up to ~ 30 km/s, and compression of matter to the densities up to ten times higher than that of normal matter. Another type of EGG was developed to provide dynamic compression of samples and the subsequent recovery of them. Spherical systems are of great value among the systems with recovery. They enable research under pressures of up to several terapascals. Analysis of the recovered samples gives information on the history of matter transformation in dynamic processes, evaluated with the help of numerical simulations. Each EGG was developed for a specific type of experiment. Some of them were used thousands of times, and other very expensive ones, were used only in several experiments. We'll touch upon just some of them.

One of the early examples of EGG is well known as gas-dynamic lenses, which transform a divergent detonation wave initiated by a detonator into a convergent or plane one. They are usually used to generate a plane or convergent spherical detonation wave in the adjoining layer of high explosives of similar configuration. Usually both lenses and HE layers are constituents of a more complex EGG designated for shock-wave (SW) research. The most widespread ones are plane SW generators. The simplest ones have a layer of material under study attached directly to the HE layer. The intensity of SW can be decreased by emplacement of an intermediate layer or increased with the help of some special layer or set of layers (pusher), accelerated by DP release, and hit against the layer of material under study. A set of such generators was developed by L.V. Altshuler and his colleagues at Sarov in late 1940s and early 1950s [viii]. A similar scheme can be used in the spherical geometry. However, in this case, the enhanced intensity of the convergent spherical flow helps to additionally increase the intensity of shock waves in the layers under study.

Shock waves provide especially valuable opportunity to study thermodynamic properties of dense materials at high pressure and multiple compression of matter. Laws of mass, momentum, and energy conservation have an algebraic form at the shock front. The first parameter to be measured is usually a shock front velocity D. It can be measured rather easily by recording the times of front arrival at certain points or surfaces. The second parameter can be mass velocity u, pressure P, or material density ρ. The two parameters, when known, determine all parameters of the flow after the jump:

$$\rho/\rho_0 = D/(D-u), \ P = \rho_0 u D, \ \varepsilon = (P/2)(\rho_0^{-1} - \rho^{-1}), \quad (0.1)$$

ρ_0 is initial density of material. On the other hand, these equations establish a relationship between the thermodynamic parameters of matter (except temperature T) known as shock adiabat or Hugoniot. A single experiment (or set of experiments for the same SW intensity) gives a single point for this law. A set of experimental points is usually used to determine parameters of phenomenological EOS of material under study.

To extend the study area of matter properties, it was proposed to compress porous materials. There will be a similar relationship for them, but initial density ρ_{00} in this case is not equal to material density at normal conditions. Studying matter within a wide range of porosities $k = \rho_0/\rho_{00}$ produces data for a wide range of thermodynamic parameters. However, porosity growth results in the decrease of pressures which can be reached with the developed experimental assemblies.

Direct measurement of the second parameter often faces difficulties. Depending on parameters and technique of the measurement, various experimental methods were developed. For relatively low pressures an artificial-spall method was developed, which doubles mass velocity of the free surface $w_{fs} = 2u$ on the SW arrival. It enables measurements of up to 50 GPa for materials such as iron.

The so-called method of obstacles became the next step. It is based on soft (without essential changing of initial state) acceleration of a layer (projectile) made of the material under study and its subsequent collision with a layer made of the same material. Accurate measurements of the projectile velocity $w_p = 2u$ before the collision, and SW velocity D in the sample give the result. The constraint is not to change the states during acceleration. For experimental assemblies with HE w_p should be less than velocity of DP outflow in vacuum, which is about $5 \div 6$ km/s. To meet this requirement, it was necessary to build large experimental assemblies with an intermediate layer between HE and projectile. In the case of spherical systems (even if the projectile layer is accelerated without destruction) the material changes state during the convergent movement. Some improvements can be introduced through adding a cascade of layers. This enables pressure of up to 1 TPa for iron. However, two cascade semispherical systems are rather complex and expensive. Experiments with them are rather rare. Two-cascade light gas guns overcome this restriction. Their limit for w_p is about 10 km/s. It is reasonable to use such facilities to accumulate accurate data for Hugoniot at high pressures (in particular, for quartz in the melting region).

The method of obstacles is rather expensive to be frequently used. Fortunately, there is one very valuable option, which facilitates data acquisition in SW research. If one has any material with the well known equation of state (standard material), it is possible to get data on SW compressibility of a new material by the measuring of SW velocity in just two layers, one of the standard material, another of the material under study, arranged one after another along the shock wave path. Measured D_e in the standard material and the known EOS give complete information about the material state at the shock front and determine loading and unloading $P_{e,l/u}(u)$ curves. Intersection of $P_{e,l/u}(u)$ and line $P_s = \rho_0 u_s D_s$ (subscript "s" here means the material under study) in P, u – plane gives the final state in the material under study. This method is called impedance-match or reflection (the latter is more frequently used in Russian).

There are two factors limiting application of this method: existence of standard material and feasibility of its application, and capability of an experimental assembly to reach the necessary level of loading. Besides, uncertainties in the EOS of the standard material contribute to an error. However, simplicity of the experimental technique made this method popular among the SW researchers and a great deal of SW data was acquired with its help.

What kind of data on the properties of dense matter is needed now?

When dealing with the theoretical models of matter at high pressures, it is reasonable to separate contributions from electrons and ions. As for an electron component, the quantum-statistical model with temperature and quantum correction (TFQ) [xviii] gives good thermodynamic data. V.P. Kopyshev [xix] proposed to account for contribution from ions at high temperatures in the approximation of non-ideal one-component plasma. However, there was a large gap between the data of these models and data of the laboratory experiments. So, one of the goals for further experimental study was to obtain SW data to cover this gap including compression of porous materials. For some materials, that gap was covered through the SW measurements carried out during underground nuclear explosions. However, there is still a need for similar data for other materials.

One more area for future studies is related to relatively low pressures of about $10 \div 100$ GPa. Most dense materials have polymorphic phase transitions in addition to melting and evaporation. Kinetics of these transitions become essential in dynamic processes. It makes shock-wave structures and relaxation processes behind the front pronounced, and sometimes breaks a single front into a set of fronts. Today there are no reliable theories to describe these processes. Phenomenological models are usually used based on some simple theoretical observations. To use these models, it is necessary to have lots of experimental data to define constants in the phenomenological relations. Fortunately, it turns out that at relatively low pressures; analogue methods can be developed to measure varying SW parameters: pressure, mass velocity, and velocity or displacement of free surface. Section 5 of the Conference discusses it in detail.

3. SHARPENING (OR CUMULATIVE) FLOWS

Problems of fast compacting of dense material drew scientists' attention to flows with central symmetry such as the convergence of spherical layers and focusing of spherical shock waves. The first is related to collapse of a spherical cavity in incompressible liquid studied by Rayleigh in 1917. An essential new point was in understanding that matter became compressible due to pressure growth during convergence (see, [i]). It was also realized that even a weak spherical acoustic wave during convergence increases its amplitude causing compression of matter. Therefore, compression of matter becomes essential for intensive flows of both types. At high-intensive flows, even dense matter behaves as a perfect gas in high energy density states.

There are self-similar solutions for cavity collapse and shock wave convergence [xx]. Their characteristic features are time-dependent linear scale and amplitudes of parameters, and conservation of profiles if normalized to the amplitudes and linear scale. In particular, matter density and pressure are tending to infinity in the case of cavity collapse, as well as temperature and pressure – in the case of shock wave convergence. It is clear that there should be physical limits for the growth of actual parameters, at least, due to the discrete structure of matter and maybe some other physical reasons, like the instability of specific flow and transition from hydrodynamic mechanism of energy transfer to another one. However, capability to attain high values of some physical parameters could be very useful in scientific research, and many researchers strive for that. These problems were deeply analyzed by E.I. Zababakhin [xxi].

Self-similar solutions to hydrodynamic equations were developed by K. Bechert [xxii] in 1940. Being asymptotic, they are not used in the designing of experimental assemblies. However, at the time when there were no computational techniques, they played an important role in understanding of the processes involving high energy density, pressure, and temperature. Moreover, a strong need for such solutions stimulated discovery of new classes of self-similar solutions. In particular, E.I. Zababakhin studied self-similar flows in periodic layered media, which were used in some experimental systems [xxiii]. Self-similar solutions with uniform deformation were discovered and studied, which appeared to be very valuable for some astrophysical studies [xxiv].

Self-similar solutions for cavity collapse and SW convergence are self-sharpening flows. Sharpening of the profiles is an internal property of the flow. In the recent decades, the world witnessed development of powerful experimental facilities with high energy flows: laser systems, pulsed electric current facilities, and high-current charged-particle accelerators. The facilities have some capability to vary energy pulse shape. This gave a push to studies of the flows with external sharpening driven through the boundary conditions. Such self-similar flows were studied, in particular, in [xxv]. They are also used as asymptotes for real flows.

Self-similar solutions were used to study an influence of real properties of matter on the evolution or destruction of the sharpening flows. In particular, influence of viscosity on cavity collapse was analyzed in [xxvi]. Both regimes (growth of amplitudes, and destruction) were shown to be feasible in viscous liquids. Shock wave convergence in materials with nonlinear heat conductivity was analyzed in [xxvii]. Non-linear heat transfer, if occurred, does not eliminate sharpening, but transforms it. Account of such effects in real applications is done usually with the help of rather sophisticated physical models and numerical codes. However, their complexity can conceal possible errors. So, the exact solutions and analytical studies still remain valuable for testing codes and models and for comprehension of general results.

4. STRONG EXPLOSIONS AND STUDY
OF MATTER PROPERTIES

A new stage in high energy density research was opened by use of the processes accompanying nuclear explosions. Nuclear reactions are characterized by a high level of energy released (about 1 MeV or higher per nucleon). If these reactions are extremely numerous (as in nuclear explosive devices), reaction products transfer released energy to nearby atoms, and subsequently, very high temperatures are reached in matter. Therefore, follow-on energy transfer to ambient matter goes through emission of hard X-rays. In dense materials this radiation reaches equilibrium with matter very fast. Non-linear thermal conductivity becomes a main mechanism of energy transfer and the transfer itself has the form of a non-linear thermal wave [xxviii]. Intensity of the wave traveling from the center outwards declines and at some distance $r_{T \to G}$ from the energy-release region, heat transfer becomes inferior to hydrodynamic energy transfer. At this stage shock wave goes ahead and further movement becomes adiabatic.

Development of the thermal stage depends strongly on non-linear heat conductivity, or averaged length of radiation path. Theoretical models, though complex and not too perfect, are the main source of this data [xxix]. So, experimental data is of great value [xxxiii].

To make a rough estimate of an event with high energy yield E_0, one can assume that the energy source is a point one with an instantaneous energy release. If coefficient of thermal conductivity depends on temperature as a power function, the heat wave evolution can be described with a self-similar solution [xxviii]. The estimates for $r_{T \to G}$ based on this solution are

$$r_{T \to G}(\text{m}) \simeq 4.70 \cdot \left[E_0(\text{kt}) \right]^{0.315} \quad \text{for air,}$$
$$r_{T \to G}(\text{m}) \simeq 0.02 \cdot \left[E_0(\text{kt}) \right]^{0.315} \quad \text{for typical rock (granite).}$$

(0.2)

In the general case this stage should be accurately described with the help of rather complex numerical codes.

After the change in the energy transfer mechanism, the hydrodynamic motion becomes prevailing. Taking into account compressibility of dense matter, it would be better to call this mechanism as gasdynamic one. Therefore, we will call these explosions strong. We would like to emphasize that they can be caused also by absorption of powerful pulsed laser beam or other very intensive energy sources.

At relatively large distances from the source, when mass of matter involved in the flow becomes much higher than that of the source, the process develops similarly to a point instantaneous explosion. For an explosion in real medium with non-linear heat transfer, the region heated with the thermal wave can be treated as the region of efficient energy release. In this case, matter flow becomes close to that induced by a point explosion at distances $r \gtrsim 3 r_{T \to G}$. If the medium is air or another gas, matter will have an EOS of simple gas form. And for many gases EOS of perfect gas gives good approximation at some efficient value of Poisson index γ. At the initial stage of SW motion, pressure P_{sw} is much higher than the initial gas pressure P_0 and matter is compressed to a limiting value $h = (\gamma+1)/(\gamma-1)$. In the 1940s it was shown [xxx] that evolution of the SW induced by a point instantaneous explosion if self-similar. At large distances deviation from this solution can be caused by the influence of initial pressure $P_0 = \rho_0 c_0^2/\gamma$, where c_0 is a velocity of sound in the gas at the initial state. And characteristic linear scale for such influence is

$$r_d = \left(E_0/\rho_0 c_0^2\right)^{1/3} \qquad (0.3)$$

usually called dynamic radius of the strong explosion. For air it comprises $r_d \simeq 300\,\mathrm{m/kt}^{1/3}$ at normal conditions. Comparison of this value with $r_{T \to G}$ shows that there is a wide range of distances $r_{T \to G} < r < r_d$ where shock wave follows the self-similar solution.

What do we have for dense matter? Equations of states for dense materials can be presented in the following form

$$P = P_c\left(\rho\right) + P_T\left(\rho,T\right); \quad \varepsilon = \varepsilon_c\left(\rho\right) + \varepsilon_T\left(\rho,T\right), \qquad (0.4)$$

where subscripts c and T mean cold and thermal components of pressure and internal energy ε, respectively. The cold components result from quantum interaction of atoms. Strong interaction of atoms in condensed matter leads to high values of matter density and sound velocity. It also causes low dynamic radius of explosion, e.g., $r_d\left(\mathrm{m}\right) \approx 3 E_0^{1/3}\left(\mathrm{kt}\right)$ for typical quartz-containing rocks. SW induced by a point instantaneous explosion in such medium begins to essentially deviate from SW described with self-similar solution at $r > 0.1 r_d$. Comparison of this value with $r_{T \to G}$ shows that there is actually no space for application of self-similar motion in such medium. This means that well developed mathematical models and codes should be used to describe accurately dense matter flow caused by the strong explosion, and theoretical models should include accurate wide-range equations of state for specific materials and rocks.

Construction of such equations of state becomes an important task. In general, such EOS should include evaporation, melting and for many materials polymorphic phase transition as well. The most difficult part is to include polymorphic transitions. As it was mentioned they often display rather complex kinetics in dynamic processes.

A great deal of theoretical and experimental work was carried out to study strong-explosion-driven shocks in different natural materials (quartz, granite, dolomite, tiff, tuff, rock salt, loams, water, etc.). Those efforts enabled scientists to estimate effects of melting and polymorphic kinetics in some materials (quartz, granite). Those data helped to develop more accurate models of natural materials. However, there is still need for accurate experimental information for materials with complex composition (as granite) in relatively low pressure range to construct good models of polymorphic kinetics. And for some rocks, it would be reasonable to get precise data on shock compression to cover the gap between low pressures in laboratory studies and high pressures in theoretical models.

Underground nuclear explosions provided a good opportunity for such studies. The shock wave in surrounding rocks was used to load experimental assemblies similar to those developed for laboratory experiments. Experimental methods were modified to account for new conditions and the reflection method became the most popular one

[xxxi]. That technology was applied to obtain data for important metals, rocks and some other materials in the range unattainable in the laboratory experiments, and to study accurately shock-driven compressibility of porous and composite materials.

However, to apply that technology efficiently it was necessary to conduct high yield explosions. Such experiments were rather expensive and rare. To overcome this difficulty, another approach was developed. The flow intensity on the thermal stage was controlled with the help of heat conducting channels, and assemblies under study were placed not far from boundary of transition from thermal to gasdynamic flows. That approach gave SW data in the highest studied pressure range 10-50 TPa for Al, Pb and some other materials [xxxii]. They showed the influence of electron shell structure on normal Hugoniot behavior, proving that theoretical models needed improvement.

New methods were developed on the basis of nuclear explosion peculiarities, such as neutron heating of material and subsequent unloading, mass velocity measurement based on Doppler shift of neutron scattering resonance, mass velocity measurement based on registration of gamma-ray benchmark movement, etc [xxxiii].

Large linear scales of high intensive flows facilitated measurements of effects caused by phase transitions. In particular, split of shock wave front in quartzite into two fronts at quartz-stishovit transition was directly observed. In the laboratory experiments this effect was infeasible to observe.

All data obtained during underground experiments are intensively used now to improve theoretical models and phenomenological equations of state.

5. DEVELOPMENT AND APPLICATION
OF NEW POWERFUL PULSED FACILITIES

Ban of nuclear tests gave an additional impetus to the development of various types of powerful laboratory facilities for HEDP research. The most well known ones are pulsed powerful laser facilities, high intensive electric current installations with explosion of conductors, charged particle accelerators with intensive beams. Their development and applications deserve individual overviews. We would only give brief remarks comparing these new approaches with traditional ones reviewed above and some thoughts on their possible applications.

The declared primary goal for all these directions is research into inertial confinement fusion (ICF) [xxxiv]. It looks very attractive to initiate thermonuclear burning in small mass pellets with a yield of several gigajoules. However, a very high compression of pellets with thermonuclear material is needed in the correspondent flows with high sharpening of profiles. Though intensive efforts and huge resources have been applied, work on ICF still faces great difficulties, including difficulties on experimental creation of convergent flows and with development of instabilities and turbulence during convergence.

However, in 1990 a unique example of a very high quality of convergence was discovered during acoustic collapse of a small bubble in liquid. Unfortunately, it takes place only in the systems with rather low energy supply. It appears that a single small bubble (with radius of $3 \div 5$ µm) can periodically grow and collapse if located in a center of resonant acoustic field applied to liquid. During a collapse, energy density increases by a factor of $10^{10\div11}$. In modified experiments, a cluster of bubbles in deuterium-contained acetone produced a neutron yield feasible in the case of utmost enhancement of flow intensity [xxxv]. So, these experiments show that there are some opportunities for progress in dynamic concentration of energy.

One more application of the mentioned facilities is to study material properties at high energy density and high intensive processes. There is a principle opportunity to carry out shock wave research. However, for this purpose it is necessary to precisely measure short time intervals. This is still a restriction to the current experimental technique and new reliable data has not yet been obtained. New facilities furnish better capabilities for studying high intensive processes, such as turbulent mixing, interaction of matter and intensive magnetic fields, non-equilibrium radiation emergence and transport. They can be especially valuable for measuring optical properties of plasma.

6. EXPLOSION-DRIVEN ACTION
ON SMALL SPACE OBJECTS

In addition to traditional physical studies, high energy density physics opens new horizons for astrophysical research. To illustrate this, consider two very different examples. The first one is related to small space objects – asteroids, and comets and their fragments, which can impact the Earth.

Scientific investigations in recent decades have shown that such impacts take place at the current stage of Solar system evolution. Intervals between the events are much longer than a human life, comparable with the time of civilization evolution and small if compared to geological periods. Depending on dimensions of space objects, the collisions can cause local, regional or global disaster including the worst ones – elimination of human civilization or even mankind extinction. Frequency of collisions decreases inversely to a square of typical space object dimensions. By now most asteroids posing global threat and only a small fraction of smaller objects have been discovered. With time, observational programs widen. Several asteroids having a small probability of heating the Earth were discovered. Apophis, discovered in 2004, is an asteroid which will be the first to threaten the Earth. In 2029 it will fly at the distance of about 30,000 km from Earth. If an unfavorable deviation of its trajectory occurs, collision with the Earth will happen in 2036.

In the present-day world full of densely populated areas and dangerous technological facilities, the collision even with small Tunguska-like object (with energy release of about 15 Mt) can cause a severe disaster. If Apophis, with characteristic dimension of about 350 m hits the Earth, the energy released will be about 1.5 Gt TNT. No matter where the impact occurs, it will cause a catastrophe of huge regional scale with potentially severe environmental consequences for the whole planet. Analysis of the potential consequences of such events is based on description of high velocity impact and accompanying phenomena with the help of HEDP scientific technology described above. There is one more important application of this technology: for prevention of dangerous impacts.

Analysis of possible alternatives to avert collisions of dangerous objects shows that at the modern technology level the most promising method is to use nuclear explosive devices. Their energy can be in the range sufficient to change the velocity of object or to disperse it. Due to high energy density, these devices are small enough to be delivered to the threatening object by up-to-date rockets and spacecrafts. And they can be efficiently deployed and placed on the surface of the object to ensure a required effect. Some useful experience for such applications was accumulated in programs on peaceful nuclear explosions.

Processes and mechanisms (radiation, gas-dynamic and neutron flux) of energy transfer and formation of correcting momentum were analyzed for various locations of explosive devices with regard to the object surface (above-surface, surface and subsurface explosions) [xxxvi]. It was shown that for objects composed of strong rocks subsurface explosions are the most efficient ones. However, changing the trajectory for the fragmented objects or ones composed of porous materials can require more sophisticated applications of nuclear explosions.

Depending on warning time (several years or more for large objects of global threat, several days or month for small objects of local or regional threat), two possible approaches can be developed to avoid collision.

In the case of long warning time, the technology of "soft" correction of the object trajectory can be applied. In particular, this scenario is applicable for Apophis. Nuclear devices, their positions, firing synchronization and energy distribution should be chosen in such a way that to efficiently transfer correcting momentum and to avoid destruction of the object. To have reliable predictions, it is necessary to collect rather complete information about properties of matter at different scales and global properties of object. Such data can be accumulated during a set of exploratory missions. In some cases reliable and environmentally friendly correction will require several sets of explosions.

Another scenario should be developed for an object with short warning time. To avoid a collision in this case, it is necessary to impart a strong momentum to the object. This can be done only with the help of powerful explosive action, which will cause fracturing and dispersion of the object. Major part of object mass will be in the form of its fragments having sufficiently high velocity to pass by the Earth at the safe distance. Minor part of mass will form an outflowing jet of vaporized matter consisting of gas and dust. It will interact with radiation belts and atmosphere of the Earth, can damage satellites and cause some undesirable atmospheric phenomena. The explosive action should be optimized to reliably change trajectories of the main fragments and to weaken unfavorable impact of small particles on the radiation belts, satellites and atmosphere.

Similar conceptual approach can be employed to avert collisions with long-period comets. However, properties of comets are known even worse. There is an apprehension that the explosive action can cause disintegration of the comet nucleus. Orbits of comets have large dimensions and part of them move along parabolic or near-parabolic trajectories. The explosion-driven correction can be applied to comet orbits. However, taking into account their high velocity inside the Solar system, this interference should be done at a great distance from the Earth. There are significant uncertainties in the information about the comet orbits. Therefore, dealing with a potentially threatening comet, the world community should be prepared to send combined exploratory and defense missions long before colliding trajectory of the comet becomes obvious. The missions should be equipped by a wide set of exploratory instruments, and have a flexible research program, which can be changed depending on new information acquired. The defense subsystem should be capable of diverse explosion-driven action based on the information obtained with the help of the exploratory subsystem. And there should be a capability to cancel the correction program and ensure safe destruction of the explosive devices.

At the current stage of the development of our civilization, nuclear-explosion-driven action is the only instrument which can be applied to avoid catastrophic collisions of space objects with the Earth. And this could become the best rationale for the enormous resources spent on the development of such devices, and the social and political dramas, which accompanied their existence.

7. THERMONUCLEAR BURSTS ON THE SURFACE OF NEUTRON STARS

Although explosions in stars are short-time events in their long evolution, the star population of our Galaxy and other close galaxies is large, that is why observations of various explosions in space are not rare. Development of the experimental techniques essentially extended the variety of the observed phenomena. Their essential role in the evolution of star-like objects became clear. Each type of them is of great interest as phenomena of high energy density physics. Let us consider just one example: a regularly observed phenomenon of X-ray bursts on neutron stars (NS).

Progress of X-ray astronomy during the three recent decades has opened good opportunities for observation of NS, which are in the active stage of their evolution. It is possible to identify a class of objects called X-ray bursters. It has turned out that these neutron stars are components of so called low-mass binaries with mass of several masses of the Sun ($M_\odot$). The second component is an ordinary star or white dwarf. Star-companion moves usually not far from the NS, and some of its matter flows onto the NS surface at a rate of $\dot{M} \approx 10^{-8} M_\odot$ a year. Caused by gravitation, significant energy release (about 200 MeV per nucleon) provides bright luminosity in X-ray spectral range $\sim 10^{37}$ erg/s. This luminosity is not constant, which can be explained by mutual motion of stars resulting in variation of accreted matter flux. A number of objects show pulsed increases of brightness by more than ten times. Pulse rises in about a second and goes down in several seconds or tens of seconds. Intervals between bursts are from several hours to a day. Ratio of energy released during the burst to integral energy of the "quiet" stage shows that bursts are of thermonuclear nature.

Some pulses show modulation of intensity of the growth and fall stages at a frequency of 300÷600 Hz. Indirect observation points out that this could be the frequency of NS rotation. It is natural to suggest that modulation is caused by non-homogeneity of thermonuclear burning and subsequent cooling along the NS surface. It was necessary to find out the mechanism, which cause the spreading of burning wave. Traditional mechanisms – thermal conductivity and convection at the burning front– gave unacceptably low values for the wave velocity. It seemed valuable to solve this problem basing on the experience accumulated during the experimental study of thermonuclear detonation [xxxvii].

The accreted matter forms a surface layer, which is often called the atmosphere. On accreting new portions of matter, previous ones go deeper and their composition changes because of thermonuclear and picnonuclear reactions. This leads to stratification of the atmosphere. In particularly, in typical neutron stars at the depth of 10^7 g/cm^2 a layer consisting mainly of helium is formed. Gradual increase of matter density and temperature in this layer will cause thermal instability due to the reactions of fusion of three helium nuclei. For slow accreting rates there will be rather homogeneous conditions along the surface layer, and thermonuclear burning can cover the entire surface at the high velocity of about 10^8 cm/s due to adiabatic gravitational wave. In observations it looks as simultaneous ignition. Development of this phenomenon can be described by 1D simulation. Essential peculiarity of the phenomenon is a strong effect of convection-driven turbulence on the released energy transfer during the initial active stage of helium burning. Radiative non-linear heat transfer becomes important only during the subsequent cooling stage [xxxviii].

If the accretion rate is high enough, matter distribution along the surface becomes inhomogeneous. In this case, helium burning begins at the most prepared location and then spreads along the surface for approximately one second. Fast rotation of NS (with frequency up to 300 Hz and more) will lead to periodical moving (often partial) of the burning region away from the field of observation that results in modulation of the observed intensity during the pulse growth stage. When burning covers the entire surface (or its main part), cooling of the atmosphere will be non-uniform and will cause modulation of the pulse fall stage.

Burning propagation was simulated with the help of two 2D gas-dynamic codes TIGR and APM (Advanced Particle-in-cell Method). The simulation showed that turbulent flow lifts the burning products to the top layers of the atmosphere, where they spread covering the cold adjacent regions. This additional thickening of the atmosphere results in an increase of pressure and temperature in the portions of the helium layer adjoining the burning region, and makes the burning front to move. These simulations enable estimating velocity of burning as $(3 \div 6) \cdot 10^6$ cm/s which is well agreed with observations.

This example of thermonuclear burning is interesting not only as space exotics. It shows the mechanism of self-sustained thermonuclear "detonation", which used to be a dream of researchers who worked on similar "terrestrial" systems several decades ago.

CONCLUSION

Exploration of high intensive processes and extreme states of matter has a relatively short (about 60 years) but bright history. Even at the initial stage using a high explosive, it initiated studies of compression of condensed matter, which for a long time was perceived as incompressible. The brightest result of that stage was the creation of nuclear explosive devices, which produced a dramatic impact on the evolution of human community far beyond the realm of science. The use of energy of nuclear explosions opened a new era in high energy density physics. It enabled studies of intensive flows with high temperature radiation and their application for thermonuclear processes. Creation of nuclear and thermonuclear systems extended capabilities of scientific research. Experimental data was obtained on shock-induced compression up to the pressures of several hundreds terapascals, on dense plasma opacity, and on polymorphic phase transitions in large-scale dynamic processes. These results stimulated the development of theoretical models.

A desire to carry out similar studies in laboratories stimulated the development of powerful pulsed laboratory installations based on different principles (lasers, electric explosions, charged particle accelerators). These activities expanded application of HEDP and opened new opportunities for promising research. In particular, it seems feasible to burn a small amount of thermonuclear material with the help of pulsed compression of small pellets in the near future.

The HEDP results and corresponding scientific technique are applied efficiently for astrophysical research, as we tried to show with the help of two examples. In the first case of near-earth asteroids and comets they can be used to study the evolution of these objects and to develop the technology to avoid their catastrophic collisions with Earth. In the case of neutron stars in low-mass X-ray binaries, a very interesting mechanism was discovered of self-sustainable thermonuclear burning along the surface due to additional compression of cold material by hot ashes, which spread in the upper layers of the NS atmosphere.

We believe that exploration of high intensive processes and extreme states of matter has a great promise for basic research and technological applications.

[i]. Rhodes, R., The making of the atomic bomb, Simon & Schuster, 1988.

[ii]. Grib, A.A. in journal Applied mathematics and mechanics (in Russian), v.8, pp.148-169, 1944; Doctorate thesis, June 1940 г.

[iii]. Zeldovich, Ya.B., in Journal of experimental and theoretical physics (in Russian), v.10, № 5, pp.542-567, 1940; , v. 12, № 9, p. 389, 1942; also in book Theory of shock waves and introduction in gasdynamics (in Russian), Moscow, Leningrad, published of the USSR Academy of Sciences, 1946; also in the book Zeldovich, Ya.B., Kompaneets, A.S., Theory of detonation (in Russian), Moscow, GITTL, 1955.

[iv]. J. Von Neiman, Progress Report on the theory of detonation waves, N.Y., 1942, (Office of Scientific Research and Development Report # 549 Division B, Section B-1, Serial # 238).

[v]. W. Döring, Uber der detonation vergang in gases, Ann. Phys., 1943, Bd. 43, # 5, s. 421-436.

[vi]. A. Schmidt, Ztschr. Schiess- und Sprengstoffw., 1936, Bd. 5, N 6, s. 12.

[vii]. Landau, L.D., Stanukovich, K.P., in Reports of Academy of Sciences of USSR (in Russian), v.46, pp. 399-402, 1945; also in book Physics of explosion, ed. Orlenko (in Russian), Moscow, Fizmatlit, 2002, т.1; also in book Stanukovich, K.P., Non-stationary movements of continuum medium (in Russian), Moscow, Nauka, 1971.

[viii]. Altshuler, L.V., in book Shock waves and extreme states of matter, Eds. Fortov, V.E., Altshuler, L.V., Trunin R.F., Funtikov, A.I., Development of methods for high pressure study, (in Russian), Moscow, Nauka, 2000.

[ix]. Altshuler, L.V., Zhuchenko, V.S., Levin, A.D., Detonation of condensed matters, in book Shock waves and extreme states of matter, Eds. Fortov, V.E., Altshuler, L.V., Trunin R.F., Funtikov, A.I., Development of methods for high pressure study, (in Russian), Moscow, Nauka, 2000, pp. 43-74.

[x]. Driomin, A.N., Savrov, S.D., Trofimov, V.S., Shvedov, K.K., in book Detonation waves in condensed matters (in Russian), Moscow, Nauka, 1970,.

[xi]. Johansson, C.H., Persson P.A., Detonics of high explosives, Academic Press, London and New York, 1970.

[xii]. Zubarev V.N., in journal Applied mechanics and technical physics (in Russian), PMTF, №2, 1965, pp.54-61.

[xiii]. Physics of explosion, ed. Orlenko (in Russian), Moscow, Fizmatlit, 2002, т.1.

[xiv]. W. Fickett, W.C. Davis, Detonation. – Berkley: Univ. California Press, 1976.

[xv]. Kanel, G.I., Razorenov, S.V., Utkin, A.V., Aortov, V.E., book Shock-wave phenomena in condensed matter Moscow, Yanus-K, 1996.

[xvi]. Aminov, Yu.A., et al, in journal Physics of burning and explosion (in Russian), FGV, № 1, pp. 103-108, 1995; Aminov, Yu.A., et al, in journal Physics of burning and explosion (in Russian), FGV, v. 33, № 1, pp. 94-97, 1997.

xvii. Grebionkin, K.F., in journal Letters to Journal of technical Physics (in Russian), v. 24. issue. 20, p.1, 1998; Grebionkin, K.F., Zherebtsov, A.L., Tarannik, M.C., Shnitko, A.S., Tsarenkova, S.K., in Abstracts of VII Khariton Scientific Talks , Sarov, 2005, p. 18.

[xviii]. Kalitkin, N.N., Kuz'mina, L.V., Preprint of Institute of applied mathematics (in Russian), Preprint N35, Moscow, Institute of Academy of Sciences of USSR, 1975.

[xix]. Kopyshev, V.P., in journal Numerical methods of mechanics of continua (in Russian), ChMMSS, v.8, issue 6, pp.54-67, 1977.

[xx]. Brushlinskiy, K.V., Kazhdan, Yu.B., in journal Progress of mathematical sciences (in Russian), UMN, **18**, issue 2 (109), pp. 3-23, 1963.

[xxi]. E. Zababakhin, I. Zababakhin, Unlimited cumulation phenomena, Nauka publishers, Moscow, 1990.

[xxii]. K. Bechert, Zur Theorie ebener Störungen in reibungsfreien Gasen, Annalen der Physic, **37**, 89-123, 1940; **38**, 1-25, 1940; **39**, 169-202, 1941; Differentialgleichungen der Wellenausbreitung in Gasen, **39**, 357-372.

[xxiii]. Zababakhin, E.I., in Journal of experimental and theoretical physics (in Russian), **49**, №2(8), pp. 643-645, 1965.

[xxiv]. Sedov, L.I., book Method of similarity and dimensions in mechanics (in Russian), Moscow, Nauka, 1965.

[xxv]. Zababakhin, I.E., Simonenko, V.A., in journal Applied mathematics and mechanics (in Russian), PMM, **42**, № 3, p. 573, 1978.

[xxvi]. Zababakhin, I.E., Nechai, M.N., in journal Applied mathematics and mechanics (in Russian) Забабахин Е.И., PMM, v. 24, issue 6, 1960.

[xxvii]. Zababakhin, E.I., Simonenko, V.A., in journal Applied mathematics and mechanics (in Russian), PMM, v. 29, № 2, 1965.

[xxviii]. Zeldovich, Ya.B., Raizer, Yu.P., book Physics of shock waves and high temperature hydrodynamic phenomena, KMoscow, Nauka, 1966.

[xxix]. Uvarov, V.B., Novikov, B.K., Nikiforov, A.F., book Quantum-statistic models of high temperature plasma (in Russian), 2000.

[xxx]. Sedov, L.I., in Reports of Academy of Sciences of USSR (in Russian), **52**, № 1, (1946); in Journal Applied mathematics and mechanics (in Russian), PMM **10**, № 2, (1946); in book Method of similarity and dimensions in mechanics (in Russian), Moscow, Nauka, 1987; Stanukovich, K.P., in book Non-stationary movements of continuum medium (in Russian), Moscow, Gostekhizdat, 1955; Taylor G., Proc. Roy. Soc., **201**, 175 (1950).

[xxxi]. Trunin, R.F., in journal Progress in Physical Sciences (in Russian), Shock Compression of Condensed Matter by Strong Shock Waves of Underground Nuclear Explosions (in Russian), UFN, **164,** No.11, pp. 1216-1237, 1994.

[xxxii]. Arorin, E.N., Vodolaga, B.K, et al, in Journal of experimental and theoretical physics (in Russian), ZhETF, v. 93, issue 2(8), pp.613-626, 1987.

[xxxiii]. Nuclear Explosions: physical studies, Avrorin, E.N., Simonenko, V.A., Schibarshov, L.I., in journal Progress in Physical Sciences (in Russian), UFN, **176** (4) 449 (2006).

[xxxiv]. Nuclear fusion with inertial confinement, book, ed. Sharkov (in Russian), Moscow, Fizmatlit, 2005.

[xxxv]. Taleyarphan, R.P., West, C.D., Chao, J.S. Jr., Nigmatulin, R.I. Evidence for nuclear emission during acoustic cavitation, Science **298**, 1868 (2002).

[xxxvi]. Simonenko V.A., Nogin V.N., Petrov D.V., Shubin O.N., and Solem J.C. Defending the Earth against Impacts from Large Comets and Asteroids в монографии Hazards Due to Comets and Asteroids, Ed., T. Gehrels, The University of Arizona Press, 1994.

18

[xxxvii]. Simonenko, V.A., Neutron stars and nuclear explosions, in book Problems of modern technical physics, (in Russian), published by RFNC-VNIITF, Snezhinsk, 2002.

[xxxviii] Simonenko, V.A., Gryaznykh, D.A., Lykov, V.A., Karlykhanov, N.G., Shushlebib, A.H., Gryaznykh, A.I., Litvinenko, I.A., et al, in Proceedings of Zababakhin Scientific Talks 2003, (in Russian), published by RFNC-VNIITF, Snezhinsk, 2002.

Nanoscale Bubble Thermonuclear Fuson in Acoustically Cavitaded Deutorated Liquid

R.I. Nigmatulin

K. Marx Street, 6, Ufa, Russia, 450000
Fax (7-3472)-73 35 69, Email: nigmar@anrb.ru

In ORNL with the participation of the author the experiments with ultrasound acoustic cavitation in liquid deutorated acetone (C_3D_6O) were held. At the moments of periodical acoustic compression of light and fast neutron (2,5 MeV) emissions and tritium (T) nuclei production by intensity $Q \sim (0,4 - 0,6) \times 10^6 \, s^{-1}$ were found. By the authors' opinion (R. Nigmatulin, R. Taleyarkhan, R. Lahey) it was a sequence of thermonuclear fusion D + D in the bubbles at the moments of implosions like in "micro-hydrogen thermonuclear bombs".

Because of the critics at ORNL the experiments were made under a control of the special commission. These measurements confirmed our measurements and the proper paper "Additional evidence ..." [2] was published in Physics Review (E). Then more detailed analysis of the experiments was published in the papers [3-6], and large paper at greater length has been accepted by Physics of Fluids. In this paper a detailed theoretical analysis and numerical calculations for the phenomena are presented

The author of the paper and his principal collaborators (R. Nigmatulin, R. Lahey, R. Taleyarkhan) presented the experimental data and their analysis in many scientific conferences and seminars as plenary or invited lecturers. In particular, we lectured at special session of the American Nuclear Society (Miami, Florida, USA, 2002), at 145 Conference of American Acoustic Society (Nashville, Tennessee, USA, 2003), at special workshop of DARPA (Arlington, Virginia, USA, 2003) on the perspectives of sonoluminescence and sonofusion. The author of the paper presented the talks at 16-th International Symposium on Non-Linear Acoustic (Moscow, Russia, 2002), 3-rd International Conference on Transport Phenomena in Multiphase Systems (Kielce, Poland, 2003), International Conference on Combustion and Detonation - Zeldovich Memorial (Moscow, Russia, 2004), 3-rd International Symposium on Two-phase Flow Modelling and Experimentation (Pisa, Italy, 2004), 5-th International Conference on Multiphase Flow, Heat Mass Transfer and Energy Conversion (Xi'an, China, 2005), 11th International Topical Meeting on Nuclear Reactor Thermal-Hydraulics (NURETH-11, Avignon, France, 2005).

The author presented the talks at Russian nuclear centers: Kurchatov Center (2000), Zababakhin Center (2000, 2002), Dubna Center (2002), St Petersburg Institute of Nuclear Physics (2003).

Recently Y. Xu and A. Butt (2005) from Purdue University (Indiana, USA) implemented independent experiments by scheme closed to [1] and confirmed production of the thermonuclear neutrons and tritium. The paper was presented at the

CP849, *Zababakhin Scientific Talks - 2005*,
edited by E. N. Avrorin and V. A. Simonenko
© 2006 American Institute of Physics 0-7354-0345-7/06/$23.00

last annual meeting of the American Nuclear Society (California, USA, 2005) and published in Nuclear Engineering and Design [8].

At the beginning of 2006 one more our experimental paper was published [9]. Experiments were conducted in which bubbles were self-nucleated without the use of external neutrons. Four independent detection systems used, in particular a neutron track plastic detector. Neutron emission rates were in range $Q \sim 5 \times 10^3$ to 10^4 s^{-1}. This rate was 50 times less than at the experiments [1, 2] because the frequency of bubble cluster formation f_{CL} was 50 times less ($f_{CL} = 1$ s^{-1} in [9] and $f_{CL} = 50$ s^{-1} in [1, 2].

Our theoretical and numerical analyses have shown that the acoustically-forced implosion of vapor bubbles of radius $a_{max} \sim 600$ to 800 μm in a bubble cluster due to a 15 bar incident pressure around the cluster and pressure amplification within the cluster, is accompanied by the formation of a strong compression shock wave cumulating (focusing) toward the center of the bubbles. This shock wave reflects from the center of the bubble producing extremely high local velocities ($w_* \approx 1,000$ km/s), and a hot ($T_* \approx 2 \times 10^8$ K), dense ($\rho_* \approx 10$ g/cm^3), high pressure ($p_* \approx 10^{11}$ bar) plasma core of radius $r \leq 60 - 65$ nm, having $\sim 10^9$ nuclei. This extreme state lasts for only a very short time ($\Delta t_* \approx 10^{-13} - 10^{-12}$ s). If this core is comprised of a deutorated hydrocarbon vapor (e.g., D-acetone) during this state, thermonuclear D/D fusion can take place producing about 12 fast neutrons (2.45 MeV) per bubble per implosion, and an equivalent amount of tritium. In any event, our analysis strongly supports the plausibility of the experimental results on bubble fusion[1,2] and the prior speculations about sonofusion of other researchers[33].

Some important features of the bubble fusion process are:

(1) *The cold liquid effect* – where relatively small variations of the liquid pool temperature strongly influence the kinetic energy of the liquid and thus the intensity of the thermonuclear fusion reaction.

(2) *The bubble cluster effect* – where bubble cluster dynamics produces a significant amplification of the interior liquid pressure compared with the incident pressure of the impressed acoustic field.

(3) *The bubble coalescence effect* – which promotes the formation of larger bubbles within the bubble cluster, having a maximum radius of $a_{max} = 600$ to 800 μm. This allows for a high cumulation of the shock waves near the center of the bubble producing conditions in a central core region of the imploded bubbles which are suitable for thermonuclear fusion.

(4) *Non-dissociation of the liquid* – where, in spite of the high pressures experienced (10^5–10^6 bar), the liquid near the interface has insufficient time for dissociation (10 ns). This is why the liquid is much less compressible than implied by an equilibrium adiabat, which corresponds to more than a microsecond of compression. Thus extrapolation of first part of D-U shock adiabat should be used for the estimation of compressibility of the liquid. This implies stronger shock waves in the bubble since less strain energy is built up in the liquid.

(5) *"Cold" electrons* – where during the extremely short time of the ultra-high compression process (10^{-13} – 10^{-12} s), the electrons have little time to be heated by the ions and the electron temperatures are many times less than ion temperatures $T_e \ll T_i$.

21

Thus the heat capacity of the vapor is ~ 2,000 J/kg instead of the equilibrium heat capacity of a completely ionized plasma, ~ 8,000 J/kg. This causes the temperature of the ions to be about four times higher than for an equilibrium plasma, which, in turn, results in conditions suitable for thermonuclear fusion. Moreover, the "cold" electrons do not produce significant energy losses by radiation emissions.

(6) *Multiscale Phenomena* – where the energetic collapse of the bubbles is a multi-scale phenomenon with a final rapid change of the scale ("sharpening"), and during the different stages, different physical phenomena, spatial and time scales dominate the process. These physical processes are: heat transfer, evaporation, condensation, and transition from a two-phase mixture to a supercritical fluid. The transition from an incompressible liquid and a homobaric pressure distribution in the vapor (this stage occupies most of the time of the process (i.e., 41.5 µs from 42 µs) to high compression of the liquid and to shock wave phenomena in gas (0.5 µs in duration), dissociation, ionization and finally to thermonuclear fusion conditions. The spatial scales are the following: the acoustic field scale is R_{AC} ~ 10^{-2} m, the bubble cluster scale is R ~10^{-3} m, the bubble size scale is a ~ $10^{-5} - 10^{-4}$ m, the dissociated core size scale is ~ 10^{-6} m, the ionized core scale is 10^{-7} m and the thermonuclear core scale is $10^{-8} - 10^{-7}$ m. The time scales are the following: the evaporation and condensation time scale is ~ 10^{-5} s, the compression wave time scale is ~10^{-6} s, the dissociation time scale is ~10^{-9} s, the time scale for full ionization is $10^{-11} - 10^{-10}$ s, and the thermonuclear reaction time scale is 10^{-13} s. The numerical code must vary the equations to accommodate the different physical phenomena and use different size grids and different time steps (from $\Delta t = 10^{-7}$ s to 10^{-14} s). To clarify the process, in the tiny central thermonuclear core this zone should use cell sizes of $\Delta r = 10^{-10}$ m in a bubble of radius 10^{-5} m. The same problem exists for the thin boundary layers near the interface.

The thermonuclear fusion process that occurs in imploding cavitation bubbles takes place within a spatial scale of about a few tens of nanometers and a time scale of a few tenths of a picosecond. Thus we might refer to it as a "Nano-Picosecond Bubble Fusion" process.

(7) *Three-dimensional phenomena* – where multidimensional analyses of the shape of the bubble supports the assumption of a spherically symmetric flow (i.e., shock wave) creating the concentration of the energy in the interior of the imploding bubbles.

All these effects are crucial for the prediction of the thermonuclear reaction's intensity. In addition, to achieve nano-scale thermonuclear fusion it is important for the test liquid to have:

1. A high atomic fraction of deuterium atoms in the molecule (in D-acetone it is 6/10 = 60%)

2. A high molecular weight (i.e, a low sound speed in the vapor, which promotes a strong shock wave) and high condensation (accommodation) coefficient to mitigate vapor cushioning during bubble implosion (for D-acetone, $M = 64$ and $\alpha \approx 1$).

3. A low saturation pressure of vapor (which can be a property of the test fluid or can be achieved due to a low pool temperature).

4. Weak non-linear compressibility of the liquid.

5. High cavitation strength of the liquid.

It is important to note that heavy water (D_2O) is not very appropriate for cavitation bubble fusion because of its relatively low molecular weight ($M = 20$), low accommodation coefficient ($\alpha \approx 0.05 - 0.07$), relatively high nonlinear compressibility and low cavitation strength.

It is also not appropriate to use laser-generated bubbles in bubble fusion experiments because these bubbles are relatively large and non-spherical, and have comparatively large vapor mass, which does not permit the liquid around an imploding bubble to reach a high kinetic energy because of cushioning by the uncondensed vapor. Moreover, lasers do not produce suitable bubble clusters, which are essential to the achievement of thermonuclear conditions.

The perspectives of intensification of Bubble Fusion are discussed in the paper. To be a source of positive energy production Bubble Fusion should be intensified $10^3 - 10^4$ times for D-D fusion. Then for D-T fusion it will be intensified 10^6 times that will be enough to produce energy.

This investigation was made with I. Akhatov, A. Topolnikov, R. Bolotnova, N. Vakhitova, R. Lahey (Jr), R. Taleyarkhan.

REFERENCES

1. R.P. Taleyarkhan, C.D. West, J.S. Cho, R.T. Lahey, Jr., R.I. Nigmatulin and R.C. Block, "Evidence for nuclear emissions during acoustic cavitation," Science 295, 1868 (2002).
2. R.P. Taleyarkhan, C.D. West, J.S. Cho, R.T. Lahey, Jr., R.I. Nigmatulin and R.C. Block, "Additional evidence of nuclear emissions during acoustic cavitation," Phys. Rev. E 69, 036109 (2004).
3. R.I. Nigmatulin, R.P. Taleyarkhan and R.T. Lahey, Jr., "The evidence of thermonuclear fusion D + D during acoustic cavitation," Vestnik ANRB (Ufa, Bashkortostan, Russia) 4, 3 (2002).
4. R.I. Nigmatulin, R.P. Taleyarkhan and R.T. Lahey, Jr., "The evidence for nuclear emissions during acoustic cavitation revisited," J. Power and Energy 218-A, 345 (2004).
5. R.I. Nigmatulin, "Nanoscale thermonuclear fusion in imploding vapoe bubbles", Nuclear Engineering and Design (Elsevier), 235, 1079 –1091 (2005).
6. R.T. Lahey, R. Taleyarkhan, R.I. Nigmatulin and I.Sh Akhatov "Sonoluminescence and the search for sonofusion," in press, Advances in Heat Transfer, 39, Academic Press (2005).
7. R.I. Nigmatulin, I.Sh. Akhatov, A.S. Topolnikov, R.Kh. Bolotnova, N.K. Vakhitova. R.T. Lahey, Jr, R.P. Taleyarkhan "The Theory of Supercompression of Vapor Bubbles and Nano-Scale Thermonuclear Fusion", Physics of Fluids, 17, 107106 (2005).
8. Y. Xu, A. Butt, "Confirmatory experiments for nuclear emissions during acoustic cavitation", Nuclear Engineering and Design, 236, (2005).
9. R.P. Taleyarkhan, C.D. West, R.T. Lahey, Jr., R.I. Nigmatulin, R.C. Block, and Y. Xu "Nuclear emissions during self-nucleated acoustic cavitation", Physical Review Letters, 96, 034301 (2006).
10. R.F. Trunin, M.V. Zhernokletov, N.F. Kuznetsov, O.A. Radchenko, N.V. Sichevskaya and V.V. Shutov, "Compression of liquid organic substances in shock waves," Khimicheskaya Fizika 11 (4), 424 (1992) (in Russian).

Fifty Years of Quantum Electronics

O.N. Krokhin

P.N. Lebedev Physical Institute, Russian Academy of Sciences, Moscow

Undoubtedly the discovery of the maser and laser principles in 1954 was one of the most outstanding achievements of modern physics. This discovery has imparted a significant impulse to the development of civilization.

During the 50-year period of rapid development a large family of lasers of different types was created and advanced so that now lasers become habitual occurrence.

Here are some results obtained in the field of quantum electronics and laser physics, which, in my view, are very significant. They reflect both practical or applied achievements and scientific ones that are still awaiting practical implementation.

A global system of fiber-optic communications. Today, we can assert that a fiber-optic channel capacity of up to 10^{11} bit/s has been achieved, allowing simultaneous transmission of nearly 10000 television programs. It is worthy of note that there are sizable reserves for increasing the rate of transmission through updating technical means. We may obviously state that optical communication lines will fully satisfy future societal needs for information exchange.

Frequency-time standards. The contemporary methods of stabilizing the frequency of generation of laser electromagnetic oscillations make it possible to register the frequency of these oscillations with a relative accuracy of 10^{-15} that is comparable with the accuracy of passive standards, but promising further improvement. This correlates with the same accuracy in measuring time intervals; e.g., the error of time standards is equal to 1 s over an interval of 10^{15} s, that is, over 100 million years. The problem of transferring this accuracy to microwave ranges, in which the Time Service operates, is nearing a solution.

Supershort light pulse generation. Specially designed laser systems with a wide-band spectrum are capable of generating light pulses with a minimal duration of $4 \cdot 10^{-15}$ s. The pulse is extended in space over the length of one wavelength, approximately 1 μm. Such pulses make possible to "photograph" superfast atomic-molecular processes, e.g., chemical reactions, and to obtain superhigh densities of light power in focusing radiation to small volumes.

Laser cooling of atoms in a gaseous state. The process essentially consists in a small quantity of atoms, which initially have thermal velocities corresponding to normal temperature, being caught in a so-called "trap" and radiated from all sides by laser emission with a frequency correlating with resonance quantum transition. In so doing, the radiation frequency should be tuned so that absorption occurs only for atoms moving toward a laser beam (due to the Doppler effect). Light absorption is accompanied by photon pulse transmission; i.e., deceleration takes place. Then, the atom returns to its basic state and the process continues. It was thus possible to cool

CP849, *Zababakhin Scientific Talks - 2005*,
edited by E. N. Avrorin and V. A. Simonenko

atoms to a temperature of about 10^{-9} K and experimentally study quantum-statistical regularities in the particle-ensemble behavior (Bose–Einstein condensation).

Semiconductor lasers. In practical implementation of the new opportunities offered by quantum electronics, semiconductor (diode) lasers play a special role. This is due to the fact that electric current in them is directly converted into light, and, consequently, their efficiency reaches 70%. Moreover, they are space saving and, in recent years, their capacity has been tangibly increased almost up to 20 W from one diode. Diode lasers are predominant among quantum electronics items by the volume of sales and, particularly, by nomenclature. They are widely used in optical communication systems, information recording-playback devices on various types of compact disks, in printers, range finders, velocity meters, homing systems for high-precision weapons, etc. Recently, diode lasers have found application as effective sources of pumping crystalline, glass, and fiber lasers, which has actually led to a breakthrough in laser technology. It has become possible to create high-capacity complexes for large-scale machine building. Thus, the development of up-to-date diode lasers enabled one to overcome the basic difficulty in broad application of laser technology—low efficiency.

Highly efficient industrial laser complexes with "diode pumping." These have already been put into operation in the automobile industry, aircraft, railway cars, and shipbuilding. For example, around 600 welding laser units operate at Volkswagen plants. Laser welding has a number of essential advantages, namely, higher accuracy, an even and clean weld, and the absence of a large heating zone and, consequently, deformation. It is effective also in the manufacture of "composite" blanks in mechanical engineering, e.g., sheet products of variable thickness or materials with different mechanical parameters. Lasers also have a wide application in pattern cutting of sheet materials, including textiles, drilling of thin holes (e.g., in engine injectors), cleaning of surfaces, rapid and high-quality cutting of glass, etc.

Mass data recording and storage. It seems that today everyone knows what a compact disk—CD—is. An enormous amount of these products is sold in a network of trade outlets, ranging from conventional music disks to huge-capacity carriers of video, graphic, and text information. Here, I would like to emphasize that similar hardware is developing and improving in plain sight. This is associated with enormous technological potentialities that have not been assimilated yet. A move to shorter-wave lasers (the spectrum blue region) on gallium nitride diodes will bring about still greater recording densities, and the implementation of a multilayer recording in several different wavelengths will increase the information recording density and the speed of information readout.

Active systems of precise positioning, location, and high-precision weapons. Lasers make it possible to create active range finders, object velocity counters, and locators. It is unquestionable that all of them, such as active missile target homing systems, find their place in military equipment.

Lasers for medicine. The widespread introduction of endoscopic equipment makes it possible to use lasers in various types of major surgery. In many cases, the use of this equipment yields good results. Another field of laser application is ophthalmology, where lasers have become standard instruments for operative treatment of eye diseases.

Stimulation of thermonuclear reactions. The application of lasers to solve problems of nuclear fusion is still at the stage of research. The main scientific problem lies in finding ways to overcome gas-dynamic instabilities that accompany compression and heating of small amounts of hydrogen isotopes—mixtures of deuterium and tritium. The low efficiency of laser facilities is the original cause of the technical difficulties in solving these problems. It seems that the difficulties may be removed with a transition to "diode pumping." Nevertheless, the cost of a project (given the present-day technological level) may turn out to be high, because a "fusion" laser, unlike an industrial one, must develop a huge pulse power. There is hope that rapid development of laser technology and a parallel search for more effective methods to "ignite" fusion "fuel" will lead to the solution of this problem in the future.

Since my report is devoted to the 50th anniversary of the advent of quantum electronics, it is necessary to return to the time when this remarkable discovery took place and try to recall the history of events that brought about this outstanding result.

FIGURE 1. The 1964 Nobel Prizewinners in Physics (left to right): A.M. Prokhorov, C. Townes, and N.G. Basov at the Physical Institute. October 1965.

In 1964, the Nobel Committee for Physics conferred its prize on N.G. Basov, A.M. Prokhorov, and C. Townes (Fig. 1) "for fundamental work in the field of quantum electronics, which has led to the construction of oscillators and amplifiers based on the maser-laser principle." The prize was divided into two equal parts (one went to Basov and Prokhorov and the other to Townes). It is noteworthy that investigations associated with quantum electronics won three more Nobel Prizes in subsequent years. In 1981, N. Bloembergen and A. Schawlow became Nobel Prize laureates for their research into laser spectroscopy; in 1997, the prize went to S. Chu, C. Cohen-Tannoudji and W.D. Phillips for the development of laser methods for atom cooling; and, in 2000, Zh.I. Alferov became a Nobel prizewinner for the development of heterostructures, including for semiconductor diode lasers.

Thus, in the early 1950s, at the Mandel'shtam and Papaleksi Laboratory of Oscillations of the Lebedev Physical Institute (LPI), on the initiative of Prokhorov, researchers began spectroscopic investigations of molecules in the radio frequency band of electromagnetic radiation. Basov, who came to the FIAN shortly before graduation from the Moscow Engineering Physics Institute, had become actively involved in the work. At that time radio spectroscopy was a novel and rapidly advancing field of physics, in which several groups of scientists in different countries were working and among which we single out, first and foremost, the team led by Townes of Columbia University in the United States. The objective of radio spectroscopy is to study the structure of molecules. What specific features and, consequently, problems did the pioneers of radio spectroscopy face?

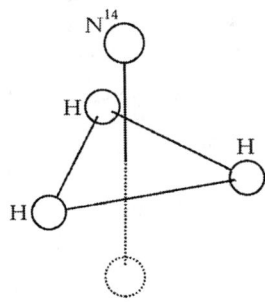

FIGURE 2. Ammonia (NH₃) molecule structure. The vertical axis corresponds to the molecule axis of rotation, which forms rotational quantum levels; the nitrogen atom may jump from the upper to the lower position (dotted line) and inversely ("inversion'"), which results in splitting of each rotational level into two sublevels between which there may occur transition with absorption or induced radiation used in a maser.

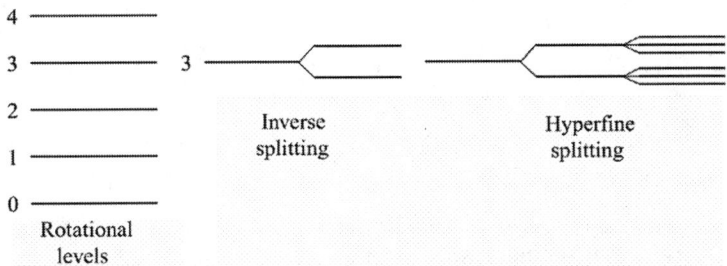

FIGURE 3. Spectrum of an ammonia molecule's rotational levels. Everywhere $kT >> E_n - E_m$; population $N_n \sim \exp(-E_n/kT)$ is equal for many lower levels down to $E_n \sim kT$.

One subject of research in the radio frequency band is quantum transitions between rotational levels of molecules (Figs. 2 and 3). The frequencies of these transitions, e.g., for ammonia molecules, lie in the field of the submillimeter wave bands (the maximum value of the quantum energy is $2 \cdot 10^{-3}$ eV and $\lambda \sim 0.5$ mm). Since the quantum energy is less than the value of $k_B T$ (T is temperature and k_B is the Boltzmann constant), several lower rotational levels of the molecule are usually filled at normal temperature. This reduces the absorption of incident radiation at the transition frequency, because it is the balance of the acts of absorption during the transition from the lower level to the upper one and induced emission in the reverse process, that is, proportionate to the difference

$N_1 - N_2$, where N_1 and N_2 are the populations of the lower and upper levels, respectively. As a result, the spectroscopy method based on radio wave absorption measurement becomes less sensitive.

Another circumstance that also impairs the accuracy of determining the frequencies of quantum transitions of molecules in a gaseous state is absorption line broadening due to the Doppler effect. Doppler line broadening is present in all variants of gas spectroscopy, including, naturally, the optical range.

Both problems can be solved by using a beam of molecules instead of gas. These molecules pass through a high-frequency resonator in the direction where the type of field oscillations in the resonator is almost maximal, that is, has a very high phase velocity in this direction. It is obvious that, since the Doppler shift of the transition frequency is proportional to v/c_{ph}, where v is the molecule thermal velocity and c_{ph} is the wave phase velocity, then with $c_{ph} \rightarrow \infty$ the frequency shift will be small (Fig. 4).

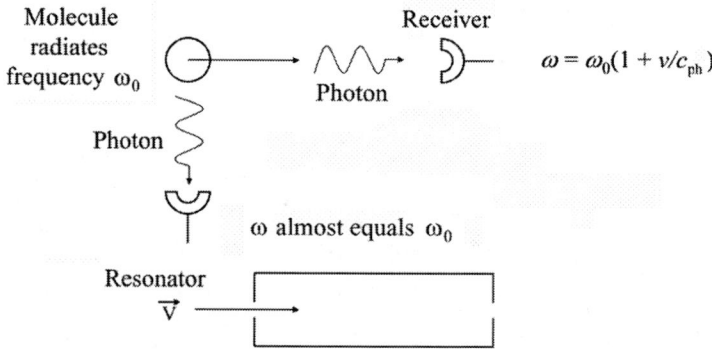

FIGURE 4. Doppler line broadening and exclusion of this effect. The frequency of an electromagnetic wave emitted by the molecule moving to the right at velocity v is shifted relative to the frequency of transition to the value of $\omega_0 v/c_{ph}$ for the receiver arranged to the right and is almost not shifted for the receiver arranged at the bottom; an effective (phase) velocity $c_{ph} \rightarrow \infty$ can be provided in the resonator to rule out the frequency shift; field $E \sim E_0 \cos \omega t$; if E_0 does not depend on the axis coordinate, this means that $c_{ph} \rightarrow \infty$ and $\omega \rightarrow \omega_0$.

The use of a molecular beam instead of gas also allows increasing the absorption efficiency owing to the "sorting" of molecules over the levels with a nonuniform electric or magnetic field. In particular, if we suppose the molecules to be "sorted" by a nonuniform electric field (for example, a quadrupole capacitor consisting of four metal rods, along the axis of which the beam propagates), then, at the input to the sorting system of the molecule that fills several quantum rotational levels as a result of thermal excitation (in a "mixed up" state), the electric field "mixes up" these states. A dipole moment arises, and, correspondingly, a force that acts on the molecules in the direction transverse to that of the beam motion in the direction of the nonuniform field gradient. In this way, the molecules located at the levels near the transition under study are focused toward the system axis or ejected from the beam (Fig. 5). Thus, application of molecular beams allowed solution of two problems important for spectroscopy, namely, ruling out Doppler broadening of the transition line and making absorption more effective due to "sorting" of molecules. The principle of maser creation was now around the corner.

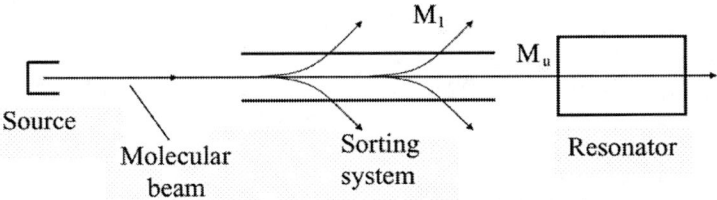

FIGURE 5. "Sorting" molecules by rotational levels in an electric field.
The electric field on the system axis is minimal: the molecules located in the upper level (M_u) pass to the resonator; the molecules in the lower level (M_l) are ejected from the system.

Really, why not use radiative transitions instead of absorbing ones for the purposes of radio spectroscopy? If we try to follow the logic of Basov and Prokhorov, on the one hand, and that of Townes, G. Gordon, and H. Zeiger, on the other, we will be convinced that both teams of researchers in the field of radio spectroscopy wrote in their 1954 publications primarily about an increase of radio spectroscopes' resolving power owing to the use of induced emission instead of absorption, because this emission, due to regeneration, should bring about the transition line narrowing (Fig. 6). These two studies marked the onset of a new era in radio physics—the application of quantum systems for generation of electromagnetic radiation.

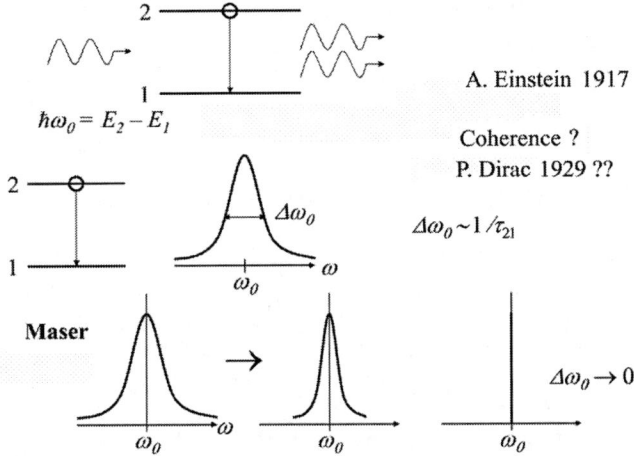

FIGURE 6. Induced radiation "line forms." At a single act. the probability of radiating the preassigned frequency is given by a bell-shaped function with a spectrum width $\Delta\omega_0 \sim 1/\tau_{21}$, where τ_{21} is the time of radiation from the second level to the first; with multiple induced radiation, the spectral probabilities are multiplied by as many times, and, within a great number of transitions, the spectrum will tend to be monochromatic (generation in laser); the induced radiation is coherent (identical) to an incident wave, which is corroborated by the discovery of a maser effect.

29

At the end of 1963, when it was already clear that the discovery had resulted in the formation of a new trend in science and technology that came to be known as "quantum electronics" and the scientific community viewed it as an outstanding achievement, D.V. Skobel'tsyn presented it for a Nobel Prize. In his letter to the Nobel Prize Committee, he wrote, "Investigating this problem, one can remember that the first considerations about the use of induced radiation for coherent amplification and generation of electromagnetic waves were stated independently by these researchers in the early 1950s at conferences, the proceedings of which, unfortunately, were not published later.

However, the first publications related to quantum oscillators appeared in 1954. In January 1954, Basov and Prokhorov contributed the article "Application of Molecular Beams for Radio Spectroscopic Investigation of Molecules' Rotational Spectra" to the *Journal of Experimental and Theoretical Physics,* where it was published in Volume 27, pp. 431-438, in October 1954. In May 1954, Townes and his colleagues sent the article "Molecular Microwave Oscillator and New Hyperfine Structure in the Microwave Spectrum of NH_3" to the journal *Physical Review,* where it was published in Vol. 95, pp. 282-284, in July 1954."

Skobel'tsyn ends his letter as follows: "That diversity of studies on quantum oscillators that we have had up to now reflects in one way or another the new ideas put forward and formulated at the same time independently by Townes, Basov, and Prokhorov. I hope that my arguments are well founded enough for the Nobel Committee to consider my reasoning outlined in this letter."

Let us now discuss those first papers on quantum electronics. An article by Townes and his colleagues Gordon and Zeiger, "Molecular Microwave Oscillator and New Hyperfine Structure in the Microwave Spectrum of NH_3," stated that

... an experimental facility has been developed and is now operating that can be used as a high-resolution microwave spectrometer, microwave amplifier, or very stable oscillator. This device, in which an inversion spectrum of the ammonia molecule is used, is based on energy emission inside the resonator with a high Q factor by the beam of ammonia molecules ...

This was the first information about the development of a molecular oscillator—a maser.

Basov and Prokhorov wrote the following in their article "Application of Molecular Beams for Radio Spectroscopic Study of Molecules' Rotational Spectra":

Using a molecular beam where there are no molecules in the lower state of the transition under review, one can construct a "molecular oscillator." The principle of operation of the molecular oscillator consists in the following.

A sorted molecular beam that lacks molecules in the lower state of the transition under review is passed through a cavity resonator. Over the time of the molecules' passage through the cavity resonator, some of the molecules pass from the upper state to the lower one, giving off energy to the cavity resonator. Should the power of losses inside the resonator be less than that of the molecules' radiation, there self-excitation begins, at which the resonator power grows to the value determined by the effect of saturation. Thus, self-excitation begins if....

Furthermore, they have the formulas determining the oscillator basic parameters, and, eventually, the Q factor needed to self-excite the molecular oscillator is evaluated. For the CsJ molecule, which was investigated by Basov and Prokhorov, the Q factor should be at least $7 \cdot 10^6$ (!). This gave reasons to state that self-excitation is impossible in this case (the Q factor falls short by three orders).

Basov and Prokhorov started developing a facility with a more intensive molecular beam—hypothetically, formaldehyde—rather than with a beam of CsJ molecules. However, after Townes' publication, they took up ammonia and soon obtained generation. Basov's doctoral dissertation, which was presented for approval in 1956, contained a detailed description of a molecular oscillator operating with an ammonia molecular beam (Fig. 7). In his review of the dissertation, Prokhorov wrote, in particular, that "It was N.G. Basov who first pointed to the possibility of creating a molecular oscillator in 1952."

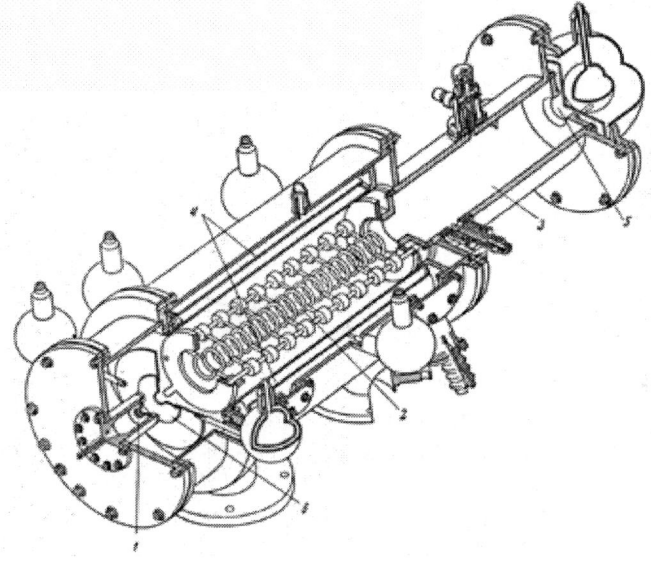

FIGURE 7. Molecular oscillator design.
(1) Molecular beam source, (2) sorting system, (3) resonator, (4) vacuum jacket cooled by liquid nitrogen, (5) end trap of spent molecular beam, and (6) diaphragm to form molecular beam.

The report by Basov and Prokhorov was the only item that they have managed to find on this subject in the archives. The report was contained in the verbatim record of the Conference on the Magnetic Moments of Nuclei held on January 22-23, 1953 (RAS Archives, Fund 1522, Inventory 1, File 59, pp. 36-47). The report delivered by Basov contains two basic points directly relating to the subject under study. It said, in part,

After we have elucidated, in general, the theoretical potential for sorting the molecules, it was natural to switch immediately to the new method of production (the actual text—O.K.). Namely, to consider radiation rather than absorption of microwaves, that is, sort out the molecules that are not in the lower rotational state but single out the molecules that are in the upper rotational state. We are to observe already the spectrum of molecular radiation rather than the absorption spectrum .... In order to observe the induced radiation of such a molecular beam, it is passed through a resonator. With time, if the resonator's Q factor is adequately great, the energy stored in the resonator grows and the probability of molecules emitting the energy tends to unity.

Another interesting statement made in the same report says that

As A.M. Prokhorov has shown, it is not imperative to monochromatize the beam by velocities, because one can use a high frequency field so high that the Doppler expansion (the actual text—O.K.) will not be obtained. Given that the beam is passed along the waveguide in such a direction in which wave phase velocities, say, that of the E wave, equal infinity, there is no shift due to the Doppler effect, because the frequency shift is determined by the ratio of the molecular beam velocity to the wave phase velocity in the direction of the beam propagation.

I believe it unavoidable to observe that the existence of the process of induced radiation was postulated by A. Einstein in 1917 as a result of analyzing the thermal process of setting equilibrium between radiation and the atomic system. The conclusion that the quantum of induced radiation must be equal to the quanta that caused the radiation follows from quantum field theory and is usually associated with P. Dirac. However, it is not all that simple to bridge the notion of coherence in the classical understanding of this word (a type of the field oscillation in a resonator) with a contention about the identity of the quanta (they lack the notion of phase). It is precisely the establishment of coherence in molecular ("quantum") oscillators operating due to a stimulated emission that is a nontrivial fact of this outstanding discovery (see Fig. 6).

Among the studies of the early period of quantum electronics, note the proposal to "sort" particles by the method of pumping active media with electromagnetic radiation, a so-called three-level diagram. It has been successfully used in creating masers— low-noise amplifiers of the microwave range in a ruby crystal—and, subsequently, in the construction of lasers. This suggestion was in the paper by Basov and Prokhorov "Possible Methods of Obtaining Active Molecules for a Molecular Oscillator," published in the *Journal of Experimental and Theoretical Physics* in February 1955 (Vol. 28, pp. 249-250).

The discovery of new methods of generation of electromagnetic radiation boosted research in the field of quantum electronics and yielded, in a number of cases, fantastic results, which in those bygone 1950s were simply impossible to imagine. Primarily, we should point out implementation of the principles of a molecular oscillator in the optical range of frequencies, in other words, the development of lasers.

In June of 1958, Prokhorov contributed the article "A Molecular Amplifier and Oscillator in Submillimeter Waves" to the *Journal of Experimental and Theoretical Physics* (Vol. 34, pp. 1658- 1659) (sent to the editors in April 1958). It presents a so-called open resonator, where a high Q factor is provided due to a short radiation wavelength. "Two plane-parallel mirrors may be used to create a molecular oscillator as a resonator. If the distance between the mirrors is l and the coefficient of reflection from the mirror is k (bearing in mind that plane wave energy losses occur only upon reflection from the mirror), then the Q factor of such a system is equal to

$$Q = \frac{2\pi l}{\lambda}(1-k)^{-1} \ .$$

As is known, open "multimode" resonators (with a great number of various types of field oscillations having close frequencies) are an indispensable attribute of lasers.

In December of that same year, Schawlow and Townes published the large article "Infrared and Optical Masers" (it was sent to the editors in August 1958) in *Physical Review* (Vol. 112, pp. 1940-1949). The article said that "The extension of maser techniques to the infrared and optical region is considered. It is shown that by using a resonant cavity of centimeter dimensions, having many resonant modes, maser oscillation at these wavelengths can be achieved by pumping with reasonable amounts of incoherent light."

And, finally, in August 1959, the *Journal of Experimental and Theoretical Physics* (Vol. 37, pp. 586, 587) carried an article "Quantum-Mechanical Semiconductor Oscillators and Amplifiers of Electromagnetic Oscillations" by Basov, B.M. Vul, and Yu.M. Popov (received by the editors in May 1959 and registered with the Committee for Inventions and Discoveries under the USSR Council of Ministers on July 07, 1958). The article dealt with "a possibility of using electron transitions between the conductivity zone (valence zone) and donor (acceptor) admixture levels of a semiconductor to produce electromagnetic radiation with the aid of an induced radiation mechanism, just as it occurs in a molecular oscillator."

All three articles attempted for the first time to extend the molecular oscillator operation principles to the region of infrared and optical frequencies. They, of course, only initiated such investigations and attracted the attention of the scientific community to the new scientific trend.

I have already said what we have received from all this. Here I consider it expedient to note that 50 years of the development of quantum electronics are a brilliant example of the importance and potential of basic science or a fundamental discovery for practical applications useful to society and the economy. Quantum electronics has actually revealed its potential over the life span of one or two generations.

A tremendous amount of the results obtained with the methods of quantum electronics in Russia has become possible thanks to much attention on the part of a number of ministries: the Ministries of Defense, the Electronic and Radio Industry, and Medium Machine Building and the USSR Academy of Sciences. It is obvious that a great many research institutes and design organizations with highly skilled specialists and prominent scientists were involved in the implementation of programs that had appeared thanks to this discovery. Of course, it would be rather risky to try to give all the names.

Nevertheless, I have in mind two prominent scientists who have created a scientific school of coherent and nonlinear optics at Moscow State University (MSU): R.V. Khokhlov and S.A. Akhmanov. This scientific school, which now combines the Departments of Quantum Electronics and General Physics and Wave Processes, as well as the MSU International Laser Center, has made a sizable contribution to the study of parametric processes and the processes of harmonic generation, fundamental regularities of high-power light propagation in media, and the development of methods of nonlinear spectroscopy.

Also noteworthy is the team of researchers at the Vavilov State Optical Institute (SOI)—the leading Soviet and Russian center in the field of optics. At this institute, actually at the same time with the LPI, the first Russian ruby lasers were put into operation and, in subsequent years, intensive research was conducted in the field of quantum electronics. The Institute of Laser Technology, which separated from the SOI several years ago, is successfully continuing these investigations. Note the tremendous role of the Leningrad Institute of Precise Mechanics and Optics (nowadays, St. Petersburg Technical University) in the advancement of quantum electronics. The Leningrad Optical-Mechanical Association has done much to upgrade technology and engineering in the field of optical and laser instrument-making, particularly in the implementation of large projects.

As was noted above, the 2000 Nobel Prize in Physics was conferred upon Alferov, who in the early 1960s successfully obtained heterostructures based on gallium-aluminum arsenide and, later on, more complex four-component compounds. It is precisely due to the assimilation of heterostructures that semiconductor laser diodes have become up-to-date devices with high efficiency and power. Their parameters will not stop at the present stage, and I hope we will witness impressive progress.

The Institute of Laser Physics of the RAS Siberian Division is deeply involved in high-precision spectros-copy and methods of creating highly stable optical oscillators. The development of standard frequencies for the Time Service calls for providing devices to transfer frequencies from the optical to the radio range. The Institute, jointly with the LPI and other institutions, is successfully solving this problem. The Institute of Laser Physics, Institute of Applied Physics, and RAS Institute of Spectroscopy are conducting in-depth research into lasers with pulse femtosecond duration and their application to high spatial resolution spectroscopy. The Institute researchers are giving top priority to the problem of laser cooling of atoms.

The Prokhorov Institute of General Physics and the RAS Institute of Problems of Laser and Information Technology have done much for medical applications of lasers.

The "Polyus" Research Institute (now Stel'makh State Unitary Enterprise "Polyus") has made a tangible contribution to the development of laser technology. Multifunctional laser systems and components, as well as laser radiation control systems, are now produced on an industrial scale and are available to many organizations thanks to the "Polyus" developments.

The "Astrofizika" Scientific and Production Association and the All-Russia Research Institute of Experimental Physics (Sarov) have been successfully developing high-power laser technology, in particular the Iskra-5 laser unit for thermonuclear studies. The unit utilizes a photodissociation method to produce an active medium on iodine atoms (energy of 20 kJ). A more powerful Iskra-6 system is being designed now. Institutes and industrial enterprises in the Republic of Belarus, where the scientific and production complex of the Soviet optical industry were created in the past, are actively developing laser technology and its many applications.

We are indebted to scientists and engineers from the above-mentioned institutions and to many others for the fact that, for over half a century, domestic science has been at a high level of achievement and made a tremendous contribution to quantum electronics.

Investigation of Neutron Emissions from D(d,n)³He and T(d,n)⁴He Reactions in a 10 TW Picosecond Laser Facility SOKOL-P

A.V. Andriyash, V.V. Andryushin, O.V. Chefonov, M.N. Chizhkov,
D.A. Dmitrov, A.G. Kakshin, I.A. Kapustin, A.V. Levin, E.A. Loboda,
V.A. Lykov, V.A. Pronin, V.G. Pokrovskiy, A.V. Potapov, V.N. Sanzhin,
V.N. Saprykin, A.A. Ugodenko, D.A. Vihklyaev, A.L. Zapysov,
Yu.N. Zuev

*Federal State Unitary Enterprise "Russian Federal Nuclear Center —Zababakhin All–Russia
Research Institute of Technical Physics", Snezhinsk, Chelyabinsk region, Russia*

Abstract. Experimental results on fast neutron generation in $D(d,n)^3He$ and $T(d,n)^4He$ reactions in the SOKOL-P laser facility are presented. Solid targets were irradiated by 1.054 μm, s- or p-polarized laser pulses of energy 5-8 J on target and duration 0.85-2 ps. The peak laser intensity was 0.5-$2 \cdot 10^{18}$ W/cm². Flat deuterated plastic $(CD_2)_n$ targets and $TiD_{0.5}T_{0.5}$ targets were used in experiments. Some experiments were carried out with additional targets placed in front of and behind the laser target. The used time-of-flight technique helped identify neutrons from $D(d,n)^3He$ and $T(d,n)^4He$ reactions. Yields up to 10^6 DD-neutrons and 10^7 DT-neutrons were measured. Interaction of the fast ion beam with the target can explain the observed yield.

Keywords: Laser acceleration of ions; Neutron generation
PACS: 52.38.Kd, 29.25.Dz

INTRODUCTION

For last decade significant progress in construction of laser systems generating short intense pulses is achieved. The plasma which occurs at interaction of these laser pulses with matter is a source of fast electrons and ions, X-rays.

Ions can be accelerated on the frontal or rear side of a laser target. Mechanisms of ions acceleration on the frontal side are not completely investigated yet. Ions which leave the target can be studied by means of the charged particles detectors.

The ions accelerated on the frontal side into the target, are slowed down by the target. If the target is thick ions do not leave the target, and their detection is impossible. If the target is thin ions which get through the target are influenced by the electromagnetic field on the rear side. We can diagnose the accelerated ions by nuclear reactions which they initiate in the target. For example, the deuterons accelerated by the laser and directed into the target can interact with cold deuterons of the target, producing neutron in $D(d,n)^3He$ reaction [1-3]. On the neutron spectra measured at different angles, energy distribution and angular dependence of the accelerated ions

CP849, *Zababakhin Scientific Talks - 2005,*
edited by E. N. Avrorin and V. A. Simonenko
© 2006 American Institute of Physics 0-7354-0345-7/06/$23.00

can be restored. This information can be useful to studying of acceleration mechanisms. In Russian Federal Nuclear Center - VNIITF 10 TW SOKOL-P CPA Nd:glass laser facility was put into operation [4]. Some experiments on neutrons generation have been carried out. The main purpose of these experiments was investigation of the ions accelerated into the target.

EXPERIMENTAL SETUP

Experiments on the irradiation of flat solid targets were carried out on SOKOL-P laser facility [4]. SOKOL-P produced pulses with a duration of 0.85 to 2 ps and an energy of 5 to 8 J. A peak laser intensity on target was $(0.5-2) \cdot 10^{18}$ W/cm^2. The intensity contrast of the main pulse was varied in the range from $5 \cdot 10^5$ to 10^9. Figure 1 illustrates the experimental setup.

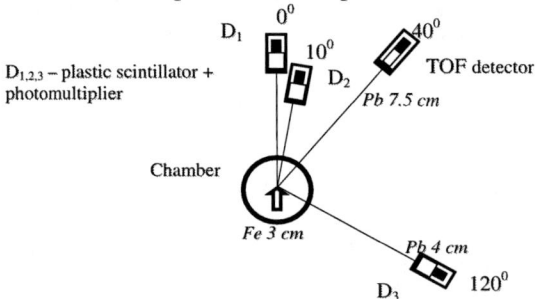

FIGURE 1. Experimental setup.

Three detectors measured neutron yield at different angles. A current mode (TOF) time-of-flight detector determined neutron energies. All detectors were shielded by lead to protect from X and γ rays. The target was placed into a steel vacuum chamber. In experiments, s- or p-polarized laser light irradiated flat deuterated plastic targets $(CD_2)_n$ with thickness of 5-300 μm and targets containing tritium $TiD_{0.5}T_{0.5}$ with thickness of 5 μm placed at Cu- substrate (See Fig.2 a, b). In some experiments, we used additional targets – catchers placed in front of and behind the laser target (Fig.2 c, d). The purpose of using the additional targets was to increase the neutron yield.

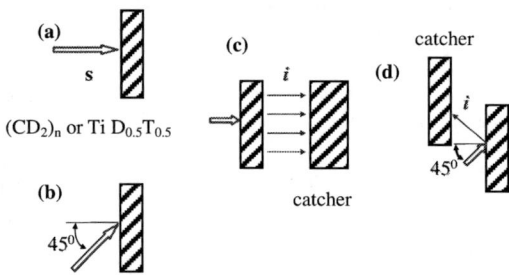

FIGURE 2. Types of targets.

36

Similar experiments at higher laser intensity were published earlier in [1-3]. The particular feature of this work is use of targets with tritium, and also use of catchers.

EXPERIMENTAL RESULTS

At an irradiation of deuterated plastic targets, the maximum yield was about $8 \cdot 10^5$ neutrons and average neutron yield was about $3 \cdot 10^5$. In experiments with single targets containing tritium, we detected up to $2 \cdot 10^6$ neutrons. Experiments with tritium containing catchers, which were placed behind of the 100 μm $(CD_2)_n$ target, have given approximately tenfold increase of neutron yield in comparison with single $(CD_2)_n$ target experiments. Additional $TiD_{0.5}T_{0.5}$ targets placed in front of the $(CD_2)_n$ targets have given about 50-fold increase in neutrons yields. In the experiments with the $(CD_2)_n$ catchers placed in front of the $(CD_2)_n$ laser target neutron yield was 3-5 – times more than without catchers. In the experiment with two tritium-containing catchers, we measured more than 10^7 neutrons. The used time-of-flight technique helped identify neutrons from $D(d,n)^3He$ and $T(d,n)^4He$ reactions (See Fig.3).

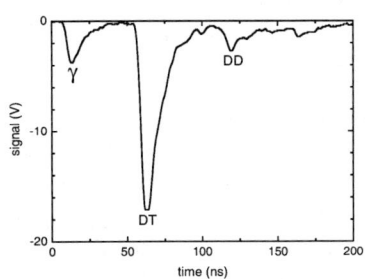
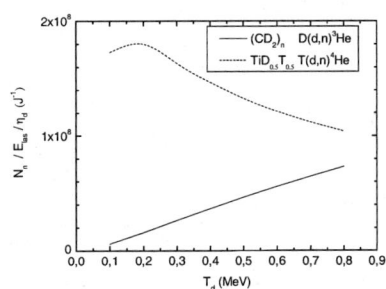

FIGURE 3. TOF signal.
FIGURE 4. Theoretical normalized neutron yield versus temperature of fast deuterons.

To analyze experimental result we used a simple model. We assumed the energy distribution of fast ions to be exponential with a temperature T_d:

$$\frac{dN_d}{dE_d} = N_d / T_d \cdot \exp\left(-E_d / T_d\right) \qquad (1)$$

The neutron yield per unit of laser energy depends on the ion temperature and the efficiency of energy transfer to ions η_d (See. Fig.4):

$$\frac{N_n}{E_{las}} = \eta_d \cdot n_d / T^2 \int_o^\infty dE_d \, \exp\left(-E_d / T_d\right) \int_o^{E_d} dE \, \sigma(E) / |dE / dx| \qquad (2)$$

where n_d is deuterium atomic density, σ is fusion reaction cross section, and $|dE/dx|$ is stopping power of deuteron in the target. Our approach is a possible alternative to Monte-Carlo calculations which were used in work [3]. At the irradiation of targets with tritium the significant contribution to the neutron yield give deuterons with energy less than 1 MeV unlike $(CD_2)_n$ targets (See. Fig.5).

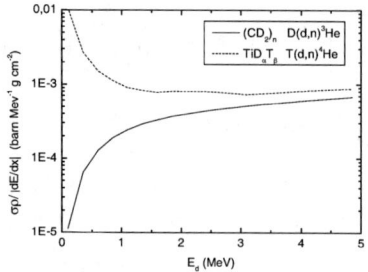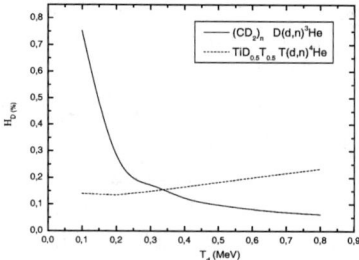

FIGURE 5. Ratio of reaction cross section to deuteron stopping power for $TiD_{0.5}T_{0.5}$ and $(CD_2)_n$ targets.
FIGURE 6. The efficiency of laser energy transfer to deuterons versus deuterons temperature that was obtained with using of neutron yields registered in the SOKOL-P experiments.

The calculated efficiency of energy transfer to deuterons that is needed to explain the observed neutron yield from single targets versus deuteron temperature is presented in Fig 6. At fast ion temperature T_d about 100-500 keV, 0.1-0.8% of laser energy must be transferred to deuterons accelerated into the laser target to obtain the experimental yield from $(CD_2)_n$ targets and 0.1-0.2% to obtain the yield from $TiD_{0.5}T_{0.5}$ targets. Total laser energy transferred to fast ions (into 4π) is about 3 times higher.

CONCLUSION

The neutron generation from $D(d,n)^3He$ and $T(d,n)^4He$ reactions at intensity of 10^{18} W/cm^2 was investigated at the SOKOL-P laser facility. The highest yield from single deuterated targets was about $8 \cdot 10^5$ neutrons. The maximum yield from targets containing tritium was about $2 \cdot 10^6$ neutrons. Yields up to 10^6 DD-neutrons and 10^7 DT-neutrons were detected in experiments with additional targets. The theoretical interpretation of neutron measurements have given the efficiency of laser energy transfer to fast deuterons accelerated into the laser target about 0.1-1% at laser intensity of 10^{18} W/cm^2.

REFERENCES

1. P.A. Norreys, A. P. Fews, F.N. Beg et al. Plasma Phys. Control. Fusion 40, 1998, p.175.
2. L. Disdier, J-P. Garconnet, G. Malka et al. Phys. Rev. Lett 82, 1999, p.1454.
3. N. Izumi, Y. Sentoku, H. Habara et al. Phys. Rev. E 65, 2002, 036413.
4. D.A. Dmitrov, L.A. Fomichev, A.G. Kakshin et al. "10 TW picosecond Nd:glass laser facility Sokol-P" in *Book of Abstracts of 28th ECLIM*, Rome, Italy, 2004, p.187

Multi-Component Multiphase Compressible Hydrodynamics: Current State

V.F. Kuropatenko

Federal State Unitary Enterprise "Russian Federal Nuclear Center —Zababakhin All–Russia Research Institute of Technical Physics", Snezhinsk, Chelyabinsk region, Russia

Abstract. The dynamic loading of multi-component materials disturbs them from equilibrium and generates relaxation processes in which components interact with each other exchanging momentum and energy (and mass if chemical reactions occur). The description of exchange processes through pair interactions allows considering the individual properties of interacting components (particle size, surface roughness, adhesion etc.). The paper discusses a new type of interaction, namely a cluster interaction which gives a new force tensor and new energy fluxes whose dependences on the characteristics of each component and on barycentric velocity are proposed and validated. The author introduces a concept of non-equilibrium kinetic energy for a component and proposes a new equation for volume concentrations which closes the system of conservation equations along with an equation of state for the i-th component, and does not impose additional restrictions on mixture properties. The model admits state and phase changes in each component.

INTRODUCTION

Pure materials are very few in nature. Mixtures are most often. If a small volume $d\theta$ contains several ($N > 1$) materials, each of the materials is called a component and the medium in the volume is called multi-component. Impacts on a multi-component medium lead to displacements of its components, to their mixing or separation, heating and deformation, phase transitions or changes in the state of aggregation (i.e., thyxotropy), chemical reactions in the components and other changes.

The history of multicomponent models goes back to the middle of the XIX century but their theory is still to be completed. Necessary information is still being collected through study and application of particular models. In what all models are weak is the description of interactions between components. Two fundamental problems remain unsolved:

- Derivation of conservation laws for a mixture from the conservation laws for its components; and
- Closure of the system of equations for the i-th component.

Both have been challenging for science during several decades. Below are their solutions.

CP849, *Zababakhin Scientific Talks - 2005*,
edited by E. N. Avrorin and V. A. Simonenko

1. MULTICOMPONENT MODELS

All multicompoenent models can be classified into diffusion models and multivelocity continua models in which the behavior of each component is defined by conservation laws for macro-scale quantities. This means that we have passed from micro- to macro-scale quantities and each i-th component is a continuum physically characterized by a pressure P_i, a temperature T_i, a density ρ_i, a specific internal energy E_i, a velocity $\bar{U}_i$, an entropy S_i etc. Thermodynamic parameters of the i-th component are related through an equation of state.

Consider a small volume $d\theta$ containing a mixture of N component of a mass dM. Divide dM and $d\theta$ between all components:

$$dM = \sum_{i=1}^{N} dM_i, \quad d\theta = \sum_{i=1}^{N} d\theta_i.$$ (1)

The ratios

$$\eta_i = \frac{dM_i}{dM}, \quad \alpha_i = \frac{d\theta_i}{d\theta}$$ (2)

are called [1-5] mass and volume concentrations of the i-th component. Density is defined as a mass of material per unit volume. Therefore

$$\rho_i = \frac{dM_i}{d\theta_i} \text{ and } \rho = \frac{dM}{d\theta},$$ (3)

where ρ_i is the density of the i-th component and ρ is the density of the mixture. Assume that the mass dM_i is "smeared" over the entire volume $d\theta$. The quantity

$$\alpha_i \rho_i = \frac{dM_i}{d\theta}$$ (4)

is virtual and called a partial density of the i-th component. It follows from (2)-(4) that $\alpha_i \rho_i$ and ρ relate as

$$\alpha_i \rho_i = \eta_i \rho.$$ (5)

For the specific volume $V = 1/\rho$, the equation (5) takes the form

$$\alpha_i V = \eta_i V_i.$$ (6)

It follows from (1), (3) and (4) that the density of the mixture is a sum of the partial densities:

$$\rho = \sum_{i=1}^{N} \alpha_i \rho_i . \tag{7}$$

Each component is characterized by a momentum $\bar{U}_i dM_i$. The law of momentum conservation at a fixed time t gives

$$\bar{U} dM = \sum_{i=1}^{N} \bar{U}_i dM_i .$$

Replacing dM_i by $\eta_i dM$ with (2) and canceling dM give that the mixture velocity $\bar{U}$ is a sum of the partial velocities $\eta_i \bar{U}_i$:

$$\bar{U} = \sum_{i=1}^{N} \eta_i \bar{U}_i . \tag{8}$$

It follows from (5) and (8) that the specific momentum $\rho \bar{U}$ of the mixture is a sum of specific momenta for components:

$$\rho \bar{U} = \sum_{i=1}^{N} \alpha_i \rho_i \bar{U}_i . \tag{9}$$

The velocity $\bar{U}$ defined by (8) is called the barycentric velocity [1, 2].

The specific internal energy E is energy per unit mass. The law of internal energy conservation at a fixed time gives

$$E dM = \sum_{i=1}^{N} E_i dM_i .$$

Dividing by dM and using (2) give that the specific energy of a mixture is a sum of the partial specific internal energies for components:

$$E = \sum_{i=1}^{N} \eta_i E_i . \tag{10}$$

Similar manipulations with the specific energy give the following equation for the total specific energy ε:

$$\varepsilon = \sum_{i=1}^{N} \eta_i \varepsilon_i . \tag{11}$$

Using (6), rewrite (10) and (11) in the forms

$$\rho E = \sum_{i=1}^{N} \alpha_i \rho_i E_i \quad \text{and} \quad \rho \varepsilon = \sum_{i=1}^{N} \alpha_i \rho_i \varepsilon_i . \tag{12}$$

$E(V,S)$ and $E_i(V_i,S_i)$ are thermodynamic potentials and their differentials are

$$dE = \left(\frac{\partial E}{\partial V} \right)_S dV + \left(\frac{\partial E}{\partial S} \right)_V dS \quad \text{and} \quad d_i E_i = \left(\frac{\partial E_i}{\partial V_i} \right)_{S_i} d_i V_i + \left(\frac{\partial E_i}{\partial S_i} \right)_{V_i} d_i S_i . \tag{13}$$

41

For $S = \text{const}$ and $S_i = \text{const}$, differentiate (6) and (10) with respect to the thermodynamic variables V, S and V_i, S_i bearing in mind that η_i and α_i are independent of V, S and V_i, S_i. We obtain two equations

$$\left(\frac{\partial E}{\partial V}\right)_S dV = \sum_{i=1}^{N}\left(\frac{\partial E_i}{\partial V_i}\right)_{S_i} \eta_i d_i V_i \text{ and } \alpha_i dV = \eta_i d_i V_i$$

which give

$$\left(\frac{\partial E}{\partial V}\right)_S = \sum_{i=1}^{N}\left(\frac{\partial E_i}{\partial V_i}\right)_{S_i} \alpha_i . \tag{14}$$

Since

$$P = -\left(\frac{\partial E}{\partial V}\right)_S \text{ and } P_i = -\left(\frac{\partial E_i}{\partial V_i}\right)_{S_i},$$

the equation (14) means that the total pressure is a sum of the partial pressures:

$$P = \sum_{i=1}^{N}\alpha_i P_i . \tag{15}$$

This is a general equation. For the case of ideal gases with identical temperatures $T_i = T$, it was proved by Dalton and was called Dalton law.

P_i, ρ_i, E_i, T_i, $\bar{U}_i$ and others are macro-scale quantities of the i-th component which describe it as a continuum. However, each component is a structural element of the mixture. So, these parameters are meso-scale quantities of the mixture. The macro-scale quantities characterizing the behavior of the mixture are derived from the meso-scale quantities with the equations (7), (8), (10), (11), and (15).

2. COMPONENT INTERACTIONS

A multicomponent medium may not be in equilibrium. Mixture equilibrium conditions are

$$P_i = P_j, \quad T_i = T_j, \quad \bar{U}_i = U_j .$$

If at least one of them is not satisfied, the mixture is not in equilibrium; it tries to attain equilibrium through a number of relation processes in which mixture components exchange momentum and energy.

For a long time, multicomponent models have considered only pair interactions in which the i-th and j-th components interact independently of all others [1-4]. If $\bar{U}_i \neq \bar{U}_j$, the exchange of their momenta is most often defined by the vector

$$\bar{R}_{ji} = a_{ij}(\bar{U}_j - \bar{U}_i)/\tau_{ij}^U .$$

The functions a_{ij} and τ_{ij}^U depend on the degree of heterogeneity, on the properties of the i-th and j-th components, on the size of their particles, on sound velocity, compressibility, equations of state for the components, on their states of aggregation, adhesion and other properties in such a way as to satisfy Onsager reciprocity condition

$$a_{ij} = a_{ji}, \quad \tau_{ij}^U = \tau_{ji}^U,$$

due to which $\bar{R}$ satisfies the condition

$$\bar{R}_{ij} = -\bar{R}_{ji}.$$

Specific forms for the dependencies are established from the conditions of each particular problem. They are often validated using a unit cell of two components. The order of the indices ij indicates that the j-th component acts on the i-th one. The action of all N components (i.e., the mixture) on the i-th one is accomplished through summation with respect to j

$$\bar{R}_{Si} = \sum_{j=1}^{N} \alpha_j \bar{R}_{ji} . \tag{16}$$

From (16), $\bar{R}_S$ acting on the mixture from all components vanishes:

$$\bar{R}_S = \sum_{j=1}^{N} \sum_{i=1}^{N} \alpha_i \alpha_j \bar{R}_{ij} = 0 . \tag{17}$$

Similar reasoning applies to a scalar function Φ_{ij} which describes energy exchange between the i-th and j-th components. As a rule [4], the function is taken in the form

$$\Phi_{ji} = \frac{b_{ji}}{\tau_{ji}^P}(P_j - P_i) + \frac{c_{ji}}{\tau_{ji}^T}(T_j - T_i) , \tag{18}$$

where

$$b_{ij} = b_{ji}, \quad c_{ij} = c_{ji}, \quad \tau_{ij}^P = \tau_{ji}^P, \quad \tau_{ij}^T = \tau_{ji}^T, \quad \Phi_{ji} = -\Phi_{ij} .$$

The flux of energy to the i-th component from all others results from summation:

$$\Phi_{Si} = \sum_{j=1}^{N} \alpha_j \Phi_{ji} .$$

The energy flux acting on the mixture from all components is zero:

$$\Phi_S = \sum_{i=1}^{N} \sum_{j=1}^{N} \alpha_i \alpha_j \Phi_{ij} = 0 . \tag{19}$$

The equations (17) and (19) are fundamental to the pair interaction models.

With the equations (7), (8), (10), (11) and (15) for the macro-scale quantities ρ, $\bar{U}$, E, ε and P, we can introduce a new type of interaction in which each i-th component interacts with the mixture. Call it a **cluster** interaction. In this interaction, momentum and energy changes in the i-th component are controlled by the parameters P, ρ, E, $\bar{U}$, and T for the mixture and by the parameters P_i, ρ_i, E_i, $\bar{U}_i$, and T_i for the i-th component. To express the cluster interaction, introduce a force tensor and an energy

43

flux. Let F_{Si} stand for the tensor of forces to the i-th component from the mixture and $\bar{Q}_{Si}$ for the energy flux. They satisfy the cluster interaction conditions

$$F_S = \sum_{i=1}^{N} \alpha_i F_{iS} = -\sum_{i=1}^{N} \alpha_i F_{Si}, \quad \bar{Q}_S = \sum_{i=1}^{N} \alpha_i \bar{Q}_{iS} = -\sum_{i=1}^{N} \alpha_i \bar{Q}_{Si}. \tag{20}$$

3. CONSERVATION LAWS FOR COMPONENTS

Write the conservation laws for mass, momentum and energy as

$$\frac{\partial}{\partial t}(\alpha_i \rho_i) + \nabla(\alpha_i \rho_i \bar{U}_i) = 0, \tag{21}$$

$$\frac{\partial}{\partial t}(\alpha_i \rho_i \bar{U}_i) + \nabla \alpha_i P_i + \sum_{k=1}^{3} \frac{\partial}{\partial x_k}(\alpha_i \rho_i \bar{U}_i U_{ki} + \alpha_i \bar{F}_{kSi}) - \alpha_i \bar{R}_{Si} = 0, \tag{22}$$

$$\frac{\partial}{\partial t}(\alpha_i \rho_i \varepsilon_i) + \nabla(\alpha_i \bar{U}_i(P_i + \rho_i \varepsilon_i)) + \sum_{k=1}^{3} \frac{\partial}{\partial x_k}(\alpha_i \bar{F}_{kSi} \bar{U}_i) + \nabla \alpha_i \bar{Q}_{Si} - \alpha_i \Phi_{Si} - \alpha_i A_{Si} = 0. \tag{23}$$

Here A_{Si} is the work of the vector $\bar{R}_{Si}$ defined as

$$A_{Si} = 0.5 \sum_{j=1}^{N} \alpha_j \bar{R}_{ji}(\bar{U}_i + \bar{U}_j).$$

Add to the equations (21)-(23) an equation of state in the form

$$P_i = P_i(\rho_i, E_i), \quad T_i = T_i(\rho_i, E_i)$$

and an equation for ε_i in terms of E_i and $0.5\bar{U}_i^2$:

$$\varepsilon_i = E_i + 0.5\bar{U}_i^2. \tag{24}$$

Compared with [1-3], these equations contain several new quantities, namely a vector $\bar{F}_{kSi}$ created by elements of the k-th line of F_{Si}, a vector $\bar{Q}_{Si}$, and a scalar A_{Si}.

The macro-scale quantities P, ρ, E, $\bar{U}$, ε, and T characterize a continuum (a mixture). Write mass, momentum and energy conservation laws for the mixture with the cluster interaction in the form

$$\frac{\partial \rho}{\partial t} + \nabla \rho \bar{U} = 0, \tag{25}$$

$$\frac{\partial}{\partial t}(\rho \bar{U}) + \bar{U}\nabla(\rho \bar{U}) + \rho(\bar{U}\nabla)\bar{U} + \nabla P + \sum_{k=1}^{3} \frac{\partial}{\partial x_k} \bar{F}_{kS} = 0, \tag{26}$$

$$\frac{\partial}{\partial t}(\rho \varepsilon) + \nabla \bar{U}(P + \rho \varepsilon) + \sum_{k=1}^{3} \frac{\partial}{\partial x_k}(\bar{U}\bar{F}_{kS}) + \nabla \bar{Q}_S = 0. \tag{27}$$

$\overline{F}_{kS}$ contains elements of the k-th line of $\overline{F}_S$ defining the force to the mixture from all components. F_S and F_{Si}, $\overline{Q}_S$ and $\overline{Q}_{Si}$ are related through the cluster interaction condition (20).

4. FORCE TENSOR F_{Si}

Let $\delta \overline{U}_i = \overline{U}_i - \overline{U}$ be velocity oscillations round a mean $\overline{U}$. Substitute

$$\overline{U}_i = \overline{U} + \delta \overline{U}_i$$

in the momentum equation for the i-th component (22) and sum with respect to i. We obtain

$$\frac{\partial}{\partial t}(\rho \overline{U}) + \overline{U}\nabla(\rho \overline{U}) + \rho(\overline{U}\nabla)\overline{U} + \nabla P + \sum_{i=1}^{N}\sum_{k=1}^{3}\frac{\partial}{\partial x_k}(\alpha_i(\overline{F}_{kSi} + \rho_i \delta \overline{U}_i \delta \overline{U}_{ki})) = 0. \quad (28)$$

The first four terms in this equation coincide with those in the equation (26). Equation (28) coincides with the equation (26) if

$$\sum_{i=1}^{N}\sum_{k=1}^{3}\frac{\partial}{\partial x_k}(\alpha_i(2\overline{F}_{kSi} + \rho_i(\overline{U}_i - \overline{U})(U_{ki} - U_k))) = 0.$$

Zeroing each of the summands, integrating with respect to x_k, and requiring that $\overline{F}_{kSi} = 0$ at $\overline{U}_i = \overline{U}$ gives the following equation for $\overline{F}_{kSi}$:

$$\overline{F}_{kSi} = -0.5\rho_i(\overline{U} - \overline{U}_i)(U_k - U_{ki}). \quad (29)$$

5. ENERGY FLUX $\overline{Q}_{Si}$

The specific kinetic energy of velocity oscillations is

$$H_i = 0.5(\overline{U}_i - \overline{U})^2. \quad (30)$$

It follows from the conservation laws for each type of energy at $t = \text{const}$ that

$$\rho H = \sum_{i=1}^{N}\alpha_i \rho_i H_i. \quad (31)$$

Substituting (30) into (31) and using (7) and (9) yield

$$\rho H + 0.5\rho \overline{U}^2 = \sum_{i=1}^{N}0.5\alpha_i \rho_i \overline{U}_i^2. \quad (32)$$

Using (12), express the mixture parameters $P\overline{U}$, $\rho\varepsilon$, $\rho\varepsilon\overline{U}$, $\overline{Q}_S$, $\overline{F}_{kS}$, and A_S in terms of partial quantities

45

$$P\bar{U} = \sum_{i=1}^{N} \alpha_i P_i \bar{U}_i + \sum_{i=1}^{N} \alpha_i P_i (\bar{U} - \bar{U}_i), \quad \rho \varepsilon \bar{U} = \sum_{i=1}^{N} \alpha_i \rho_i \varepsilon_i U_i + \sum_{i=1}^{N} \alpha_i \rho_i \varepsilon_i (\bar{U} - \bar{U}_i),$$

$$\bar{Q}_S = -\sum_{i=1}^{N} \alpha_i \bar{Q}_{Si}, \quad \bar{F}_{kS} = -\sum_{i=1}^{N} \alpha_i \bar{F}_{kSi}, \quad -\sum_{i=1}^{N} \alpha_i A_{Si} = 0, \quad -\sum_{i=1}^{N} \alpha_i \Phi_{Si} = 0$$

and substitute in (27). We obtain

$$\sum_{i=1}^{N} \left[\frac{\partial}{\partial t} (\alpha_i \rho_i \varepsilon_i) + \nabla(\alpha_i \bar{U}_i (P_i + \rho_i \varepsilon_i)) + \right.$$

$$\left. + \sum_{k=1}^{3} \frac{\partial}{\partial x_k} \left(\alpha_i \bar{F}_{kSi} \bar{U}_i \right) + \nabla \alpha_i \bar{Q}_{Si} - \alpha_i \Phi_{Si} - \alpha_i A_{Si} + B_i \right] = 0, \qquad (33)$$

where

$$B_i = \nabla(\alpha_i (P_i + \rho_i \varepsilon_i)(\bar{U} - \bar{U}_i)) - 2\nabla \alpha_i \bar{Q}_{Si} - \sum_{k=1}^{3} \frac{\partial}{\partial x_k} \left(\alpha_i \bar{F}_{kSi} (\bar{U} + \bar{U}_i) \right).$$

Each i-th summand in (33) coincide with the energy equation (23) for the i-th component if $B_i = 0$. Using (24) and (29), rewrite it as

$$\nabla(\alpha_i ((\bar{U} - \bar{U}_i)(P_i + \rho_i E_i) - 2\bar{Q}_{Si})) = 0. \qquad (34)$$

Integrating (34) with respect to x_k and finding integration constants from the condition that $\bar{Q}_{Si} = 0$ at $\bar{U}_i = \bar{U}$ give the following expression for the energy flux:

$$\bar{Q}_{Si} = 0.5(\bar{U} - \bar{U}_i)(P_i + \rho_i E_i). \qquad (35)$$

With the equations (29) and (35) for $\bar{F}_{kSi}$ and $\bar{Q}_{Si}$, the conservation laws (21)-(23) for the i-th component are invariant to Galilean transformation and have the property that the summation of conservation laws for components with respect to i gives corresponding conservation laws for the mixture. The first of the above two problems is thus resolved.

6. COROLLARIES TO CONSERVATION LAWS

Before we start to manipulate the conservation laws (21)-(23) and (25)-(27), consider components of the total specific energy ε. Multiply ε_i (24) by $\alpha_i \rho_i$ and sum with respect to i:

$$\sum_{i=1}^{N} \alpha_i \rho_i \varepsilon_i = \sum_{i=1}^{N} \alpha_i \rho_i E_i + \sum_{i=1}^{N} 0.5 \alpha_i \rho_i \bar{U}_i^2.$$

After substituting (12) and (32) in this equation and canceling ρ we obtain

$$\varepsilon = E + 0.5\bar{U}^2 + H.$$

Thus the total specific energy of the mixture, ε, is the sum of the specific internal energy of the mixture, E, the specific kinetic energy of the mixture, $0.5\bar{U}^2$, and the specific non-equilibrium kinetic energy of the mixture, H (specific energy of velocity oscillations). As velocity relax, H goes into E.

Applying identical transformations to the conservation laws (21)-(23) for the i-th component and (25)-(27) for the mixture gives equations of motion, equations for the specific internal energy, and equations for specific entropy:

$$\alpha_i\rho_i\frac{d_i\bar{U}_i}{dt}+\nabla\alpha_iP_i+\sum_{k=1}^{3}\frac{\partial}{\partial x_k}(\alpha_i\bar{F}_{kSi})-\alpha_i\bar{R}_{Si}=0,$$

$$\alpha_i\rho_i\frac{d_iE_i}{dt}+\alpha_i\left(P_i\nabla\bar{U}_i+\sum_{k=1}^{3}\bar{F}_{kSi}\frac{\partial\bar{U}_i}{\partial x_k}+\bar{R}_{Si}\bar{U}_i-A_{Si}-\Phi_{Si}\right)+\nabla\alpha_i\bar{Q}_{Si}=0,$$

$$\alpha_i\rho_iT_i\frac{d_iS_i}{dt}-P_i\frac{d_i\alpha_i}{dt}+\alpha_i\left(\sum_{k=1}^{3}\bar{F}_{kSi}\frac{\partial\bar{U}_i}{\partial x_k}+\bar{R}_{Si}\bar{U}_i-A_{Si}-\Phi_{Si}\right)+\nabla\alpha_i\bar{Q}_{Si}=0, \quad (36)$$

$$\rho\frac{d\bar{U}}{dt}+\nabla P+\sum_{k=1}^{3}\frac{\partial\bar{F}_{kS}}{\partial x_k}=0,$$

$$\rho\frac{dE}{dt}+\rho\frac{dH}{dt}+P\nabla\bar{U}+\sum_{k=1}^{3}\bar{F}_{kS}\frac{\partial\bar{U}}{\partial x_k}+\nabla\bar{Q}_S=0,$$

$$\rho T\frac{dS}{dt}+\frac{\partial\rho H}{\partial t}+\nabla(\rho H\bar{U})+\sum_{k=1}^{3}\bar{F}_{kS}\frac{\partial\bar{U}}{\partial x_k}+\nabla\bar{Q}_S=0. \quad (37)$$

For $V=\text{const}$ and $V_i=\text{const}$, and for η_i and α_i independent of thermodynamic quantities, the equations

$$dE=\sum_{i=1}^{N}\eta_id_iE_i, \quad dE=\left(\frac{\partial E}{\partial S}\right)_V dS, \quad d_iE_i=\left(\frac{\partial E_i}{\partial S_i}\right)_{V_i}d_iS_i, \quad T=\left(\frac{\partial E}{\partial S}\right)_V, \quad T_i=\left(\frac{\partial E_i}{\partial S_i}\right)_{V_i}$$

give

$$TdS=\sum_{i=1}^{N}\eta_iT_id_iS_i.$$

Using (6), the equation can be written as

$$\rho TdS=\sum_{i=1}^{N}\alpha_i\rho_iT_id_iS_i.$$

Dividing by dt gives

$$\rho T\frac{dS}{dt}=\sum_{i=1}^{N}\alpha_i\rho_iT_i\frac{d_iS_i}{dt} \quad (38)$$

47

which expresses entropy production for the mixture as the sum of entropy productions for components.

Substituting (37) and (36) in (38) gives the equation for the volume concentration of the i-th component:

$$P_i \frac{d_i \alpha_i}{dt} + (\bar{U} - \bar{U}_i) \nabla \alpha_i P_i - 2 \nabla \alpha_i Q_{Si} = 0. \tag{39}$$

This equation closes the conservation laws for the i-th component. Thus the second problem is also resolved.

7. MODEL PROPERTIES

The new force F_{Si} (29) and energy flux $\bar{Q}_{Si}$ (35) contain quantities with the subscript i which characterize components, or structural elements of a multicomponent medium and only two macro-scale quantities, namely density ρ and velocity $\bar{U}$ of the mixture. The force F_{Si} and the flux $\bar{Q}_{Si}$ are universal; they do not contain empiric constants and do not depend on component characteristics or properties which control relaxation times. This is the property in which the cluster interaction fundamentally differs from the pair one.

The force F_{Si} and the flux $\bar{Q}_{Si}$ vanish as velocity equilibrium establishes. It follows from (39) that at $\bar{U}_i = \bar{U}$ the volume concentrations of components remain constant along trajectories.

Conservation laws for a multicomponent medium result from the summation of the laws for components. This property of the model is no more than an argument in its favor because the behavior of the i-th component can be described without these laws. Equations for the i-th component include the conservation laws (21)-(23), equations for A_{Si}, $\bar{F}_{Si}$, $\bar{Q}_{Si}$, $\bar{R}_{Si}$, Φ_{Si} and ε_i, the equation (39) for α_i, equations of state, and the equations (8) and (9) for ρ and U. So, the number of equations and the number of functions are identical in the full system of equations which is closed with no additional hypotheses specifying the mixture. The model presented makes it possible to describe concurrently a lot of phenomena in multicomponent media and even increase the accuracy of predictions for systems not yet studied experimentally.

The work was supported by Russian Foundation for Basic Research under Project 04-01-00050.

REFERENCES

1. A.N. Kraiko, R.I. Nigmatulin, V.K. Starkov, and L. B. Sternin, Itogi Nauki Tekh., Ser.: Gidromekh. **6**, 93 (1973).
2. R.I. Nigmatulin, *The Essentials of Heterogeneous Medium Mechanics* (Nauka, Moscow, 1978) [in Russian].
3. N.N. Yanenko, R.I. Soloukhin, A.N. Papyrin, and V.M. Fomin, *Supersonic Two-Phase Flows Under the Conditions of Velocity Nonequilibrium of Particles* (Nauka, Novosibirsk, 1980) [in Russian].
4. V.F. Kuropatenko, Mat. Model. **1** (2), 118 (1989).
5. V.F. Kuropatenko, A multicomponent model. *Proceedings of RF Academy of Sciences*, 2005, #6, P. 761-763.

SECTION 1

CUMULATIVE PHENOMENA AND HIGH-INTENSIVE PROCESSES

Development of A595 Explosion-Resistant Container Design. Numerical, Theoretical and Experimental Justification of the Container Design Parameters

A.I. Abakumov, I.V. Devyatkin, V.Yu. Meltsas, A.L. Mikhailov,
G.F. Portnyagina, V.N. Rusak, V.P. Solovyev, M.A. Syrunin,
S.M. Treshalin, A.G. Fedorenko

*Russian Federal Nuclear Center – All-Russian Research Institute of Experimental Physics, 607190
Sarov Russia*

Abstract. The paper presents the results of numerical and experimental study on the AT595 metal-composite container designed in VNIIEF within the framework of international collaboration with SNL (USA). This container must completely contain products of an 8-kg-TNT detonation cased in 35 kg of inert surrounding material. Numerical and theoretical studies have been carried out of the containment capacity and fracture of small-scale open cylinder test units and container pressure vessel models subjected to different levels of specific explosive load (beneath, equal to and above the required design load defined for this container), and two AT595 containers have been tested for the design load and a higher load.

INTRODUCTION

One of the approaches to protect the environment against the impacts of an explosion is to contain the detonation in an explosion-resistant vessel. Containment vessels are required for evacuation of terrorist explosive devices that may be found in residential areas, of damaged munitions, accidental or special devices containing HE or other hazardous products. These containers can also be used for scientific research or other activities involving explosives.

It is clear that improving the design quality and reliability of such containers and reducing their cost is of essential importance now under the high threat of terrorism, and more efficient new approaches, including technical means, must be developed to resolve this problem. A key issue in the development of such container designs is to minimize their dimensions and weight without decreasing the containment capacity, their affordable cost (i.e. the minimum cost among commercially available similar designs) being also very important.

1. AT595 CONTAINER DESIGN

The container AT595 (see figure 1) designed in VNIIEF in the framework of international collaboration with the SNL (USA) is a cylindrical container with

CP849, *Zababakhin Scientific Talks - 2005*,
edited by E. N. Avrorin and V. A. Simonenko
© 2006 American Institute of Physics 0-7354-0345-7/06/$23.00

hemispherical end caps and a two-layer pressure vessel [1-3]. The inner steel lining (steel 12X18H10T) ensures leak tightness, and the outer layer is made of basalt composite and bears the load. Loading mouths are located at the end caps that are hermetically closed by steel lids. Packages of tied-wire fabric are used in the vessel as shrapnel protection [4], while the mouth lids are protected from shock-wave and fragment impacts by steel cylinders filled with polyfoam and covered with inner lids [2,3,5]. Roving RB9-1250-4C of basalt fiber 9μm in diameter is used for the basalt plastic layer. This layer is made by combined winding of ribbons of epoxy-binder-impregnated braids, in interchanging double spiral and ring layers with about equal thickness ratio [6, 7].

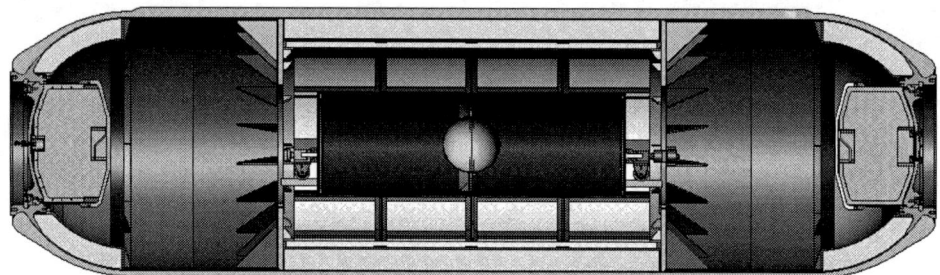

FIGURE 1. The AT595 container design schematic.

The container working space where the HE cased in inert material (the hazardous object) is placed, is limited by throttles designed to mitigate the impacts of shock wave and gaseous products resulting from HE detonation on the container end caps [5]. In addition, the throttles protect the peripheral areas of the end caps against fragments resulting from HE detonation.

The total length of the container is 3320mm, the outer diameter is 985mm and the working diameter of the mouth is 420mm. The dangerous load consists of a spherical HE charge, 8 kg in TNT equivalent, surrounded by a 35kg cylindrical inert casing.

2. COMPUTER CODES USED IN NUMERICAL SIMULATIONS

The computer codes B71 [8] and DRACON [9] were used to perform numerical analyses of the container performance under detonation load. The code B71 is based on S.K.Godunov's method and intended for gas dynamic flow simulations. The code DRACON is used to model stress state in the elements of complex 3D structures taking into account the non-linear behavior of material and contact interactions under intense mechanical and thermal load in 2D and 3D geometries.

Shrapnel protection made of tied-wire fabric with 2x2mm cells and 0.5mm wire thickness is installed in the container. Owing to this, detonation products, as shown experimentally, accelerate the net as they are moving through it, thus decreasing its permeability for detonation products. In addition, the net, due to its branched structure, efficiently reduces the products temperature. The presence of tied-wire fabric in the container required joint simulations of gas dynamics and the response of the container elements.

Two-component two-velocity material model is used to model the behavior of shrapnel protection net under detonation load in the AT595 container. Gas flow inside and outside of the net package, as well as the heat transfer between gas and net are calculated by B71, while DRACON is used for the net compression and elastoplastic deformation. The communication model that was developed for this particular purpose solves the problem of contact interaction at the interface between gaseous products and deformable solids and implements mutual transfer of necessary information to other modules.

Taking into account the peculiarities of the behavior of AT595 container subjected to detonation loading, the computer code DRACON was augmented by modules for:

- Anisotropy of mechanical properties of basalt plastic shell made by spiral-ring winding;
- The fracture of basalt plastic layer;
- Strain rate effects on steel shell strength parameters;
- Shrapnel protection net deformation under HE detonation product impact;
- Interaction between HE inert casing and shrapnel protection net and load transfer to steel shell inner surface.

3. THE TESTING OF THE AT595 CONTAINER

Two AT595 containers were fabricated to experimentally confirm the required containment capacity and to evaluate the container strength margin. The loading regimes for the design load and the increased load were chosen based on the results of the open cylinder tests and numerical predictions.

The two containers were tested by the detonation of HE contained in the dangerous load inside the container, see fig.2. The dangerous load designed for explosive tests on containers includes the following elements:

- Inert casing with the actual weight of ~38 kg made as a four-layer (lead, plastic, aluminum, steel) hollow cylinder, closed by four-layer lids at its edges, with aluminum hemispherical beds mounted in the container center where the HE charge is placed;
- The HE charge is spherical and is made of 40:60 trotyl-hexogen alloy with specific caloric value of 4.93 MJ/kg, the actual weight of 7.6 kg (experiment 1) and 10.35 kg (experiment 2). Scaled to TNT, the caloric value of this is equivalent to 8.86 kg and 12.06 kg TNT, while its impulse action is equivalent to 8.2kg and 11.17kg TNT [2].

FIGURE 2. The AT595 container with the dangerous load and the HE charge.

Streak camera, tensometry, electrocontact measurements [2, 3] were used in the container tests to record fast phenomena. These techniques were used to measure strain and displacements in the key elements of the AT595 containers.

3.1. The First Container Test

The explosive test performed on the first AT595 container showed that this vessel was able to contain the internal load created by the detonation of the dangerous load, thus confirming the containment capacity of the AT595 container for the design level of explosive load.

The detailed history of the phenomena occurring under explosive loading of the container can be provided by numerical modeling. Fig. 3 presents the product pressure field and the location of the dangerous load shells at different moments of time.

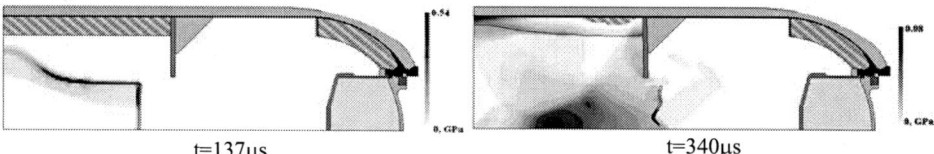

t=137µs   t=340µs

FIGURE 3. Detonation product pressure field and the location of the dangerous load shells at different moments of time.

The analysis of the container deformed state carried out based on strain measurements revealed that it is the most severely loaded central section of the pressure vessel that determines the vessel strength, as was predicted numerically before the test. The distribution of peak strain achieved on the basalt plastic shell outer surface throughout the loading process is shown in fig. 4. This figure also provides experimental data recorded by different strain measurement techniques. It is seen from fig.4 that the maximum level of strain that takes place over the detonation center is 1.8-1.9%. It should be noted that the numerical results are in good agreement with the experimental data.

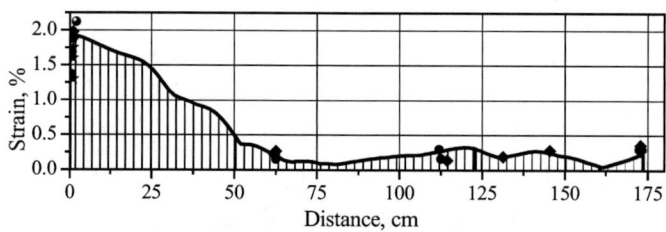

FIGURE 4. Peak hoop strain distribution on the outer surface of basalt plastic shell. Comparison to experimental data.

The shrapnel protection layer of tied-wore fabric inserted in the container was considerably damaged, but it was able to protect the container inner surface against the fragment impact. The shrapnel protection of the hemispherical end caps received almost no damage. Severe damage, i.e. a through hole was observed in the damper inner lid that was not shrapnel-protected.

Judging by the damage character on the container inner parts, the severest damage was caused by fragments of both destroyed inert casing fixture and the casing itself.

3.2. The Second Container Test

The test performed on the second AT595 container with dangerous load containing a spherical HE charge of 10.35kg cased in 37.2 kg of inert material demonstrated that the AT595 container preserves its bearing capacity under a load from explosive and shrapnel impacts that exceeds the design value by a factor of ~1.4.

Though the container preserved its integrity in this case, the basalt plastic layer shows considerable damage in the two opposite zones in the central section (fig.5). This damage was caused by the impact of the fragments of the cart where the dangerous load was mounted. In addition to basalt plastic, local micro damage was found in the steel lining.

FIGURE 5. The second AT595 container after the performed test (views from the opposite sides).

Similarly to the first container test, the inner lid of the damper was severely damaged. Taking into account the results of the previous test, a package of tied-wire fabric was installed in the damper compartment in this experiment. No other methods could be used to reinforce the inner lid, since the container had been already manufactured.

The container performance in this experiment with a higher load can be seen from the numerical results. Fig.6 shows detonation product pressure distribution field and the location of the dangerous load shells at different moments of time. It is seen here that at the time 335µs the bottom of the dangerous load has developed a shape that would promote the lid perforation. The maximum velocity of the lid pole point of the dangerous load lid was 3.3km/sec. The character of inner lid perforation observed in the experiment is shown in fig.7.

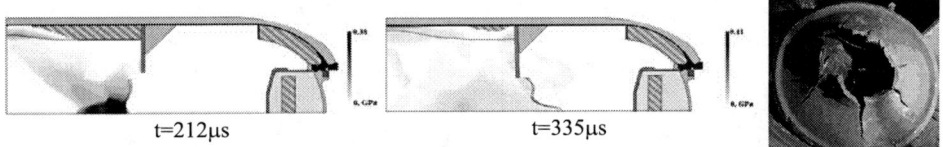

t=212µs   t=335µs

FIGURE 6. Product pressure distribution fields and the location of the dangerous load shells at different moments of time.

FIGURE 7. A post-test view of the inner lids.

The distribution of the maximum achieved strain over the outer surface of the basalt plastic shell throughout the loading process is shown in fig.8. This figure also presents experimental data. It is seen from fig.8 that the peak strain level over the detonation center is 2.8-2.9%. Here, similarly to the previous experiment, the results are in rather good agreement with the experimental data.

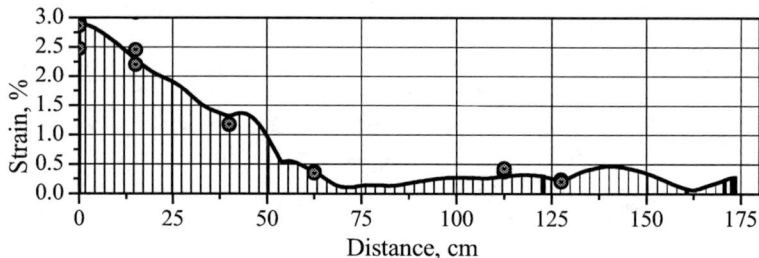

FIGURE 8. Peak hoop strain distribution on the outer surface of the basalt plastic shell. Comparison to experimental data.

It can be concluded based on the obtained results that despite the damage that developed in the test, the container vessel preserved its integrity and, therefore, its bearing capacity. Therefore, the AT595 container design has a margin of not less than 1.4 and, at the same time, there is still room for further optimization to increase its strength margin within practically the same weight and dimension limits.

The work was performed under Contract #12831 between SNL (USA) and VNIIEF (Russia)

REFERENCES

1. Editorial Board: RAS Academician R.I.Ilkaev et. al., The main developments of the RFNC-VNIIEF, 2003, Sarov, 2004.
2. Edited by Prof. A.G.Ivanov, *Explosive fracture of objects of different scale*, Monograph, RFNC-VNIIEF, Sarov. 2001, 482p
3. . Syrunin M.A., Fedorenko A.G., Ivanov A.G., "Response to loading and strength of a fiberglass plastic container subjected to internal detonation loading", *Fizika gorenya i vzryva*, vol.38, 3, 2002, p. 127-136.
4. Belov G.V., Dyakin E. P., Protasov S. A. et al. "Penetration of compact steel projectiles into heterogeneous metal targets of tied-wire fabric (TWF) type", *International Journal of Impact Engineering*. 23, 1999, c. 63-66.
5. Ryzhanskiy V.A.., Rusak V.N., Zaikin S.N. "Explosion-Resistant Container". *Patent 1793790, Russian Federation. Byul. Izobr* 4, 1995. C.257.
6. Rusak V.N, Fedorenko A.G., Syrunin M.A., Sobol L.A., Sukhanov A.V., Popov V.G. "Ultimate deformability and strength of basalt plastic shells subjected to internal detonation load", *Prikladnaya Mehanika I Tehnicheskaya Fizika*,43, 2002, 1, p.186-195.
7. Ivanov A.G., Syrunin M.A., Fedorenko A.G. "The influence of winding structure on the ultimate deformability and strength of shells made of oriented fiberglass plastic subjected to internal detonation loading", *Prikladnaya Mehanika I Tehnicheskaya Fizika*, 4. 1992, p. 130-135.
8. S.K.Godunov. *Numerical Solution of Multidimensional Gas Dynamics Problems*. Moscow, Nauka, 1976.
9. A.I.Abakumov, P.N.Nizovtsev, A.V.Pevnitskiy, V.P.Solovyev. "The DRACON Code for Calculations of Elastoplastic Flows under Shock-Wave Load in 2D and 3D". *Proceeding of the International Conference (IV Zababakhin's Scientific Conference)*. 1995, pp.89-90.

Shock and Explosive Interaction of Charge in Shell with Combined Barrier

A.V. Gerasimov, N.S. Alexeev, V.N. Mikhailov,
SV. Pashkov

Research Institute of Applied Mathematics and Mechanics, Tomsk, Russia

Abstract. The problems of shock and explosive interactions of monolithic and layered barriers with flattening and penetrating shells, filled in with charges of plastic or ordinary explosives, are considered in two- and three-dimensional positions. Barriers can include several layers of porous or easily deformable material. Detonation initiation proceeds both during shock interaction and just after it. Barrier piercing by an explosive-charged shell, barrier and shell fragmentations are considered taking into account natural heterogeneity of the material.

Keywords: **Impact, fragmentation, shell, charge**

PACS: **02.60.Cb, 02.70.Bf, 62.20.MK.**

INTRODUCTION

Heterogeneity of real material structures is one of the factors determining fracture pattern, since it affects distribution of physico-mechanical characteristics through material volume. One used random distribution of strength property deviations from nominal value to simulate initial faulted structures.

Basic relations, describing motion of porous strong retractable and perfectly elastoplastic medium, are based on laws of conservation of mass, impulse and energy and are closed by Prandtl-Reuss relations on Mises flow condition. Equation of state is used in Tate and Mi-Gruneisen form [1]. The given set of equations is supplemented with a kinetic equation describing pore growth and compression [2]. Attainment of limit value for equivalent plastic deformation was used as a fracture criterion at intensive shear deformations [3]. Attainment of critical value for porosity was used to simulate fractures of spall type. Detonation products are considered as non-viscous and non-heat-conducting gas; Landau-Stanyukovich equation was used as an equation of state [1].

Wilkins method was used to solve two-dimensional problems, whereas to calculate three-dimensional flows one used the procedure realized on tetrahedral meshes and based on combined application of Wilkins method intended to calculate inner body points [4] and of Johnson method intended to calculate contact interactions [5]. For flattening shells studied were the effects of physico-mechanical characteristics of the

CP849, *Zababakhin Scientific Talks - 2005*,
edited by E. N. Avrorin and V. A. Simonenko
© 2006 American Institute of Physics 0-7354-0345-7/06/$23.00

material of interacting bodies, number and geometrical design of barrier layers on the pattern of barrier stressed-strained state and fracture.

RESULTS

Numerical simulation of ring fracture problem was carried out as a test calculation [6]. Limit value of equivalent plastic deformation was distributed in meshes of the estimated range according to normal distribution law with 10-percentage deviation variance. As the ring expands one could observe the formation of large fragments (Figure 1); the comparison of the calculated (circles) and experimental (crosses) data [6] testified to their completely satisfactory fit (Fig.2).

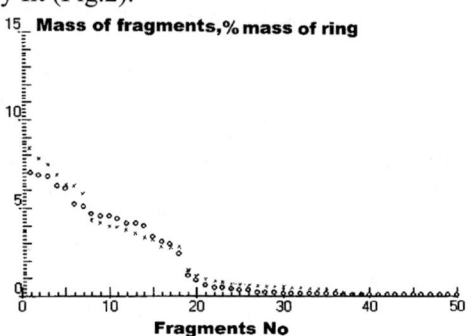

FIGURE 1. Ring fragmentation

FIGURE 2. Fragment spectrum

Fig. 3 presents system configurations for two points of time, where normal barrier piercing and shell fragmentation is shown. The shell was made of steel and barrier – of copper; trotyl was used as explosive. Impact rate was equal to 1000 m/s. Outer radius of the shell was 0.03 m and inner one - 0.015 m, length - 0.12 m, barrier thickness – 0.01 m and that of shell bottom – 0. 015 m and barrier radius was - 0.12 m. While calculating one realized delay of explosive detonation initiation, which amounted to 100 μs. As a result shell fragmentation started after its penetration through a barrier, as is clearly seen in Fig. 3, b.

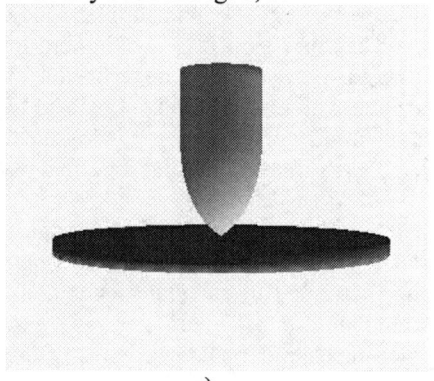

a)
b)
FIGURE 3. Shell-barrier interaction at normal impact: a)0 μs; b)180 μs

At shell and barrier fractures two groups of fragments are formed, which interact with each other. A mesh size in detonation products, comparable with that in shell material, in case of Lagrange approach to the description of continuum motion makes it impossible to describe gas break through the forming cracks at earlier stages of fracture. However towards destruction point shell acceleration has been generally completed and one can neglect pressure loss because of detonation product outflow. The process of shell and barrier fragmentations (explosive initiation proceeds at the bottom at t=100 μs) is presented in Fig. 3; probabilistic nature of fragments formation is observed on the external shell surface, where the fragments of different sizes are formed. Owing to probabilistic nature of barrier fracture at the points of time t >0 the problem is no longer axially symmetric and becomes spatial. It is clearly seen from the figures, where cracks appear on the back surface of the barrier and fragment field is formed. One might note that the fragments have different shapes and sizes relative to the initial axis of symmetry of interacting barrier and shell. Such character of fragmentation is more adequate to real processes of barrier piercing, than the results of the studies considering the piercing process without taking into account probabilistic nature of real body fragmentation.

The process of barrier piercing at angel of 15^0 with explosive-charged shell and further detonation of the explosive, are presented in Fig. 4. The delay in explosive charge initiation is 100 μs. In this case the initial asymmetry of impact process is reflected on barrier fracture. The processes of barrier and shell fractures are defined by the interactions of deterministic and probabilistic factors, which are observed under actual fracture conditions.

a)   b)

FIGURE 4. Shell-barrier interaction at oblique impact: a)0 мкс; b)163 μs

The impact of shell, charged with plastic explosive, with a laminated plate is considered below (Fig. 5). The plate dimensions are as follows: thickness 4 cm and radius - 11 cm; shell dimensions: outer radius – 3.5 cm, inner radius - 3 cm, length – 5.5 cm, thickness of bottom contacting with plate is 0.5 cm and that of external bottom – 0.6 cm; explosive charge height is 4.4 cm and its radius - 3 cm. The plate is made of steel and shell – of copper. The density of explosive charge $\rho_0 = 1.6$ g/cm^3 and detonation rate D =6000 m/c.

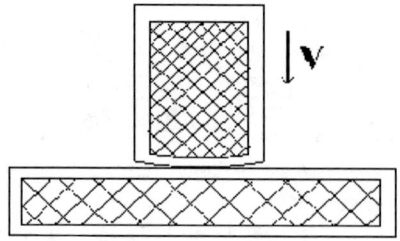

FIGURE 5. Scheme of charged shell interaction with barrier

Figures 6 and 7 present calculations for the interaction of explosive-charged shell with monolithic and laminated barriers, respectively. As is seen from the calculations

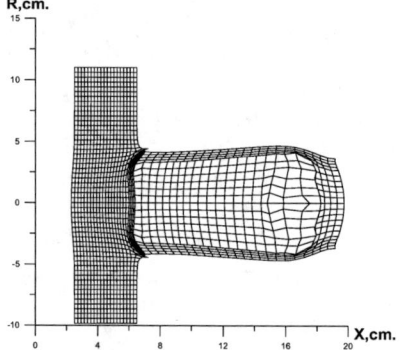

FIGURE 6. Plate-shell system - at point of time t=26 μs   (monolithic barrier)

presented in Fig 8 in case of monolithic barrier porosity amounts to critical value on a back surface and therefore spallation of barrier pieces is possible. The injection of a porous layer (steel, 0.5 cm) into the barrier (Fig.9) decreases damage level below critical one and thereby prevents from spall effect.

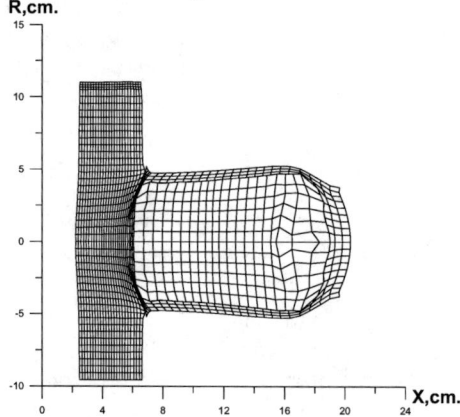

FIGURE 7. Plate-shell system - at point of time t =35 μs (barrier with a porous layer)

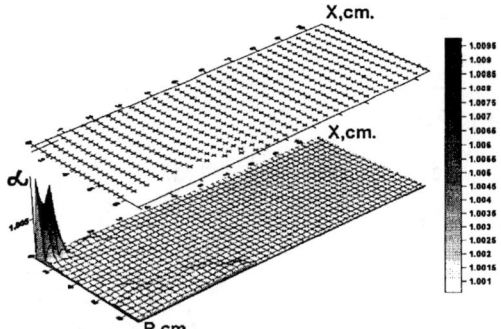

FIGURE 8. Porosity δ distribution in a monolithic barrier (t=26 μs).

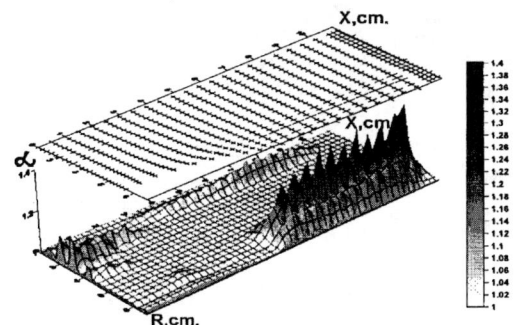

FIGURE 9. Porosity δ distribution in the barrier with a porous layer (t=35 μs).

CONCLUSION

Due to the developed technique intended to solve fragmentation problems one can most entirely, from physical view point, reproduce the processes of solid fragmentation in three-dimensional position under explosive and shock loads. Moreover it is also possible to affect the character of a fragment spectrum for the construction under consideration varying physico-mechanical parameters and structures of the material. The work is made at financial support of the Russian Fund of Basic Researches (grant № 05-08-01196a).

REFERENCES

1. Stanyukovich, K. P. et al., Physics of explosion. Moscow. NAUKA Publishers, 1975, 704 p.
2. Johnson, J. N., Dynamic fracture and spallation in ductile solids, J. Appl. Phys. 52 (1981), p. 2812
3. Kreinhagen, K.N. et al., A determination of a ballistic limit at impact with structural targets, AIAA J. 8 (1970), p. 42.
4. Wilkins, M.L., Computer simulation of dynamic phenomena. Berlin-Heidelberg-New-York, Springer, 1999, 246 p.
5. Johnson, G.R., Colby, D.D., Vavrick, D.J., Tree-dimensional computer code for dynamic response of solids to intense impulsive loads, Int. J. Numer. Methods Engng. 14 (1979), p.1865.
6. Diep, Q.B., Moxnes, J.F., Nevstad, G., Fragmentation of projectiles and steel rings using numerical 3D simulations, 21th Int. Symp. of Ballistics-2004, Adelaide, Australia, 2004, p. 752.

Structural Changes and Energy Cumulation in an Iron-Nickel Alloy upon Quasi-Spherical Explosive Loading

V. I. Zel'dovich*, I. V. Khomskaya*, N. Yu. Frolova*, A. E. Kheifets*, V. M. Gundyrev*, B. V. Litvinov[r], and N. P. Purygin[r]

* *Institute of Metal Physics, Ural Division, Russian Academy of Sciences, ul. S. Kovalevskoi 18, Ekaterinburg, 620041 Russia*
[r] *All-Russia Research Institute of Technical Physics, Russian Federal Nuclear Center, Snezhinsk, Chelyabinsk oblast, 456770 Russia*

Abstract Ball samples of the Fe-31.8 wt % Ni-0.05 wt % C iron--nickel alloy one of which was in an austenitic state and the other was in a martensitic-austenitic state were subjected to quasi-spherical shock-wave loading under identical conditions. A comparison of the results obtained under the same loading conditions on the samples of the same alloy in two different initial states made it possible to establish the influence of the initial phase composition on the structural changes and on the effect of energy cumulation.

INTRODUCTION

After cooling from a high temperature to room temperature, the iron-nickel alloys with a nickel concentration in the vicinity of 30% are in an austenitic state. After an additional cooling into the negative-temperature range, they undergo a $\gamma \to \alpha$ martensitic transformation. The subsequent warming to room temperature retains the obtained structure of martensite with some amount of retained austenite. Thus, these alloys can be obtained in two structural states differing in phase composition. The $\alpha \to \gamma$ transformation occurs under the action of a high static or dynamic pressure, and martensite is transformed into a denser phase, austenite [1]. This circumstance makes it possible to investigate the role of phase transformation in the formation of a structure that arises under the action of shock waves and to clarify "in a pure form" the problem brought forth by Zababakhin: how the phase transition influences the effect of energy cumulation upon loading of materials with convergent shock waves [2]. The purpose of this work was to consider this problem.

Phase transformations in iron-nickel alloys under the action of shock waves were studied repeatedly [3,4]. However, distinctions in the loading schemes did not allow one to unambiguously clarify the influence of the initial phase composition on the formation of macro- and microstructure, the process of formation of the internal cavity, and the effect of energy cumulation.

CP849, *Zababakhin Scientific Talks - 2005*,
edited by E. N. Avrorin and V. A. Simonenko
© 2006 American Institute of Physics 0-7354-0345-7/06/$23.00

EXPERIMENTAL

Ball samples of alloy Fe-31.8 wt.% Ni-0.05 wt % C diameter 20 mm was placed in steel holder. The external diameters of construction was 60 mm. The first sample had austenitic structure, second sample had martensitic-austenitic structure, martensite in it was about 50%. The converging shock waves were created by means of the explosion of spherical charge, around construction. Explosion initiated on a surface of a charge in several points uniform arrangemetnt regular on the sphere. The special conditions for delay of unloading were undertaken with the purpose of preservation of samples under loading. The calculation of pressure on a surface of construction showed 46 GPa. After loading, the samples were taken out of the cans to be subjected to further investigations. The sample structure was studied in diametrical sections; if necessary, other cuts could be made. The structural investigations were performed with the help of metallography, transmission electron microscopy, and X-ray diffraction analysis.

EXPERIMENTAL RESULTS

When an originally austenite sample is examined metallographically, the initial stages of the formation of an internal cavity are observed; three elongated sections of the cavity converging to the center are formed in the central part of the sample (Fig. 1a). Two sections were joined together to form an obtuse-angled pair, the third section was located separately. The nonequiaxed, elongated shape of the cavity sections and their relative positions indicate that each section was formed under the action of tensile stresses upon the reflection of neighboring convergent shock waves, before the reflected waves emerged onto the outer surface. This means that the initial geometry of the experiment (dodecahedral symmetry) was not forgotten to the time instant of focusing; the shock-wave motion remained three-dimensional. The cavity sections are partly filled with metal that has a dendritic structure. The narrow sections adjacent to the cavity also have a dendritic structure (Fig. 1b). The "regular" structure of the dendrites and their extent indicate that they were formed upon crystallization in the process of cooling of the molten metal. This means that in these sections the residual temperature after unloading was higher than the melting temperature of the alloy and the cavity arose in the liquid state. Near the sections that underwent melting, a zone in which recrystallization occurred is located (Fig. 1b). Further toward the sample periphery, large-size deformed regions in which no recrystallization occurred were observed. The main deformation mechanism is twinning, which is characteristic of this type of loading. Twinning in different grains

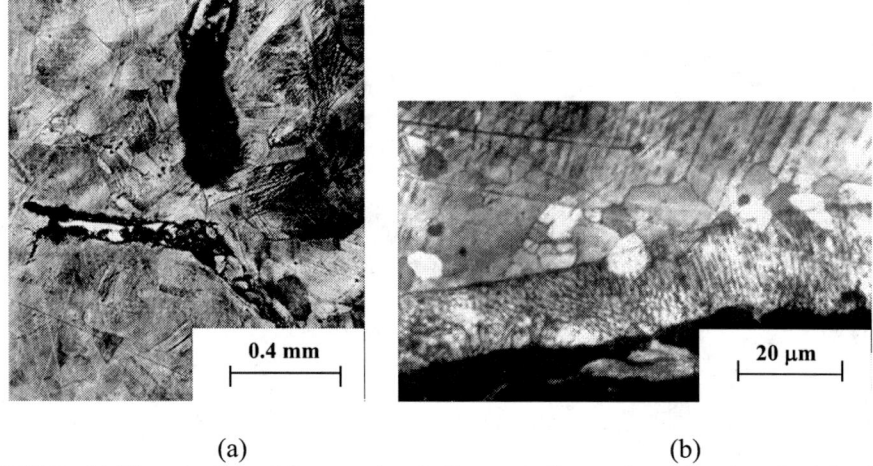

(a)   (b)

FIGURE 1. (a) Microstructure of the central part of an austenite sample subjected to shock loading (general view) and (b) dendrites and a recrystallized structure near the internal cavity.

occurs on one, two or three systems. The high rate of loading causes the simultaneous action of several twinning systems. Figure 2a display electron micrograph of such twins. Twins are frequently grouped into packets. The thickness of individual twins is very small (5-10 nm). The thickness of twin packets may be larger by an order of magnitude. It may be concluded that the twinning arises from a very large number of nucleation centers and that separate twins merge into conglomerates. It is seen in the dark-field micrographs that twins consist of weakly misoriented parts.

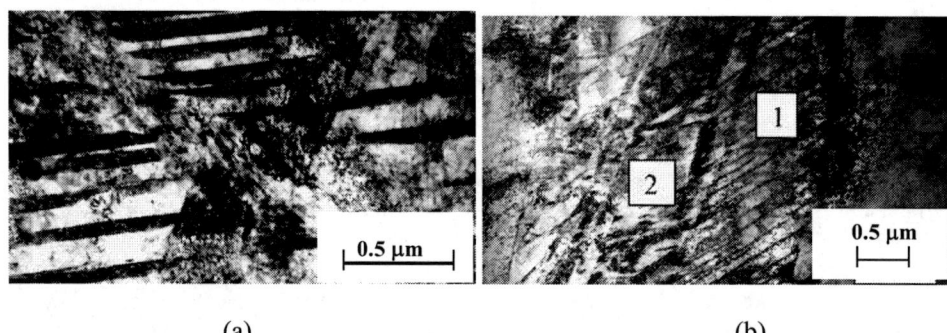

(a)   (b)

FIGURE 2. Microstructure of austenite (a) and martensite-austenite sample (b) after shock loading. The twins (a) and structure of reverted-austenite crystals (b).

The microstructural study showed that in the мartensite-austenite sample neither a cavity nor a dendritic structure or recrystallization is observed. Diffraction patterns taken from different parts of the sample showed that an almost complete transformation of martensite into austenite took place under the action of shock waves. It is known from experiments on the effect of static and dynamic pressure on the reverse martensitic transformation in alloys of the given composition [4, 5] that the transition starts at a pressure of 7-10 GPa and terminates at 17-20 GPa. Consequently, the pressure in this experiment was 20 GPa or more.

The transformation of each martensite crystal into austenite occurs from a large number of nucleation centers. Figure 2b displays the electron micrograph of a part of a reverted crystal (former lens-type martensite crystal that underwent the $\alpha \rightarrow \gamma$ transformation upon shock-wave loading). It is seen that the crystal consists of two regions differing in structure. The structure of region 1 consists of thin unidirectional bands. Dark-field and diffraction analysis showed that this region has a structure of reverted austenite with a single orientation ($[123]_\gamma$ zone axis), while thin (~10 nm) extended bands represent slip bands in the reverted austenite (Fig. 2b). A great amount of disperse austenite crystals differing in orientation were formed in region 2. Two different structure mechanisms of austenite formation can be realized within one lens-type martensite crystal.

Thus, in two samples of one alloy having different initial phase composition, structural changes caused by the action of shock waves (under the same loading conditions) are substantially different.

DISCUSSION OF RESULTS

The observation of structural changes caused by the action of shock waves suggests that in the first sample after unloading the residual temperature at the external boundary of the dendritic structure was equal to the melting temperature of the alloy (1720 K). At the internal boundary, the temperature was even higher. In the second sample, the residual temperature was lower than the recrystallization temperature of the deformed alloy. If the recrystallization temperature of the alloy is taken to be equal to 970 K [6], we come to the conclusion that the residual temperature in the second sample was less than 970 K. From the residual-temperature values, we estimated the magnitude of pressure at the center of the first sample and the upper pressure limit in the second sample as this was done in [7].

Figure 3 displays the shock adiabat (curve 1), unloading isentropes (curves 2), and a T-P diagram of iron in the austenitic state borrowed from [8]. The T-P diagram is represented by the pressure dependence of the melting temperature (curve 3). The isentropes were plotted so that the residual temperatures were equal to the melting temperature and the recrystallization temperature. As is seen, the maximum pressure in the first sample reached 180 GPa or more, and the maximum pressure in the second sample was certainly less than 100 GPa. The obtained distinctions in the structure of the two

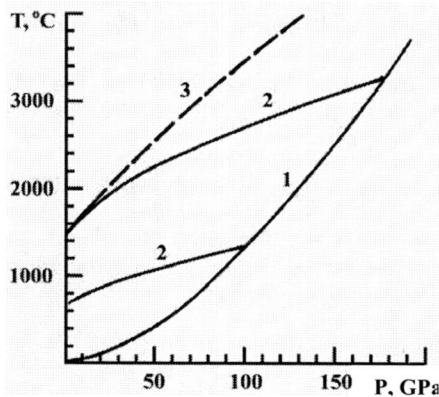

FIGURE 3. Shock adiabate of the γ–iron and unloading isentropes combined with a T-P diagram.

samples and the pressure values that differ strongly at the center indicate that the transformation of martensite into austenite, which occurs with energy absorptance, weakens the effect of energy cumulation. This result confirms Zababakhin's hypothesis that the enhancement of shock waves upon convergence is "limited by phase transitions" [2] and also agrees with the data of Kozlov on the role of the α→ε transformation in steels in the effect of energy cumulation [9].

Let us estimate what fraction of the shock-wave energy is spent on the transformation of martensite into austenite. We restrict ourselves to simple energy calculations disregarding the kinetics of the process. The latent heat of the phase transition is on the order of 10^3 cal/mol [10], or $6 \cdot 10^8$ J/m^3. When the transformation occurs in an alloy containing 50% martensite, the heat of transition is $3 \cdot 10^8$ J/m^3. The shock-wave energy transferred to the material is $E_v = 0.5 P \Delta V$. The table contains pressures, changes in volume, and volume energies in shock wave for the 18Cr18Ni10Ti stainless alloy which is close in the properties to the above alloy but exhibits no phase transition. Data on the pressure and volume changes are borrowed from [11], and the last column is obtained using the above formula.

TABLE Pressure, volume changes, and energy in shock wave

P, GPa	ΔV, m^3	$E_v \cdot 10^8$ J/m^3
6.45	0.037	1.2
14.03	0.074	5.2
19.43	0.099	9.6
24.67	0.121	14.9

It is seen from a comparison of the latent heat of the phase transition with the data given in the table that the energy of a shock wave with a pressure of 6.45 GPa is insufficient to transform 50% martensite present in the sample into austenite. At a pressure of 14 GPa, the energy input per cubic meter of the material is 520 MJ, of

66

which 300 MJ is spent on the transformation. The remaining 220 MJ are spent on compression and heating. Thus, in the presence of a phase transition and a moderate pressure at the shock-wave front (15- 25 GPa), the pressure produced is substantially lower than in the absence of a phase transition. At high pressures, however, the energy spent on the phase transition becomes low as compared to the shock-wave energy and the role of phase transition in the cumulation effects becomes insignificant.

Thus, the distinctions observed in the structure of the two samples and the strongly differing pressure values indicate that under given conditions of experiment the transformation of martensite into austenite, which occurs with energy absorption, weakens the effect of energy cumulation.

ACKNOWLEDGMENTS

This work was performed in the framework of the "Thermal Physics and Mechanics of Intense Energy Treatment" Program of the Presidium of Russian Academy of Sciences.

REFERENCES

1. R. W. Rohde, Acta Metall. 18 (8), 903-913 (1970).
2. E. I. Zababakhin and I. E. Zababakhin, Phenomena of Unlimited Cumulation, Nauka, Moscow, 1988.
3. V. I. Kolomytsev, V. A. Lobodyuk, and G. I. Savvakin, Metallofizika 3 (6), 69-75 (1981).
4. V. A. Teplov, V. M. Schastlivtsev, E. A. Kozlov, et al. Phys. Met. Metallogr. 92, 411-420 (2001).
5. V. D. Sadovskii, V. A. Teplov, D. I. Tupitsa, et al. Phys. Met. Metallogr.56 (4), 775-784 (1983).
6. V. I. Zel'dovich, I. G. Komarova, and V. D. Sadovskii, Phys. Met. Metallogr. 44 (2), 301-310 (1977).
7. V. I. Zel'dovich, B. V. Litvinov, N. P. Purygin, et al., Dokl. Akad. Nauk 343 (5), 621-624 (1995).
8. E. A. Kozlov, G. V. Kovalenko, A. I. Uvarov, and V. A. Teplov Phys. Met. Metallogr. 88, 605-611 (1999).
9. E. A. Kozlov Metals and Minerals Research in Spherical Shock-Wave Recovery Experiments", ed. by B. V. Litvinov ONTI RFNC-VNIITF, Snezhinsk,1996, pp.3-11.
10. M. A. Krivoglaz, V. D. Sadovskii, L. V. Smirnov, and E. A. Fokina, Steel Quenching in a Magnetic Field Nauka, Moscow, 1977.
11. R. F. Trunin, M. Yu. Belyakova, M. V. Zhernokletov, and Yu. M. Sutulov, Izv. Akad. Nauk SSSR, Fiz.
12. Zem. 2, 99-106 (1991).

Hugoniot Adiabatic Curves
With Heat Transfer and Phase Jump

S.V. Khabirov

Laboratory for Differential Equation of Mechanics, Institute of Mechanics, Ufa Branch of Rassian Academy of Sciences, October street, 71, Ufa, 450054, Russia

Abstract. The approximate model of the shock wave moving along the heat conducting gas is given. It is possible 9 types of the shock adiabatic curves with the heat transfer and the phase jump. The problem of the stationary source is solved.

Keywords: shock wave, heat conductivity, stationary source.

INTRODUCTION

The academician R.I. Nigmatulin recommended to consider an approximate model of the heat conducting gas into a small domain bounded by the shock wave. In this domain the gradient temperature is a small value but the coefficient of the heat conductivity is a big value so that the heat current was finite.

APPROXIMATE SUBMODEL

We assume that gas is perfect $p = R\rho T$, is polytropic with the sound speed $a = \sqrt{\gamma RT}$, $R = c_p - c_V$, $\gamma = c_p / c_V$ are the gas constants. The constants γ_i, R_i are different on the shock wave sides. Hence the phase transfer is into the shock wave.

In a small heat conducting domain the equations for the spherically symmetric motions are

$$U_{1t} + U_1 U_{1r} + R_1 T_1 \rho_1^{-1} \rho_{1r} = 0, \quad T_1 = T_1(t),$$

$$\rho_{1t} + U_1 \rho_{1r} + \rho_1 (U_{1r} + 2r^{-1} U_1) = 0, \tag{1}$$

$$\rho_1 c_{V1} [T_1' + (\gamma_1 - 1) T_1 (U_{1r} + 2r^{-1} U_1)] + q_{1r} + 2r^{-1} q_1 = 0$$

where U_1 is the velocity, ρ_1 is the density, T_1 is the temperature, q_1 is the heat current.

CP849, *Zababakhin Scientific Talks - 2005*,
edited by E. N. Avrorin and V. A. Simonenko
© 2006 American Institute of Physics 0-7354-0345-7/06/$23.00

The characteristics of the system (1) are

$$dt = 0, \quad dr = (U_1 \pm \sqrt{R_1 T_1})dt.$$

It follows that the velocity of the perturbations on the moving gas is infinite or is equal $c_1 = \sqrt{R_1 T_1} < a_1 = \sqrt{\gamma_1 R_1 T_1}$ for $\gamma_1 \neq 1$.

With respect to the other side of the shock wave gas is not heat conducting

$$U_{2t} + U_2 U_{2r} + R_2 (T_{2r} + T_2 \rho_2^{-1} \rho_{2r}) = 0,$$

$$\rho_{2t} + U_2 \rho_{2r} + \rho_2 (U_{2r} + 2r^{-1} U_2) = 0, \tag{2}$$

$$T_{2t} + U_2 T_{2r} + (\gamma_2 - 1) T_2 (U_{2r} + 2r^{-1} U_2) = 0, \quad q_2 = 0.$$

The characteristics of the system (2) are

$$dr = U_2 dt, \quad dr = (U_2 \pm \sqrt{\gamma_2 R_2 T_2})dt$$

so the perturbations extend along the moving gas with the sound speed $\sqrt{\gamma_2 R_2 T_2}$.

The relations on the surface of the shock wave have the form

$$m \equiv \rho_1 (U_1 - D_1) = \rho_2 (U_2 - D_1),$$

$$R_1 \rho_1 T_1 + m \rho_1 (U_1 - D_1) = R_2 \rho_2 T_2 + m \rho_2 (U_2 - D_2), \tag{3}$$

$$m((U_1 - D)^2 + 2c_{p1} T_1) + 2q_1 = m((U_2 - D)^2 + 2c_{p2} T_2)$$

where the heat current $q_1 > 0$ is bring to the wave and the heat current $q_1 < 0$ is radiating from the wave.

TIPES SHOCK ADIABATIC CURVES

From (3) follows the Hugoniot adiabatic curve (shock adiabatic curve)

$$\frac{\gamma_2 + 1}{\gamma_2 - 1} pV - \frac{\gamma_1 + 1}{\gamma_1 - 1} p_1 V_1 + p_1 V - V_1 p - 2q_1 \sqrt{\frac{V_1 - V}{p - p_1}} = 0. \tag{4}$$

We suppose $1 \leq \gamma_i \leq 2, \ i = 1, 2$.

There are 9 types of the shock adiabatic curves depending on the parameters

$$\delta = \gamma_2 \left(1 - \frac{(\gamma_1+1)(\gamma_2-1)}{(\gamma_1-1)(\gamma_2+1)} \right), \quad v = 2q_1 \frac{\gamma_1-1}{\gamma_2+1} \gamma_2^{-1/2} V_1^{-1/2} p_1^{-3/2}, \quad \gamma = \frac{2\gamma_2}{\gamma_2+1}.$$

The polynomials $b^{\pm}(s^2) = v^2 \pm 4s^2 (s^2 \pm \gamma)(\gamma s^2 \pm \delta)$, its minimum and maximum are involved in the classification of the shock adiabatic curves.

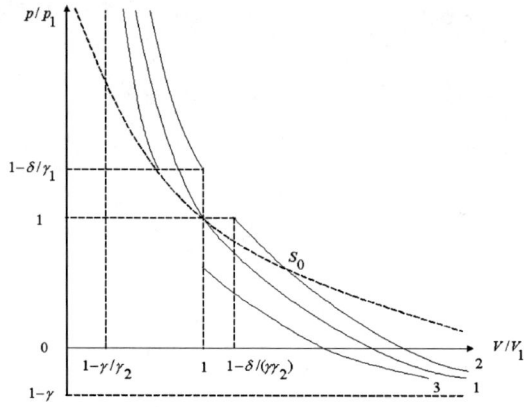

FIGURE 1. Shock adiabatic curves for $v = 0$. Type 1 ($\delta = 0$), type 2 ($\delta < 0$), type 3 ($\delta > 0$).

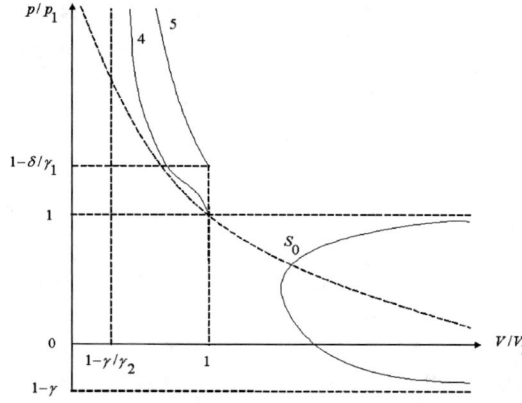

FIGURE 2. Shock adiabatic curves for $v > 0$. Type 4 ($\delta = 0$), type 5 ($\delta < 0$).

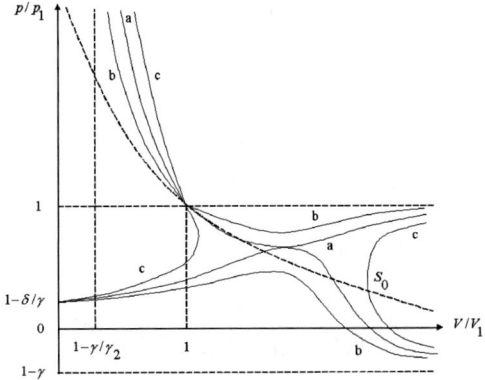

FIGURE 3. Shock adiabatic curves for $v > 0$.

Type 6. $\delta > 0$: $b^{+}_{min} = 0$ (a), $b^{+}_{min} < 0$ (b), $b^{+}_{min} > 0$ (c).

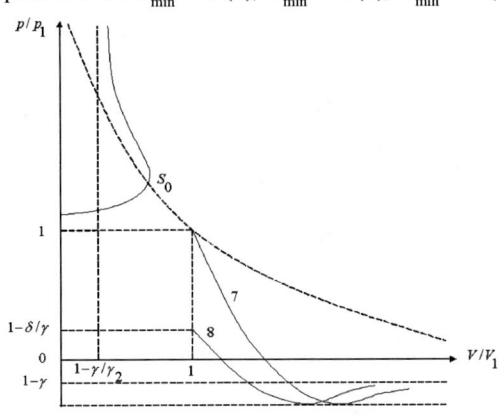

FIGURE 4. Shock adiabatic curves for $v < 0$. Type 7 ($\delta = 0$), type 8 ($\delta > 0$).

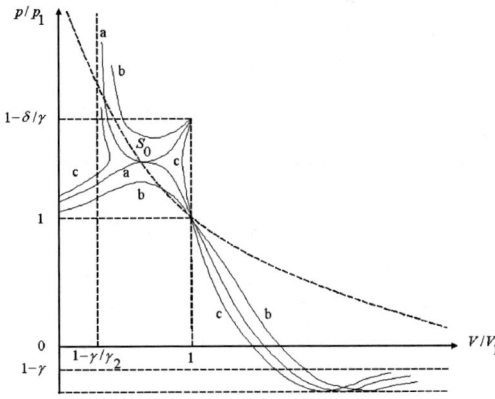

FIGURE 5. Shock adiabatic curves for $v < 0$.

Type 9 ($\delta = 0$): $b^{-}_{max} = 0$ (a), $b^{-}_{max} < 0$ (b), $b^{-}_{max} > 0$ (c).

Theorem. The entropy increase along Hugoniot adiabatic curve $p > p_1$ for the types 1, 3, 7, 8, 9 under the conditions $|v| > |\delta|$ and $|v| \geq 2\gamma^2$ or under the conditions $|v| < 2\gamma^2$, $|v| \geq 2|\delta|$.

There is the stationary point S_0 of the entropy along Hugoniot adiabatic curve for the types 2, 4 5 6 9 under the conditions $|v| < |\delta|$ and $|\delta| < |v| < 2|\delta|$.

The entropy in crease with the pressure growth when the pressure is more then one in the stationary point. So the relative velocity is suppose in front of the shock wave and one is subsonic over the front of the shock wave (Tsemplen theorem).

STATIONARY FLOW

The stationary submodel is integrable. It gives description of a nonpoint sink or a source. In the domain with the heat conduct we have

$$r^2 U_1 \rho_1 = Q_1, \; U_1^2 + 2R_1 T_1 \ln \rho_1 = A_1 Q_1^2,$$
$$r^2 q_1 - R_1 Q_1 T_1 \ln \rho_1 = B_1 Q_1,$$
$$r^{-4} = \rho_1^2 (A_1 - 2R_1 T_1 Q_1^{-2} \ln \rho_1) \equiv F(\rho_1) \leq r_0^{-4} < r_{1*}^{-4}$$

where Q_1, T_1, A_1, B_1 are constants. The flow is determined for $r \geq r_0$. The velocity is equal the sound velocity for $r = r_{1*}$. There are two solutions $\rho_i^{(1)}(r), i = 1,2$, the first of them has the velocity more then $\sqrt{R_1 T_1}$, the second of them is subsonic and has the velocity less then $\sqrt{R_1 T_1}$,

In the other hand we have

$$r^2 U_2 \rho_2 = Q_2, \; T_2 = B_2 \rho_2^{\gamma_2 - 1},$$
$$r^{-4} = A_2 \rho_2^2 - \frac{2\gamma_2}{\gamma_2 - 1} R_2 B_2 Q_2^{-2} \rho_2^{\gamma_2 + 1} \equiv G(\rho_2) \geq r_{2*}^{-4}$$

where $B_2 > 0$, $A_2 > 0$, Q_2 are constants. The flow is determined for $r \geq r_{2*}$. It is the sound sink or the sound source. There are two solutions $\rho_i^{(2)}, i = 1,2$. The first is supersonic and the other is subsonic.

On the surface of the shock wave we have

$$Q_1 = Q_2 = Q,$$
$$Q^2 r_v^{-2} = \rho_1^2 (A_1 Q^2 - 2R_1 T_1 \ln \rho_1) = \rho_2^2 (A_2 Q^2 - \frac{2\gamma_2}{\gamma_2 - 1} R_2 B_2 \rho_2^{\gamma_2 + 1}),$$

72

$$2B_1 + \frac{2\gamma_1}{\gamma_1 - 1} R_1 T_1 + A_1 Q^2 = A_2 Q^2,$$

$$\rho_1 (T_1 R_1 + A_1 Q^2 - 2R_1 T_1 \ln \rho_1) = Q^2 A_2 \rho_2 - \frac{\gamma_2 + 1}{\gamma_2 - 1} R_2 B_2 \rho_2^{\gamma_2}.$$

From that it follows the quadratic equation for ρ_2

$$\left(\frac{\rho_2}{\rho_1}\right)^2 \frac{\gamma_2 - 1}{\gamma_2 + 1} (2B_1 + \frac{2\gamma_1}{\gamma_1 - 1} R_1 T_1 + A_1 Q^2) - $$

$$- \frac{2\gamma_2}{\gamma_2 + 1} (T_1 R_1 + A_1 Q^2 - 2R_1 T_1 \ln \rho_1) + A_1 Q^2 - 2R_1 T_1 \ln \rho_1 = 0. \tag{5}$$

Let the solution $\rho_1^{(1)}(r)$, $U_1^{(1)} > c_1 = \sqrt{R_1 T_1}$ is determined in the domain $r_0 < r < r_v$, $r_v > r_{*_1}$ in which the sound speed in reached. The gas state „1" is in front of the detonation wave by Tsemplen theorem if the pressure p_2 is more then the pressure in the stationary point S_0 for the entropy on the shock adiabatic curve. It is the source $(Q > 0, \ r = r_0)$. The sign „+" stands in the solution of the equation (5). There exists the unique

Let the solution $\rho_1^{(1)}(r)$, $U_1^{(1)} > c_1$, $U_1^{(1)} < a_1$ is determined in the domain $r_0 < r < r_v'$, $r_{*_2} < r_v' < r_{*_1}$ or the solution $\rho_2^{(2)}(r)$, $U_1^{(2)} < c_1 < a_1$ is determined in the domain $r_0 < r < r_v'$, $r_v' > r_{0*}$. Then the gas state „1" is over the detonation wave if the pressure p_1 is more then the pressure in the point S_0. It is the sink $(Q < 0, \ r < r_0)$. We have the supersonic flow $\rho_2^{(1)}(r)$ in front of the detonation wave. It is possible two shock jumps with the strong wave (the sign „+" in the solution of (5)) and with the weak wave (the sign „−" in the solution of (4)).

ACKNOWLEDGMENTS

This work was supported by Russian Foundation for Basic Research (Grants OFI-a 04-01-08050, 05-01-00080, 05-01-00775-a).

Time-Dependent Compression of Cylindrical or Spherical Ideal Gas Volumes

A.N. Kraiko

Central Institute of Aviation Motors of P.I. Baranov, Russia, Moscow

Abstract. Two problems of time-dependent compression of ideal (inviscid and not heat-conducting) gas, related with the idea of controlled inertial thermonuclear fusion (CITF) are considered. In first of them for time, close to time of an acoustic wave run by an initial cylindrical or spherical volume, the gas is isentropically compressed from rest to rest. Such energy optimal compressing does not ensure, however, the required level of temperature. In opposite to it when faster compressing with "a head shock" (HS) the necessary density is not reached. In the second problem the centered compressing wave of characteristics coming to an axis or to center of symmetry (further – CS) simultaneously with HS is introduced. As a result it is possible to create the conditions required for CITF.

Keywords: Shock-free compressing from rest to rest, shock, centered compressing wave.
PACS: 28.52, 47.40.-x, 52.35.Tc, 52.57.-z.

INTRODUCTION

For realizing the CITF it is necessary to compress the substance of the thermonuclear target in $(0.5\text{-}4)\cdot10^3$ times to increase its temperature up to $(0.3\text{-}1)\cdot10^8$ K and to execute "a condition of ignition": $I \equiv \int \rho dr \geq I_*$, where I_* is known constant, ρ is density, and r is distance from CS. It is necessary to ensure these requirements when rather small energy expenditures, and to make the compressing and the heating faster as possible for decreasing of parasitic influence of different dissipative effects.

The main result of the solution [1] of a variation problem about the energy optimal compressing up to the given ratio c of initial and final volumes – is the detection of fast (for time $t^m = kt_0$) shock-free (isentropic) compressing "from rest to rest" (ICRR) capability. Here t_0 – the time of an acoustic wave run from a wall of an initial volume up to CS, t^m – the function of c, of index v, equal 2 in cylindrical and 3 in spherical cases, and (for perfect gas) of isentropic ratio γ, and k is the monotonically decreasing function of c, equal 2 when $c = 1$ and 1 when $c = \infty$. The examples of this problem solving are below presented.

The energy optimal ICRR allows to achieve the necessary for CITF values of $\rho_f/\rho_0 = n$ and the criterion of fire when low level of temperature T_f of compressed substance. Really, at any isentropic compressing $T_f/T_0 = (\rho_f/\rho_0)^{\gamma-1}$. If $\rho_f/\rho_0 = 10^4$, then

CP849, Zababakhin Scientific Talks - 2005,
edited by E. N. Avrorin and V. A. Simonenko

$T_f/T_0 = 39.8$ and 464 accordingly for $\gamma = 7/5$ and $5/3$. When $T_0 = 300$ K it is many orders less of required value $T_f = (0.3\text{-}1)\cdot 10^8$ K.

When faster compressing with a HS according to the Guderley solution (GS) [2-4] the gas density increases in a final number of times (from 33 for $\gamma = 5/3$ up to 145 for $\gamma = 7/5$). The realizing of major densities is possible if to attach to particular C^--characteristic of GS the flow with a focusing of C^--characteristics to CS in a moment when arrival there of HS. In a focusing point the gas velocity and the sound velocity will approach to infinity. Therefore for calculation by a method of characteristics of the piston trajectory implementing this flow, it is necessary to determine such singularity analytically. Earlier the attempt of the author to make it was completed by failure [5]. In the last section of the present paper the reason of the indicated failure is clarified, the required analytical solution is found and with its use by the method of characteristics the piston trajectories are designed, which implement the necessary for CITF conditions.

ISENTROPIC COMPRESSING FROM REST TO REST

Implementing the ICRR trajectory of two-dimensional, cylindrical or spherical piston, corresponds to the scheme of flow shown in rt -plane in figure 1, a. Here along with a trajectory of the piston if the C^+-and C^--characteristics are drawn. For the scales of the length, velocity, time and density are hereinafter taken the initial coordinate of the piston r_i, the sound velocity in uncompressed gas a_0, ratio r_i/a_0 and initial density ρ_0. Under the C^--characteristic $i0$ and above the C^+-characteristic c^+f the gas is at rest. Its thermodynamic parameters are constant and are determined by the state equations with specific entropy equal to initial one $s_f = s_0$ and with a density equal for selected scales to a given degree of compressing $c = r_f^{-\nu}$. So,

$$r_f = c^{-1/\nu}, \quad \rho_f = c, \quad s_f = s_0, \quad a_f = a(\rho_f, s_f) = a(c, s_0). \tag{1}$$

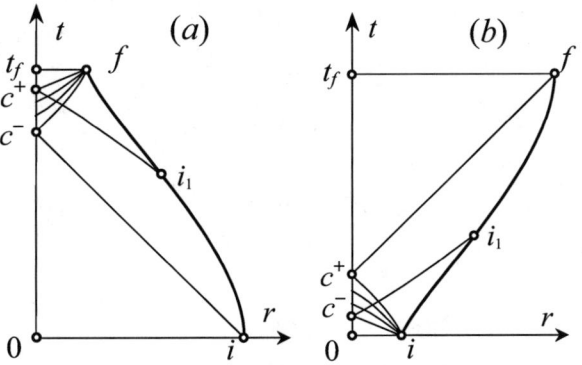

FIGURE 1. rt-cards of isentropic compressing (a) and expansion (b) from rest to rest

When ICRR the trajectory of the piston is those, that C^--characteristics, which go from its "initial" site ii_1, being reflected from the axis t as C^+-characteristics, are focalized in a point f. On the first sight, the capability of such trajectory designing is looked problematic. However, the given problem has a rather simple solution being a consequence of an invariance of equations and conditions, which describe shock-free one-dimensional flows of ideal gas regard to arbitrary shift of a time-zero and simultaneous modification of signs of time and x-component of a velocity v. After the corresponding shift and modification of signs the problem of compressing (the fig. 1, a) becomes the problem of expansion (fig. 1, b) with known from (1) constant parameters of the gas at rest on C^--characteristics ic^- and with equal to initial those in a problem of compressing known parameters of also being at rest gas on C^+-characteristics c^+f. The solving of the problem of gas expansion is reduced to the consequence solving of two standard problems of the characteristics method.

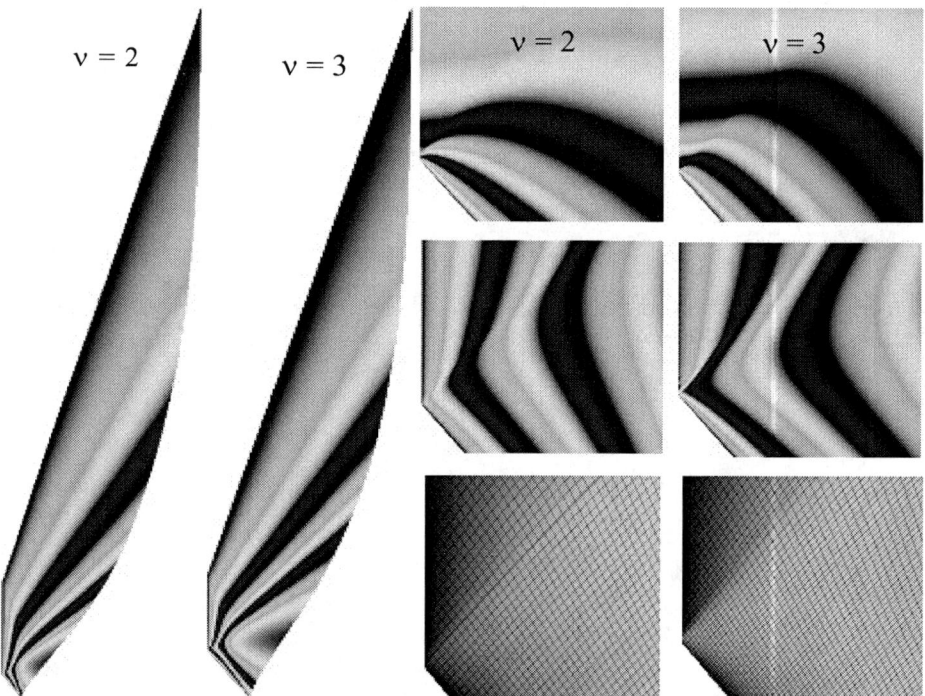

FIGURE 2. The fields of the pressure (above) and velocity and grids of characteristics when isentropic expansion of cylindrical and spherical pistons from rest to rest

In figure 2 for the perfect gas with $\gamma = 1.4$ the calculation results of cylindrical and spherical symmetrical expansions from rest to rest up to $\rho_f/\rho_0 = 0.01$ are presented. With periodic gamma the fields of pressure and velocity are shown. Two left-hand figures represent the fields of a velocity during all process of expansion. Two right columns give the information about the neighborhood of a point c^- – of arrival to time axis of initial C^--characteristics $i0$. The characteristic grids, distributions of the velocity (in the middle) and the pressure (above) are figured there. These distributions are in agreement with the theoretical analysis [6, 7]. The capability of a such expansion was rejected for many years by S. P. Bautin. His objections were reduced to the statement that in CS allegedly there is a shock wave. The Fig. 2 demonstrates the weakness of these objections.

CENTERED COMPRESSING WAVE, CONJOINED TO SINGULAR C⁻-CHARACTERISTICS OF GUDERLEY SOLUTION

As it was already noted, for $t_f < t_0$ the compressing should begin with HS. Close to HS the flow is described by the Guderley solution (GS). In agreement with the GS the density in a neighborhood of CS will increase in a final number of times [2-4]. The greater compressing is possible, if to change the piston motion above the point of going out from it the singular C⁻-characteristics of GS C_0^-, catching up the HS SI at the moment of its arrival to CS. This motion must ensure the focusing to the same point of C⁻-characteristics bundle. For a two-dimensional piston this capability is elementary. The xt-card, corresponding to it, is presented in figure 3, a. Increasing an initial velocity of the piston and the velocity of HS SI, it is possible to receive as much as small time of compressing when equal to $(\gamma + 1)/(\gamma - 1)$ the HS ratio of densities. The required as much as major density will be ensured with the centered wave of compressing $i_1 0 i_2$, catching up the HS in the CS.

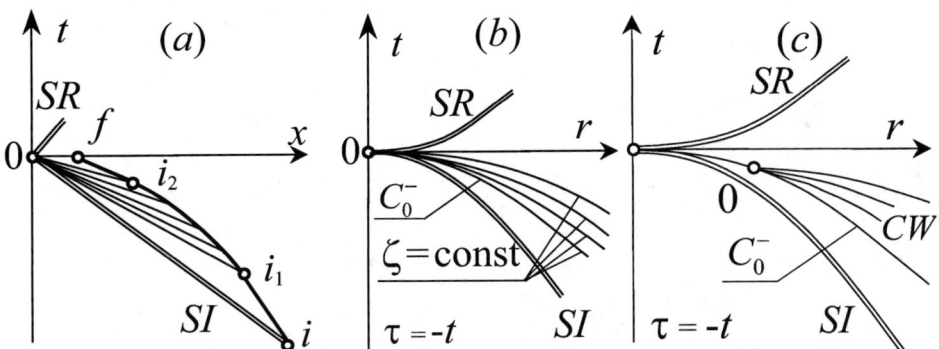

FIGURE 3. The flow cards with the centered compressing wave

Let, as in fig. 3, $\tau = -t$, the time t is read out from the moment of arrival of HS and of singular C⁻-characteristics C_0^- to CS, $u = -v$, and v is the gas velocity. If the focusing of C⁻-characteristics to the origin behind the HS (SI – in a fig. 3, b) for $v = 2$ and 3 is possible, their design should go from the origin ("points 0") $r = \tau = 0$. The first its stage should become the clearing up of flow structure near to point 0. The flow behind the HS SI and at coming to point 0 characteristics C_0^- is non-isentropic with known entropy function $p/\rho^\gamma = S(m)$, where the constant along the trajectories of particles the Lagrangian variable m is introduced by the equality

$$dm = r^{\nu-1}\rho(dx - u\,d\tau) = \frac{\rho a}{u+a}\,dz, \quad z = \frac{r^{\nu}}{\nu}. \tag{2}$$

In (2) the second expressions for dm is valid at C^--characteristics.

At C_0^-, which coincides a line of persistence of self-similar variable of GS $\xi = r/\tau^n$, in agreement with [2-4] and (2) when adequate scaling we have

$$a = m^{\alpha}, \quad u = \mu m^{\alpha}, \quad z = (1+\mu)m, \quad \alpha = \frac{n-1}{\nu n}. \tag{3}$$

Assistant here constant m and exponent n or α are known from GS [2-4], where $n < 1$ and therefore $\alpha < 0$, i.e. a and u infinitely increase at $m \to 0$. From (3) and the requirement of saving in a bundle of p/ρ^{γ} as a function of m we shall find, that in it

$$\rho = a^{1/\kappa}m^{\omega}, \quad p = \frac{1}{\gamma}a^{\gamma/\kappa}m^{\omega}, \quad \omega = -\frac{\alpha}{\kappa}, \quad \kappa = \frac{\gamma-1}{2}. \tag{4}$$

The formulas (4) predetermine the use of m as one of independent variables. Having taken as the second independent variable a persistent along everyone C^--characteristics of a bundle "characteristic" variable ζ and having eliminated with the help of formulas (4) the derivatives of p and ρ, we shall put the flow equations to a kind

$$z_m = \frac{u+a}{\rho a}, \quad u_m + \frac{1}{\kappa}a_m + \frac{\omega a}{\gamma m} + \frac{\nu-1}{\rho\nu z}u = 0,$$

$$\rho a z_{\zeta} a_m + (\gamma u_{\zeta} - a_{\zeta})u + \rho a^2 \frac{\kappa\omega}{\gamma m}z_{\zeta} = 0, \quad \tau_m = (\nu z)^{(1-\nu)/\nu}(\rho a)^{-1}. \tag{5}$$

It is possible to search for the solution of this system in the form

$$r = \delta(m)\zeta, \quad a = \psi(m)A(\zeta), \quad u = \psi(m)U(\zeta). \tag{6}$$

with functions, being a subject to definition, $A(\zeta)$, $U(\zeta)$, $\delta(m)$ and $\psi(m)$, where $\delta(0) = 0$, and $\psi(0) = \infty$. In [5] the solution of such type was under construction for six distinguished by values of ν and γ cases. In five of them the functions, possessing the required properties, were found. In sixth they have occurred the complex. However either in the first five examples the determined C^--characteristics of a bundle placed not above the singular characteristic of GS, but under it, i.e. the found solutions have appeared physically senseless. As opposed to this, in a problem of the cylindrical or spherical compressing wave conjoined to "uniform rest" ($\xi = r/\tau$, $n = 1$, $\omega = \alpha = 0$), the solution in the form (6) has coincided with the known solution of unlimited cumulative action [8, 9]. The fact that in the latter case the submissions (6) have given the exact solution, but in a problem with HS have reduced in physically senseless

78

outcome, is a paradox, without explanation of which it was impossible to rely on a success.

When analysis the reasons of the formulated paradox the attention to the key difference of two problems was attracted. At the initial characteristic of a wave conjoined to uniform rest, $a = 1$, $u = 0$, i.e. there are no singularities, and infinitely increasing "disturbances" a and u, determined by the formulas (6) are the main. On the contrary, as it was noted above, in GS the values of a and u at the singular characteristic when $m \to 0$ infinitely increase. Therefore in this case we shall present z, a and u in the form

$$z = (1+\mu)m + \varphi(m)Z(\zeta), \quad a = m^\alpha + \psi(m)A(\zeta), \quad u = \mu m^\alpha + \psi(m)U(\zeta), \qquad (7)$$

where the first items consistent with distributions at C_0^- from (3), when $m \to 0$ are much greater, than second, i.e.

$$(1+\mu)m \gg \varphi(m)Z(\zeta), \quad m^\alpha \gg \psi(m)A(\zeta), \quad \mu m^\alpha \gg \psi(m)U(\zeta). \qquad (8)$$

Substituting expressions (7) in the first three equations of a system (5) with account of (8) and that first ("main") items of these expressions satisfy to these equations, we get equations for definition $\varphi(m)$, $\psi(m)$, $Z(\zeta)$, $A(\zeta)$ and $U(\zeta)$. The solutions of these equations satisfying (8), show that in a small neighborhood of a point 0

$$a = m^\alpha(1+\zeta m^\Delta), \quad u = \mu m^\alpha \left(1 + \frac{k}{\mu}\zeta m^\Delta\right), \quad \zeta \geq 0 \qquad (9)$$

and $\varphi(m) = m^{1+\Delta}$. The characteristic variable $\zeta = 0$ at the characteristic C_0^- and increases with removal from it.

TABLE 1. Constants being included in the Guderley solution and in the formulas (9)

ν	γ	n	α	ω	Δ	$10^{-3v\Delta}$	μ	k
2	5/3	0.81563	-0.113	0.339	0.00437	0.9414	1.30225	2.39739
3	5/3	0.68838	-0.151	0.453	0.00481	0.9051	1.28192	2.16128
2	1.4	0.83532	-0.099	0.493	0.01103	0.8587	1.88677	4.09769
3	1.4	0.71717	-0.131	0.657	0.01401	0.7480	1.88509	3.76773
2	1.2	0.86116	-0.081	0.806	0.01761	0.7840	2.97599	8.48025
3	1.2	0.75714	-0.107	1.069	0.02374	0.6114	3.01599	7.96798

Determined by the described above method k and Δ, and also n, α, ω and μ, found from the solution of Guderley problem for six cases, considered in [5], are assembled in the TABLE 1. Besides there are in it describing a little of index Δ values of $10^{-3v\Delta}$. If at C_0^- $m = 1$ when $r = 1$, then when $r = 10^{-3}$ we have $m = 10^{-3v}$. Therefore values of $10^{-3v\Delta}$ show, that even at so small distances from CS the components, appeared owing to the centered compressing wave and equal to zero in CS, become the values about unity. For the same reason when calculation the flow near CS it is necessary to determine z and τ not on expansions such as (9), but by numerical integration when ζ

79

is constant of first and last equations of a system (5) for the initial conditions: $z(0,\zeta) = \tau(0,\zeta) = 0$. The functions $a(m,\zeta)$ и $u(m,\zeta)$ thus are determined by the formulas (9), and $\rho(a,m)$ – by the first formula from (4).

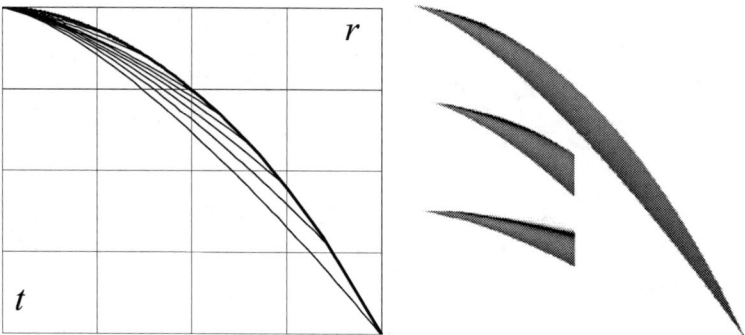

FIGURE 4. C⁻-characteristics and the trajectory of the piston (at the left) and fields of $\log_{10}\rho$ (on the right)

The found solution and the method of characteristics with the consequent calculation of C⁻-characteristics, corresponding to increasing values of ζ, allow to construct the flow up to any fixed trajectory of particles – line m = const, for example, up to $m = 1$. It can be accepted for a required trajectory of the piston. The result of one of such calculations is shown in figure 4. In its left-hand part the characteristics and trajectory of the piston are shown, and in the left-hand there are the fields of $\log_{10}\rho$. In the given example $0 \le \log_{10}\rho \le 4$, and, as well as everywhere, $\rho = 1$ at C_0^-. Along with all compressing wave neighbourhood of CS is shown with an increased scale. The evaluation of $\log_{10}\rho$ as a function of τ at the piston is given by curve 1 in figure 5.

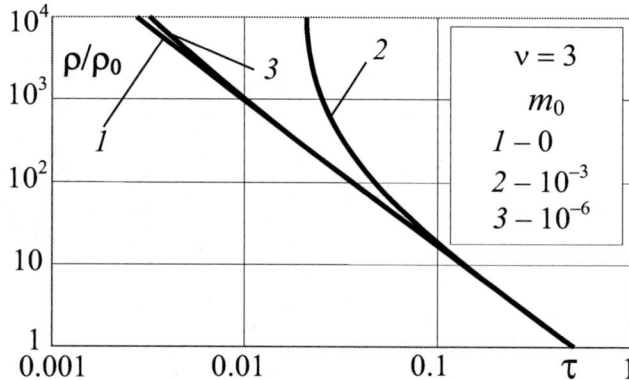

FIGURE 5. The evaluation of $\log_{10}\rho$ at the piston when three positions of a focusing point

After constructing of the solution with a bundle of compressing waves focalized in CS, it is natural to consider the solution with their focusing in a point 0, located at the same singular characteristic of GS when $m_0 > 0$. The card of such flow is shown in a fig. 3, c, and the evaluations of $\log_{10}\rho$ at the piston, adequate to two values of m_0, give curves 2 and 3 in fig. 5. When $m_0 > 0$ the flow in a point 0 is described by the formulas for two-dimensional centered wave of Riemann (requirements of persistence in this point of an entropy and "right" invariant). Though they have no anything common with the formulas (9), the curves 2 and 3 depart from a curve 1 only when approach to a focusing point.

The results, introduced above, have shown the principle capability of simultaneous compressing and heating of gas up to values of densities and temperatures necessary for CITF, for times, on the orders smaller than time of acoustic wave run by uncompressed target. The dissipative effects and actual thermodynamics inevitably will change the formulas and numerical results but hardly so, that the required conditions become unattainable.

ACKNOWLEDGMENTS

The author is grateful to V. N. Malov and N. I. Tillyaeva, which have executed the calculations by a method of characteristics, G. G. Chernyi and V. A. Simonenko – for a boosting appraisal of works and V. A. Belokogne – for useful discussions. The investigation is carried out with the support of RFFI (the projects: 02-01-00422 and 05-01-00846) and the State program of a support of the leading scientific schools of the Russian Federation (project SS-2124-2003.1).

REFERENCES

1. A. N.Kraiko, *J. Appl. Maths Mechs,* **57**, 793-808 (1993).
2. G. Guderley, *Luftfahrtforschung* **19**, 302-312 (1942).
3. G. B. Whitham, *Linear and Nonlinear Waves*, A Willey-Interscience Publication, 1974.
4. L. D. Landau, E. M. Lifshitz, *Theoretical Physics. Vol. VI. Hydrodynamics.* M.: Nauka, 1986, pp. 563-569 (in Russian).
5. A. N. Kraiko, *Proceedings of VII ZST*, Snezinsk, Russia, 2003, http://www.vniitf.ru/rig/konfer/7zst/reports/6s/6-1.pdf, 13 p. (in Russian).
6. A. N. Kraiko, *Proceedings of the Steklov Institute of Mathematics*, **223**, pp. 185-194 (1998).
7. A. N. Kraiko, *J. Appl. Maths Mechs,* **63**, pp. 909-916 (1999).
8. Ya. M. Kazdan, *J. Appl. Mech. and Technical Phys.*, pp. 23-30 (1977) (in Russian).
9. I. E. Zababakhin, V.A. Simonenko, *J. Appl. Maths Mechs,* **42**, pp. 573-576 (1977) (in Russian).

Isentropic Self-Similar Gas Compression

V.A. Simonenko, N.E. Chizhkova

Russian Federal Nuclear Center - All-Russia Research Institute of Technical Physics
PO BOX 245, Snezhinsk, 456770, Russia

Abstract. Self-similar isentropic compression of perfect gas is studied in 1D spherical and cylindrical symmetry for two cases: (1) compression of homogeneous gas by means of smooth compression wave (e.g. with the help of convergent piston) and (2) steady velocity convergence of gas shell with self-similar initial profile of gas. Main attention is paid to flows with two-stage limited compression of matter by smooth isentropic convergent wave and shock wave reflected from the center of symmetry with jump of entropy. Being isentropic the solutions present the most efficient way to dynamically compress the gas, which can be used in various experimental program on inertial confinement fusion and similar. The presented solutions can be useful to test numerical codes designed to simulate the problem with so strong sharpening of flows.
Keywords: isentropic flow, limited and unlimited compression, self-similar solution, shock wave, piston, shell.
PACS: 51.35.+a, 42.55.-f, 47.40.Rs, 47.55.Ca, 52.35.Tc, 64.90.+b

INTRODUCTION

Studying of isentropic flows of compressible media is one of the earliest directions of gas dynamics research. Pioneer work apparently belongs to Riemann [1]. The initial interest was caused by mathematical attractiveness of the exact solutions of mechanics of continuum equations. However gradually the applications of such solutions began to emerge. Well known examples are one-dimensional outflow of initially compressed gas from a pipe and flow of explosive products behind the front of plain detonation wave.

In the second half of the last century the great interest emerged to use isentropic flow to achieve high compression of matter. Such flows can be used to provide phase transitions of condensed matter, to achieve conditions for development of chain reactions of heavy nuclear fission or thermonuclear reactions of light nuclei. Isentropic processes are characterized by the minimal energy consumption and they can be considered as asymptotic ones for realistic flows. In particular, they are useful in research of optimal targets construction and regimes of compression in various modifications of inertial confinement fusion with the help of sharpening flows.

Such flows in real systems are usually far from ideal ones of exact solutions. More complex physical models are used for their description. These models include various dissipative processes. Description of such complex processes can be obtained on the base of complex physical models with the help of rather complex codes. Complexity of codes makes it difficult to estimate accuracy of simulations. Therefore the exact solutions are necessary to test such codes.

CP849, *Zababakhin Scientific Talks - 2005,*
edited by E. N. Avrorin and V. A. Simonenko

In the presented work 1D self-similar flow of isentropic compression of perfect gas is studied for spherical and cylindrical symmetry. The techniques of Riemann invariants are used. It gives symmetrical and very convenient for research appearance of self-similar equations. However this technique is applied just for isentropic flows.

The self-similar solution for isentropic compression of homogeneous gas was firstly obtained by K.P. Stanyukovich [2] for the plain case. This solution is known as a centered compression wave. It gives a convergence of all matter particles and β – characteristics simultaneously at the same moment, e.g., $t = 0$, into one point (focus). Simultaneous convergence of all particles results in unlimited compression of a chosen mass of matter. It was also paid attention to limited compression of gas by means of centered wave of compression. To have such flow a plane piston should move with acceleration as in previous case till moment t_1, and after that should move with steady velocity $u_1 = u(t_1) = const$ obtained by this moment. It this case there is a focusing of characteristics at the moment $t = 0$ on the focus plane, which follows by reflection of a shock wave from the plane. The matter compressed smoothly by compressive wave increases its density by jump into the front of reflected shock wave and comes to the rest behind the front.

Unlimited compression can be organized also for cases of 1D spherical or cylindrical symmetry, for instance, by means of strictly chosen movement of "spherical or cylindrical piston", which comes to the focus simultaneously with all particles, or by means of assignment of definite change of pressure versus time in the last particle of the matter. It can be also organized the limited compression with the reflected shock wave like for the plane case. However for spherical and cylindrical symmetry there is no coincidence of self-similar lines and β – characteristics.

Generalization of such solutions for cases of spherical and cylindrical symmetries was made by I.E.Zababahin and V.A.Simonenko [3] and Ya.M.Kazhdan [4]. A.N.Kraiko [5] and A.N.Kraiko and N.I.Tillyeva [6] used these solution to compare energy consumption for various gas dynamics flows.

There is an extension of interest to convergent smooth compression for the last caused by new types of experimental facilities dealing with cylindrical symmetry (electric explosion wire array and heavy ion beam targets). They revived interest to efficient compression of targets. It is why we decided to presents more detailed analysis of early published solutions paying more attention to the cases of more demanded finite compressions, which were just mentioned in [3]. The solutions with reflected wave are very valuable for testing of hydrocodes. It is usually used for this purpose well exact solution, proposed by Noh [7] for compression of initially cold gas. However on this reason it does not account the thermodynamics of smooth compression. The presented solutions allow performing more complete tests.

SELF-SIMILAR SOLUTIONS PRODUCTION

As in [3], we will represent the gas dynamics equations in Riemann's invariants $\alpha = u + 2c/(\gamma - 1)$ and $\beta = u - 2c/(\gamma - 1)$, where u and c are matter velocity and sound velocity:

$$\begin{cases} \dfrac{\partial \alpha}{\partial t} + \left(\dfrac{\gamma+1}{4}\alpha + \dfrac{3-\gamma}{4}\beta \right)\dfrac{\partial \alpha}{\partial r} + \dfrac{(v-1)(\gamma-1)}{8r}\left(\alpha^2 - \beta^2\right) = 0, \\[3mm] \dfrac{\partial \beta}{\partial t} + \left(\dfrac{\gamma+1}{4}\beta + \dfrac{3-\gamma}{4}\alpha \right)\dfrac{\partial \beta}{\partial r} + \dfrac{(v-1)(\gamma-1)}{8r}\left(\beta^2 - \alpha^2\right) = 0 \end{cases} \tag{1}$$

Let's choose $\tau = v_0 t / r$ as a self-similar variable where r is a distance to the center or axe of symmetry, $v = 3$ for spherical and $v = 2$ for cylindrical symmetry, time t is counted from the moment of focusing, v_0 is the velocity scale. After that we will receive

$$\begin{cases} \left(\dfrac{\gamma+1}{4}\alpha + \dfrac{3-\gamma}{4}\beta - \tau^{-1} \right)\alpha' - \dfrac{(v-1)(\gamma-1)}{8\tau}\left(\alpha^2 - \beta^2\right) = 0, \\[3mm] \left(\dfrac{\gamma+1}{4}\beta + \dfrac{3-\gamma}{4}\alpha - \tau^{-1} \right)\beta' - \dfrac{(v-1)(\gamma-1)}{8\tau}\left(\beta^2 - \alpha^2\right) = 0. \end{cases} \tag{2}$$

It is convenient to go to functions $A = \alpha\tau$, $B = \beta\tau$ and to analyze field of integral curves on a plane (A, B). Then

$$\frac{dB}{dA} = \frac{q(A,B)}{q(B,A)} \times \frac{p(A,B)}{p(B,A)}, \tag{3}$$

where $q(A,B) = \dfrac{\gamma+1}{4}A + \dfrac{3-\gamma}{4}B - 1$,

$p(A,B) = \dfrac{(v-1)(\gamma-1)+2(\gamma+1)}{8}B^2 + \dfrac{3-\gamma}{4}AB - \dfrac{(v-1)(\gamma-1)}{8}A^2 - B$.

Points of intersection isoclinal lines are special. Let's consider the most interesting ones. Direction centers are $N\left(-\dfrac{2}{\gamma-1}, \dfrac{2}{\gamma-1}\right)$ and $G(1,1)$. A saddle is the point $S\left(\dfrac{2-2\sqrt{v}}{v\gamma - v + 2}, \dfrac{2+2\sqrt{v}}{v\gamma - v + 2}\right)$ through which pass two special solutions. They have different inclinations: $K_{1,2} = -\dfrac{9}{2} \pm 2\sqrt{5}$ at $v = 2, \gamma = 3$; $K_{1,2} = -4 \pm \sqrt{15}$ at $v = 3, \gamma = 3$;

Point O is a two-critical center. All integrated curves can have any value of dB / dA.

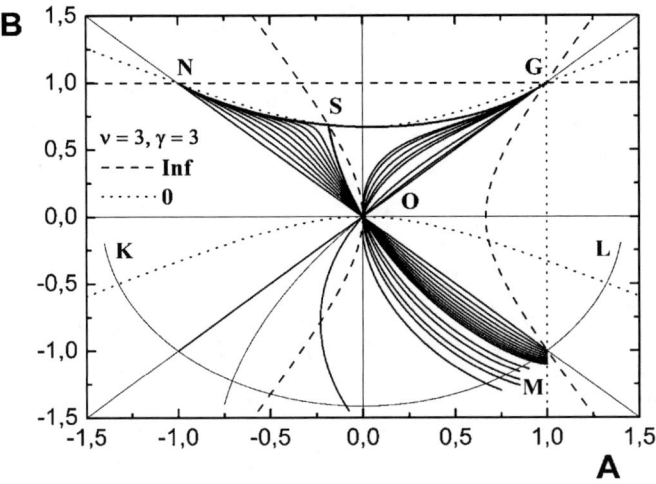

FIGURE 1. $v = 3$; $\gamma = 3$.

In the picture 1 zero and infinity isoclinal lines, some integrated curves and the percussive adiabatic KL for $\gamma = 3$ are represented.

The curve NS correlates with unlimited homogeneous gas compression by the piston, the curve NOM correlates with the limited compression by the piston, the curve GS correlates with unlimited compression profiled shell, the curve GOM correlates with the limited shell compression. We will consider these solutions in more detail.

COMPRESSION OF GAS BY THE PISTON

Unlimited and Limited Compression

Let compressible substance is ideal gas with initial values of pressure, density and sound velocity, accordingly, c_0, ρ_0, P_0. Firstly, gas do not move ($u_0 = 0$) and then it smoothly starts converges to a center / axis of symmetry under action of the outside applied pressure. The velocity scale is the value $v_0 = c_0 = 1$.

Unlimited shock-free compression of substance by the piston is the special solution. The law of the piston movement $r_p(t)$ is numerical integration of the equation $dr/dt = r(A + B)/(2t)$ (in the picture 4).

85

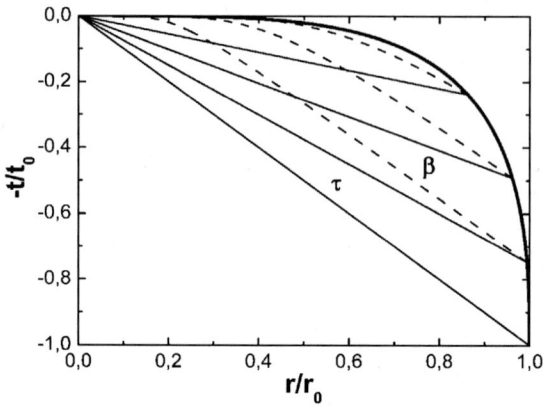

FIGURE 2: **The piston trajectory at unlimited collapse.**

The wave of compression spreads on rest gas with sound velocity. Its "head" is weak break, and "tail" goes with piston velocity u_p. Near the point S at $r \to 0$ all values unlimited grow: $r \propto t^{\eta} \; c \propto t^{-\nu(\gamma-1)\eta/2} \; p \propto t^{-\nu\eta}$, where $\eta = 2/(\nu\gamma - \nu + 2)$.

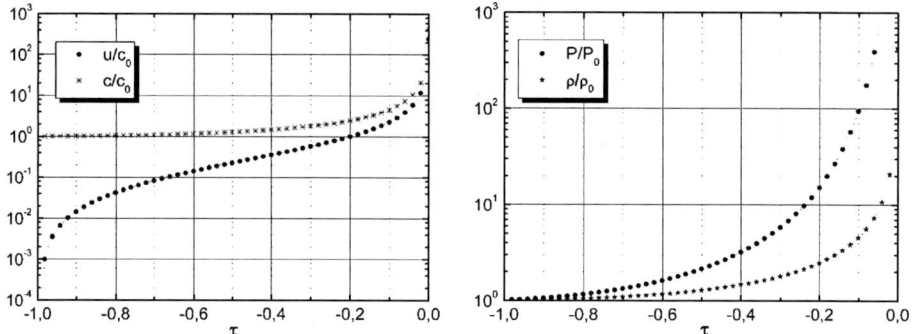

FIGURES 3,4. Dependences of substance velocity $u = u(\tau)$, sound velocity $c = c(\tau)$,

pressure $P = P(\tau)$ and density $\rho = \rho(\tau)$.

The set of integral curves filling triangle $\triangle$ NSO and coming on percussive adiabat KL corresponds to final compression of substance. Near the point O A and B variables are proportional to self-similar variable $B = qA$, $A = m\tau$ (q,m-constants). Solutions at $2/(\gamma-1) < m < \infty$ are correlated with limited compression. At the moment of focusing the substance goes with steady velocity to the center. This condition is initial one by L.I.Sedov [8]. In the picture 5 shock wave is reflected from the center and behind this wave the substance is rest.

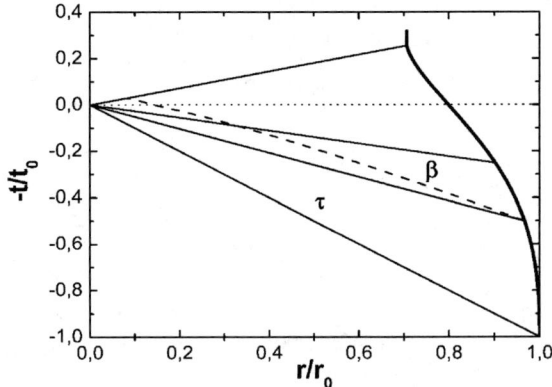

FIGURE 5. The piston trajectory for limited collapse.

At the moment of focusing pressure and density have the same values in the whole system. The velocity of shock wave is $D = 2,8 \cdot c_0$. At the self-similar variable $\tau = 0,3616$ shock wave comes on the piston and stops it.

COLLAPSE OF FINITE THICKNESS SHELL

Unlimited and Limited Collapses

The results correlates with Δ GSO in the picture 1 are qualitatively new. Separatrix GS correlates with collapse of shell. There is absolute vacuum $t_0 = -1$ $u = -1, c = 0$ inside the shell at the initial moment of time. Movement of shell occurs under action of the outside applied pressure to its external border. Let thickness Δ of shell is 0.15 from initial radius. The velocity scale is $v_0 = |u_{0b}|$, where u_{0b} is initial velocity of the internal border.

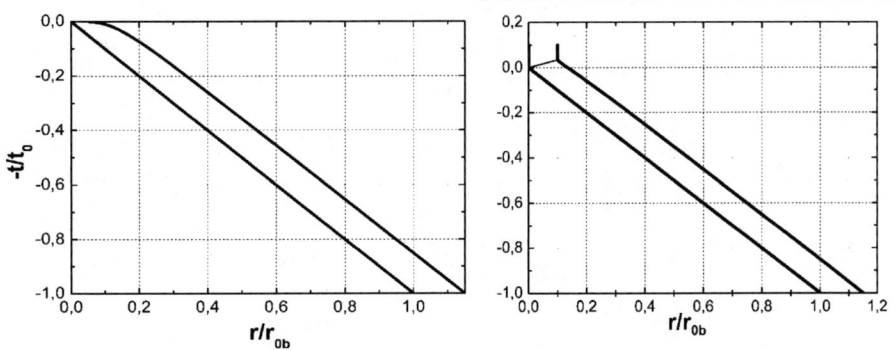

FIGURES 6,7: The trajectory of internal and external borders shell at unlimited and limited collapses.

At limited collapse at the moment of focusing the shock wave on internal border of the shell is formed. Speed of the shock wave is $D = 2,86 \cdot v_{ob}$. At value of self-similar variable $\tau = 0,3487$ the shock wave achieves external border of shell and stops it.

CONCLUSION

The constructed solutions cover all range of compression from infinity up to unit. Solutions with limited compression always contain reflected shock wave from a center or axis of symmetry. The reflected wave spreads in compressed matter preliminary "heated" by isentropic wave of compression and on this reason has finite amplitude. In this respect this solution has an advantage in comparison with exact solution proposed by Noh and popular among code designers.

Such solutions represent also significant applied interest, in particular to the problems of inertial confinement fusion. They allow receiving the shape of a pulse of pressure which provides originally smooth isentropic compression of matter and then generate reflected shock wave, in which moderate increase of compression provides essential rise of temperature. Such sequence is favorable for ignition of thermonuclear burning. In the sense of energy consumption isentropic regime of compression is the most efficient because there is no extra spending of energy for additional heating of matter.

REFERENCES

1. B. Riemann, *Uber die Fortpflanzung ebener Luftwellen von endlilicher Schwingungsweite*, Abhandl. d. Ges. d. Wiss. zu Gottengen., 1860.
2. K.P. Stanyukovich, *Neustanovivshiesya Dvizheniya Sploshnoi Sredy* (in Russian: Non-Steady-State Motion of Continuum), Moscow, GITTL, 1955.
3. I.E. Zababakhin and V.A. Simonenko, *Prikl. Mat. Mekh., Akad. Nauk SSSR*, 1978, vol. 42, no. 3, p.573.
4. Ya.M. Kazhlan, *K Voprosu ob Adiabaticheskom Szhatii Gaza pod Deistviyem Sfiricheskogo Porshnya*, (in Russian) PMTF, 1977.
5. A.N. Kraiko, *Prikl. Mat. Mekh., Ros. Akad. Nauk*, 1996, vol. 60, no 6, p.1000.
6. A.N. Kraiko and N.I. Tillyaeva, *High Temperature, Nauka*, 1998, vol. 36, no. 1.
7. W.F. Noh, Errors for Calculations of Strong Shocks Using an Artificial Viscosity and an Artificial Heat Flux, J. Comput. Phys. 72 (1987), 78.
8. L.I. Sedov, *Metody Podobiya i Razmernosti v Mekhanike* (The Similarity and Dimensions Methods in Mechanics), Moscow: Nauka, 1987.

SPH Modification Based on the Riemann Solver

Anatoly D. Zubov, Alexander M. Lebedev, Valentina L. Sokolovskaya

Russian Federal Nuclear Center – Zababakhin Institute of Technical Physics
RFNC-VNIITF,P.O. Box 245, Snezhinsk, Chelyabinsk Region, Russia 456770

Abstract. In this paper a modification of the Smooth Particle Hydrodynamics (SPH) method is considered. This modification allows increasing the accuracy of calculation of shocks and contact discontinuities in hydrodynamic flows. This paper also presents a research which shows the influence of different kernel forms used in the SPH simulations, and also the influence of particle parameter difference in distinct simulation domains. An example of a test one-dimensional Riemann problem is used to compare the computations by a «standard» SPH-method, by the A.N.Parshikov's method and also by a modified method presented in this paper. The comparison of the computation results with the exact solution in discrete analogs of the Chebyshev and L_2 norms displays advantages of the offered modification.

INTRODUCTION

Smooth Particle Hydrodynamics is a Lagrangian technique for mechanics of continua calculations. Due to its meshfree nature this method can handle complex material deformation and fragmentation. The thickening of particle in high density domains is a convenient circumstance for evaluations, which automatically provides reaching sufficient approximation in these domains. The difficulties of the method implementation include the boundary conditions specification and the initial particle distribution.

As well as many other particle methods, SPH is insufficiently competitive in comparison, for example, with finite-difference methods in 1D problems. Therefore the purpose of the present paper is to analyze the available opportunities to increase the efficiency and accuracy of 1D calculations by the methods that can be created in the framework of the SPH-methodology.

HYDRODYNAMICS EQUATIONS

The 1D hydrodynamic equations system is given below:

$$\frac{d\rho}{dt} = -\rho \nabla \cdot \mathbf{U}, \quad \frac{d\mathbf{U}}{dt} = -\frac{1}{\rho}\nabla P, \quad \frac{dE}{dt} = -\frac{P}{\rho}\nabla \cdot \mathbf{U}, \quad \frac{d\mathbf{x}}{dt} = \mathbf{U}. \tag{1}$$

Taking into account the artificial viscosity q_{ij} the SPH approximation provides for the following arrangement of the system:

CP849, *Zababakhin Scientific Talks - 2005,*
edited by E. N. Avrorin and V. A. Simonenko

$$\frac{d\rho_i}{dt} = \rho_i \sum_j \frac{m_j}{\rho_j} \left(\mathbf{U}_i - \mathbf{U}_j \right) \cdot \nabla W \left(\left| \mathbf{x}_i - \mathbf{x}_j \right|, h \right)$$

$$\frac{d\mathbf{U}_i}{dt} = -\sum_j m_j \left(\frac{P_i}{\rho_i^2} + \frac{P_j}{\rho_j^2} + q_{ij} \right) \nabla W \left(\left| \mathbf{x}_i - \mathbf{x}_j \right|, h \right)$$

$$\frac{dE_i}{dt} = -\frac{1}{2} \sum_j m_j \left(\frac{P_i}{\rho_i^2} + \frac{P_j}{\rho_j^2} + q_{ij} \right) \left(\mathbf{U}_i - \mathbf{U}_j \right) \cdot \nabla W \left(\left| \mathbf{x}_i - \mathbf{x}_j \right|, h \right)$$

$$\frac{d\mathbf{x}_i}{dt} = \mathbf{U}_i$$

$$(2)$$

where $W\left(\left| \mathbf{x} - \mathbf{x}' \right|, h \right)$ is a kernel (weighting) function, and h is a smoothing length.

Most often the artificial viscosity in SPH is used in the form that was proposed by Monaghan and Gingold [1]:

$$q_{ij} = \begin{cases} \dfrac{-k_1 \overline{C}_{ij} \mu_{ij} + k_2 \mu_{ij}^2}{\overline{\rho}_{ij}}, & \left(\mathbf{U}_i - \mathbf{U}_j \right) \cdot \left(\mathbf{x}_i - \mathbf{x}_j \right) < 0 \\ 0, & \left(\mathbf{U}_i - \mathbf{U}_j \right) \cdot \left(\mathbf{x}_i - \mathbf{x}_j \right) \geq 0 \end{cases}, \qquad (3)$$

where $\mu_{ij} = \dfrac{h(\mathbf{U}_i - \mathbf{U}_j) \cdot (\mathbf{x}_i - \mathbf{x}_j)}{(\mathbf{x}_i - \mathbf{x}_j)^2 + k_3 h^2}$, $\overline{C}_{ij} = \dfrac{C_i + C_j}{2}$, $\overline{\rho}_{ij} = \dfrac{\rho_i + \rho_j}{2}$, C is the sound

speed, and k_1, k_2, k_3 are coefficients.

The hydrodynamic equations were supplemented with the equation of perfect gas state

$$P = (\gamma - 1) \rho E . \qquad (4)$$

SPH MODIFICATIONS

A.N. Parshikov [2] offered a version of the SPH-method in which the Riemann problem solution [3] is used to calculate the particles interaction in SPH-environment, where the particle contact point velocity U and pressure P are estimated in the acoustic approximation (U_{ij}^* and P_{ij}^* respectively):

$$U_{ij}^* = \frac{U_j^R \rho_j C_j + U_i^R \rho_i C_i - P_j + P_i}{\rho_j C_j + \rho_i C_i}, \quad P_{ij}^* = \frac{P_i \rho_j C_j + P_j \rho_i C_i - \rho_j C_j \rho_i C_i \left(U_j^R - U_i^R \right)}{\rho_j C_j + \rho_i C_i}, \quad (5)$$

where $U^R = \mathbf{U} \left(\mathbf{r}_j - \mathbf{r}_i \right) / \left(\left| \mathbf{r}_j - \mathbf{r}_i \right| \right)$.

Usage of an acoustic approximation allows a more precise calculation in the neighbourhood of a contact discontinuity (and also shock wave) without regard to any artificial viscosity. In this case, the equations (2) are become as follows:

$$\frac{d\rho_i}{dt} = 2\rho_i \sum_j \frac{m_j}{\rho_j}\left(\mathbf{U}_i^R - U_{ij}^*\right)\cdot\nabla W\left(\left|\mathbf{x}_i - \mathbf{x}_j\right|, h\right)$$

$$\frac{d\mathbf{U}_i}{dt} = -\frac{2}{\rho_i}\sum_j \frac{m_j P_{ij}^*}{\rho_j}\nabla W\left(\left|\mathbf{x}_i - \mathbf{x}_j\right|, h\right)$$

$$\frac{dE_i}{dt} = -\frac{2}{\rho_i}\sum_j \frac{m_j P_{ij}^*}{\rho_j}\left(\mathbf{U}_i^R - U_{ij}^*\right)\nabla W\left(\left|\mathbf{x}_i - \mathbf{x}_j\right|, h\right)$$

$$\frac{d\mathbf{x}_i}{dt} = \mathbf{U}_i.$$

(6)

A.D. Zubov, V.L. Sokolovskaya [4] obtained that the version of the SPH-method by A.N. Parshikov worsens the accuracy of the description of a rarefaction wave in comparison with a "standard" SPH method. To achieve a better agreement with the exact solution, it was offered to use the feature of a rarefaction wave $\nabla\rho\cdot\nabla U < 0$ (at this condition the solution is based on the "standard" formulas). Also in calculation of internal energy of substance there is no need to use the acoustic approximation for pressure.

There are two different Riemann solution-based methods [6], which belong to the class of Godunov SPH methods (GSPH). According to the Godunov method the evolution of flow variables at the ith particle is calculated as

$$\rho_i^{n+1} = \rho_i^n - 2\rho_i^n \frac{U_i^a - U_i^b}{\Delta x_i}\Delta t$$

$$U_i^{n+1} = U_i^n - 2\frac{P_i^a - P_i^b}{\rho_i^n \Delta x_i}\Delta t$$

$$E_i^{n+1} = E_i^n - 2\frac{P_i^{ah}U_i^{ah} - P_i^{be}U_i^{be}}{\rho_i^n \Delta x_i}\Delta t$$

(7)

where $\Delta x_i = x_{i+1} - x_{i-1}$.

The top indexes (a) and (b) mean respectively "ahead interface state" and "behind interface state". The time step has to be evaluated according to Courant criterion [6].

Also we have made a comparison of the calculation accuracy in case of a choice of the common kernel function based on B-spline (third degree) [1,2,4] and kernel function based on Q-spline [5] (fifth degree).

91

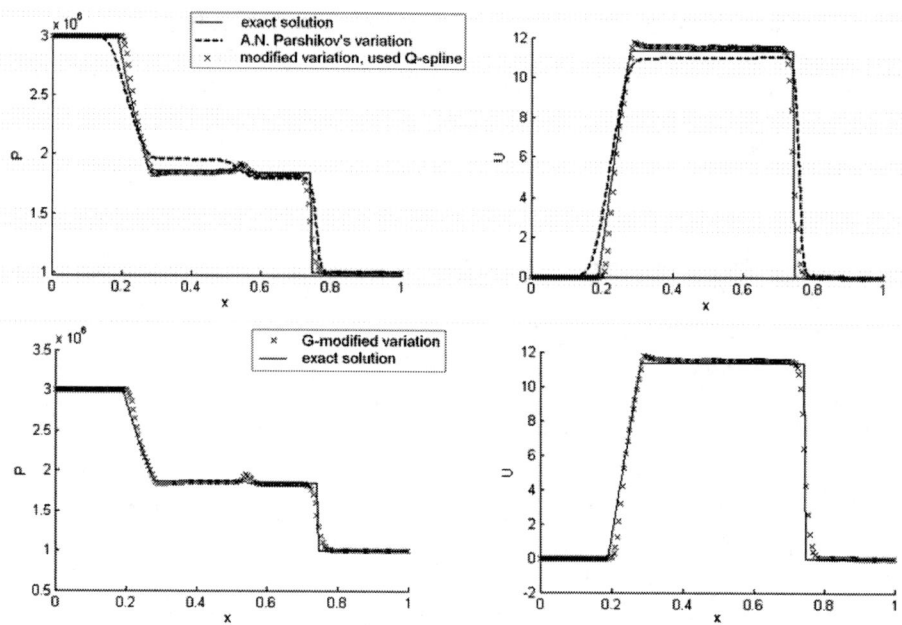

FIGURE 1. Flow parameters profiles in test 1 ($t = 0.004$).

TWO TESTS ON A RIEMANN PROBLEM

Test 1. Initial parameters of gas: $P_1 = 3 \cdot 10^6$ Pa, $\rho_1 = 1.5 \cdot 10^3$ kg/m^3, $\gamma_1 = 3$ to the left of a contact surface and $P_2 = 1 \cdot 10^6$ Pa, $\rho_2 = 1.2 \cdot 10^3$ kg/m^3, $\gamma_2 = 3$ to the right of a contact surface. Speed of the gas at the initial moment is equal to zero. The simulation domain is a line segment of length $L = 0.1$ m. An initial position of a contact discontinuity is $x_0 / L = 0.5$. In the tests presented in the figures we used 200 particles. [2,4].

TABLE 1. Relative estimation of an error in all the domain for the modified methods.

	Standart	Modified	Godunov - Modified
$\|\rho\|_C$	1	0.782	0.752
$\|E\|_C$	1	0.789	0.761
$\|P\|_C$	1	0.832	0.804
$\|\rho\|_{L_2}$	1	0.449	0.447
$\|E\|_{L_2}$	1	0.474	0.474
$\|P\|_{L_2}$	1	0.484	0.474

92

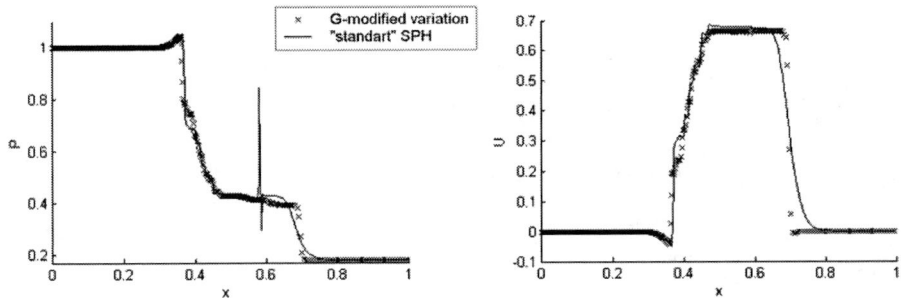

FIGURE 2. Flow parameters profiles in test 2 ($t = 0.15$).

Test 2. Initial parameters of gas: $P_1 = 1$, $\rho_1 = 1$, $\gamma_1 = 1.4$ to the left of a contact surface, $P_2 = 0.1795$, $\rho_2 = 0.25$, $\gamma_2 = 1.4$ to the right of a contact surface. All other parameters were the same as in Test 1. [5].

The flow parameters profiles in these tests are presented in Fig. 1 and Fig. 2 respectively. Relative estimation of an error (in discrete analogues of Chebyshev and L_2 norms) in full domain for the modified methods for kernel based on Q-spline is indicated in Table 1. The usage of the smoothing kernel based on Q-spline is justified at increasing of the particles number.

SUMMARY

The variation of the GSPH method using Godunov algorithm (7) allows improving the accuracy of calculations in comparison with the "standard" SPH method, especially in the domain of contact discontinuity. Using a Godunov algorithm together with the modified method has also given an increase of accuracy, though insignificant.

REFERENCES

1. J. J. Monaghan, "An introduction to SPH", *J. Comput. Phys.* 48 (1988), pp.89-96.
2. A. N. Parshikov, S. A. Medin, I. I. Loukashenko, V. A. Melikhin, "Improvements in SPH method by means of interparticle contact algorithm and analysis of perforation tests at moderate projectile velocities", *Internat. J. Impact Engng.* 24 (2000), pp.779-796.
3. R. D. Richtmyer, K. W. Morton, "Difference methods for initial-value problems", Intercience Publishers, a division of J. Wiley & Sons, New York, 1967.
4. A. D. Zubov, V. L. Sokolovskaya, "About one version SPH of particles based on the task of Riemann", preprint 1/2002, SGPhTA, Snezhinsk, 2002 (in Russian).
5. J. Hongbin, D. Xin, "On criterions for smoothed particle hydrodynamics kernels in stable field", *J. Comput. Phys.* 202 (2005), pp.699-709.
6. Molteni D., Bilello C., "Riemann Solver in SPH", Mem. S.A.It. Suppl., vol. 1, № 36 (2003), pp.36-44.

Long-living luminous objects, forming in large-scale water cavity

I.L. Veremeenko, P.I. Golubnichiy, J.M. Krutov, D.V. Reshetniak

Department of Physics, East-Ukrainian national university by Vladimir Dahl.
bl. Molodezhny, 20a, 91034, Lugansk, Ukraine

Abstract. During research of electric discharge in water the authors found a new type of cavitative luminescence. The difference in it was that the luminescence was not connected to a collapse of the cavity, and was observed during all period of pulsation of the cavity. Radiation from the cavity was registered not from all volume of a bubble, and from areas with sizes much less than the diameter of the cavity. The time of life (luminescence) of these formations by three orders surpassed the characteristic lifetime of breaking up discharge plasma, therefore the authors named them long-living luminous objects (LLO). LLO had a spherical form and were transparent. The brightness of its luminescence did not decrease monotonously, and had a flashing character. For further research of LLO a special chamber has been made, allowing one to throw out luminescence objects from water to air. We report the results of the set of experiments, which have enabled to estimate some properties of LLO in water and air.
Keywords: Electric discharge, cavitation, long-living luminous objects.
PACS. 78. 60. Mq -- Sonoluminescence.
PACS. 52. 50. Lp – Plasma production.

Long-living luminous objects (LLO) were discovered as a result of studies of disintegration of water plasma in electric discharge[1,2,3]. The objects existed inside pulsing cavities, formed in water after electric discharge. The observed formations moved independently one from another and from movement of the border of the cavity, their average size amounted to 1mm, its form was close to spherical, its colour was orange. Time of life, which was defined by time of the luminescence was $\geq 6 \cdot 10^{-3} s$ and was limited by the period of pulsations of the cavity. In some cases

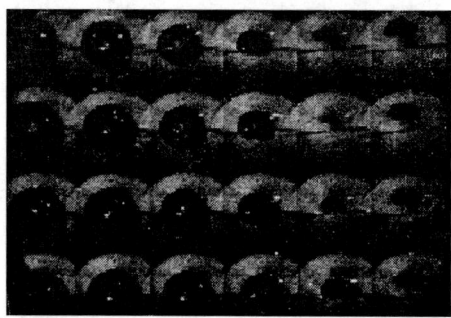

FIGURE 1. – Luminescence of objects inside cavities.

CP849, *Zababakhin Scientific Talks - 2005,*
edited by E. N. Avrorin and V. A. Simonenko
© 2006 American Institute of Physics 0-7354-0345-7/06/$23.00

LLO penetrated from cavity to surrounding liquid. On figure 1 we present multiframe photos of such events. The order of following of the frames - from top to bottom, from left to right. The exposure of each frame and interval between them 50 mks, delay from the beginning of shooting to the end of energy discharge was 1ms.

The shooting was conducted by means of an optical-mechanical camera, equipped with amplifier of brightness. For visualization of the cavities was used end-to-end translucence, the intensity of which was selected so as to shade cavity to reduce to minimum hindrance in observation of the LLO luminescence.

Analysis of optical characteristics of investigated formations, its spectrum of radiation, dynamics of the luminescence and nature of motion of these formations has shown that they can not be a particle of overheated material of electrodes, a result of chemical reactions in gas phase or clot of plasma.

A further study of LLO was prevented by water, surrounding cavity, in which they were found after the discharge. As a result of a row of tests a chamber was designed, which allowed to throw out the products of disintegration of water plasma and luminous objects to air [4]. Using the same conditions of discharge we found that average size of the objects thrown away to atmosphere increased by 3 - 5 times and their time of life (luminescence) went up to 0.1c. On fig. 2(a). is shown a photo of the ejected luminous objects to atmosphere. The shooting was maid in LLO's own light. Recording was maid by electronic-optical camera (EOC), collected on the base of electron-image tube and amplifier of brightness.

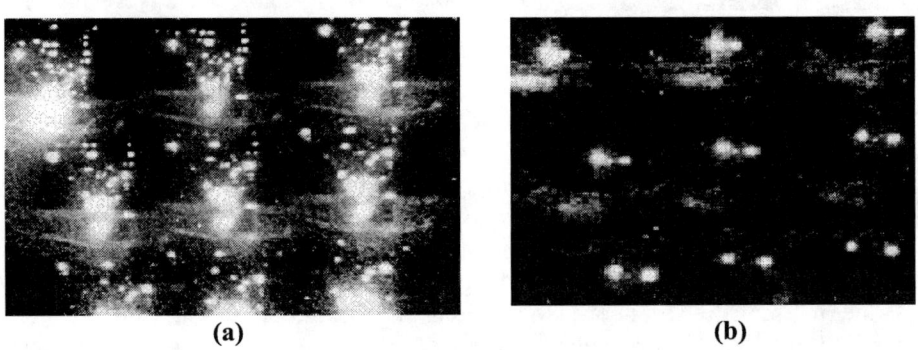

(a)   (b)

FIGURE 2. – Throw-out of LLO to atmosphere (a), "condensation" of LLO in air (b).

On picture there are nine frames, order of their following – from left to right, from top to bottom. The first frame is displaced to the left for simplification of its identification. Exposure of each frame – 5 microseconds, interval between them 200 microseconds, delay between beginning of shooting and moment of ending of the discharge 0,5 ms.

A study of LLO in air has allowed to register the process of "condensation" of objects from luminous cloud, consisting of products of disintegration of water plasma. On Fig. 2(b) a track record of this process is shown. The shooting was maid in object's

95

own light. The order of following of the frames, their nature and rate of the shooting is similar to the ones shown on fig. 2(a).

Having conducted multiple experiments with luminous objects in air we found that the most effective were conducted with influence on LLO of high-temperature fields. Once we put on nozzle of the chamber from which the objects were thrown out a flat double spiral of wire consisting of alloy of Ni and Cr. Thickness of the wires was 0.1 mm, a distance between its loops – 2.5 mm . The distance between upper and lower layers was 5 mm. Through the wire electric current was passed. The upper layer was warmed to 1000 C, the lower to 800 C. The temperature was measured by a pyrometer. In most cases LLO, having passed through the spiral, disappeared. A result of such event is shown on Fig. 3(a). Here there are six frames, the order of following is similar to previous ones.

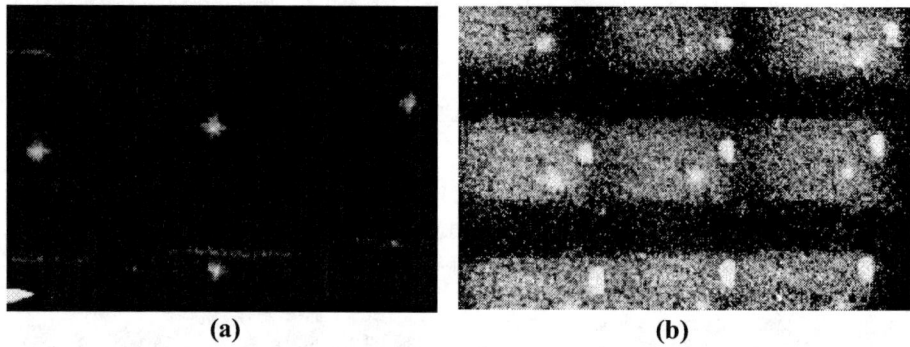

(a)   (b)

FIGURE 3. – The object passing the zone of heating (a). Process of lengthening of the object in inhomogeneous thermal field to the side of the lower temperatures (b).

For suppression of light from the heated spiral we used colour glasses, selected so as to vastly weaken radiation from it and to pass as much as possible light from LLO. On the picture the upper spiral is seen in the form of weak horizontal band, location of the lower one is shown by arrow on the fourth frame.

On the first frame the luminous object, being at a distance of 10 mm from the lower layer of the spiral has a nearly spherical shape. Approaching the zone of the heating its form changes - he looks like melting. While passing the lower heated layer LLO disappears (frame 4), appearing again between layers of the spiral on fifth frame to disappear again while passing upper layer. Screening the luminescence of the object by wire is impossible. This is proved by presence of a small luminous area on frame 6 on supposed location of LLO remains and by model experiments.

In the other experiments on place of the spiral we put a copper plate, cooled to the liquid nitrogen temperature. For visualization of space between the chamber and cooled plate we used sidewise light, with radiation dispersed on microdrops of water. On fig. 3(b) we present result of registering interesting type of the observed phenomena.

On fig. 3(b) we registered lengthening of LLO towards lower temperatures while flying. Velocity of "growing" is up to 15 m/s.

Spectral distribution of LLO radiation (fig.4) was registered by a glass monochromator. For restoration of real spectrum, distorted by absorption in optical tract and spectral sensitivity of the photocathode EOC, the spectrum under investigation was compared with spectrum of the ordinary lamp with the temperature 2850 K. Lines of radiation or absorption are absent on the photo, as is evidenced by experiment which uses diffractive monochromator. The resolution is 0,1 nm.

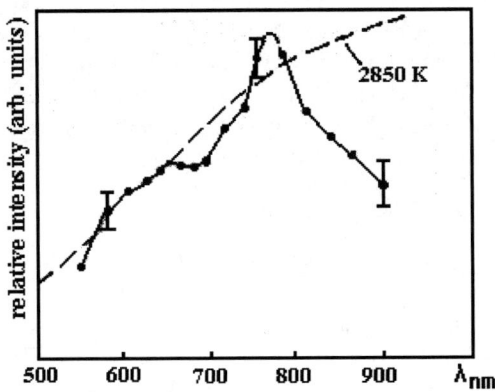

FIGURE 4. Spectral distribution of LLO radiation.

Within the range of 550 - 750 nm spectral distribution of LLO is close to distribution of an absolutely black body with the temperature 2850 K, however its luminescence intensity is more than two orders lower. This has allowed to estimate the power of radiation of LLO - $7 \cdot 10^{-4} [Wt] < N < 7 \cdot 10^{-3} [Wt]$ [5].

Transition of luminous objects from cavity to water enabled to estimate LLO's density in supposition that the object is a solid body - $\dfrac{\rho_o}{\rho_w} >> 0,67 \cdot s$. Here - ρ_o is LLO's density, ρ_w - density of water, c - a coefficient of hydrodynamical resistance (for sphere with corresponding to the experiment Reynolds number=0.4 [6]).

Observation of behavior of long-living luminous objects in high-temperature fields has enabled to estimate temperature conductivity of LLO's material in the same suggestions, as above $D \approx 10^{-2} \dfrac{m^2}{s}$. This value is by two orders more than temperature conductivity of copper and by three and a half orders higher than temperature conductivity of ice.

Using the preceding data and [5] it is possible do suppose:

1. As a result of electric discharge in water LLO are formed with time of life (luminescence) by four orders exceeding typical time of relaxation of any discharge plasma.

2. The form of the objects formed in cavities or thrown away to atmosphere was close to spherical.

3. Heating and cooling of LLO brings about suppression of their luminescence (the range of LLO existence is 100 - 1100 K).

4. Quick cooling of scattering products of the discharge (cooled barrier, exit to atmosphere) facilitates formation of the objects.

5. LLO's ability to exist and radiate in air, in water, as well as inside a pulsing cavity allows to speak of their self sustenance.

6.Conservation of LLO shape while moving with significant velocity in air and in water, as well as ability to condense from shapeless cloud or on cooled surfaces points to significant inter-particle interaction within material, forming LLO.

7. Long-living luminous objects can be a porous friable bodies, the "framework" of which consists of unusual metastable combinations of hydrogen and oxygen.

8. After transition from metastable to usual combination there must be some release of energy.

REFERENCES

1. Golubnichiy P.I. at al., DAN of the USSR, 311, 1990, pp. 356-360.
2. Golubnichiy P.I. at al., ZhTF, 60, 1990, pp. 183-186.
3. Golubnichiy P.I. at al., An abstract of the reports at 3 All-Union seminar on ball lightning, M., IVTAN, 1990, p. 40.
4. Golubnichiy P.I. at al., An abstract of the report at 15th All-Union research conferences "High-speed photography, fotonics and metrology of fast occurring processes", M., VNIIOFI, 1991, pp. 113-117.
5. Veremeenko I. L. at al., The Herald of East-Ukrainian national university, 12 ,2000, pp. 98-107.
6. Batchelor G. K. Introduction to fluid dynamics. Cambridge University Press , 1970.

Nonlinear Dynamics of the Interface of Dielectric Liquids in a Strong External Electric Field

N.M. Zubarev

Institute of Electrophysics, Ural Branch, Russian Academy of Sciences,
106 Amundsen Street, 620016 Ekaterinburg, Russia

Abstract. The evolution of the interface between two ideal dielectric liquids in a strong vertical electric field is studied. If the ratio of the permittivities of the liquids is inversely proportional to their densities, we find that there is a flow regime for which the velocity and electric field potentials are linearly dependent functions. The corresponding reduced equations for interface motion are derived. In the limit of small density ratio, these equations coincide with the well-known equations describing the Laplacian growth.

It is well known [1,2] that a flat interface between two dielectric liquids in a rather strong vertical electric field is unstable. The dynamics of nonlinear waves on a liquid-liquid interface is usually investigated in the quasi-monochromatic approximation (see [3-6] and the references therein). The applicability of this approach is limited by the condition that the slopes of the surface must be small. However, the development of the instability can violate this condition. In this connection, there arises a need to seek analytic methods for studying large-amplitude interfacial waves. In view of the complexity of the equations of motion of a liquid-liquid interface, it would be of particular value to find an effective way of reducing the number of these equations.

In this paper we suggest an approach to studying the nonlinear interface dynamics which simplifies considerably the equations of motion. The linearized equations can be split into two separate subsystems in the strong-field limit, and our goal is to extend this property to the nonlinear equations. To clarify this idea, consider the dispersion relation for linear interfacial waves. It has the following form [1,2]:

$$\omega^2 = \frac{\rho_1 - \rho_2}{\rho_1 + \rho_2} gk - \frac{E_1 E_2(\varepsilon_1 - \varepsilon_2)}{4\pi(\rho_1 + \rho_2)(\varepsilon_1 + \varepsilon_2)} k^2 + \frac{\alpha}{\rho_1 + \rho_2} k^3 ,$$

where k is the wave number; ω is the frequency; g is the acceleration of gravity; α $ is the surface tension coefficient; ρ_1 and ρ_2 are the mass densities of the lower and the upper liquid, respectively ($\rho_1 > \rho_2$); ε_1 and ε_2 are the dielectric constants of the respective liquids. The external electric field strengths under and above the interface, E_1 and E_2, are related by the expression $\varepsilon_1 E_1 = \varepsilon_2 E_2$. It can be seen from the dispersion relation that, if the electric field is strong enough,

CP849, *Zababakhin Scientific Talks - 2005*,
edited by E. N. Avrorin and V. A. Simonenko
© 2006 American Institute of Physics 0-7354-0345-7/06/$23.00

$$E_1 E_2 \gg \frac{\varepsilon_1 + \varepsilon_2}{(\varepsilon_1 - \varepsilon_2)^2} \sqrt{g\alpha(\rho_1 - \rho_2)},$$

the second term on the right-hand side of the dispersion relation dominates for the wave numbers in the range

$$\frac{g(\varepsilon_1 + \varepsilon_2)(\rho_1 - \rho_2)}{E_1 E_2 (\varepsilon_1 - \varepsilon_2)^2} \ll k \ll \frac{E_1 E_2 (\varepsilon_1 - \varepsilon_2)^2}{\alpha(\varepsilon_1 + \varepsilon_2)}.$$

Then $\omega^2 \sim k^2$ and, hence, we can separate the dispersion relation into two branches

$$\omega^{(\pm)} = \pm ick, \qquad c^2 = \frac{E_1 E_2 (\varepsilon_1 - \varepsilon_2)^2}{4\pi(\varepsilon_1 + \varepsilon_2)(\rho_1 - \rho_2)}. \tag{1}$$

For one branch, small periodic perturbations of the surface increase exponentially with characteristic times $(ck)^{-1}$, while for the other branch, these perturbations attenuate. In this situation, we can restrict our consideration to the increasing branch $\omega^{(+)} = +ick$, and thus substantially simplify the problem of describing the evolution of the interface at the linear stage of the development of instability. The buildup of perturbations of the surface inevitably transforms the system to a state in which its evolution is determined by nonlinear processes. Then, in the general case, splitting into branches becomes impossible. We will show below that, for the particular case $\varepsilon_1 \rho_1 = \varepsilon_2 \rho_2$, we can extract the separate branches from the equations of motion. This makes it possible to halve the number of equations required for describing the evolution of the boundary.

Consider the evolution of the interface between two ideal liquids of infinite depth in an external vertical electric field. In the unperturbed state, the boundary of a liquid is a flat horizontal surface. Let the z axis of the Cartesian coordinate system be normal to the unperturbed interface. The function $\eta(x, y, t)$ specifies the shape of the deformed boundary, i.e., the liquids occupy the regions $z < \eta(x, y, t)$ and $z > \eta(x, y, t)$, respectively. It is convenient for the subsequent analysis to choose the origin of coordinates so that the level of the liquids be determined by the expression $z = -vt$. In other words, the origin will move with respect to the interface at a certain constant velocity v.

Let us assume that the motion of both liquids is potential. The velocity potentials for incompressible liquids, Φ_1 and Φ_2, satisfy the Laplace equations,

$$\nabla^2 \Phi_1 = 0, \qquad \nabla^2 \Phi_2 = 0, \tag{2}$$

with the following conditions at the boundary and at infinity:

$$\rho_1 \left[\frac{\partial \Phi_1}{\partial t} + \frac{(\nabla^2 \Phi_1)}{2} \right] - \rho_2 \left[\frac{\partial \Phi_2}{\partial t} + \frac{(\nabla^2 \Phi_2)}{2} \right] = \frac{\varepsilon_1 - \varepsilon_2}{8\pi} (\nabla \varphi_1 \cdot \nabla \varphi_2), \qquad z = \eta(x, y, t),$$

$$\tag{3}$$

$$\frac{\partial \Phi_1}{\partial n} = \frac{\partial \Phi_2}{\partial n}, \qquad z = \eta(x, y, t), \tag{4}$$

$$\Phi_1 \to -vz, \qquad z \to -\infty, \tag{5}$$

$$\Phi_2 \to -vz, \qquad z \to +\infty, \tag{6}$$

100

where φ_1 and φ_2 are the respective electric field potentials in and above the liquid, and $\partial / \partial n$ denotes the derivative along the normal to the interface. The expression on the right-hand side of the dynamic boundary condition (nonstationary Bernoulli equation) is responsible for the electrostatic pressure at the interface between two ideal dielectric liquids in the absence of free electric charges [7]. The evolution of the interface is determined by the kinematic relation

$$\frac{\partial \eta}{\partial t} = \frac{\partial \Phi_1}{\partial z} - \left(\nabla_\perp \eta \cdot \nabla_\perp \Phi_1 \right), \qquad z = \eta(x, y, t), \tag{7}$$

where $\nabla_\perp$ is the 2-D gradient in the $\{x, y\}$-plane.

The electric potentials φ_1 and φ_2 satisfy the Laplace equations

$$\nabla^2 \varphi_1 = 0, \qquad \nabla^2 \varphi_2 = 0. \tag{8}$$

Since the electric field potential and the normal component of the displacement vector should be continuous at the interface, we must add the following conditions at the boundary:

$$\varphi_1 = \varphi_2, \qquad z = \eta(x, y, t), \tag{9}$$

$$\varepsilon_1 \frac{\partial \varphi_1}{\partial n} = \varepsilon_2 \frac{\partial \varphi_2}{\partial n}, \qquad z = \eta(x, y, t). \tag{10}$$

The system of equations is closed by the conditions of the electric field uniformity at an infinite distance from the surface:

$$\varphi_1 \to -E_1 z, \qquad z \to -\infty, \tag{11}$$

$$\varphi_2 \to -E_2 z, \qquad z \to +\infty. \tag{12}$$

Let us show that for certain relations between the problem parameters a flow regime is possible where the velocity and electric field harmonic potentials are linearly dependent functions. Suppose that

$$\varphi_1 = a \Phi_1 \left(4\pi \rho_1 / \varepsilon_1 \right)^{1/2}, \qquad \varphi_2 = b \Phi_2 \left(4\pi \rho_2 / \varepsilon_2 \right)^{1/2}, \tag{13}$$

where a and b are unknown constants. It is necessary to verify that the initial equations of motion (2)-(12) can be compatible with these relations.

It is apparent that the Laplace equations (2) and (8) are always compatible with condition (13). Moreover, it can readily be seen that Eqs. (4)-(6) and (10)-(12) are compatible with (13) if

$$a \left(\rho_1 \varepsilon_1 \right)^{1/2} = b \left(\rho_2 \varepsilon_2 \right)^{1/2}, \tag{14}$$

and the auxiliary parameter v takes the value

$$v = a^{-1} v_0, \qquad v_0 = E_1 \left(4\pi \rho_1 / \varepsilon_1 \right)^{-1/2} > 0.$$

The only nontrivial problem concerns the compatibility of Eq. (13) with the nonlinear dynamic (3) and kinematic (7) relations. Let us rewrite Eq. (7) so that it would not contain the function η explicitly. With the help of formulas (13) and (14), the boundary condition (9) can be rewritten as follows:

$$\varepsilon_1^{-1} \Phi_1 = \varepsilon_2^{-1} \Phi_2, \qquad z = \eta(x, y, t). \tag{15}$$

Differentiating this expression with respect to time or spatial variables, we obtain

$$\frac{\partial \eta}{\partial t} \cdot \left[\varepsilon \frac{\partial \Phi_1}{\partial z} - \frac{\partial \Phi_2}{\partial z} \right]_{z=\eta} = - \left[\varepsilon \frac{\partial \Phi_1}{\partial t} - \frac{\partial \Phi_2}{\partial t} \right]_{z=\eta},$$

$$\nabla_\perp \eta \cdot \left[\varepsilon \frac{\partial \Phi_1}{\partial z} - \frac{\partial \Phi_2}{\partial z} \right]_{z=\eta} = -\left[\varepsilon \nabla_\perp \Phi_1 - \nabla_\perp \Phi_2 \right]_{z=\eta},$$

where $\varepsilon = \varepsilon_2 / \varepsilon_1$ is the ratio of the permittivities. These relations allow us to eliminate η from (7), which, after simple rearrangement, takes the form

$$\frac{\partial \Phi_1}{\partial t} - \varepsilon^{-1} \frac{\partial \Phi_2}{\partial t} + \left(1 - \varepsilon^{-1}\right) \left(\frac{\partial \Phi_1}{\partial n}\right)^2 = 0, \qquad z = \eta(x,y,t). \tag{16}$$

Let us now transform the dynamic boundary condition (3) so that it be analogous to (16). Eliminating the electric field potentials from Eq. (3) with the help of relations (13) and decomposing the velocities of liquids into the normal $(\partial \Phi / \partial n)$ and tangential $(\partial \Phi / \partial \tau)$ components, we get

$$\frac{\partial \Phi_1}{\partial t} - \rho \frac{\partial \Phi_2}{\partial t} + \frac{\left(1 - \rho - a^2 \varepsilon^{-1} + a^2\right)}{2} \left(\frac{\partial \Phi_1}{\partial n}\right)^2 + \frac{\left(1 - \rho \varepsilon^2 - a^2 + a^2 \varepsilon\right)}{2} \left(\frac{\partial \Phi_1}{\partial \tau}\right)^2 = 0,$$

$$z = \eta(x,y,t),$$

where $\rho = \rho_2 / \rho_1$. This expression coincides with (16) and, therefore, they are compatible if

$$\rho = \varepsilon^{-1},$$
$$1 - \rho \varepsilon^2 - a^2 + a^2 \varepsilon = 0,$$
$$1 - \rho - a^2 \varepsilon^{-1} + a^2 = 2 - 2\varepsilon^{-1},$$

whence $\varepsilon \rho = 1$, i.e.,

$$\varepsilon_1 \rho_1 = \varepsilon_2 \rho_2, \tag{17}$$

and also $a^2 = 1$. The equation for the parameter a has two roots, $a^{(\pm)} = \pm 1$, which correspond to different branches of the solutions.

Thus, we have proved that the functional relation (13) can be compatible with the equations of motion if condition (17) is valid. The corresponding flow regime is described by the following equations:

$$\nabla^2 \Phi_1 = 0, \qquad \nabla^2 \Phi_2 = 0, \tag{18}$$

$$\frac{\partial n}{\partial t} = \frac{\partial \Phi_1}{\partial n} \sqrt{1 + (\nabla_\perp \eta)^2}, \qquad z = \eta(x,y,t), \tag{19}$$

$$\frac{\partial \Phi_1}{\partial n} = \frac{\partial \Phi_2}{\partial n}, \qquad z = \eta(x,y,t), \tag{20}$$

$$\rho_1 \Phi_1 = \rho_2 \Phi_2, \qquad z = \eta(x,y,t), \tag{21}$$

$$\Phi_1 \to -a^{(\pm)} v_0 z, \qquad z \to -\infty, \tag{22}$$

$$\Phi_2 \to -a^{(\pm)} v_0 z, \qquad z \to +\infty. \tag{23}$$

The reduction of the initial equations (2)-(12) to Eqs. (18)-(23) significantly simplifies the analysis of a moving interface between two liquids.

The dispersion relation for Eqs. (18)-(23) has the form

$$\omega = i a^{(\pm)} \frac{\rho_1 - \rho_2}{\rho_1 + \rho_2} v_0 k = a^{(\pm)} \frac{i \sqrt{\varepsilon_1} \left(\rho_1 - \rho_2\right)}{\sqrt{4\pi \rho_1 \left(\rho_1 + \rho_2\right)}} E_1 k.$$

It can be seen that, for the branch $a = a^{(+)} = +1$, an initial sinusoidal perturbation will increase and, for $a = a^{(-)} = -1$, it will attenuate. With regard to Eq. (17), this expression coincides with expression (1), that specifies different branches of the dispersion relation for the unreduced equations of motion.

Thus, if condition (17) is satisfied, the separation of solutions into two branches, one increasing and the other decreasing with time, is possible not only for the linearized equations, but also for the initial nonlinear equations (2)-(12).

In the formal limit $\rho_1 \gg \rho_2$, the evolution of the interface will be governed by the influence of the lower liquid. Then the set of equations (18)-(23) reduces to (compare with Refs. [8,9])

$$\nabla^2 \Phi_1 = 0,$$

$$\frac{\partial \eta}{\partial t} = \frac{\partial \Phi_1}{\partial n} \sqrt{1 + (\nabla_\perp \eta)^2}, \qquad z = \eta(x, y, t),$$

$$\Phi_1 = 0, \qquad z = \eta(x, y, t),$$

$$\Phi_1 \to -v_0 z, \qquad z \to -\infty.$$

Here, we set $a = +1$ and, as a consequence, $v = v_0$. They coincide with the equations that describe so-called Laplacian growth, viz., the motion of a phase boundary with a velocity directly proportional to the normal derivative of a certain harmonic scalar field (Φ_1 in our case). Depending on the frame of reference, this field may have the meaning of temperature (Stefan's problem in the quasi-stationary limit [10]), electrostatic potential (electrolytic deposition [11]), pressure (viscous fingering in Hele-Shaw cells, or flow through a porous medium [12,13]), etc.

Thus, we have shown that it is possible to halve the number of equations required for describing the motion of the interface of two dielectric liquids in an applied electric field. In the limit of a small liquid density ratio, the reduced equations coincide with the well-known equations that describe Laplacian growth. It is of importance to us that many exact solutions are known for these equations. They describe the evolution of the interface up to the formation of ``fingers", cuspidal dimples, and so on (see, e.g., [13-15]). In particular, these solutions can be used for testing numerical algorithms which are used to simulate the motion of a liquid-liquid interface.

ACKNOWLEDGMENTS

This study was performed within the framework of the program "Mathematical Methods in Nonlinear Dynamics" of the Presidium of the Russian Academy of Sciences. It was supported in part by the "Dynasty" Foundation and the International Center for Fundamental Physics in Moscow.

REFERENCES

1. J.R. Melcher, *Field-Coupled Surface Waves*, MIT Press, Cambridge, MA, 1963.
2. J.R. Melcher, Phys. Fluids **4**, 1348 (1961).

3. A.A. Mohamed, E. F. Elshehawey, J. Fluid Mech. **129**, 473 (1983).
4. R.K. Singla, R. K. Chhabra, S.K. Trehan, Int. J. Eng. Sci. **35**, 585 (1997).
5. M.F. El-Sayed, D. K. Callebaut, J. Colloid Interf. Sci. **200**, 203 (1998).
A. R.F. Elhefnawy, Int. J. Eng. Sci. **40**, 319 (2002).
6. L.D. Landau and E. M. Lifshitz, *Course of Theoretical Physics, Vol. 8: Electrodynamics of Continuous Media*, Nauka, Moscow, 1982; Pergamon, New York, 1984.
7. N. M. Zubarev, JETP Lett. **71**, 367 (2000).
8. N. M. Zubarev, JETP **94**, 534 (2002)].
9. J. S. Langer, Rev. Mod. Phys. **52**, 1 (1980).
10. M. Matsushita, M. Sano, Y. Hayakawa, *et al.*, Phys. Rev. Lett. **53**, 286 (1984).
11. P.G. Saffman and G. I. Taylor, Proc. R. Soc. London, Ser. A **245**, 312 (1958).
12. D. Bensimon, L. P. Kadanoff, Sh. Liang, *et al.,* Rev. Mod. Phys. **58**, 977 (1986).
13. S.D. Howison, SIAM J. Appl. Math. **46**, 20 (1986).
14. M.B. Mineev-Weinstein and S. P. Dawson, Phys. Rev. E **50**, R24 (1994).

Dynamic of a Bubble in the Field of a Short Bipolar Acoustic Pulse

D.V. Reshetnyak, P.I. Golubnichiy, Y.M. Krutov

Department of Physics, East-Ukrainian national university by Vladimir Dahl,
bl. Molodezhny 20a, 91034 Lugansk, Ukraine

Abstract. In present paper the dynamics of gas-vapor bubble, pulsing in a field of a short bipolar acoustic pulse of different polarity is analyzed. The thermodynamic parameters of the bubble interior are simulated based on a model including heat and mass exchange between bubble and ambient liquid.

Keywords: acoustic pulse, sonoluminescence, boundary layer, gas-vapor mixture.
PACS: 43.25Yv, 78.60.Mq, 43.25.+ y.

INTRODUCTION

The discovery of single-bubble sonoluminescence (SBSL) [1] stimulated a great number of papers devoted to phenomena of high energy concentration in cavitation liquid under the action of a sound field. A great number of models of the processes has appeared (see the review [2]). SBSL features greatly differ from multibubble luminescence. The actual problem is determination of conditions, under which characteristics of luminescence peculiar to SBSL in multibubble system are realized. We assume that SL investigations in pulsed [3] and pulsed-modulated [4] acoustic fields are able to make clear this problem because in this way the "clarification" of the cavitation zone effect [4] is appeared. This brings about weakening of the inter-bubble interaction keeping the stability of spherical shape. As a result, the energy concentration during the collapse increases. So, the collapse of separate bubbles must be similar to SBSL condition. Such approach also can be perspective for initialization of thermonuclear fusion reactions in collapse bubbles [5] under the specific conditions of cavitation creation and in sonochemical technology.

In connection with the aforesaid we analyze the dynamics of single spherical bubble, pulsing in the field of a short bipolar acoustic pulse (AP) of different polarity in water. We examine the dynamics of water vapor contents. Based on the model, including heat and mass exchange between bubble and liquid, the thermodynamic parameters of gas-vapor mixture (GVM) in the bubble for different gases are simulated.

CP849, *Zababakhin Scientific Talks - 2005,*
edited by E. N. Avrorin and V. A. Simonenko

GAS-VAPOR BUBBLE MODEL

The bubble radius is defined by a generalized Rayleigt-Plesset equation, which takes into account a compressed and viscous liquid in the form [6]:

$$\left(1 - \frac{U}{c}\right) R \frac{dU}{dt} + \frac{3}{2}\left(1 - \frac{U}{3c}\right) U^2 = \left(1 + \frac{U}{c}\right) \frac{P_R - P_0 - P_a}{c} +$$
$$+ \frac{R}{cc} \frac{d}{dt}(P_R - P_0 - P_a) \tag{1}$$

where U, R are the velocities of the wall and the variable radius of the bubble, c is the velocity of sound in liquid, ρ is density of liquid, $P_0 = 1$ bar is the hydrostatic pressure, P_a is the external variable acoustic pressure, $P_R = P_{gv} - 2y/R - (4\mu U/R)$ is the pressure on the bubble wall. Here σ, μ are correspondingly coefficient of surface tension and dynamic viscosity of liquid, P_{gv} is the GVM pressure in the bubble. It is defined by equation of state:

$$P_{gv} = \frac{(\rho_g R_g + \rho_v R_v)T}{1 - (b_g \rho_g + b_v \rho_v)}, \tag{2}$$

where ρ_g, ρ_v, R_g, R_v, b_g, b_v are the density, the gas constants and van der Vaals constants of gas and vapor accordingly, T is the GVM temperature.

The variable external pressure was set as plane sinusoidal AP of the different polarity:

$$P_a = \mp P_m \sin\left(\frac{2p}{\phi}t\right) F(t), \tag{3}$$

where P_m, τ are correspondingly amplitude and duration of the pulse, $F(t) = \{1 + exp[(t - \phi)/(0.01\phi)]\}^{-1}$ is Fermi function; the sign (-) corresponds to the pulse of the expansion-compression (E-C), sign (+) to the compression-expansion pulse (C-E).

The processes of evaporation, condensation and heat transfer were taken into account in the present paper by the boundary diffusion layer which is in thermal equilibrium with liquid [7]. The equations of molecules vapor flow and thermal conductivity through the bubble surface in used theoretical model are:

$$\frac{dN_v}{dt} = 4\pi R^2 D_{gv} \frac{n_{v0} - n_v}{l_{diff}}, \quad l_{diff} = min\left(\sqrt{\frac{RD_{gv}}{\left|\frac{dR}{dt}\right|}}, \frac{R}{p}\right), \tag{4}$$

$$\frac{dT}{dt} = \frac{1}{C_v}\frac{dQ}{dt} - \frac{P_{gv}}{C_v}\frac{dV}{dt} + \left[4T_0 - 3T - T\sum_j \left(\frac{u_{vj}/T}{\exp(\theta_{vj}/T)} \right) \right] \frac{k}{C_v}\frac{dN_v}{dt}, \qquad (5)$$

where D_{gv} is the binary diffusion coefficient, n_{v0}, n_v are the equilibrium and current concentration of the vapor in the bubble, l_{diff} is the thickness of diffusive boundary layer.

In the equation (3) $D_{gv} = D^0_{gv}(n_0/n_{gv})$ [8], where D^0_{gv}, n_0 are the binary diffusion coefficient and concentration of gas under normal pressure, n_{gv} is the current concentration of the mixture. In the equation (4): C_v is the mixture heat capacity, T_0 is the ambient liquid temperature, θ_{vj} is the characteristic oscillation temperatures of the water molecules and k is the Boltsmann constant. The heat flow through the surface is described by the equation:

$$\frac{dQ}{dt} = 4\pi R^2 \kappa_{gv} \frac{T_0 - T}{l_{th}}, \quad l_{th} = min \left(\sqrt{\frac{R\chi_{gv}}{\left|\frac{dR}{dt}\right|}}, \frac{R}{p} \right), \qquad (6)$$

where κ_{gv}, χ_{gv} are the thermal conductivity and thermal diffusivity coefficients of the mixture in the layer, l_{th} is the thickness of thermal boundary layer.

Depence of mixture heat conductivity on the pressure (density) in present paper is defined by the expression:

$$\kappa_{gv} = \kappa^0_{gv} + \Delta\kappa_g + \Delta\kappa_v, \qquad (7)$$

where κ^0_{gv} is the mixture thermal conductivity at low pressure, calculated by the Lindsey-Bromli method [9], $\Delta\kappa_g$, $\Delta\kappa_v$ is the additional thermal conductivity components as function of their density; they were set as analytical functions, obtained by the tabular data approximation [10].

RESULTS AND DISCUSSION

The numerical integration of the ODE system (1), (4), (5) was maid by Runge-Kutt 5 order method. The values of AP calculation parameters were defined by (2) from the paper [3]: $P_m = 6$ bar, $\tau = 7$ μs. We assume the values of temperature is $T_0 = 293$ K, initial radius of bubble $R_0 = 1.5$ μm. The quantity of such gas micro-bubbles in water is $10^3 - 10^4$ cm^{-3} [11].

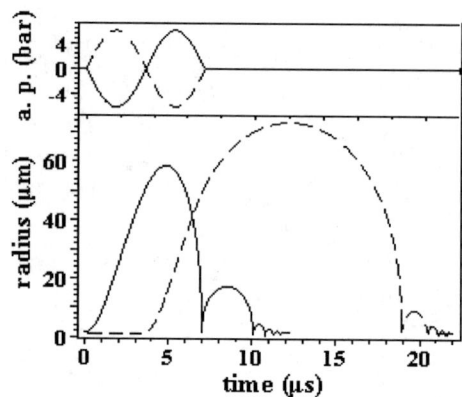

FIGURE 1. Dynamics of bubble. a. p.– acoustic pressure.

The results of gas-vapor bubble (gas – argon) dynamics calculation are represented in Figure 1, in the field of the E-C pulse (the solid line) and C-E (the dashed line). In the first case the bubble expands to a bit smaller size, for it undergoes the action of compression wave at a moment $\tau/2$, preventing its further rise. Calculated values of the minimum radius, the percent of vapor content, peak temperature and pressure of the mixture at a moment of the first collapse of bubble, containing different noble gases or air are given in Table 1. In the last column we present coefficients $D^0{}_{gv}$ for gas-vapor mixtures.

The more high temperature in the collapsing bubble at C-E pulse in comparison with such bubble in the case of E-C pulse is explained by the different quantitative GVM composition in these cases. The bubble collapses only in the field of the hydrostatic pressure at C-E pulse therefore average velocity near the border of the wall will be smaller, than at E-C pulse. Consequently there will be less molecules of vapor in C-E pulse at the collapse moment, as it has a time to condense. Simulations show that with of the decreasing molecular mass noncondensable gas decreases the part of vapor, and increases the peak temperature and pressure. The possible explanation of the processes is the coefficient of the binary diffusion increasing with the molecular mass of gas decreasing (see Table 1) and accordingly vapor is out of the bubble faster. Therefore for the bubbles containing light noncondesable gases the collapse will be deeper. GVM in them will has smaller heat capacity that provides more extreme conditions at the collapse moment. The same is for C-E pulse.

TABLE 1. Parameters of GVM in collapse.

Gas	R_{min}, μm		n_v/n_{gv}, %		T_{max}, K		P_{max}, 10^5 bar		
	E-C	C-E	E-C	C-E	E-C	C-E	E-C	C-E	
Kr	0.96	0.63	98.9	95.8	4900	5100	0.77	0.95	20.21
Ar	0.80	0.44	98.0	88.7	5400	6100	1.01	1.49	24.99
Ne	0.46	0.27	90.2	56.0	7400	10500	2.13	2.80	37.53
He	0.24	0.22	31.0	19.0	19000	21700	4.73	3.48	81.41
air	0.82	0.47	98.2	90.0	5280	5540	0.98	1.41	24.44

We can conclude that for increasing the energy concentration for investigations of cavitation phenomena in pulsed fields of the moderate amplitude it is reasonable to saturate water with helium and use AP C-E polarities.

REFERENCES

1. D.F. Gaitan, L.A. Crum, C.C. Church and R.A. Roy, *J. Acoust. Soc. Am.* 91, 3166-3183 (1992).
2. M.P. Brenner, S. Hilgenfeldt and D. Lohse, *Rev. Mod. Phys.* 74, 425-484 (2002).
3. I.L. Veremeenko, P.I. Golubnichiy, Yu.M. Krutov and D.V. Reshetnyak, *Proc. of Scientifically-practical Conference "Perspective developments of the science and technology"*, Belgorod, 2004, pp. 35-38.
A. Francescutto, P. Ciuti, G. Iernetti, N.V. Dezhkunov, *Europhys. Lett.* 47(1), 49-55 (1999).
4. R.P. Taleyarkhan, J.S. Cho, C.D. West et al., *Phys. Rev. E.* 69, 036109 (2004).
A. Prosperetti and A. Lezzi, *J. Fluid. Mech.* 168, 457-478 (1986).
5. R. Toegel, B. Gompf, R. Pecha and D. Lohse, *Phys. Rev. Lett.* 85, 3165-3168 (2000).
6. X. Lu, A. Prosperetti, R. Toegel and D. Lohse, *Phys. Rev.* 67, 056310 (2003).
7. R.C. Reid and T.K. Sherwood, *The Properties of Gases and Liquids,* Khimiya, Leningrad, 1971, pp. 522-523.
8. N.V. Vargavtik, *Manual on Thermophysical Properties of Gases and Liquids,* Nauka, Moscow, 1972.
9. A.S. Besov, V.K. Kedrinskii and E.I. Pal'chikov, *Pis'ma v JTF* 10, 240-244 (1984).

Experimental Researches of High–Speed Hydrodynamic Processes

N.V. Velichutin, A.L. Kartashev, Y.M. Kovalev, V.G. Lupanov

The Chelyabinsk State University, Kashirinyh Brothers street, 129, Chelyabinsk, Russia

Abstract. One of the major problems hydro and gas dynamic is research of high–speed processes in fluids and gases. During motion of bodies in fluids appears a different physical effect, such as, a wave motion of fluids, cavitations phenomena, generation of flaws of different forms around of streamlined body, etc. In gases particular interest presents research of shock - wave structures, and also jet flows. Feature of considered processes is their rapidity that produces special requirements to methods of recording used during researches and data-acquisition equipment.

For motion study of bodies in a fluids the experimental equipment (Figure 1) consisting of reservoir with a fluids, a set of bodies (Fig. 2) which motion is explored, and registration–measuring system will be used. Motion of bodies in a fluid is implemented by their free fall by gravity on a surface of a fluid.

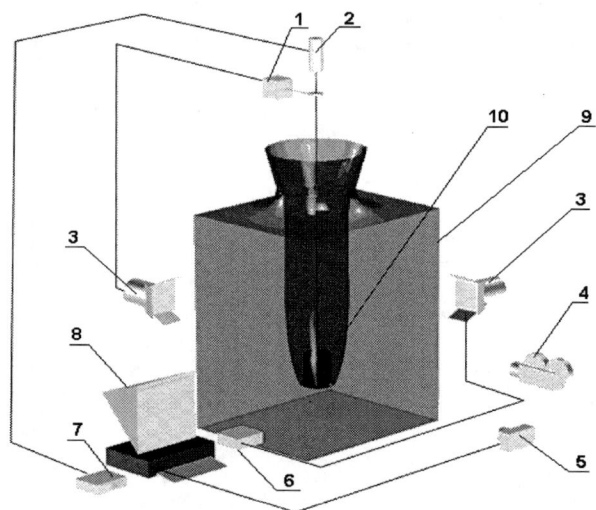

FIGURE 1. Experimental equipment: 1 – synchronizer; 2 – sensor of pressure; 3 – source of light; 4 – fast video camera; 5 – electron-optical gated digital camera; 6 – unit of synchronization; 7 – oscilloscope THS 730A; 8 – computer; 9 – reservoir with fluid; 10 – body.

CP849, *Zababakhin Scientific Talks - 2005*,
edited by E. N. Avrorin and V. A. Simonenko
© 2006 American Institute of Physics 0-7354-0345-7/06/$23.00

FIGURE 2. Set of bodies.

The recording system of kinematics and power characteristics of explored bodies is founded on obtaining of the measuring information with the help of oscillograph THS730A and the personal computer. Registration of dynamic characteristics is made with usage of tens meter, piezoelectric, capacitive and induction sensors. Visualization of proceeding processes is provided with shooting with nanosecond exposure time with the help of electron–optical gated digital camera NANOGATE GC–1, a video shooting and photographing. Processing of the visual information is carried out with usage of the personal computer.

Researches of drag to a linear motion of bodies of the different form in a fluid with the help of complex registration of kinematics and power characteristics are carried out. The drag force from the direction of a fluid, acting on a body in a direction an inverse to a driving direction, is registered by a measuring system. As fluids the mixture of water with glycerin in different ratio will be used.

Photographs and video films of process of motion of bodies in a fluid with registration of wave processes and process of generation of a flaw in a fluid around of a driving body are received. Oscillograms of change of a drag force of a fluid to a linear motion of bodies of the different form are received. Results of experimental investigations are shown on Fig. 3, 4, 5, and 6.

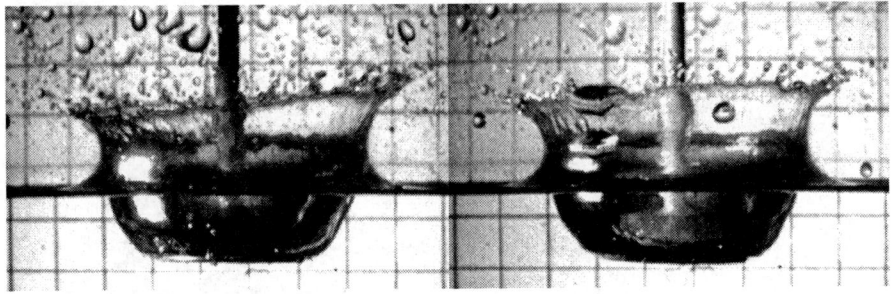

FIGURE 3. Penetration of a round plate in fluid.

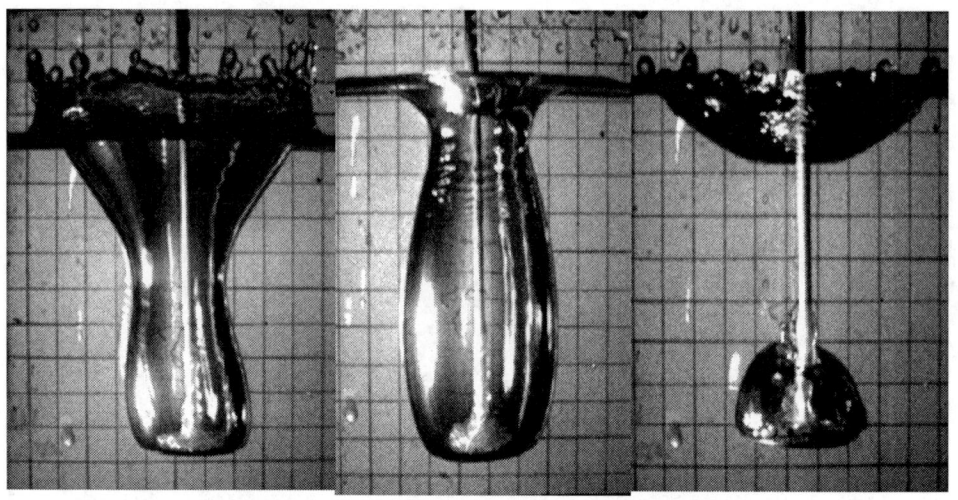

FIGURE 4. Penetration of a round plate in fluid under different initial velocity at one time.

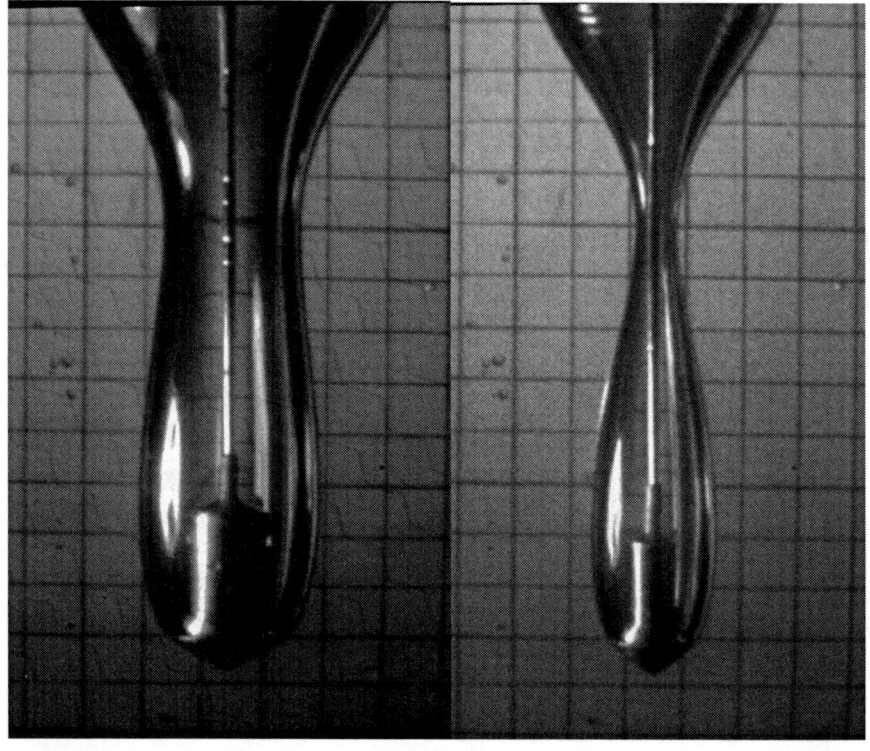

FIGURE 5. Penetration of a body with conical nose in fluid under different initial velocity at one time.

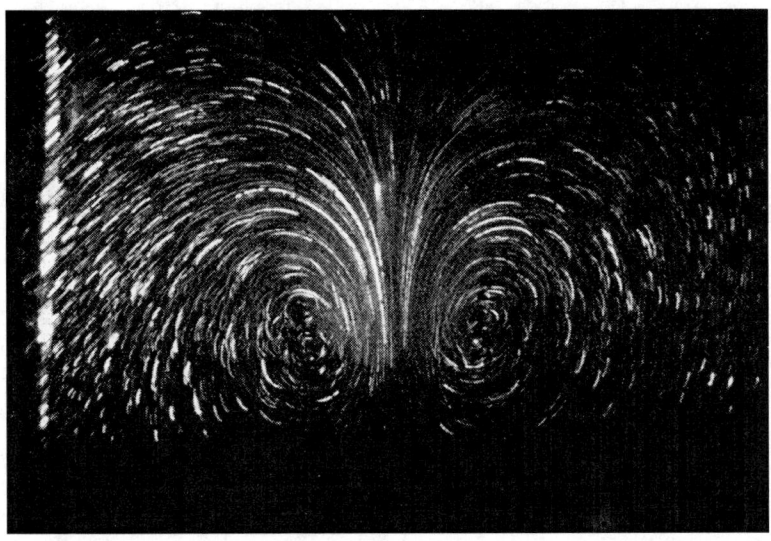

FIGURE 6. Trajectory of a trace particle under motion body in fluid with generation of cavity.

Researches of origination and motion of shock waves in gases are carried out. With the help of electron-optical gated digital camera NANOGATE GC–1 photographs of the shock wave structures arising in air jet, effusing in the flooding area are received (Fig. 7).

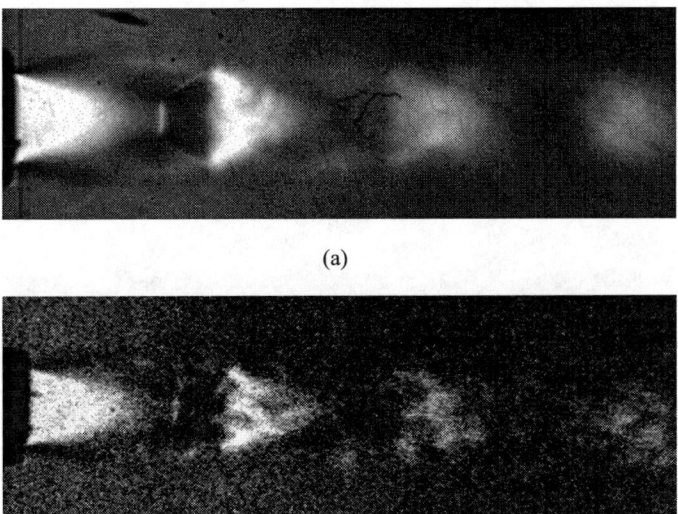

(a)

(b)

FIGURE 7. Shadow photographs of air jet; time of exposition: 100 microseconds (a), 0.5 microseconds (b).

The approved technique of registration of kinematics and power characteristics of motion bodies in a fluid and shock wave structures in gases can be used at research hydro and gas dynamic processes in different technical devices.

The work is executed at support of International Science and Technology Center, project # 0904–2003.

The Synthesis of Materials
in a Cumulative Jet

O.L. Pervukhina, L.B. Pervukhin

ISMAN, Chernogolovka, Russia

Abstract. In the present work processes taking place in welding gap ahead of a point of contact are used for synthesis of new materials in extreme conditions of interaction of a cumulative jet and compressed gas. Carried out researches allow to assume an opportunity of realization of purposeful synthesis of materials with their simultaneous drawing on a surface of a target.

Keywords: Explosion welding, cumulative jet, titanium, oxides, synthesis of materials
PACS: 02.60.Cb, 02.70.Bf, 47.11.-j.

INTRODUCTION

It is known that connection received by explosion welding can have the wavy form. Inclusions are formed on tips of waves and have intermediate structure between welded materials. Intermetallides, oxides, carbides etc. are formed on tips of waves also. Inclusions are named "a white phase", "vortex-type zones", "melting". Structure and properties of inclusions differ from structure and properties of initial materials. Sizes of inclusions depend on a mode of welding and usually do not exceed 0.1 mm^2. It is shown at [1], that inclusions are formed ahead of a point of contact in a cumulative jet. Quantity of inclusions grows during process of titanium - steel explosion welding with moving from starting point. On a distance of 1 m they cover all surfaces. It complicates explosion welding of large-sized sheets of titanium and steel.

In the present work processes taking place in welding gap ahead of a point of contact are used for synthesis of new materials. To prove an opportunity of synthesis we shall consider development of process ahead of a point of impact contact of titanium plates in explosion welding mode.

The area of compressed gas and cumulative jet, which consists of superficial layers of welded materials, is formed as a result of collision of plates ahead of a point of contact in a welding gap. The sizes of area increase with increase in distance from the starting point of collision. The cumulative jet cooperating with compressed gas forms disperse particles cloud.

Thus, compressed gas with saturated dispersed particles of a cumulative jet moves ahead of a point of contact. Brightness temperatures of gas behind front of the shock wave between steel plates at their collision measured at [2] reveal its significant excess in comparison with calculated shock adiabatic curve. The shock wave at point of contact velocity $V_K = 3500$ km/s compresses and heats up gas up to temperatures

CP849, *Zababakhin Scientific Talks - 2005*,
edited by E. N. Avrorin and V. A. Simonenko

of the order 5000°C. Particles are considered as the additional piston and lead to increase in temperature of gas and velocity of a shock wave. Influence of this stream leads to heating up of metal and fusion of its surface of impact. On big enough distance from a place of initiation the energy that has acted from gas in metal becomes comparable with the energy allocated in a zone of a seam as a result of collision. The condition of gas ahead of a point of contact can be estimated in one-dimensional approach by the equations of the air shock wave raised in a pipe by the moving piston. The front of such shock wave will have speed in 1.3-1.5 times higher than the velocity of a point of contact, and gas behind front of a shock wave can heats up to temperature of the order 3000-6000 °C. Metal with increase in distance from the initial point of process of impact stood longer time under influence of a stream of energy from compressed gas. Calculation of heating up of metal by this energy source shows, that a layer there were thickness 10 microns on distance of 1 m from the beginning of process of welding at speed of a point of contact of 2500 km/s can get heat up to temperature of the order 500 °C, Al and Cu - up to 300 °C, and the Ti - up to 900 °C [3].

Let's consider balance of heat ahead of a point of contact:

$$Q = Q_1 + Q_2 + Q_3 \qquad (1)$$

Here Q_1 is amount of heat generated during gas compression, Q_2 is amount of heat generated in the process of commutation and air-dynamical braking of particles in compressed gas, Q_3 is amount of heat generated by chemical reaction.

Calculation of a thermal emission has been made under conditions considered at [3]. Comparison of kinetic energy of thrown plates and kinetic energy of ejected particles and internal energy of ejected particles shows that Q_1 allocated at shock compression of gas makes 17.3 % kinetic energy of impact of thrown plates.

Total quantity of heat brought to a surface of a body mass m during aerodynamic braking defined [4] under the formula:

$$Q_2 = \frac{mV_0^2}{2} \left(\frac{1}{2} \frac{c_\tau}{c_\tau + c_x} \right) \qquad (2)$$

Here V_0 is initial velocity of a body, c_τ is resistance of viscous friction of gas to a surface of a body and c_x is resistance of forces of normal pressure.

Calculations show, that Q_2 is insignificant and makes 0.22 % from kinetic energy of thrown plates.

It is necessary to note, that Q_1 and Q_2 for all materials practically equally. Calculation of thermal emission Q_3 due to chemical reaction between oxygen and nitrogen shows, that in welding gap it is not enough air for full combustion of particles of the titanium. Welding gap depending on saturation of particles has share of the reacted titanium 9-25 %. Calculation of heat of combustion of particles and the allocated energy of other metals leads to a similar results. Fact tells that for the majority of metals at explosion welding of greater areas problems do not arise. The situation sharply varies if we consider unique ability of the titanium to absorb oxygen and nitrogen. If in such metals as Fe, Cu and Zr, the maximal solubility of oxygen is $2 \cdot 10^{-3}$, $4 \cdot 10^{-2}$ and 8 at.% accordingly, in Al oxygen practically do not dissolve, the titanium is capable to comprise up to 30 at.% of oxygen and 20 at.% of nitrogen [5]. The sites of a surface located away from welded plates have an opportunity to absorb more gases, than the weldings located on an initial site. On some length L gasing a superficial layer reaches specified above limiting values. When the point of contact comes to this place saturated by oxygen and nitrogen of a particle start to take off in heat.

Burning of a surface of a plate of the titanium begins, as the temperature of particles is high enough. Such ignition is the most probable as the surface of particles in addition heats up at contact to air. Since the big thermal emission there is a sharp rise in temperature of a particle down to its fusion, and internal burning due to the oxygen dissolved in a particle and nitrogen begins. Calculation of thermal effect shows [3] that due to internal burning kinetic energy of impact allocation of heat is 52-150%.

The temperature of particles in welding gap increases due to heat exchange with hotter air surrounding them and due to their braking. Ignition of dispersed particles in welding gap leads to additional allocation of heat and growth of temperature that creates conditions of simultaneous avalanche reaction of burning in the certain zone. Hence, conditions for synthesis in extreme conditions of new materials are created ahead of a point of contact in welding gap. The composition and structure of the synthesized materials will be defined by processes cumulation and composition of an atmosphere in welding gap.

EXPERIMENTS

From practical experience it is known that the effect of fusion of a surface can arise at welding sheets longer, than 1 m. In this case the amount of energy of gas acted in metal comes nearer to amount of the energy allocated as a result of impact. The technique of carrying out experiments provided installation with a welding gap of plates from the titanium thickness of 6 mm and length 2 m with hermetic sealing of a backlash on the long parties (Fig. 1).

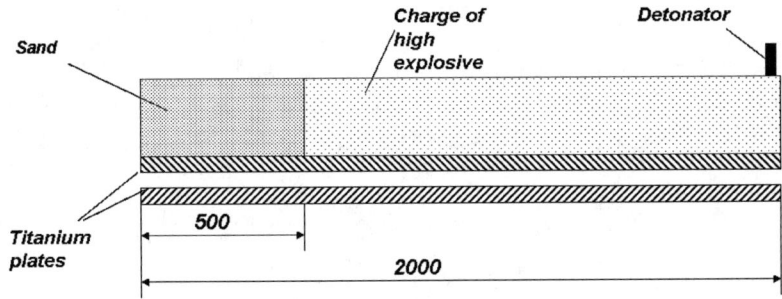

FIGURE 1. The scheme of experiment.

For fixation of products of synthesis at the last 500 mm of plates there is no charge. The plates were protected from influence of products of detonation by sand. After explosion surfaces of the ends of both plates were cut out to make samples for research of structure and a chemical compound of the synthesized materials.

RESULTS

On a surface of the bottom plate a continuous brilliant covering have revealed. Extent of a covering has made 250мм from a place of the termination of welding, and then the covering passed in separate jets. On the top plate extent of a covering has made 100 mm, which then also passed in separate jets. The covering is strongly connected to a surface of titanium plates and has not collapsed at machining.

Studying of structure of a covering (Fig. 2) and zones of connection has shown that thickness of a covering on the bottom plate has made 60-70 microns, on top - 20-40 microns. Microhardness of a covering has made 540 - 600 MPa (the titan -180 MPa).

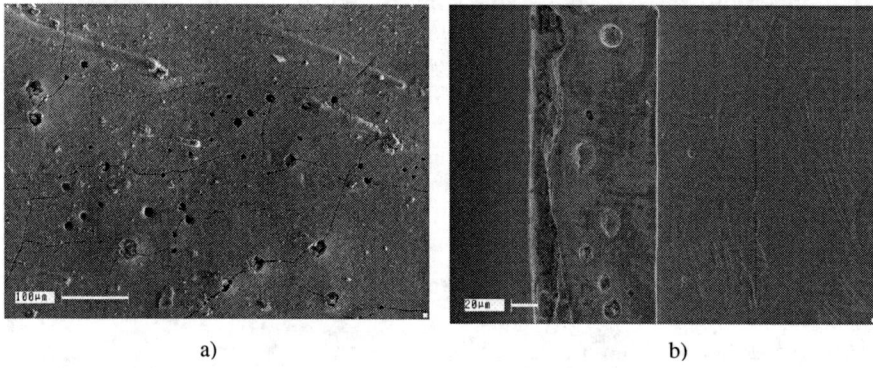

a)   b)

FIGURE 2. Surface of a covering from the synthesized material (a) and cross-section microstructure of sample that has been cut out from plates of the titanium (b).

On a surface of a covering have revealed presence of a grid of microcracks, the melted off round cavity. The covering represents monolithic structure in which grains are not looked through, the border has undressed between a covering and by the titanium is brightly expressed. In the titanium did not observe traces of significant plastic deformation.

The analysis of chemical and phase structure of a covering (Fig.3) has shown, that in it there is no pure titanium and its solid solutions. The covering consists of a mix of TiN, TiO, Ti_2O, Ti_3O. The X-ray of a covering has not been completely deciphered, as complex chemical compounds of the titanium with gases are formed, which in normal conditions are not fixed.

DISCUSSION

After studying structure, chemical and phase composition of the received materials the following mechanism of their synthesis is offered. In the beginning of process due to cumulation particles of metal are injected into area of compressed gas. Due to small extent of area of a particle have small temperature, which not sufficient for the beginning chemical reaction of oxidation. Burning of particles is not observed, that proves to be true structure of cast inclusions and their chemical compound. In process of development of process time of interaction of particles with gases increases, due to a heat transfer from compressed gas and braking of flying particles the temperature of particles raises. Mild burning begins them with formation oxides, nitrides and allocation of heat. Small time of process and low heat conductivity of the titanium interfere with absence of heat removal from a zone of reaction in an environment. It raises temperature in compressed gas promotes significant heating up of surfaces of titanium plates before a point of contact down to their fusion. Hence, the temperature of particles of the titanium formed as a result of cumulative process increases, and velocity of reaction and a thermal emission on a surface of particles increases. Conditions are created for course of avalanche reaction between the titanium, oxygen and nitrogen and the stream of energy on welded surfaces sharply increases. The increase in velocity of reaction is promoted by the fact that in the titanium of superficial layers of welded metal enters reaction. It represents a solid solution of gases of an atmosphere in the titanium, with the layer of molecules of oxygen adsorbed on it and nitrogen. High velocity of movement of the synthesized particles oxides and nitrides of the titanium at collision leads to their strong connection with the top and bottom surfaces of plates.

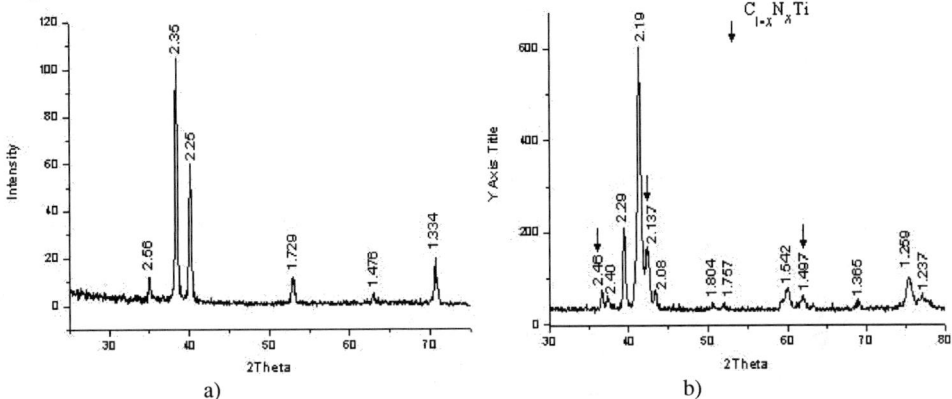

FIGURE 3. The X-ray of the titanium (a) and the synthesized material (b).

CONCLUSIONS

Conditions for synthesis of new materials are created in extreme conditions of interaction of a cumulative jet and compressed gas during impact of plates in a mode of explosion welding in welding gap ahead of a point of contact.

Carried out researches allow to assume an opportunity of realization of purposeful synthesis of similar materials with their simultaneous drawing on a surface of a target.

ACKNOWLEDGMENTS

The authors are gratitude A. A. Berdychenko and D.V. Oleinikov for their help in experiments.

REFERENCES

1. A.A. Deribas, Physics of hardening and welding by explosion, Science, Novosibirsk, 1980, 221 p.
2. S.N. Ishutkin, V.I. Kirko and V.A. Simonov, Combustion, Explosion, and Shock Wave, V. 16, № 6, (1980), 69-73 p.
3. A.A. Berdychenko, L.B. Pervukhin, A.A. Shterzer and B.S. Zlobin, Combustion, Explosion, and Shock Wave, V. 39, № 2, (2003), 128-136 p.
4. Yu.V. Polezhaev, F.B. Yurevich, Thermal protection, Energy, Moscow, 1976, 398 p.
5. Fromm E., Gerbhard E. Gaza and carbon in metals, Metallurgy, Moscow 1980, 711 p.

Calculation Technique for Kinematic Characteristics of Penetration of Combined Striker with Oblong Core Part Considering Possible Destruction of the Latter

E.V. Antsiferova, V.V. Bogdanov, E.V. Derebenko,
A.V. Lagutina, E.A. Khmelnikov

Nizhnetagilsky Technological Institute USTU – UPI, Nizhny Tagil, Russia

Abstract. The up-to-date development of the armored vehicles conditions complication of armor constructions and increased slope of shell armored plates. Combined strikers (C/S) can be used to destroy armored vehicles. We can increase total weight of the core part to increase the striker's power. However, the increase of core part diameter is limited by body dimensions. Thus, we can increase core part weight by increasing its length. Because of C/S interaction with the barriers at large deviation angles, C/S's mechanical trajectory sparks in the barrier. This results in bending stress which occurs in the core part. Because of large deviation angles, the impact of the side surface of oblong core part against the cavity edge occurs. This increases the probability of core part destruction.

The calculation technique for oblong core part penetration into different types of barriers is presented. The large number of factors can be calculated using this technique. It is assumed that the core part is destroyed when the tail part impacts against the cavity in the section where specific impact energy exceeds the critical value. Impact elasticity and destruction at bending stress were selected to be destruction criteria. The following core part destruction scenarios were investigated and calculated: (i) core head part is slightly destroyed but tail part of cylindrical shape penetrates deeper; (ii) core tail part is slightly destroyed but head part penetrates deeper, mass loss is taken into account; and (iii) after the impact, the core part is splitted up into two parts, then both of them penetrate into the barrier, one part is of ogival shape, the other is of cylindrical one.

This calculation technique was applied to computational program, then critical angles at which core part side surface is still in contact with cavity surface, and the angles at which core part destruction occurs were calculated. Depths of core part penetration for different destruction scenarios were calculated.

Different types of armor are used in armored equipment. Single-layer, multi-layer, separated and combined armors are widely used. Moreover, up-to-date armored equipment is characterized by increased slope of front armor plates that ensures angle deviation of shell axis to be about 75°. Thus, the striker's penetration trajectory is deviated that results in ricochet from the shell.

Combined strikers (C/S) can be used to destroy armored vehicles. We can increase total weight of the core part to increase the striker's power. However, the increase of core part diameter is limited by body dimensions. Thus, we can increase core part

CP849, *Zababakhin Scientific Talks - 2005*,
edited by E. N. Avrorin and V. A. Simonenko
© 2006 American Institute of Physics 0-7354-0345-7/06/$23.00

weight by increasing its length. This increases the probability of side surface impact against cavity edge, and thus, core part destruction. The calculation technique for kinematic characteristics of penetration of combined strikers into multi-layer barriers considering possible destruction of core part is presented.

Penetration of the striker into the barrier is assumed to be two-dimensional and symmetrical if compare with its flat movement. The following are the initial conditions: speed of impact of the striker against the barrier, angles of shell axis deviation to the barrier θ, geometrical properties of the barrier and striker, and physical and mechanical properties of the materials they are made of (Figure 1).

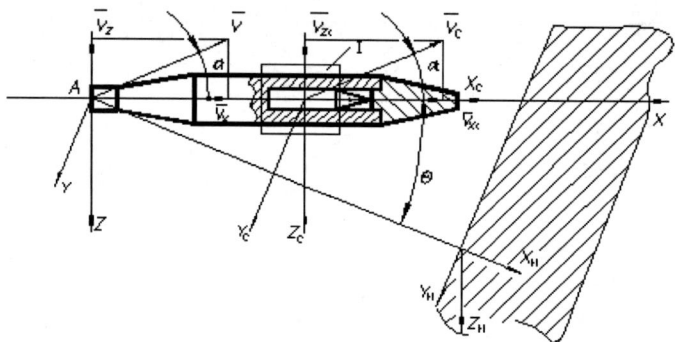

FIG. 1. Set of experiment.

Penetration process consists of three stages:

1. From the impact of the striker against the barrier (the body starts wearing out) to the moment when strength of barrier material enters the play that affects on the striker head part;

2. Non-wearing core part leaves the body; and

3. Further penetration of the striker parts into the barrier till their complete stuck in the barrier or barrier destruction.

Equation sets were used for the body and non-wearing core part in each stage of penetration, and strength factors were calculated. These equation sets are based on the system of axes when the body is at A point and the core part in the centre of mass, and the fixed system of axes related to the barrier (Fig. 1). Changes in kinematic characteristics of the striker and coordinates of striker's penetration into the barrier with time were calculated. These parameters were recalculated for each time step.

As it was already mentioned, the increase of slope of armored shell plates results in deviation of the striker's penetration trajectory. This increases the probability of the side surface impact against the cavity edge. This causes strength localization over the core part. The core part can be destructed during the further loading. The probability of striker destruction increases with the increase of the striker's length.

Cylindrical part of the striker is broken up into elementary areas to check the contact between the side surface and cavity edge. The radius vector ρ_i, the unit vector of normal n_i near the central point of each elementary area, its size and the view of the normal unit vector. The coordinates of the points and their speed are defined in the fixed system of axes (Fig. 2). The cavity edge is defined as the straight segments which were formed by linear approximation. The impact occurred when normal vector component of the point speed is directed to the cavity edge and the point outside the edge. After that, we checked if the core part made of heavy alloy was destructed based on destruction criteria.

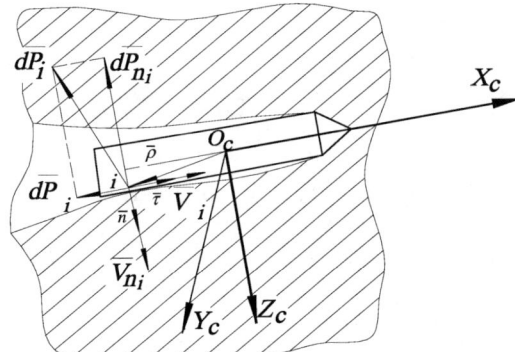

FIG. 2. Impact of side surface against cavity edge.

Impact elasticity and destruction at bending stress were selected to be destruction criteria. Impact elasticity was calculated using A_d performance (material destruction performance). The following loading mechanisms were used in the calculations involving destruction at bending stress: 1) the tail part is confined and the moment of resistance force affects the head part; and 2) the side surface of the striker impacts against the cavity edge.

The striker will be destroyed when specific impact energy or bending stress exceed the critical values. We reckon calculated three scenarios of core part destruction (Fig. 3):

1. The core head part is slightly destroyed, but the cylindrical-shaped tail part penetrates deeper;

2. The core tail part is slightly destroyed, but the head part penetrates deeper because of mass loss;

3. After the impact, the core part is splitted into two parts which penetrate into the barrier independetly; one part is of ogival shape, the other one is of cylindrical shape.

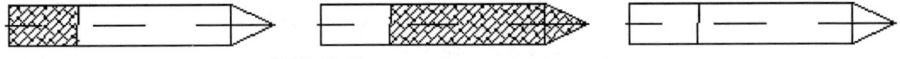

FIG. 3. Core part destruction scenarios.

The scenarios were analyzed using the calculation technique. We determined the influence of the striker destruction on penetration depth of the striker into the barrier as well as the influence of geometric parameters of the striker on striker destruction. In the analysis the model of bullet (ZBM-25) with doubled-length striker was used. The physical and mechanical properties of barrier materials corresponded to those of 52S steel and fiberglass. In the analysis, the angle between the barrier normal and the bullet axis was assumed to be 30°.

We have calculated the critical angles when impact of the striker side surface against the cavity edge was observed (Fig. 4). The striker length was varied: $l = 85$ mm, $l = 100$ mm, $l = 107$ mm, $l = 115$ mm, $l = 130$ mm (the core part was located in the bullet head part).

The impact occurs when deviation angle is rather small, and the destruction occurs when deviation angle is larger. Regularity is evident when impact elasticity is used a destruction criterion: critical deviation angles increase with the decrease of core part length (Fig. 4a).

If speed is 900 m/s, the impact of the striker tail part occurs at the end of the striker's penetration trajectory into the barrier. Thus, its kinetic energy is not sufficient to destroy the striker (impact elasticity does not excess its critical value). Destruction of the core part occurs when the angle between the barrier normal and the bullet axis is getting closer to the ricochet angle value.

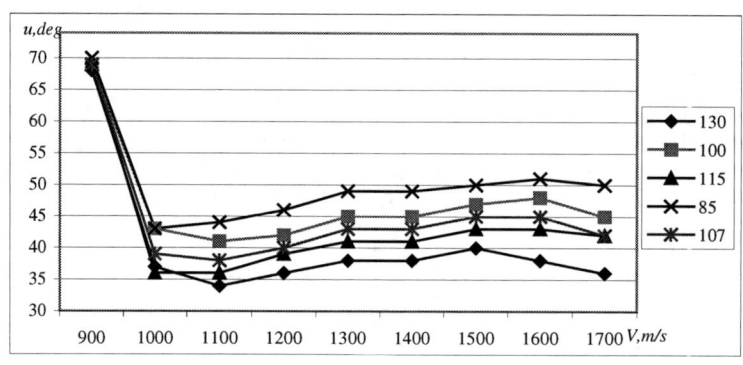

a

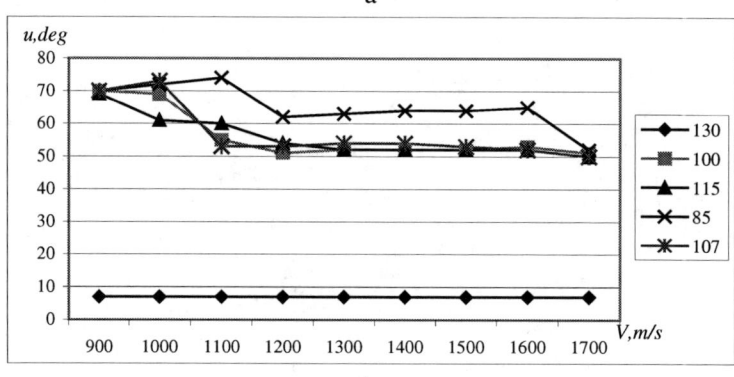

b

FIG. 4. Angle between the barrier normal and the bullet axis at core part destruction. a - impact elasticity, b - destruction at bending stress.

124

The resistant force moment is the main factor in the case of destruction at bending stress. Destruction at bending stress is directly proportional to this force (Fig. 4b). Bending stress drops with the increase of the core part length.

The striker's penetration into the semi-infinite barrier was calculated to control influence of destruction on penetration. Impact elasticity was selected to be a destruction criterion.

The analysis of the results showed that (i) core tail part destruction occurred when it was in the striker head part (Fig. 5); and (ii) core head part destruction occurred when it was in the striker tail part (Fig. 6). This can be explained by the effect of the barrier resistance force which was stronger in (i) case. In (ii) case, the body imparted a significant angular velocity to the core part, thus, destruction occurred in the striker head part.

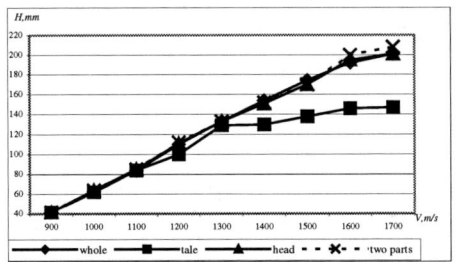

FIG. 5. Depth of penetration into the semi-infinite barrier for different destruction scenarios (core part is in the striker head).

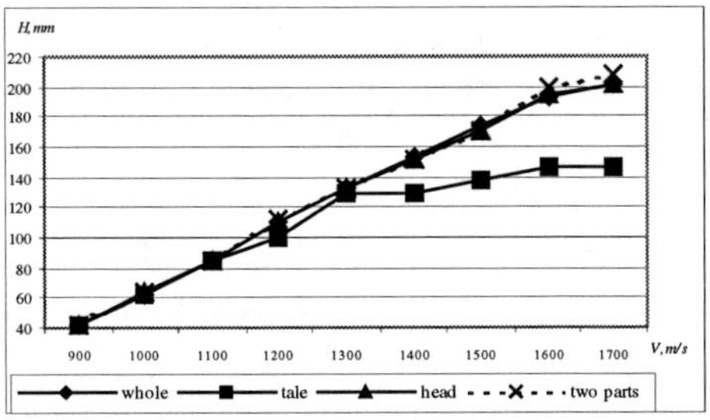

FIG. 6. Depth of the penetration into the semi-infinite barrier for different destruction scenarios (core part is in the striker head).

When the striker is splitted into two parts, they penetrate into the barrier as two stand-alone strikers. The analysis showed that penetration depth was the same when the striker was undamaged since mass and kinetic energy were constant.

Penetration into the three-layer barrier was investigated. The core part was in the striker head. Penetration depth depends on thickness of the first barrier layer (Fig. 7). The total thickness of the barrier was constant, and deviation angle of the shell axis to the barrier was 60°.

If the first barrier layer is thin, destruction of the core part starts in the third layer (Fig. 7a). Destruction of the core part occurs in the first barrier layer if barrier thickness is 100 mm and the speed is low (Fig. 7b). Destruction of the core part in the first layer results in mass loss and kinetic energy decrease. This causes the decrease of depth penetration.

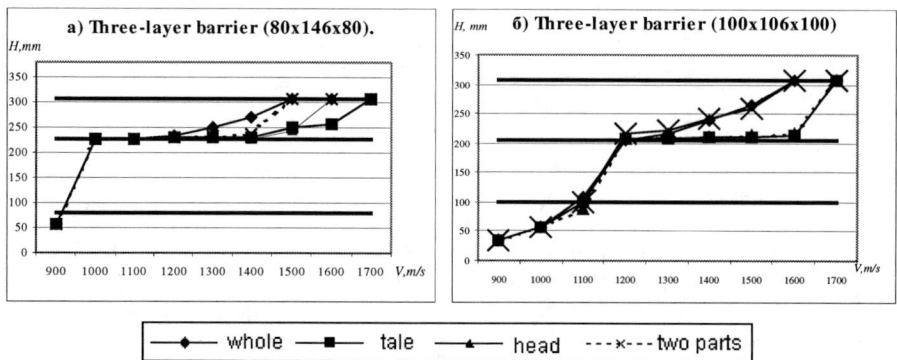

FIGURE 7. Depth of penetration into the three-layer barrier for different destruction scenarios (the core part is in the striker head).

If the core part is in the striker tail, the destruction occurs in the third layer because the core part remains in the striker for a long time (Figure 8).

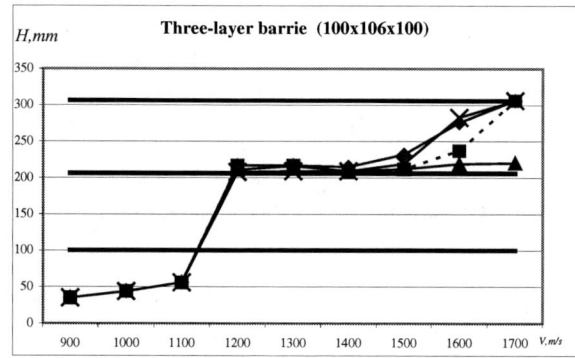

FIGURE 8. Depth of penetration into the three-layer barrier for different destruction scenarios (core part is in the striker tail).

CONCLUSION

1. The calculation technique for kinematic characteristics of penetration of combined strikers considering possible destruction of the latter was developed.

2. The present technique allows calculating a number of various parameters such as geometric, physical and mechanical characteristics of the striker and barrier as well as the conditions when the striker approaches to the barrier.

3. A computer program was developed. It helps to fit the optimal gage of the barrier.

4. The analysis of the results showed the effect of striker destruction on penetration depth.

SECTION 2

EXPLOSION AND
DETONATION PHENOMENA

Origin And Propagation Characteristics Of The Explosive Decomposition Chain Reaction In Heavy Metal Azides

E.D. Aluker[*], G.M. Belokurov[*], A.G. Krechetov[*], A.Yu. Mitrofanov[*], A.S. Pashpekin[*], B.P. Aduev[†]

[*]Kemerovo State University, Krasnaya st. 6, Kemerovo, 650043, Russia
[†]Kemerovo Branch of Institute of Solid State Chemistry and Mechanochemistry, Sovetsky av. 18, Kemerovo, 650099, Russia

Abstract. Pre-explosive luminescence has been used to visualize the nascence and propagation of a chain reaction over the sample. Results of investigations on nascence and propagation of pre-explosive luminescence in silver azide using streak-camera (space resolution 50 μm, time resolution ~ 1 ns) are as follows: under homogeneity initiation by a laser pulse and electron accelerator pulse luminescence originates in discrete centers and then propagates over a sample. From the data obtained authors hypothesized the following: a) discrete center nature of nascence of the reaction considered is provided by stochastic nature of initiation process and is not associated with unsoundness of a sample; b) propagation of the reaction over the sample is provided by diffusion of electrons.

Keywords: silver azide, pre-explosive luminescence, initiation
PACS: 47.40.-x; 82.33.Vx

EXPERIMENTAL

Under laser initiation of high-quality AgN_3 whisker crystals, pre-explosive luminescence (and hence, an explosive decomposition chain reaction [1]), has been previously shown to originate in spatially discrete centers (nascent centers) [2]. Within several tens of nanoseconds, the nascent centers have been observed to expand and overlap, forming a continuous luminescent area. It should be emphasized that the homogeneity of the initiating laser beam and the high quality of the irradiated crystals ensured that the observed nascent centers were not related to beam non-homogeneity and macro-defects of the sample [2]. Emergence of the nascent centers could be associated with several underlying causes, viz. (1) small accumulations of point defects having a high coefficient of laser radiation absorption could have induced non-homogeneous excitation (initiation); or (2) absorption of laser radiation in the sample was homogeneous but for some unknown reason, the chain reaction originated more easily in certain micro-areas of the sample (nascent centers). To determine the proper explanation, a similar experiment having an alternative initiation method that would guarantee homogeneous absorption of the initiating pulse energy in the sample was desirable. That objective became the purpose of the research presented here.

CP849, *Zababakhin Scientific Talks - 2005*,
edited by E. N. Avrorin and V. A. Simonenko
© 2006 American Institute of Physics 0-7354-0345-7/06/$23.00

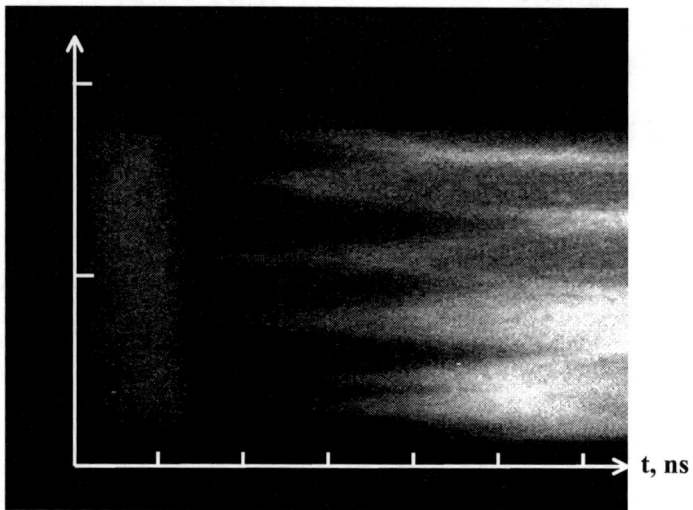

FIGURE 1. Time-dependent picture of luminescence along the length of the sample under initiation by an electron accelerator pulse. Note that the luminescence first appears in discrete points (nascent centers) and only then propagates throughout the sample (homogeneous light emission seen along the left margin of the picture (t ~ 40 ns) is the radio-luminescence of the sample excited by the initiating pulse).

A high-current pulsed electron accelerator served as the initiating pulse source (average energy of the accelerated electrons – 200 keV, current density of the beam - 1.2. kA/cm^2, pulse duration at half height – 20 ns, non-homogeneity of the beam across the area of the sample does not exceed 5%). Silver azide (AgN$_3$) whisker crystals having typical dimensions of 100×100×1000 mm^3 were used as samples. Because the track length of the accelerated electrons (~ 300 mm) exceeded the transverse dimensions of the sample, and taking into account the mechanisms of primary radiation processes in condensed matter [3], the homogeneity of the initiation of the sample was guaranteed.

An optical block was used to project an enlarged image of the sample onto the input screen of a Vzglyad-2A streak-camera operating in time-scanning mode. The spatial resolution of the method is ~ 50 μm. The axis of the image of the sample on the output screen was perpendicular to the time-scanning direction. Hence, the resulting image showed the time dependence of the spatial distribution of light emission through the length of the crystal. Measuring techniques were the same as those reported in [2]. Figure 1 presents the results obtained for one of the samples; as in the case of laser initiation [2], luminescence originated in discrete nascent centers that expanded into luminescent regions and eventually spanned the sample. Based on the divergence angle of the luminescent cones shown in Fig. 1, the propagation rate of the reaction across the sample could be estimated. The propagation rate, averaged over 20 samples, was 1300±300 m/s, which was in good agreement with the results obtained using laser initiation [4].

The initiating pulse energy was too weak to homogeneously heat the sample to temperatures sufficient to cause the onset of explosive decomposition. Taking into account the homogeneity of the excitation and the physics of the dissipation of ionizing radiation energy in solids [3], it is also unlikely that such temperatures could be reached in localized points (hot spots). Therefore, the underlying mechanism of reaction nascence, rather than the specifics of the absorption of the initiating pulse energy, accounts for the emergence of the observed nascent centers.

DISCUSSIONS

As previously discussed in [1], the chain reaction could originate around dislocations. The research results presented here did not disprove that explanation; however, numerous additional assumptions would be required to explain the higher numbers of nascent centers observed when initiating pulse energies were increased [2]. Thus, it seems appropriate to consider another plausible cause for the localized nature of the onset of the reaction.

We propose that the reaction originates in discrete nascent centers simply as a result of the stochastic nature of the process. First, the low (near-threshold) initiating pulse energies should be considered; the nascent centers have been previously clearly observed at such low energies [2]. In this case, the probability is quite low that the initial event (i.e. the consecutive trapping of electrons and holes by paired vacancies [5]) will occur and thus trigger reaction nascence; consequently, the reaction starts off only in a few random points in the sample. An electron-and-hole multiplication chain reaction commences in the vicinity of these random points, ensuring expansion of the reaction zone by diffusion of electrons and holes. The increasing reaction rate and expanding reaction zone cause luminescence intensity to increase, facilitating visual observation of the reaction.

Luminescence of the nascent centers has been observed to reach detectable levels after $\tau \sim 10^{-8}$ s elapsed time after the initiating pulse (Fig. 1). During this time interval, the reaction zone propagates over a distance $r \approx V\tau$, where $V \sim 10^5$ cm/s [2] is the propagation rate of the front edge of the reaction, and $\tau \sim 10^{-8}$ s is the time of detectable luminescence onset in the nascent centers (Fig. 2). Thus, the estimated propagation distance $r \sim 10^{-3}$ cm agrees in order of magnitude with the spatial resolution of the method used in the experiment. For typical concentrations of paired vacancies in the samples, i.e. $\sim 10^{16}$ cm^{-3}, each luminescent nascent center contains in its volume ($r^3 \sim 10^{-9}$ cm^3) approximately 10^7 active points, which is sufficient to ensure detectable luminescent intensity in the nascent centers. Higher initiating pulse energies increase the probability of the consecutive trapping of electrons and holes by paired vacancies [5], thus increasing the number of nascent centers [4]. For fairly high initiating pulse energies, the number of reaction origination points might increase to such levels that overlapping of the nascent centers would occur in less time than the time resolution of the experimental method and thus appear as quasi-homogeneous reaction onset [2].

In conclusion, it should be noted that a similar phenomenon is well known in radiation physics [3]. Particles having small specific ionization (e.g. β particles) exhibit tracks consisting of well-defined ionization areas. For particles having high specific ionization (e.g. α particles), these locations overlap, forming a continuous cylindrical ionization area.

The research was supported by RFBR.

REFERENCES

1. M. M. Kuklja, B. P. Aduev, E. D. Aluker et al. *J. Appl. Phys* **89**, 4156–4166 (2001).
2. B. P. Aduev, E. D. Aluker, A. G. Krechetov, and A. Y. Mitrofanov, *Combust., Explos., and Shock Waves* **39**, 581–584 (2003).
3. C. Lehmann, *Interaction of radiation with solids and elementary defect production,* Oxford: University Press, 1977.
4. B. P. Aduev, E. D. Aluker, A. G. Krechetov, and A. Y. Mitrofanov, *Combust., Explos., and Shock Waves* **39**, 701–703 (2003).
5. B. P. Aduev, E. D. Aluker, and A. G. Krechetov, *Combust., Explos., and Shock Waves* **40**, 209–213 (2004).

Investigation Of PETN Monocrystals Initiation By Electron Beams

Loboiko B.G., Garmasheva N.V., Filin V.P., Gromov V.T., Shukailo V.P., Stryakhnin V.L., Nesterov O.V., Khruliova O.V., Alekseev A.V., Gagarin A.L., Taybinov N.P.

All-Russian Federal Nuclear Centre – Zababakhin Institute of Technical Physics, 456770, Snezhinsk, Chelyabinsk region, Russia

Abstract. Electron beam initiation of PETN monocrystals depending on their size and defectiveness as well as electron beam parameters and environmental acoustic stiffness, was investigated. The length of PETN monocrystals was from 1 mm to 30 mm. The experiments used pulsed accelerator of electrons GIN-540 with the average beam-current value of ~1kA, pulse length of $\tau_{0.5} \approx 10$ ns, the average electron energy of ~250keV. The experiments showed that the low level of fluence of electrons led to the appearance of additional defects in PETN crystals. When fluence of electrons increased the different experimental results were observed: crystal destruction, initiation of crystal explosive decomposition, sample detonation to form a mark on witness plate. The fixed parameters of electron action showed the dependence of experimental results on acoustic stiffness of a reference plate material. Keywords: **PETN, high explosive (HE), monocrystals, electron beam, pre-explosive processes**
PACS: **98.62.Bj, 79.20.Fv, 82.33Vx**

INTRODUCTION

Studying the early stages of HE initiation under external actions is especially important, since it helps to understand the nature of HE detonation. Most important is to study processes taking place in HE during and promptly after the external actions.

Papers [1, 2] describe investigations of pre-explosion processes ongoing in filiform crystals of heavy metal azides under the pulsed electron or laser exposures. The method of optical pulsed spectrometry with high time resolution (~1 ns) developed to investigation fast processes is the result of this work. In paper [3], a similar approach was tested to investigate the processes ongoing in PETN crystals under the pulsed electron beam. Continuation of this work required improved methods of samples preparation, investigation into characteristics of these samples, as well as improvement of experimental set-up. This report contents some results of these works.

CP849, *Zababakhin Scientific Talks - 2005*,
edited by E. N. Avrorin and V. A. Simonenko
© 2006 American Institute of Physics 0-7354-0345-7/06/$23.00

1 PREPARATION OF SAMPLES

PETN monocrystals were prepared using raw PETN crystallization from oversaturated acetone solution by slow cooling (<5⁰C/hour). The length of obtained PETN monocrystals ranged from 1mm to 30 mm. (see Fig. 1, (a)).

The material was identified by FTIR spectra obtained on IR-Fourier spectrometer SYSTEM-2000, Perking Elmer. Samples for investigations have been pressed from mix of PETN powder with KBr to form pellets. Basic FTIR characteristic lines of PETN after crystallization coincide with the published data [4].

Earlier in [5], we obtained the experimental curve of PETN monocrystals transmittance (%) of electromagnetic radiation depending on the wavelength (the transparency curve). Measurements were made on spectrophotometer SF-26 in the wave range from 200 nm to 1200 nm. It has been shown that PETN single crystals are practically nontransparent in UV-range (see Fig. 1, (b)). The edge of fundamental absorption was observed to fall within 300 nm - 330 nm. Based on the above results, the optical width of the prohibited zone for PETN single crystals was estimated to be 3.8-4.1 eV.

In this work, the transmittance (%) of PETN single crystals in the range of wavelength 400-600 nm was used as the relative characteristic of sample defectiveness in the direction of external actions. Defectiveness of the used PETN crystals varied from 30 to 80%.

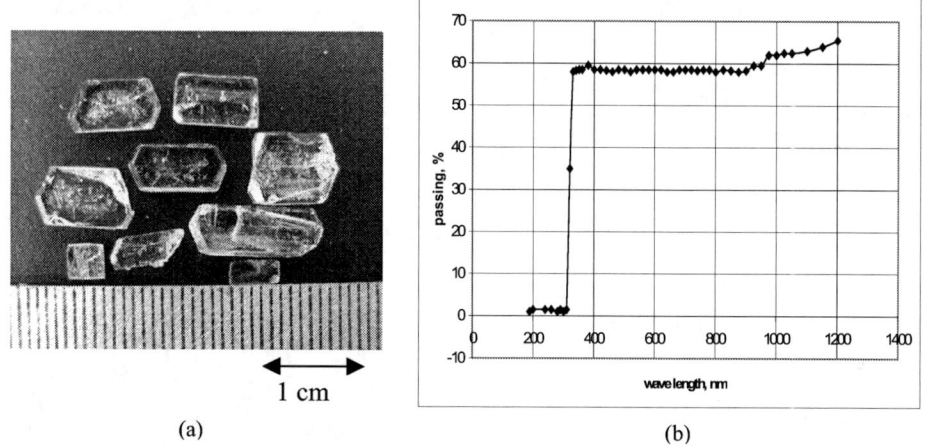

(a)   (b)

FIGURE 1. PETN monocrystals (a) and the curve of PETN monocrystals transmittance (%) of electromagnetic radiation depending on the wavelength (b)

2 ELECTRON BEAM INITIATION OF PETN CRYSTALS

The experiments used the pulsed electron accelerator GIN-540 with the average beam current of ~1 kA, pulse length of $\tau_{0,5} \approx 10$ ns, which provide the average electron energy of ~250 keV.

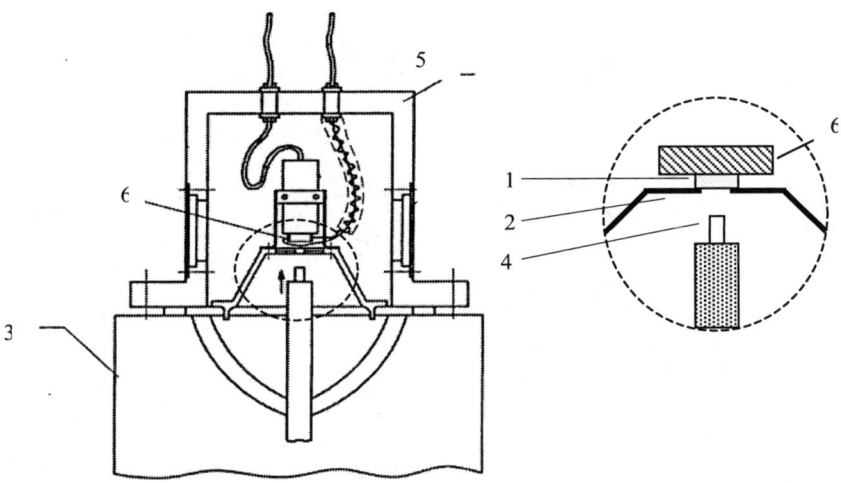

FIGURE 2. Experimental set-up based on pulsed electron accelerator GIN-540 for PETN monocrystals investigation

For irradiation, PETN crystals (1, see Fig.2) were placed above the hole (∅3.5 mm) in the foil of the anode (2) in the accelerator GIN-540 (3) (see Fig.3). Electrons action onto the crystals was varied by changing the anode-to-cathode (2-to-4) and crystal-to-anode distances and due to metal foils arranged between the accelerator anode and the sample. The protective cover (5) sealing the working chamber of the accelerator closed the assembly. During experiments, air pressure in the working chamber of the accelerator was not more than 0.53 Pa. Fig. 3 shows pictures of the experimental set-up at different steps of preparation procedure.

Experiments demonstrated that variations in parameters of the electron beam action (corresponding to the changes from 0 to $8 \cdot 10^{-7}$C in the charge the sample was exposed to) decreased the transmittance of PETN single crystals in the visible range (T,%). The shape of the transmittance curve in the 200-1200 nm range repeats the previously obtained one (See Fig.1, b). The results serve the basis for the conclusion that the above level of electron beam impact mainly causes the onset of additional defects in PETN crystals.

 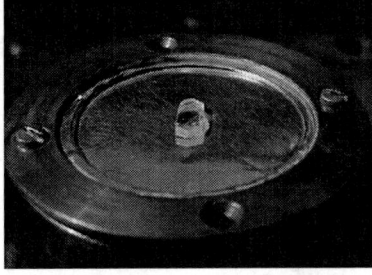

FIGURE 3. Anode of GIN-540 (a) and PETN monocrystal (b) prior to electron beam impact

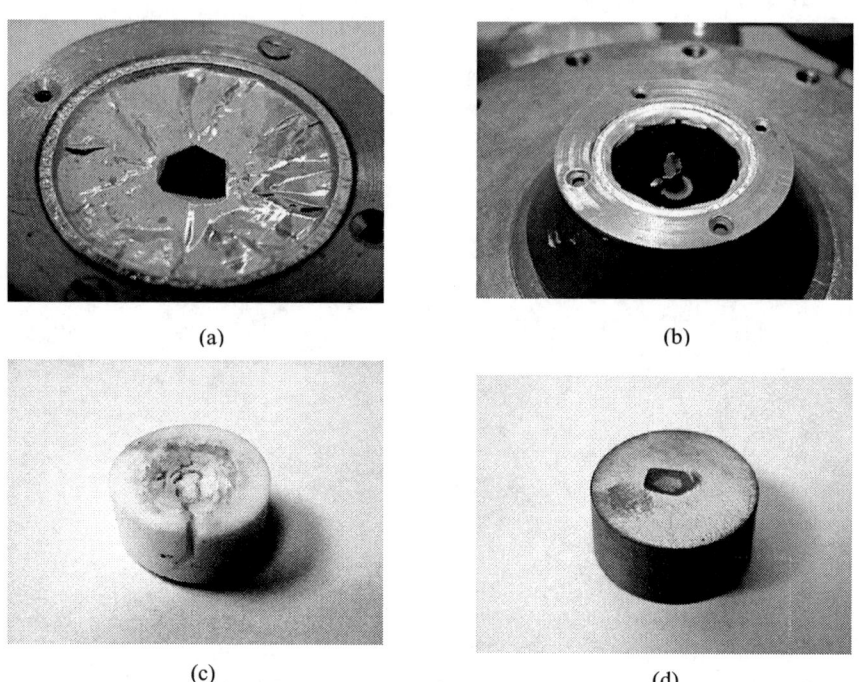

(a)   (b)

(c)   (d)

FIGURE 4. Anode and cathode of GIN-540 (a, b) at different impacts on PETN monocrystal; reference fluoroplastic (c) and copper (d) plates after detonation of PETN samples.

Depending on the level of electron beam exposure (action) on PETN crystals, experiments registered:

1 - decrease of the transmittance of samples;

2 - onset of cracks in samples;

3 - spraying of the crystal;

4 - partial destruction of the anode and cathode (Fig.4, a);

5 - complete destruction of the cathode (Fig.4, b) and/or formation of the trace on the reference plate (Fig.4, c, d).

3 DYNAMIC PROCESSES IN PETN MONOCRYSTALS

The next series of experiments reference plate with 20-mm diameter and 10-mm thickness made of different materials were placed on the studied crystals (6, see Fig.2). All experiments used the same experimental setup and geometry of samples exposure when the electric charge, the sample was exposed to, was $\sim 1.2 \cdot 10^{-5}$ C. Consideration was given to PETN crystals having approximately the same transmittance (60-70%) and dimensions in the direction of electron beam action were from 1 to 2.3 mm. At fixed parameters of the electron beam action, experimental results were observed to depend on the additional plate material. Experimental results are summed in Table 1.

TABLE 1. Experimental results for PETN single crystals (exposure ~1.34·10-5 C)

Plate material	PETN single crystal		Results
	Size*, mm	Mass, g	
Copper	1.03 – 1.73	0.07 – 0.08	Detonation (Fig.4, d)
Fluoroplastic	1.20 – 1.45	0.040-0.07	Crystal destruction, partial destruction of anode and cathode (Fig.4, a)
	1.45 – 1.95	0.07 – 0.12	Detonation (Fig.4, b, c)
	1.95 – 2.03	0.12 - 0.13	Crystal destruction, partial destruction of anode and cathode (Fig.4, a)
Foam plastic	1.17 – 1.92	0.06-0.21	Crystal destruction, partial destruction of anode (Fig.4, a)

* In the direction of electron beam impact

These results can be explained by making the following assumption. The pulsed electron beam initiates, in a PETN crystal, processes resulting in the generation of acoustic waves and their propagation through the crystal. After these waves reach the HE/plate interface, several patterns are possible; their onset depends on acoustic stiffness of materials adjacent to the sample. Table 2 gives the dynamic compressibility of the pressed high-density PETN and materials of plates used in experiments.

TABLE 2. Dynamic compressibility of some materials

Material	Density ρ, g/cm^3	$D=C_0+\lambda U$ C_0, km/s	λ	Reference
PETN	1.75	2.87	1.69	[6]
Copper	8.93	3.91	1.51	[7]
Fluoroplastic	2.16	1.95	1.67	[7]
Foam plastic	0.6	1.19	1.35	[8]

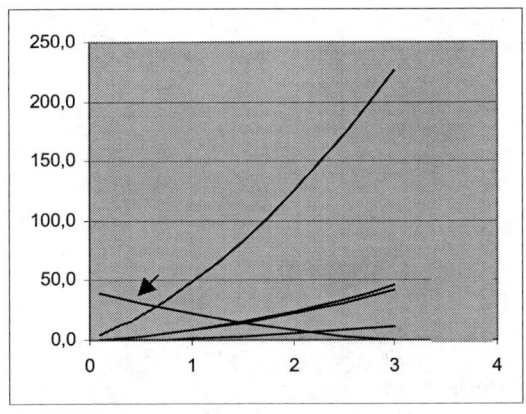

FIGURE 5. Wave processes on PETN crystals: P-U diagrams of copper (1), explosive (2), fluoroplastic (3), foam plastic (4) and wave output on explosive/plate boundary (5)

Fig.5 shows the P-U diagram of studied inert materials (copper (1), fluoroplastic (3), foam plastic (4)) and PETN (2, at ρ=1.75 g/cm^3) [6].

This diagram also considers the case when the acoustic wave propagating through the HE crystal reaches the HE/plate interface (5).

137

After the acoustic wave reaches the HE/copper interface, the shock wave, whose amplitude is significantly larger than that of the incident wave, begins to propagate inside HE. Apparently, just this shock wave initiated detonation in PETN crystals in experiments with copper plates. If the acoustic wave leaves the HE crystal and gets the HE/foam plastic interface, the rarefaction wave would propagates back to the HE crystal and destructs it.

Acoustic stiffness of foam plastic is close to that of HE. In this case, observed experimental results, i.e. the crystal thickness range within which detonation took place, can be explained if the above hypothesis is supplemented by the assumption that additional energy release is observed in the incident wave. Actually, if electron beam initiates pre-explosion processes in a PETN crystal, then certain time (and sample size, respectively) is needed for these processes to develop into detonation. On the other side, beginning from certain sizes of samples (in the direction of the arising wave propagation) rarefaction wave from lateral sides of the sample will have the determining influence on the onset of detonation process.

CONCLUSIONS

The results allow the following conclusions.
1.   The edge of fundamental absorption for PETN single crystals is shown to fall within 300-330 nm; this gives the prohibited zone width, which is 3.8-4.1 eV.
2.   Pulsed electron accelerator with the average beam current of ~1kA, the pulse length of ~10 ns and the average electron energy of ~250 keV allows early stages of PETN single crystal initiation to be studied.
3.   In PETN monocrystals, pulsed electron beam initiates pre-explosion processes accompanied by energy release, generation and propagation of acoustic waves, which are capable of causing sample detonation under certain conditions.

ACKNOWLEDGMENTS

Authors are appreciated L.N.Filina and S.G.Sazonova for the help on the sample preparations. Investigations were supported by ISTC Project №2180.

REFERENCES

1.   B.P.Aduev, E.D.Aluker, G.M.Belokurov, Yu.A.Zakharov, A.G.Krechetov "Explosive decomposition of heavy metal azides", *ZhETF*, **116**, №5(11), 1676–1693 (1999).
2.   Yu.A.Zakharov, E.D.Aluker, B.P.Aduev, G.M.Belokurov, A.G. Krechetov, Pre-explosion phenomena in heavy metal azides, "Khimmash", Moscow, 2002
3.   B.P.Aduev, G.M.Belokurov, N.V.Garmasheva, S.S.Grechin, E.V.Tupitsin, V.N.Shvayko, *"Spectral-Kinetic Characteristic of PETN Luminescence Under Initiations by Electron Beam" in* Proceedings of the 7th Seminar "New Trends in Research of Energetic Materials", Pardubice, 2004, pp.414-416
4.   Pristera F., Halik M., Castelli A., Frederics W. Analysis of Explosives Using Infrared Spectroscopy / Anal.Chemi. 1960, v.32, 3, pp. 495-508
5.   O.V.Khruleva, V.P.Filin, N.V.Garmasheva, L.N.Filina, B.G.Loboiko Optical properties of PETN crystal / VII Zababakhin Scientific Talking, 2003

6. LASL Explosive Property Data/ Ed. Tr.Gibbs, I.A.Popolato, University of California Press, Berkly - LosAngeles - London, 1980
7. Yu.N.Zhugin, K.K.Krupnikov, N.A.Ovechkin, E.V.Abakshin, M.M.Gorshkov, V.T.Zaikin, V.M.Slobodenyukov Some specific features of dynamic compressibility of quartz / j. Physics of Earth, 1994, №10, pp.16-22
8. I.P.Dudoladov, V.M.Rakitin, Yu.N.Sutulov, G.S.Telegin Shock compressibility of expanded polystyrene with different initial density / j. PMTF (Applied Mechanics and Technical Physics), 1969, №4

Temperature-Based Macrokinetic of Shock Wave Initiation of Detonation

K.F. Grebenkin, A.L. Zherebtsov, M.V. Taranik,
C.K. Tsarenkova, A.S. Shnitko

*Russian Federal Nuclear Center – All-Russian Institute of Technical Physics,
456770, Snezhinsk, Russia*

Abstract. Basing on the semiconductor model of detonation, a macrokinetic model of TATB-based composition initiation by shock waves has been developed. The model describes the influence of the initial HE temperature on its sensitivity.

Keywords: detonation, TATB, semiconductor model, hot spots.
PACS: 47.40.Rs, 82.33.Vx

INTRODUCTION

A hot spot concept serves as a basis for practically all present-day kinetic models of detonation initiation in heterogeneous HE. According to this concept, a heterogeneous HE burning rate may be represented as a production of three factors [1]:

$$-d\xi/dt \sim N^{1/3}\left(P_f\right)\cdot D(P,T)\cdot F(\xi), \tag{1}$$

where o is the HE concentration, N - the hot spots density, D – the rate of the burning wave propagation from the hot spots, $F(o)$ – a geometric factor describing a change of the burning topology after meeting of the burning waves, propagating from the neighboring hot spots, - pressure at the front of the first shock wave.

Several models of macrokinetics of plasticized TATB detonation have been proposed within the framework of the relation (1). In the given report, a new model differing from the previous ones by approach to the selection of the relation (1) factors is represented. An attempt of estimating these functions, basing on analysis of micro- and meso-physical processes as well as introducing the model parameters having a clear physical meaning, has been made in this work. We believe that such approach can open the way to further accuracy increasing by means of detailed elaboration of the models and refinement of its parameters on the basis of ab initio calculations and measurement of meso- and micro- characteristics of the explosives.

THE MODEL

The macrokinetic model presented here is based on the semiconductor model of detonation, which suggests that the energy is transferred from the hot spots by electron thermal conductivity [2]. The modeling of the burning wave propagation rate

CP849, *Zababakhin Scientific Talks - 2005*,
edited by E. N. Avrorin and V. A. Simonenko
© 2006 American Institute of Physics 0-7354-0345-7/06/$23.00

dependence on temperature of the compressed but unreacted TATB beyond the hot spots have been carried out [3]. The calculated dependence turned to be almost linear, and the logarithmic derivative of D was estimated as

The dependence of the geometric factor on HE concentration $F(o)$ was evaluated in simplified calculations of the growth of several hot spots, randomly distributed over space. Analysis of the results has shown that two stages of macrokinetics, qualitatively differing in the geometric factor, are observed. At the first stage, when the burning waves do not overlap, the cell model [4] is applicable. This stage duration was estimated as one third of the total HE decomposition time $\Delta\tau$: $t_1 \approx \Delta\tau/3$. Under the assumption that the rate of the burning wave propagation is constant, the HE burning-out at the end of the first stage may be estimated according to the cell model as

i.e. about 3 %. Therefore, as the first approximation, one may neglect the energy release at this stage and model as the time delay t_1 since coming the shock wave front.

The duration of the second stage, when the intensive HE burning occurs, is about twice as much as that of the first stage $t_2 \approx 2*\Delta\tau/3$. The model calculations have shown the geometric factor to change slightly with time at this stage, and thus it may be considered as a constant.

Finally, let's consider the dependence of the hot spots density on the initiating shock wave pressure. Initiation of the TATB-based explosive composition occurs at high pressures. In this case a hydrodynamic mechanism, namely the pores collapse, plays a determining role in the hot spots formation [5]. Microstructure of the pressed TATB-based (97%) explosive composition with a polymeric bonding (3%) was studied [6], and it was found that it possesses pores of typical size on a micron scale. We have carried out the 2D hydrodynamic calculations and revealed that the collapse of the single pore, after passing the shock with pressure of 15 GPa, results in the temperature increase up to ~ 4000 K in a zone of one order of magnitude lesser size than the initial pore size was. When decreasing the pressure at the shock wave front to 10 GPa, the temperature of the hot region reduces by ~ 1000 K. Extrapolating these calculation results to lower pressures region lead to conclusion that the critical temperature of burning of the hot spots of $\sim 0.1\mu$m in diameter (~ 2000 K according to [7]), originated from micron-size pores collapse, occurs under pressures of ~ 5 GPa at shock wave front. So, the step function at ~ 5 GPa can be used as a simplified form of the hot spots density factor $N^{1/3}(P_f)$.

More realistic dependence of the hot spots density factor on pressure should take into account the fact that pores differ in size [8]. Assuming that the pore size variations are near $\sim 1\mu$m, the hot spots size variations may be estimated as $\sim 0.1\mu$m. This lead to the variations of the hot spot critical temperature by ~ 200 K [8], and the variations of the initiating pressure by ~ 2 GPa corresponds to such variation of hot spots size.

Thus the dependence of the hot spot density factor on the pressure at the front of the shock wave must look like a smoothed step function at pressures near P $\sim$ (5–10) GPa. In figure 2 the hot spot density factor selected so to reproduce the results of the experiments on shock wave initiation of LX-17, the plasticized TATB, is given.

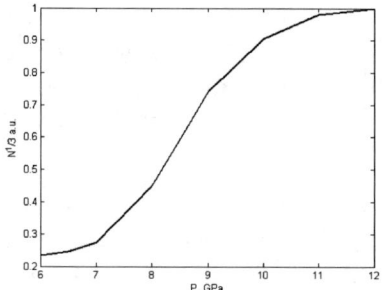

FIGURE 1. Dependence of the hot spots density factor on pressure at the front of the first shock wave.

Based on the above considerations, the HE macrokinetics equation may be written as follows (t=0 corresponds to the shock arrival moment):

$$-\frac{d\xi}{dt} = \begin{cases} 0, & \text{when } 0 \le t \le t_1, \\ \exp\{F(T)\}, & \text{when } t > t_1 \end{cases} \tag{2}$$

Function $F(T)$ is given in the fig. 2. It was estimated by the results of calculations of the burning wave rate-temperature dependence [3] and then was adjusted so to describe the results of experiments on LX-17 detonation initiation. We should note that this function is close to the linear one and its derivative varies within range $A = dF/dT = d\ln D/dT \approx (8\pm2)*10^{-3} K^{-1}$, that agrees with this parameter evaluation given above.

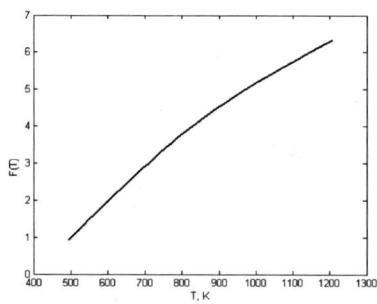

FIGURE 2. Function $F(T)$ in the burning rate equation (2); the unit of time is equal to 10^{-5} s.

RESULTS

Numeric modeling of some experiments on detonation initiation in LX-17 (92.5 % TATB) was carried out aiming the model verification. Equations of state of Me-Grunizen type for the unreacted explosive [9], equation of HOM type for the explosion products [10] and the model [11] for HE - explosion products mixture were used in calculations, which were carried out by the 2D code [12]; the grid step was equal to 0.4 mm that provides ~5 intervals per reaction zone in the stationary detonation wave. When modeling the heated explosive composition it was assumed that its porosity didn't change under the heating.

The calculated dependences of the run to detonation distance in LX-17 at normal as well as increased up to 250 C initial temperatures are represented in fig. 3. One can see the experimental results for different values of the initial temperature of explosive composition are reproduced in the calculations.

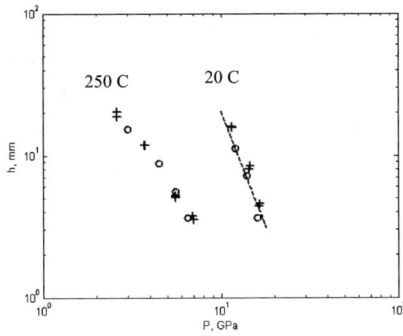

FIGURE 3. Dependence of the run to detonation distance in LX-17 on pressure: dotted line – the experimental dependence [13], crosses – the experimental values from [14], circles – the calculation results.

The critical diameter calculations were carried out in the following way. The initial geometry of the system was a truncated cone of 500 mm in length, which left and right diameters were 30 mm and 2 mm. The initiating triangular pressure pulse with 30 GPa amplitude and 1 microsecond total duration was applied from the side of the bigger diameter. As detonation wave "forgets" the initiation conditions after passing some distance, such simplified initiation model is reasonable. At normal temperature, the detonation interrupts at the distance of 31.5 cm from the initiating plane, while at T = 250 C – at distance of 45 cm; so, the charge of critical diameters equal to ~ 12 mm and ~ 5 mm correspond to these interruption distances. Thus, the experimental value of LX-17 critical diameter equal to 12 mm at normal conditions [15] is reproduced in calculations. Critical diameter of heated LX-17 was not measured, but for PBX-9502, similar TATB-based composition, the experimental value of critical diameter at $T_0 = 250$ C is about 4 mm [16].

Numeric modeling of the experiment on initiation of detonation by a flying thin plate made of mylar of 1.27 mm in thickness [17] was carried out. In contrast to Plot-Plot type measurements, the initiating pressure pulse is very short in this experiment. In calculations the detonation is initiated at the plate critical velocity ~ 3.6 km/s (run to detonation distance ~ 10 mm), that agrees well with the experimental value of the critical plate velocity equal to 3.6 ± 0.2 km/s.

CONCLUSION

The new physical model of macrokinetics of the shock wave detonation initiation in plasticized TATB was proposed. Numeric simulation of some experiments on shock wave detonation initiation in LX-17, including the experiments with preliminarily heated HE, was carried out. Calculation results are in agreement with the experimental data.

In conclusion, we should mention that the model gives the reasonable estimation of the explosion time for the case of the low velocity impact initiation, too. Considerable influence of initial HE temperature on the explosion time is predicted by the model; it decreases by order of magnitude when TATB initial temperature increases from 20 C up to 250 C. Measurement of dependence of the TATB explosion time on its initial temperature at the low velocity impact condition is of interest for the model verification.

REFERENCES

1. V.F. Lobanov. Fizika Gorenia i Vzriva (in Rus). 1980, N 6, p. 113.
2. K.F. Grebenkin. Bull. Am. Phys. Soc. 2005. V. 50, N. 5. p. 59.
3. K.F. Grebenkin, A.L. Zherebtsov, V.V. Popova, M.V. Taranik. Proc. of Int. Conf. V Khariton Scientific Reading (in Rus). Sarov. Russia. 2003, p. 189.
4. Fizika Vzriva (in Rus). Ed. L.P. Orlenko. Vol 1. Moscow. Fizmatlit. 2002.
5. R. Menikoff. Granular Explosives and Initiation Chemistry. LA-UR-99-6023. 1999.
6. G. Demol, P. Lambert, H. Trumel. Proc. 11th Deton. Symp. 1998, p. 309-316.
7. C.M. Tarver, S.K. Chidester, A.L. Nichols. J.Phys.Chem., 1996, 100, 5794-5799.
8. Yu.A. Aminov, N.S. Es'kov, Yu.R. Nikitenko. Tech. Papers of 12th Int. Deton. Symp. 2002. p. 40.
9. Yu.A. Aminov, A.V. Vershinin, N.S. Es'kov. e.a. Fizika Gorenia i Vzriva (in Rus). 1995, N 1, p. 103.
10. K.F. Grebenkin, A.L. Zherebtsov. Proc. of Int. Conf. V Zababakhin Scientific Reading (in Rus). Snezhinsk. Russia. 1998. Part 1. p. 200.
11. W.C. Davis. Proc. 11th Deton. Symp. 1998. p. 303.
12. V.A. Suchkov, A.S. Shnitko. Third Joint Conference on Computational Mathematics. Los Alamos, NM, USA, January 23-27, 1995.
13. R.L. Gustavsen, S.A. Sheffield, R.R. Alcon e.a. Shock Compression of Condensed Matter-2001. AIP. 2002. p. 1019.
14. J.C. Dallman, J. Wackerle. Proc. 10-th Deton. Symp. 1993, pp.130.
15. T.D. Tran, C.M. Tarver, J. Maienschein e.a. Proc. 12-th Deton. Symp., 2002. p. 239.
16. B.W. Asay, J.M. McAfee. Proc. 10-th Deton. Symp. 1993, p. 485.
17. C.A. Honodel, J.R. Humphrey, R.C. Weingard e.a. Proc. 7-th Int. Deton. Symp., 1981. p. 425.

Use the Numerical Experiment for Analysis of Increasing Explosive Efficiency under Initiation by High Velocity Impact

S.L. Neverov

249 apt., Pogrebnyaka str., 12 b., Krivoy Rog 50025 Ukraine

Abstract. There are revealed the mechanism and law of increment of integral impulse in detonation wave initiated under high velocity through an inert screen. Numerical solutions were performed according to problem definition that is impossible to realize in a real experiment. It is shown that an impulse in detonation wave depends on total thickness of a screen and liner by the same law as in case of initiation by direct impact with a liner. Therefore it is possible to name them as dynamical piston (simple or complex).

Keywords: Detonation, numerical solution.

In this article are represented results of the numerical modeling of explosive detonation, initiated in three ways: by direct impact of a liner (scheme 1), by detonation wave through an inert screen (scheme 2) and by the impact through an inert screen (scheme 3). As the author's development shows, use of such methods of initiation ensures considerable technological effect. To investigate mechanism of this effect there was the need of the additional research of detonation wave which was initiated by above mentioned methods.

Calculations were carried out by the program made with use of the Wilkins algorithm [1]. The model of explosive behavior chosen for use was "three-dimensional combustion" model (Meyder [2]) modified to allow to calculate explosive initiation of any intensity shock wave. For this purpose there was entered the "forced afterburning" mode as a numerical procedure. That equation of state is represented as a mixing percussive adiabat at the detonation front and as a polytrope in the case of detonation product unloading. The RDX was taken as explosive and steel as material of a liner and screen.

For comparison of results the calculations were performed by using the same initiating charge (mass of explosive for acceleration of a liner, in this case, for instance, 30 mm). That mass is taken into account in calculations for estimation of efficiency of the whole explosive mass using in the examined schemes as compared with regular initiating charge.

There was detailed enough in previous works [3] how a liner and (or) screen presence changes the profile of detonation wave pressure and was shown that these changes could be in principal important for realization of necessary physical processes in loaded materials. In any of the examined schemes a liner and a screen compress and slow down the detonation products unloading upwards and so it is possible to consider

CP849, *Zababakhin Scientific Talks - 2005*,
edited by E. N. Avrorin and V. A. Simonenko
© 2006 American Institute of Physics 0-7354-0345-7/06/$23.00

them as pistons. What is more, it is necessary to name them as "dynamic pistons" because unlike merely theoretical rigid wall and "pistons, moving on determined law", they are direct participants of shock-wave flow, so their behavior is not assigned directly but depends on flow of all other mediums.

Except for changing of detonation wave form each of these ways of initiation noticeably increases integral impulse of pressure $I = \int_0^\infty p(t)dt$ in detonation wave.

It turned out that it is the most suitable from universal physical values for estimation of general loading intensity and comparison of explosive mass using efficiency in each scheme. It also became known, that as detonation wave spread along the charge the impulse increases in each scheme exactly in the same way as in a regular initiated charge, according to classical solution [4] but on higher level. However, it turned out that if there was an initiation with a liner through an inert screen, an impulse behaves in a non-monotone way depending on thickness of a liner.

Therefore research of exactly the third scheme is conducted in this work, and on the basis of the results is made general conclusion for all three schemes. Analysis of calculation results of third scheme have shown that in the particular cases the screen moves away from a liner towards initiated explosive charge after impact depending on correlation between thicknesses of a liner and screen and speeds up upwards in a certain time under influence of detonation products and hits against a liner from below. As a result of that in all materials are generated additional waves of compression with the complex form and amplitude of several tens of Kilobar, which causes non-monotone behavior of the integral impulse. Since a liner and screen in this case play the different roles, was analyzed how they affect upon the result separately and influence of the secondary impacts as well.

For this purpose were performed numerical experiments, in which an algorithm was changed in such a way that, starting from a moment of the separation a liner is forced to speed up upwards with deliberately high velocity to exclude the secondary impacts of a screen against a liner. In those cases when separation does not occur in a 'natural' way a liner is conducted upwards forcedly during the loading waves meet on their contact surface.

After separation of the liner a screen only is a dynamic piston, making an 'undersetting' influence upon detonation products. An integral impulse calculated in these numerical experiments increases monotonously with increase of screen thickness (Figure 1).

(All figures are presented in Lagrange coordinates for thickness of initiated charge equal 10 mm to show the parameters on determined depth of explosive, where hereinafter must be situated the loaded material). So that the bigger liner thickness, the asymptotically closer these dependences approach to dependences of the impulse from liner thickness for scheme 1, rather then to dependences of the impulse from screen thickness for scheme 2. So in fact a screen itself in the third scheme influences upon parameters of detonation wave as a liner in scheme 1. Smaller efficiency of its influence it is possible to explain that, unlike direct initiation by a liner, not whole

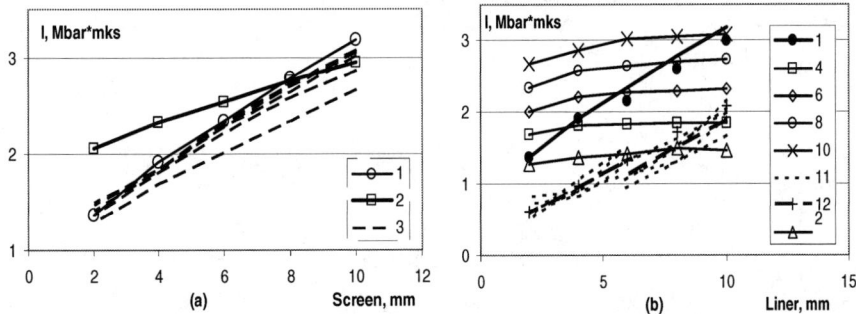

FIGURE 1. Dependences of an impulse in detonation products without secondary impact from thickness of a screen (a), 1 – in scheme 1; 2 – in scheme 2; 3 – in scheme 3 for liner thicknesses 2, 4, 6, 8, 10 mm; and from thickness of a liner (b), 1 – in scheme 1; 2, 4, 6, 8, 10 – in scheme 2 for corresponding screen thicknesses; 11 – impulses of secondary impacts; 12 – model (1) of impulses of secondary impacts.

mass of the screen moves at a moment of initiation, but only its part, which is spanned by shock wave that depends on thicknesses of a liner. So than the thicker a liner, the more a screen influences upon an impulse like a liner. To take into account this fact, it is possible to offer the mass velocity coefficient, for example as the ratio of linear momentum, which a screen is spanned by at passing the shock wave through, to linear momentum, which a screen could have moving as a whole with its mass.

These data would be presented as dependence of an impulse from liner thickness (Fig. 1 b) and then it is seen that in fact liner thickness influences upon an impulse weakly. It is interesting to note that each dependence asymptotically tends to the value, which in scheme 1 is created by a liner, whose thickness is equal to thickness of a screen. Such presentation of data allows to estimate a contribution of repeated impacts of a screen against a liner to an integral impulse. They are computed as difference between the full impulse and the impulse got after separation of the liner and it appears that practically they do not depend on screen thickness. This can be explained thereby that duration of the repeated waves of compression is defined by liner thickness (it is the less, the less screen thickness in examined cases with separated liner is). The regression equation

$$I_{add} = 0,2257 + 0,1863 * \Delta_{liner} \qquad (1)$$

describes influence of a liner on this additional part of integral impulse with high enough reliability.

In cases of loading by third scheme, when there is no separation of a screen (its thickness is less then thickness of a liner), it is possible to consider non-separated liner and a screen together as a united piston with total thickness. Dependence of an impulse from total thickness (i.e. as united piston), is shown on Figure 2.

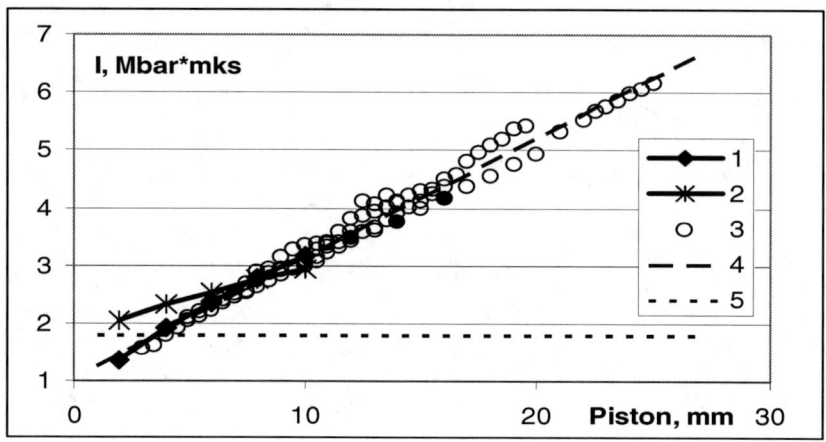

FIGURE 2. Dependences of an impulse in the detonation wave from total thickness of dynamic piston, 1 – in scheme 1; 2 – in scheme 2; 3 – in scheme 3 for liner thicknesses 2, 4, 6, 8, 10 mm; 4 – model (2); 5 – an impulse in regular initiated charge at depth of 40 mm.

This fact (a liner is a united piston) allows to represent results of initiation by direct impact by a liner (according to scheme 1) on this Figure 2 also. If to represent the full impulses made by third scheme as dependence on sum of thicknesses (as united pistol) on the same Figure 2 it is possible to draw a common conclusion. It turned out that in scheme 1 and 3 where a liner is used dependence of integral impulse in detonation wave from united thickness of a liner and a screen in scheme 3 (despite there are constituents with different nature) and from liner thickness in scheme 1 is described by the common equation. The regression equation describing this dependence (line 4) is

$$I = 1,039 + 0,2089 * \Delta_{sum} \tag{2}$$

Dependence from thickness of a screen (scheme 2, line 2) is described by different equation because of differences in nature of shock-wave flow. If to compare them we see that value of an integral impulse in case of loading by attached charge through a screen (scheme 2) is equal to one in case of impact by a liner (scheme 1) if a screen and a liner have the same thickness 8 mm. An impulse acting into total amount of regular initiated explosive at depth of 40 mm (30 mm + 10 mm) is shown as dotted line 5 for comparison.

It is necessary to note that there are some difficulties in realization such numerical experiments. Namely: in the cases without natural separation of a screen from a liner, attempts to do such a separation forcedly without taking into account characteristics of shock-wave flow could cause results to be wrong. In this case this is principal difficulty because contra directional waves of unloading on contact surface 'liner-screen' have considerable duration and it is difficult to define a moment of forced separation.

So are drawn the following conclusions.

1. The mentioned schemes allow to generate detonation wave with form different from one that exists in regular initiated charge.
2. A liner in scheme 1 and liner with screen in scheme 3 act on detonation products the same manner as "dynamic pistons". So they enlarge an impulse in detonation wave proportionally to thickness of piston by the same law.
3. According to [3] each millimeter of explosive thickness causes increasing of an impulse on 0.0447. So it is possible to say that each piston's millimeter is equivalent nearly 4 millimeters of additional explosive.

REFERENCES

1. M.L. Wilkins, Numerical methods in hydrodynamics [in Russian translation], Moscow, 1967.
2. Ch. Mader, Numerical modeling of detonation [in Russian translation], Moscow, 1985.
3. S.L. Neverov, Estimate of explosive use efficiency at different initiation scenarios using statistical processing of computational solution results. XIII Combustion and Explosion Symposium, Chernogolovka, 2005.
   (http://orel3.rsl.ru/nettext/russian/gor_i_vzr/content/Neverov.pdf)
4. Explosive physics [under edition of K.P. Stanyukovich], Moscow, 1975.

Thermoplastic Explosive Compositions on the Base of Hexanitrohexaazaisowurtzitane

V.P. Ilyin, S.P. Smirnov, E.V. Kolganov, Yu.G. Pechenev

FGUP «GosNII"Kristall"», Dzerzhinsk, 606007 Nizhegorodskaya obl., Russia

Abstract. Hexanitrohexaazaisowurtzitane is an azostructural compound known as CL-20. We performed a series of experiments with CL-20 synthesized in Russia to evaluate the possibility to use it in pressed high explosive compositions. We used it in thermoplastic compositions both with an inert binder and energetic binder. The compositions were conventionally named CL-20И and CL-20A. It was determined that the thermoplastic compositions had the most high detonation parameters and a level of sensitivity to mechanical effects acceptable to allow their processing. Their detonation characteristics were compared with that of some known foreign compositions based on CL-20.

Keywords: explosives, CL-20, hexanitrohexaazaisowurtzitane, thermoplastic compositions, explosive characteristics.

Among investigations of synthesis of explosives, which meet different operational requirements, particular interest is paid to the works concerning new high explosives with significantly higher energetic characteristics than that of HMX, capabilities of which practically completely realized both in high explosive and in solid propellant compositions.

From the achievements of last decades it is worth to pay attention to Nilsen synthesis of skeleton azo compound – hexanitrohexaazaisowurtzitane (HNIW), well known as CL-20, in 1987 in the USA. CL-20 molecular structure is a space isowurtzitane lattice in which to each bridge atom of nitrogen one nitro group is attached. Without going into details we note that HNIW synthesis is a labour-consuming process usually consisting of glyoxal condensation with benzyl amine followed by debenzylation through catalytic hydration in the presence of acetic anhydride followed by nitration in two steps. The most successful in practical realization of CL-20 production process were Thiokol Corporation (USA), SNPE (France) and Bofors Explosives AB (Sweden).

Owing to flexible conformational structure CL-20 molecules are capable to form a number of rather stable polymorphs. On the base of investigations of CL-20 six nitro groups orientation symmetry relative to stable isowurtzitane skeleton and existing steric difficulties foreign researchers assume that there exist 12 conformation forms. From derived pure polymorphs there were reliably identified four polymorphs of CL-20 (Table 1) among which ε-form with mono crystal theoretical maximum density of 2.044 g/cc was of practical interest. According to the available results of foreign investigations real density of CL-20 ε-polymorph particles measured by flotation

CP849, *Zababakhin Scientific Talks - 2005*,
edited by E. N. Avrorin and V. A. Simonenko
© 2006 American Institute of Physics 0-7354-0345-7/06/$23.00

TABLE 1. CL-20 Polymorphs Properties.

Polymorphs	Space groups	ρ, g/cc	ΔH_f, kcal/mol	Increase of thermo-dynamic stability
β	Pb2$_1$a	1.99	-	$\downarrow$
α (hydrate)	Pbca	1.97	82	
γ	P2$_1$/n	1.92	-	
ε	P2$_1$/n	2.044	95	

method is within the range of 2.033-2.038 g/cc. It is considered that this CL-20 polymorph (monoclinic space group P2$_1$/n) is the most stable thermodynamically in normal conditions.

CL-20 tendency to have polymorphs is its significant disadvantage. Non-reversible transformations of ε-polymorph to γ-polymorph were registered both at high (136-140°C, 152-183°C) and at rather low temperatures (55-58°C, 63-65°C) depending on crystals sizes, their defects, presence or absence of a solvent, rate of heating and time of standing. Such transformations, a polymorph kind and particle size depend also on conditions of CL-20 crystallization from saturated solutions: temperature, nature and purity of used solvents and precipitants. It is worth to note that CL-20 α-polymorph exists only as crystalline hydrate and a number of researchers found some amount of hardly removable water even in ε-polymorph.

CL-20 is rather thermostable: active solid phase self-increasing decomposition of its γ-polymorph is observed within the temperature range of 225-235°C and it is nearly 30-35°C lower than that of HMX. At low pressures CL-20 flameless combustion is observed with nearly 700°C temperature and formation of structural solid residue.

Characteristic properties of this explosive are high detonation velocity and pressure in combination with high crystal density and favorable enthalpy of formation (Table 2).

TABLE 2. Comparative calculated properties of PETN, RDX, HMX and CL-20.

Name	Formula	α	ρ, g/cc	ΔH_f, kcal/mol	D, m/s	Q, kcal/kg	Pdet, kbar
PETN	(structure)	0.860	1.778	-126	8200	1390	290
RDX	(structure)	0.667	1.820	16	8760	1320	326
HMX	(structure)	0.667	1.904	20	9000	1340	364
CL-20	(structure)	0.800	2.044	95	9530	1430	444

151

TABLE 3. Granulation of tested CL-20 sample.

Particles retained on a screen, %				Passing screen particles, %
250 μ	160 μ	100 μ	50 μ	50 μ
0.4	1.8	36.2	53.0	8.6

TABLE 4. Experimental Parameters Of Tested Explosives.

Name	Real density of crystals, g/cc	Lower level of impact sensitivity,mm	Rate of explosions,%	Lower level of friction sensitivity,MPa	Critical thickness of detonation wave, mm
PETN	1.77	50	100	150	0.22
RDX	1.80	70	72	270	0.48
HMX	1.89	50-70	80	200-250	0.55
CL-20	2.04	<50	80	100	0.20

We carried out a number of experiments to determine the possibility to use CL-20 in high explosive compositions applicable for pressing. In the experiments CL-20 synthesized in Russian Federation was used with the following particle size (Table 3).

At first CL-20 X-ray crystal structure was determined that showed its correspondence with ε-polymorph and then CL-20 sensitivity to external mechanical effects.

The results in Table 4 show, that real crystal density of this CL-20 sample corresponds to that of its ε-polymorph too. Levels of sensitivity to external mechanical effects and of detonation wave critical thickness are close to that of PETN and acceptable to process CL-20 with observing safety rules. At the same time CL-20 is a stable explosive and compatible with usual components of compound solid propellants and explosive compositions.

On the base of CL-20 thermoplastic compositions were prepared both with an inert binder (acrylic polymer and wax) and an energetic binder (TNT and mixed vinyl polymer). The compositions were named CL-20И and CL-20A. CL-20И was prepared using water-emulsion process and CL-20A by water-suspension process with an organic solvent. The results of tests for critical thickness of detonation wave, sensitivity to mechanical effects of the compositions, their pressing ability at different temperatures, detonation velocity of charges with 10mm diameters are shown in Table 5.

It is seen that the model thermoplastic compositions have at present the highest detonation parameters and sensitivity to mechanical effects acceptable for processing. And it is worth to note that detonation velocities of CL-20И and CL-20A are evidently not ultimate because they were determined with the charges of small diameters. In this connection the calculations were made of these compositions detonation velocities in the charges with obtained densities taking into account real maximum values and calculations of amour penetration (L) and acceleration impulse (η) relating to the most powerful foreign thermoplastic composition LX-14 on the base of HMX (Table 6).

TABLE 5. Experimental values of CL-20 model thermoplastic compositions characteristics.

Composi-tion name	Content of components %	Lower level of impact sensitivity, mm	Lower level of friction sensitivity, MPa	Density of charges at $P_{sp.}$=200MPa and pressing temperature g/cc	Critical thickness of detonation wave, mm	Detonation velocity of Ø10mm charges (at density, g/cc), m/s
CL-20И	CL-20 – 98 Inert binder – 2	70	200	20°C–1.875 (6.5) 95°C-1.966 (1.9)	0.20	9170 (1.966)
CL-20A	CL-20 – 98 Energetic binder – 2	50	150	20°C-1.850 (8.7) 95°C-1.999 (1.3)	0.19	9230 (1.999)

TABLE 6. Comparative properties of high explosive thermoplastic compositions.

Name	Components %	ρ, g/cc	D, m/s	Q, kcal/kg	$P_{sp.,}$ kbar	Amor penetra-tion L, %	η, % Radial propelling	η, % End propelling
LX-14	HMX – 95.5 Estane – 4.5	1.835	8790	1210	315	100	100	100
CL-20И	CL-20 – 98 Inert binder – 2	1.966	9190	1380	380	111	107	108
CL-20A	CL-20 – 98 Energetic binder – 2	1.999	9350	1390	400	115	109	110
LX-19	CL-20 – 95.8 Estane – 4.2	1.929	9130	1350	360	108	105	106
PATHX-2	CL-20 – 95 Estane – 5	1.923	9120	1330	355	108	104	105
PBXC-19	CL-20 – 95 Ethyleneviny l acetate – 5	1.896	9080	1330	345	106	103	104

The results of Table 6 confirm the advantages of the model compositions CL-20И and CL-20A, as well as some foreign compositions based on CL-20, according to their detonation properties in comparison with LX-14. Substitution of HMX with CL-20 (kind and quantity of a binder, conditions of making charges leaving without change) increases detonation velocities of the charges by 300-350m/s and explosion heat by 100-150kkal. In every case of substitution it is necessary to take into account a sufficiency of CL-20 energetic advantages as well as available information about its operational characteristics when formulating high-energy compositions.

153

Experimental Determination of Thermodynamic Parameters for "Explosive Products + Air" Mixture in Closed Chambers

E.E. Lin, A.V. Sirenko

Russian Federal Nuclear Center – VNIIEF, 607190, Sarov, Nizhni Novgorod Region, Russia

Abstract. We study the influence of explosive products (EP) burnout under aerial environment on thermodynamic parameters of gases mixture in closed chambers with plasticized PETN based high explosive (HE). The results are adduced for measurements of steady-state pressure $\bar{p}$ and sound velocity $\bar{c}$ in "EP + air" mixture, when varying the mean charge density in the range of $\rho_0 = 0.2 - 11$ kg/m³. The obtained data can be useful for solving tasks connected with the application of explosives in physical experiments.

Key words: explosive products, air, closed chamber, steady-state pressure, sound velocity, polytropic index.

PACS: 82.40Py

INTRODUCTION

Early in [1] we determined experimentally the thermodynamic parameters of explosive products for plasticized PETN based high explosive in closed evacuated chambers when varying the mean charge density in the range $\rho_0 = 0.75 - 11$ kg/m³. It has been stated that the value of steady-state (hydrostatic) pressure p grows proportionally to ρ_0 in accordance with empirical dependence

$$\bar{p} = 0.8\rho_0, \qquad (1)$$

where $\bar{p}$ is measured in MPa. The stated sound velocity does not depend on ρ_0 in the studied range and is equal to $\bar{c} = 1$ km/s. These results make it possible to determine the value of an effective polytropic index for EP at the stage of their intense expansion in 10^2-10^3 times as $k_0 = 1.17 \pm 0.02$. The data obtained in [1] have been used subsequently in [2] for numerical calculations of gas-dynamic flow in the explosive driven shock tube at the quasi-acoustic stage of shock wave motion in air.

It is interesting to study an influence of EP burnout for the considering HE composition in the closed air-filled chambers on thermodynamic parameters of "EP + air" mixture. For this purpose, in the present work the measurements were carried out

CP849, *Zababakhin Scientific Talks - 2005*,
edited by E. N. Avrorin and V. A. Simonenko
© 2006 American Institute of Physics 0-7354-0345-7/06/$23.00

for steady-state pressure p and sound velocity $\bar{c}$ under variation of the mean density of shot loading in the range of $\rho_0 = 0.2 - 11$ kg/m³.

EXPERIMENTAL METHOD AND RESULTS

The same as in [1], the experiments were carried out with the explosive chambers of the two types (see figures 1, 2). The inner diameter of the chamber of the first type was 0.19 m and the length was 0.2 m; the same parameters for the chamber of the second type were 0.09 m and 0.5 m correspondingly. The chambers were filled by air at atmospheric pressure. The pressure oscillograms are presented in figures 3, 4. The steady-state pressure was determined by plotting the middle curve between oscillations caused by circulations of pressure wave between the closed ends of camera. The sound velocity was determined by measurements of circulation period and of amplitudes of pressure wave (see [1]). The dependences $p(\rho_0)$ and $\bar{c}(\rho_0)$ are adduced in figures 5, 6.

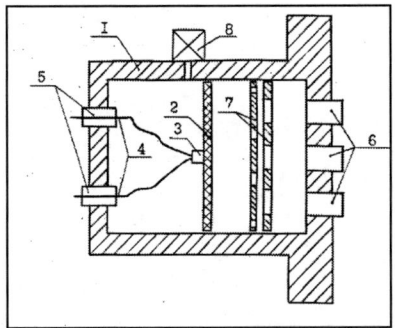

Figure 1. The scheme of experiments with the camera of type 1: 1 – the camera body, 2 – plasticized shot, 3 – electric detonator, 4 – isolators, 5 – electrodes, 6- piezoelectric gauges, 7 – throttles, 8- valves.

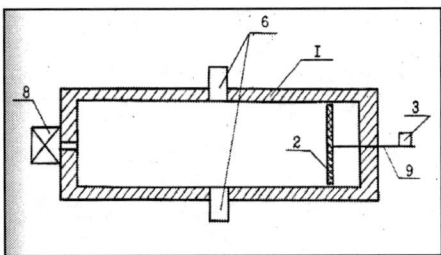

Figure 2. The scheme of experiments with the camera of type 2: the positions 1-3, 6 and 8 are similar to those in figure 1, 9 – detonating fuse.

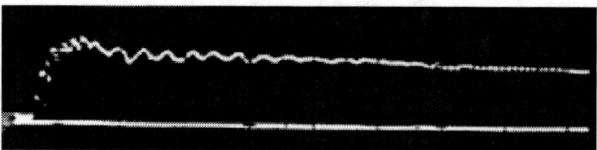

Figure 3. Pressure oscillogram for camera of type 1 (the duration is 10 ms).

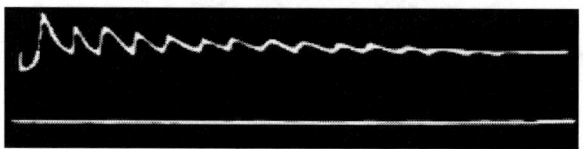

Figure 4. Pressure oscillogram for camera of type 2 (the duration is 10 ms).

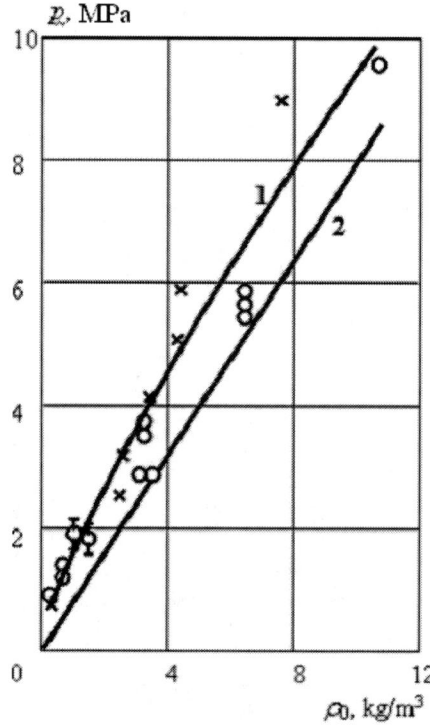

Figure 5. The dependence of steady-state pressure on the density of shot loading: 1 – mixture (EP + air), the circles correspond to the camera of type 1, the crosses correspond to camera of type 2; 2 – "pure" EP [1]

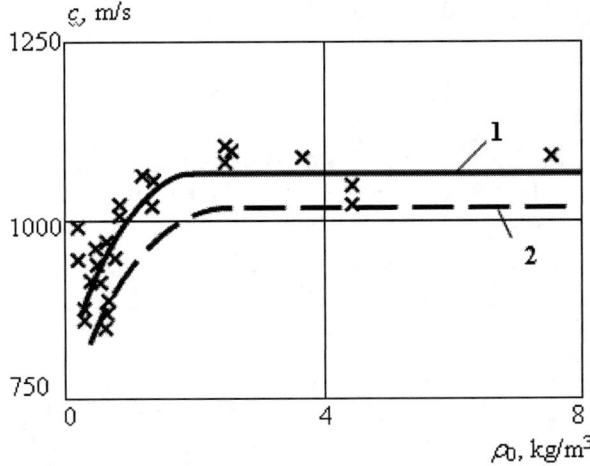

Figure 6. The dependence of sound velocity on the density of shot loading: 1 – the measurements in camera of type 2, 2 – calculations.

At the mean charge density, which is less or is approximately equal to initial density of air in chamber, the hydrostatic pressure in the studying mixture exceeds the hydrostatic pressure in "pure EP" (i.e., in evacuated chamber) approximately by the factor of 2. The discrepancy decreases with the increase of ρ_0 and at $\rho_0 = 11$ kg/m^3 is near 15 %, that does not exceeds the limits of measurements error. Most probably, the observed exceeding of steady-state pressure in the mixture "EP + air" in comparison with that parameter for pure EP is connected with EP burning down under environment of hot air oxygen, which was heated by blast wave. As a relative content of oxygen in the mixture is decreased with the augmentation of density of shot loading, a relative part of additional energy release owing to EP burning down is decreased. Therefore, the relative discrepancy of pressure values in the mixture under consideration and in pure EP is reduced.

INTERPRETATION OF RESULTS

An analysis shows that in the studied range of the mean charge density the dependence $p(\rho_0)$ in air-filled chambers can be presented in a form, which is similar to the dependence for the volume-distributed shot of plasticized PETN based HE [3]:

$$\bar{p} = 1.4\rho_{00}(\rho_0/\rho_{00})^{0.8},$$ \hfill (2)

where $\rho_{00} = 1.3$ kg/m^3 is an initial density of air (the pressure is measured in MPa). To determine the effective polytropic index k we use the relationship for steady-state pressure $\bar{p} \approx (k-1)\rho_0 E^*$, where E^* is specific energy release with taking into account of EP burning down. For pure EP we have the relationship $\bar{p} \approx (k_0 - 1)\rho_0 Q$, where Q is explosion heat for the HE. If we divide these expressions when using corresponding empirical dependences (1) and (2), we obtain the following relationship for k and k_0:

$$(k-1) \approx (k_0 - 1)1{,}75(\rho_{00}/\rho_0)^{0,2} \frac{E^*}{Q}. \tag{3}$$

It is evident that at the small density of shot loading $\rho_0 \ll \rho_{00}$ the energy release E^* is approximately equal to heat Q^* of HE combustion, and at $\rho_0 \gg \rho_{00}$ $E^* \approx Q$. Let us set the relationship E^*/Q in the following form, which conforms to these limiting conditions:

$$\frac{E^*}{Q} \approx 1 + \frac{Q^* - Q}{Q} \frac{1}{1 + (\rho_0/\rho_{00})} \tag{4}$$

From the relationships (3) and (4) we obtain that in the studied range of ρ_0 the value of effective polytropic index of the mixture decreases from 1.31 to 1.19.

The dependence for the sound velocity $\bar{c} = ((k-1)kE^*)^{1/2}$, which was calculated on the basis of expressions (3) and (4), has been dotted in the figure 6. The character of calculated curve is similar to the experimental one. The largest quantity discrepancy in the order of 20 % takes place at $\rho_0 < 1$ kg/m^3. That is connected, apparently, with inexactitude of approximation for energy release, given by the expression (4). The discrepancy decreases with the increase of the density of shot loading, and at $\rho_0 > 1$ kg/m^3 it does not exceed 10 %. The established mutual agreement for results of measuring pressure and sound velocity confirms their reliability.

The obtained data can be useful to solve the tasks applied for the application of explosives in physical experiments.

REFERENCES

1. E. E. Lin, A. V. Sirenko, and A. I. Funtikov, *Combust., Expl., Shock Waves*, **16**, 472-474 (1980).
2. V. E. Shemarulin, V. Yu. Mel'tsas, G. F. Kopytov, et. al., *Combust., Expl., Shock Waves*, **18**, 478-482 (1982).
3. A. F. Baryshnikov, V. Yu. Kainov, E. E. Lin, et. al., *Combust., Expl., Shock Waves*, **15**, 827-828 (1979).

Simulation of the Combustion of Melted High Energy Materials

V.E. Zarko[1], L.K. Gusachenko[1], A.D. Rychkov[2]

[1]*Institute of Chemical Kinetics and Combustion, Novosibirsk 630090, Russia*
[2]*Institute of Computational Technologies, Novosibirsk 630090, Russia*

Abstract. The types for the combustion instability of melted high energy materials (HEMs) are discussed. It is shown that application of the Zeldovich-Novozhilov model in the domain beyond the stability boundary is not justified. New type of the combustion instability of HEM is revealed which results from instability of subsurface reaction zone. The finding is made on the basis of both numerical and analytical calculations. This instability type allows making plausible explanation of the phenomena of "cross-surface waves" and "negative erosion" under tangential blowing the HEM.

Keywords: subsurface heat release, instability, intrinsic turbulence, negative erosion.
PACS: 47.70.Pq, 82.33.Vx

INTRODUCTION

Available experimental data [1-5] show that typical high energy materials (HEMs) have pronounced exothermic reactions in the condensed phase. This results in specific type of the combustion instability which is not described by the Zeldovich-Novozhilov phenomenological combustion model. Consequently, rather simplified approaches based on ideas of classical theory of HEMs combustion by Belyaev and Zeldovich have to be modified. The paper deals with theoretical analysis of the combustion instability arising due to thermal instability of surface reaction zone with developed exothermic reactions.

PECULIARITIES OF SIMULATION WITHIN INSTABILITY DOMAIN

Transient combustion of homogenized HEMs under conditions of varied in time "external" parameters (pressure, heat flux, gas flow velocity) can be satisfactory described by the phenomenological combustion model [6, 7]. It has to be noted that the model is not eligible for description of the burning rate auto oscillations if the parameters of the burning system belong to instability domain of steady-state combustion regimes. For eligibility, it is necessary that oscillations arise and exist over the whole surface of the burning HEM. Such situation can be formally described within the framework of one-dimension approach [8-10] exploited by the phenomenological model. However, real HEMs do not exhibit uniform pattern of

CP849, *Zababakhin Scientific Talks - 2005*,
edited by E. N. Avrorin and V. A. Simonenko

condensed phase reacting because of numerous reasons for existence of non-uniform distribution of component concentration, temperature disturbances, etc., on the HEM burning surface. As a result, the real HEM combustion is characterized by arbitrary values of the oscillation phase for neighboring sites on the burning surface. This leads to non-uniform gas release over the surface leading to creation of "intrinsic turbulence" near the surface. The characteristic time of the gas flow oscillations is of the order of the magnitude of burnout time for preheat layer in the condensed phase. Thus, the burning rate becomes dependent not only on the instantaneous values of pressure and surface temperature gradient (as it is in the Zeldovich-Novozhilov model) but also on the parameters of the gas flow above the burning surface. It means that now it becomes necessary to develop multy dimensional model describing processes in "quasi-stationary" gas phase.

LOCALIZATION OF THE BURNING RATE CONTROL ZONE

The burning rate control zone can be localized in different parts of the combustion wave depending on the type of HEM and combustion conditions [11]. The control zone localization site can be determined experimentally via analyzing the measured thermal profile. When processing the thermocouple data, one may find the values of surface temperature T_s and temperature gradient from the gas phase side $(\partial T/\partial x)_{gs}$. Note that in stationary combustion wave the heat needed to supply into the condensed phase equals $\Delta H = c\,(T_s - T_o) + Q_m$, where c is the specific heat and Q_m is the melting heat. The ratio $\varepsilon = \lambda_g \left(\dfrac{\partial T}{\partial x}\right)_{gs} / m\Delta H$ gives estimation for the contribution of gas phase heat feedback into the heat balance on the burning surface (m is the mass burning rate). Consequently, the contribution of the condensed phase heat release into the heat balance equals $(1 - \varepsilon)$. When $\varepsilon \ll 1$, the burning rate control zone belongs to the condensed phase exothermic reactions. If one takes into consideration the evaporation on the burning surface, the above inequality becomes stronger. With valid condition $\varepsilon \ll 1$ the stationary burning rate can be easily calculated by simple Zeldovich's formula:

$$(m\Delta H)^2 = 2\lambda_c Q \frac{RT_s^2}{E} W(T_s) \qquad (1)$$

Here Q, E, W(T) are the condensed phase reaction heat, activation energy and rate of the subsurface reaction, respectively. Analysis of micro thermocouple data [1-5] allows making conclusion about localization of the burning rate control zone in the condensed phase for neat RDX, HMX, ADN and mixtures RDX/HMX + HTPB or RDX/HMX + GAP. According to experiment, the ε – value for listed HEMs comprised $0.1 - 0.3$ at atmospheric pressure and became even less with pressure rise. In the case of HNF [12] the data are available only for low pressures. The value of ε equals 0.6 at P = 0.04 MPa and 0.4 at P=0.1 MPa. It is expected that at elevated pressures the contribution of condensed phase reactions will become higher. Similar conclusion can be made in the case of CL-20 where experiments at P = 0.3 MPa [13] demonstrated rather high contribution of exothermic reactions in the condensed phase into heat balance on the burning surface.

INSTABILITY OF SUBSURFACE REACTION ZONE IN THE CASE OF CONDENSED PHASE BURNING RATE CONTROL ZONE

Earlier the authors formulated the model of transient combustion of melted HEMs [14]. The combustion model was intended to describe the propagation of the combustion wave in RDX and HMX. For those HEMs the burning rate control zone locates in the gas phase at low pressures and in the condensed phase at high pressures [1, 2]. In the problem formulation the first order exothermic reaction have been considered in the condensed phase and two exothermic reactions of varied order in the gas phase. Within the framework of that model the calculation showed establishment of stationary combustion regimes at low pressures and unstable combustion behavior at elevated pressures. The latter was characterized by a sort of thermal explosion in subsurface reaction layer. Special analytical study [15] confirmed that the revealed in numerical calculations combustion instability is not caused by the calculation procedure. Instead, it reflects physical phenomenon of unstable thermal behavior which arises at elevated pressures due to strengthening the condensed phase reactions and formation of maximum on the temperature profile.

The instability boundary corresponds to the zero temperature gradient on the burning surface from the condensed phase side. In this case the evaporation of non-decomposed HEM on the reacting surface proceeds on the expense of heat flux from the gas phase:

$$\lambda_g \left(\frac{\partial T}{\partial x} \right)_{gs} = m(1-\alpha)L \qquad (2)$$

Here α is the mass fraction of decomposed HEM, L is the HEM latent heat of evaporation. When expression (2) is valid, all expenses on heating the condensed phase from T_o to T_s are covered by the heat release in the condensed phase. It is assumed that the products of decomposition exist mainly in the gaseous form and they filtrate or diffuse through subsurface layer without heat losses.

Let us also assume that the instability behaves in accordance with Le Chatelier's principle and counteracts against the cause, which creates it. This prevents further developing the instability and provides conditions for preservation of the dominant role of the condensed phase reactions in the heat balance on the burning surface. As a consequence, one may expect appearance of small amplitude high frequency burning rate oscillations. They may not be synchronized over the whole burning surface and can exist in the form of intermittent reaction spots or surface crossing combustion waves. It has been reported in [16-19] that such kind oscillations of the combustion wave were observed for several types of HEMs and they existed also in the stable combustion domain calculated by the Zeldovich-Novozhilov approach. Note that "surface crossing" combustion waves do not transfer heat, mass and information. The oscillations of the parameters differ in various points of the surface by the phase value and observed "wave" is indeed the wave of the phase change. As is known, the velocity of the phase change wave can exceed the speed of light that illustrates the fact that this sort of wave does not transfer any real matter.

161

The considerations related to non-synchronous oscillations of the combustion wave in different points of the burning surface allow formulating plausible mechanism of the combustion stabilization with negative feedback (as follows from Le Chatelier's principle). Namely, one may assume that non-synchronous (disordered) oscillations of the gas efflux generate turbulence in the gas phase, which enhances the effective magnitude of heat- and mass-transfer. Consequently, the heat feedback from the gas phase also increases.

Increased transport properties of gas (and heat feedback) can be qualitatively evaluated. Let the reaction of zero order proceeds in the liquid subsurface layer: $\rho_c (\partial\alpha / \partial t) = W(T)$, where α is the degree of HEM decomposition. With pressure rise the penetrated through the burning surface heat flux q_{cs} becomes less and starting from the moment, when $q_{cs} = 0$, the oscillations of the burning rate arise with characteristic time $t_{osc} = \rho_c/W(T_s)$. These oscillations, disordered in phase, generate near the burning surface the turbulence characterized by the mixing coefficient $D_{eff} = const (\Delta v_g)^2 t_{osc}$. Here Δv_g is the fluctuation of the flow velocity of gas released from the surface. In this case the negative feedback forms, which restricts the amplitude of auto-oscillations. With occasional enhancement of the mass burning rate Δm (and Δv_g) the mixing in the gas phase increases and gas heat feedback enhances. This results in increasing the mean burning rate and restricting the Δm value.

Another possible mechanism of the combustion stabilization relates to existence of the solubility limit for gaseous decomposition products in the liquid subsurface layer. When reaching the limit, the explosion-type gas release occurs. In this case part of heat is spent on the substance state changing while the rest of heat is lost from the subsurface reaction zone in the form of enthalpy of released gas or condensed dispersed products, if dispersion takes place.

Instability of combustion under emerging maximum on the temperature profile in subsurface layer is caused by non-stationary processes in the reaction zone and could not be described by the Zeldovich-Novozhilov model, which assumes the quasi stationary behavior of that zone. The results of [14, 15] oblige one to evaluate critically the statements of other works dealing with transient HEM combustion modeling and taking into account the effects of evaporation and subsurface exothermic reactions.

"NEGATIVE EROSION" AS A RESULT OF THE COMBUSTION INSTABILITY

So called HEM erosion combustion is stationary combustion under action of turbulence created by tangential blowing the burning surface. Let us assume that near the combustion stability limit the described above "intrinsic turbulence" takes place. Combustion generated turbulence near the burning surface has been previously discussed in [20-21]. However, that type turbulence was caused by the gas phase disturbances appearing due to non-uniform distribution of the components over the composite propellant surface.

Let us now assume that tangential gas flow destroys the "intrinsic turbulence" in the gas layer adjusting the burning surface. Consequently, the contribution of this part

162

of heat feedback to the burning surface will decrease with the velocity of tangential gas flow. At the same time the contribution of heat feedback due to "external turbulence" created by growing in velocity tangential gas flow will increase. Therefore, with increase of velocity of tangential blowing the total heat feedback to the burning surface will pass through the minimum and burning rate will first decrease and then increase. Thus, the considerations related to "intrinsic turbulence" allow proposing alternative mechanism of negative erosion.

REFERENCES

1. A.A. Zenin. *Journal Propulsion and Power*, 1995, V. 11, No. 4.
2. A.A. Zenin, V.M. Puchkov, S.V. Finyakov. *Combustion, Explosion, and Shock Waves*, V. 34, No. 2, 1998.
3. A.A. Zenin. AIAA-99-0595.
4. A.A. Zenin, S.V. Finyakov. In: Proceedings of 31st international annual conference of ICT, Karlsruhe, FRG, 2000, P 132.
5. A.A. Zenin, S.V. Finyakov. Proceedings of 32st international annual conference of ICT, Karlsruhe, FRG, 2001, V 8.
6. Ya.B. Zeldovich. *Zhurnal Experimentalnoi i Teoreticheskoi Fiziki*, 1942.V.12, p.498.
7. B.V. Novozhilov. The nonstationary burning of solid rocket propellants. Nauka, Moscow, 1973.
8 V.I. Zemskih, B.V. Novozhilov, A.V. Timchenko. *Khimicheskaya Fizika*, 1990, V. 9, No. 12.
9. B.V. Novozhilov. *Khimicheskaya Fizika*, 2004, V. 23, No. 5, p. 68.
10. A.A. Belyaev, Z.I. Kaganova, B.V. Novozhilov. *Combustion, Explosion, and Shock Waves*, 2004, V. 40, No. 4.
11. L.K. Gusachenko, V.E. Zarko. Combustion, Explosion, and Shock Waves, 2005, V. 41, No. 1.
12. V.P. Sinditskii, V.V. Serushkin, S.A. Filatov, V.Yu. Egorshev / Combustion of energetic materials, edited by K.K. Kuo, L.T. DeLuca, Begell House Inc., New York, 2002, pp. 576
13. V.P. Sinditskii, V.Yu. Egorshev, M.V. Berezin, V.V. Serushkin, Yu.M. Milekhin, S.A. Gusev, A.A. Matveev. *Khimicheskaya Fizika*, V. 22, No.7, 2003, pp. 69-74.
14. V.E. Zarko, L.K. Gusachenko, A.D. Rychkov / Challenges in Propellants and Combustion - 100 Years after Nobel, Kenneth K. Kuo, ed. Begell house, inc. New York, Wallingford (u.k.), 1997, 1014-1025.
15. L.K. Gusachenko, V.E. Zarko, A.D. Rychkov. *Combustion, Explosion, and Shock Waves*, 1997, V. 33, No. 1.
16. A.V. Anan'ev, A.G. Istratov, Z.V. Kirsanova, V.N. Marshakov, G.V. Melik-Gaikazov / XII Symphosium (Russian) on Combustion and Explosion, part I. RAS, Chernogolovka, 2000.
17. V.N. Marshakov, A.G. Istratov, V.M. Puchkov. *Combustion, Explosion, and Shock Waves*, 2003, V.39, No. 4, pp.100-106.
18. V.N. Marshakov, A.G. Istratov / Progress in Combustion and Detonation / Edited by A.A. Borisov, S.M. Frolov, A.L. Kuhl / Intern. Conf. on Combustion and Detonation. Zel'dovich Memorial, 30.08-03.09 2004, Moscow, Russia, M: TORUS PRESS Ltd.,2004, CD- disk, Paper W2-2, 11 p.
19. V.N. Marshakov / Condensed Systems Combustion. IX Symphosium (Russian) on Combustion and Explosion, Chernogolovka, 1989, pp. 47-51.
20. S.S. Novikov, P.F. Pokhil, Yu.S. Ryazantsev. *Zhurnal Prikladnoi Mechaniki i Tehnicheskoi Fiziki*, 1968, V. 3, pp. 128-133.
21. V.Ya. Zyryanov, V.M. Bolvanenko, O.G. Glotov, Yu.M. Gurenko. *Combustion, Explosion, and Shock Waves*, 1988, V. 24, No.6, pp. 17-26.

Structure, Thermal Properties, and Combustion Behavior of Plasma Synthesized Nano-Aluminum Powders

A. Pivkina[*], D. Ivanov[*], Yu. Frolov[*], S. Mudretsova[†], J. Schoonman[¶]

[*]Semenov Institute of Chemical Physics, Russian Academy of Science,Kosygin st. 4,
1199991 Moscow, Russia
[†]Moscow State University, Department of Chemistry,Lenin Hills, 119992 Moscow, Russia
[¶]Delft University of Technology, Delft Institute for Sustainable Energy, P.O. Box 5045, Delft 2600 GA,
The Netherlands

Abstract. The plasma electro-condensation process was used to synthesize nano-sized aluminum powders. Adding different chemicals modified the physical and chemical properties of these powders. To characterize the nano-sized powders, X-ray diffraction, TEM, BET analyses, and simultaneous TG/DSC analyses were performed. TG/DSC analyses revealed a dramatic degradation of the aluminum oxide layer after storage of the aluminum powder in air for a period of several months. The burning rate of the model solid propellant with nano-sized aluminum was experimentally examined. The combustion behavior of nano-sized aluminum will be presented and will be compared with the combustion behavior of the micron-sized powders.

Keywords: Aluminum Powder, Combustion, DSC, Nano-Particles.
PACS: 65.80.+n; 81.07.-b; 82.33.Vx.

INTRODUCTION

In the past, the transition from the millimeter-scale of particles to the micrometer-scale was of interest, whereas nowadays the nano-sized components are of interest in order to achieve high performance in rocket propellants and pyrotechnics. Nano-aluminum represents an example of such a material. It is expected that in comparison with the micron-sized particles, nano-sized aluminum powder will increase the burning rate, and will considerably decrease the agglomeration enhancing the specific impulse of solid rocket propellants. Metal particle agglomeration at the combustion surface is considered to be a reason for the metal particle ignition delay, chemical incompleteness of combustion, and the total efficiency losses. Experimental data show that the burning rate is strongly dependent on the component's sizes, and could be increased considerably by going down in size to the nano-scale [1-6].

Aluminum nano-particles can be prepared using a variety of techniques, including dynamic gas condensation [7], the cryomelting process [8], and by the plasma explosion process [9]. A thin native aluminum oxide layer on the nanoparticles occurs in air. This passivating oxide layer prevents fast aluminum oxidation during

CP849, *Zababakhin Scientific Talks - 2005*,
edited by E. N. Avrorin and V. A. Simonenko

combustion, as well as agglomeration problems. To increase the reactivity of the oxide layer, different types of doping materials were used (Zn, Cu, Ni, Cr, Zr, alkali-earth metals, etc.). Characterization of the particle diameters, size distributions, oxide layer thicknesses, morphology of the particles, thermal behavior, and combustion parameters is important in predicting performance for specific applications. This paper reports on the experiments and the evaluation of the oxidation of nano-aluminum synthesized by the plasma electro-condensation technique. Different types of nano-aluminum powders are investigated, i.e., barium-doped nano-powder, benzine and silicon rubber stabilized powders. Barium-doped nano-aluminum is studied *as received* and aged during several months. Comparison of thermal and combustion behavior is made to the conventional Al micron-sized powder.

EXPERIMENTAL SECTION

Materials

Micron-sized aluminum powder with a spherical particle of average diameter of 10 μm was used as a reference powder and as a precursor for the plasma-synthesis process, which results in nano-sized Al powder. During the plasma-synthesis process, the argon gas flow delivers the precursor powder with additives to an argon-filled reactor, where aluminum powder evaporates in a high-temperature plasma zone. Details of the plasma-synthesis process have been reported elsewhere [1]. To increase the aluminum chemical reactivity, 1.5at% of barium, which is equivalent to 7.5mass%Ba, was added to the green powder mixture. Subsequent condensation of metal vapor represents nano-sized aluminum powder fabrication. For the current research three types of nano-particles were produced and investigated (Table 1). Samples 1, 2 contain stabilizing compounds, which were added after condensation.

Additionally, nano-aluminum powder containing 7.5% Ba was aged at 69% relative humidity, as listed in Table 2. The thermal behavior of aged samples was monitored continuously throughout the period. Different heating rates were applied to study the thermal behavior of samples 1.2 and 1.3, i.e., $10^0 min^{-1}$ and $2^0 min^{-1}$, respectively. Special experiments were performed to study the process of the nano-aluminum oxidation in the presence of Pt-foil to study the oxidation catalysis in a nano-sized powder.

TABLE 1. Composition, specific surface, and active aluminum content of the investigated *as-received* samples

Sample number	Composition, mass. %	Specific surface, $m^2 g^{-1}$	Active Al content, %
1	92,5% Al+7,5% Ba	22,7	70.0
2	98,5% Al+1.5% Benzine	9,1	88.3
3	98,5% Al+1.5% Silicone rubber	9,1	75.0

TABLE 2. Composition, specific surface, and active aluminum content of the investigated *as-received* samples

Sample number	Composition	Aging period, months
1	Al+7,5% Ba	As-received
1.1	Al+7,5% Ba	3
1.2	Al+7,5% Ba	4
1.3	Al+7,5% Ba	4
1.4	Al+7,5% Ba +Pt-foil	4

Experimental techniques

Wet chemical analysis was performed to find the amount of active aluminum. It is a selective reaction between aluminum and water in the presence of sodium hydroxide. From the volume of hydrogen measured by displacement of water in a burette, the amount of active metal is calculated.

The Brunauer–Emmett–Teller (BET) method was used to determine the specific surface area [10].

X-ray diffraction patterns were obtained at room temperature using a Rigaku "Geigerflex" X-ray diffractometer, employing CuK_α radiation. Samples were finely ground; the diffraction angle of 2θ was scanned at a rate of $2^0 min^{-1}$. The average particle size was calculated by using the Sherrer's equation.

Transmission electron microscopy (TEM) microanalysis (JEM-2000 EX-II) was used as an independent method for characterization of the chemical composition and particle morphology. Analyses were performed at acceleration voltage of 200 kV. Various electron microscopy methods – bright- and dark-field TEM and microdiffraction were applied.

TG/DSC experiments were carried out using a NETZSCH Simultaneous Thermal Analyzer STA 409. Experiments were conducted in static air. Approximately 10-20 mg of sample and reference material (α-Al_2O_3) were placed in separate alundum crucibles. Heating rate studies were conducted; samples were heated from 20 to 1100°C at 2 or 10°C min^{-1}. Calibrations of TG mass, DSC baseline, and temperature were conducted before the experiments.

The experimental investigations of burning rate and combustion of stoichiometric compositions of aluminum with ammonium perchlorate (AP) were conducted using a constant volume bomb pressurized with nitrogen. Samples were pressed tablets made of micron-sized ammonium perchlorate mixed with (i) micron-sized aluminum (sample 4), (ii) aged Ba-doped nano-aluminum, and (iii) benzine or silicon rubber stabilized nano-aluminum. The relative density of pellets was 90-98% of theoretical maximum density. The size of the pellets was 8 mm in diameter and 10-15 mm in length. Three cylindrical tablets were glued together to form one sample to increase the measurement accuracy. Data from three different experiments were averaged. In order to produce linearly burning tablets, the sides of each sample were inhibited with two-part epoxy. An electric match was taped on the top of the ignition mixture, which was painted onto the top of the sample. The calculated average accuracy of the burning rate measurements is ±5%.

The ignition was conducted using an electrically heated Ni-Cr wire set on top of the pellet. The combustion wave propagation was recorded with a video camera through a transparent quartz window. A data acquisition board L-154 was used to collect the measuring data to PC.

RESULTS AND DISCUSSION

Chemical Purity, Particle Size and Morphology

Table 1 presents the BET specific surfaces and the active aluminum contents of the investigated powders. As revealed by X-ray analysis, sample 1 contains the only one crystalline phase – aluminum metal. Considering the immediate reaction of barium to form an oxide layer BaO in air, barium oxide and barium were expected to be present in the sample. Nevertheless, no indications of any crystalline Ba compounds were found, which is shown in Figure 1. The active Al content is 70% (Table 1), therefore, 30% of *as-received* powder is supposed to be amorphous material comprising alumina and barium oxide. The average sizes of Al nanoparticles, determined from XRD line broadening, were 45 nm (111), and 41 nm (200).

The nano-sized Al particles were visualized by TEM. A picture from sample 1 is shown in Figure 2a; and the particle size distribution calculated from twelve TEM images is plotted in Figure 2b. The particles are ideally spherical with a Gauss particle-size distribution revealing an average particle diameter of 43 nm. Assuming that particles within a particular size fraction are coated by a passivating oxide layer of the equivalent thickness (ΔR_0), the ΔR_0 value was found to be 3,3 nm, which is close to the literature value of 2 to 4 nm [11-13]. Note that according to Auger spectroscopy, the real structure of the passivating oxide layer on the surface of an Al particle at room temperature is more intriguing. Under the aluminum oxide layer of about 3 nm thickness, a transition layer of about 10 nm thickness was detected, where the aluminum oxide molecules were found along with the Al metal atoms.

TG-DSC Results

As-Received Powders
TGA scans of *as-received* aluminum nano-powders in static air are shown in Figure 3a. Initial weight loss due to adsorbed moisture of a few percent is seen below 300-400°C. Smaller particles of doped nano-aluminum would be expected to react at temperatures of 450°C, where the slow weight gain starts. The weight gain for sample 1 increases dramatically around 510°C: nano-particles oxidize very fast, in a regime similar to a thermal explosion. Indeed, the oxidation rate (DTG) for sample 1 is quite high, whereas for the samples 2,3,4 this value is much lower, as illustrated Figure 3b. Noticeable oxidation of micron-sized aluminum (sample 4) starts above 800°C.

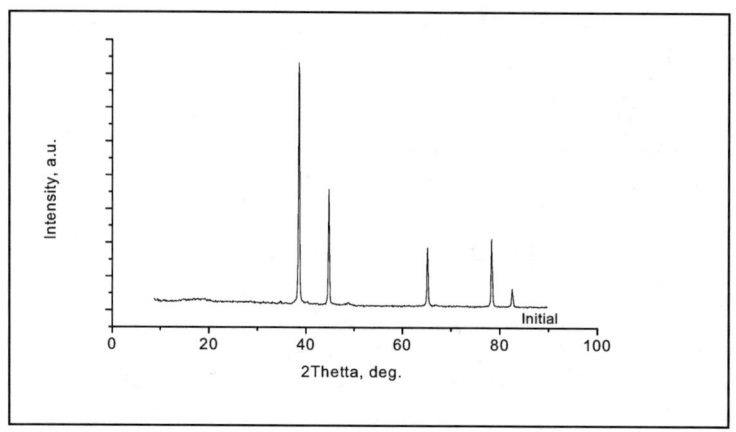

FIGURE 1. X-ray diffraction pattern of sample 1 (nano-aluminum powder doped with 7.5mass% Ba, *as received*)

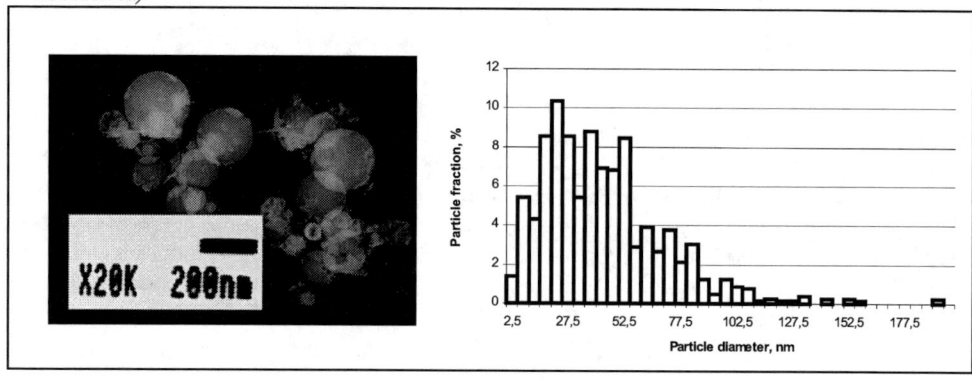

FIGURE 2. TEM images of sample 1 (nano-aluminum powder doped with 7.5mass% Ba, *as received*) with the particle size distribution obtained by TEM.

   Table 3 summarizes TGA results, *i.e.,* the first oxidation onset temperature (T_1), the maximum oxidation rate temperature (T_{max}), the weight gain at the first peak (m_1), the amount of active metal reacting at the first oxidation step (m_{1Al}), the onset temperature for the second oxidation step (T_2), the total weight increase (m^*), and the amount of active metal reacting at temperatures below 1100^0C (m^*_{Al}). The total amount of oxidized aluminum m^*_{Al} of *as-received* nano-aluminum samples is found to be equivalent for *as-received* nanopowders (about 30%), but the thermal behavior of these powders is completely different. Indeed, sample 1 shows the very fast one-step oxidation at 510^0C, followed by a TG-plateau until 950^0C, whereas samples 2,3, continuously react slowly with oxygen upon increasing temperature.

   The fast oxidation of nano-aluminum doped with Ba at 500^0C could be attributed to rapid diffusion of oxygen through the oxide layer after barium oxide's fast exothermical oxidation to form barium peroxide [14]. Arising considerable structural changes along with the additional heat release, may cause the oxide layer to crack at this temperature, thus causing a rapid increase in oxidation.

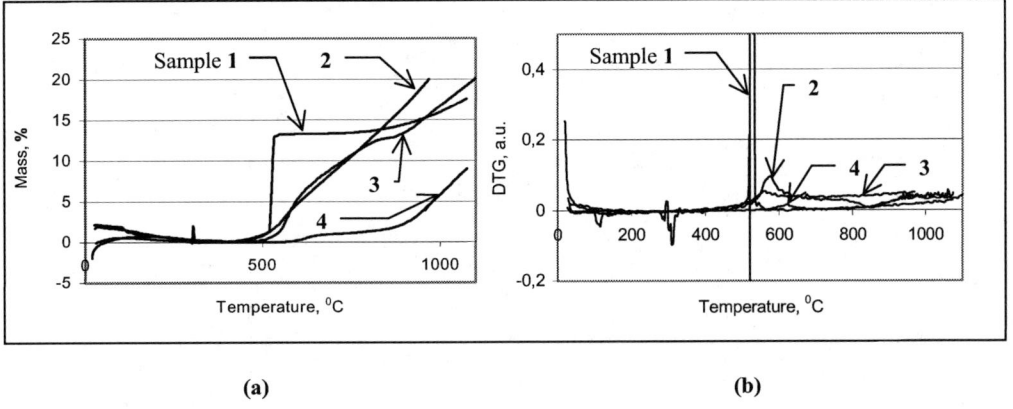

FIGURE 3. TG (a) and DTG (b) traces of *as-received* aluminum powders showing rapid weight gain above 500⁰C for doped powder: sample 1 – Al+7.5% Ba; 2 – Al+1.5% benzine; 3 – Al+1.5% silicon rubber; 4 – micron-sized Al.

Assuming that after the first oxidation step (below the melting point 660⁰C) a "new" passivating oxide layer on the particles of sample 1 has an equivalent thickness (ΔR_1) for the particles within a particular size fraction, the ΔR_1 value was calculated to be of 2,0 nm. Thus the total oxide layer thickness is 5,3 nm.

Aged nano-aluminum doped with Ba

Nano-aluminum powders doped with barium were aged at room temperature and 69% relative humidity in air using a closed can. The aging process of four months was followed using a TG/DSC thermal analysis. Results are compared in Figure 4, where TGA and DTG curves are presented. After 3 months of aging, the doped nano-aluminum (sample 1.1) oxidizes continuously and the total amount of reacted metal (72,89%) is much higher than that of *as-received* sample 1. This indicates that an aluminum oxide layer degradates during storage dramatically, which could be caused by active interaction between barium oxide and humid air components.

Further storage of doped nano-aluminum (sample 1.2) leads to increasing of the oxidation rate, and obvious appearance of the second oxidation step at temperature around 800⁰C, which is shown in Fig. 4b. To study the first oxidations step in more detail the TG/DSC experiment was repeated with a lower heating rate of 2⁰min⁻¹. Two exothermal peaks were extracted from the peak of the first oxidation step; the temperature of 561⁰C corresponds to the bending point, as was found by DDSC analysis. In this sample particles that are larger or have thicker oxide layers oxidize later than smaller ones.

To study the chemical composition of the first oxidation step products, sample 1.2 was heated up to 630⁰C and than cooled very fast. Using a computer code, the crystal phase of the reaction product was found to comprise 59.7% of metal Al, and 40.3% of γ-Al_2O_3. The amounts present were quantified by thermogravimetric and X-ray diffraction methods and are presented in Table 4.

TABLE 3. TGA results of aluminum nano-powder oxidation

Sample number	Heating rate, ^{0} min^{-1}	$T_1 {}^0C$	$T_{max}, {}^0C$	$m_1\%$	$m_{1\,Al}\%$	$T_2 {}^0C$	$m^*, \%$	$m^*_{AL}\%$
1	10	516	527	12,97	20,89	820	17,6	28,35
1.1	10	512	567	14,47	23,31	640	45,25	72,89
1.2	10	530	580	27,79	44,76	666	51,43	82,85
1.3	2	490	540	30,81	49,63	636	52,78	85,02
1.4*	10	495	540	41,13	60,00	-	-	-
2	10	530	566	6,65	8,49	650	24,51	31,28
3	10	560	580	8,12	12,21	831	19,15	28,78
4	10	571	610	1,09	1,65	825	7,09	11,88

* - sample was fast cooled at 630^0C

It seems, therefore, that the aging of Ba-doped nano-aluminum powder in humid atmosphere increases its oxidation rate at temperature around 500^0C. This fact reflects the degradation process of the passivating oxide layer covering the nanoparticles.

TABLE 4. X-ray analysis and TGA results of aged aluminum nano-powder oxidation (sample 1.2)

Temperature, ^0C	Active Al content, %	Amorphous phase content, %	γ-Al$_2$O$_3$, %	α-Al$_2$O$_3$, %
20 (XRD)	70	30	-	-
630 (XRD)	41,8	30	28,2	-
630 (TGA)	38,7	In total 61,3 (crystallinity could not be defined)		

Pt-catalyzed oxidation

As an oxidation catalyst, a platinum foil was used being crushed to form pieces and mixed with the powder to be investigated. Fig. 4a shows TG results of the catalyzed nano-aluminum oxidation (sample 1.4). At temperature of 540^0C a very fast "thermal explosion" reaction is observed; the weight increase being found to be 41.13%, which corresponds to the oxidation of 60% of the active metal. The heat release at this particular temperature is extremely high – 11,4 KJ/g. To study the melting of nanoparticles below 660^0C, the heating process was interrupted at 630^0C, and the sample was fast cooled down to room temperature. No changes in the platinum foil were detected, indicating the catalytic character of the activated oxidation, and the absence of liquid aluminum on the surface of nano-particles at temperatures below the bulk melting temperature.

Obviously, the catalyzed oxidation regime needs to be further investigated in detail, because of the great interest for the solid propellants combustion, where the condensed phase temperature is close to $400-500^0C$.

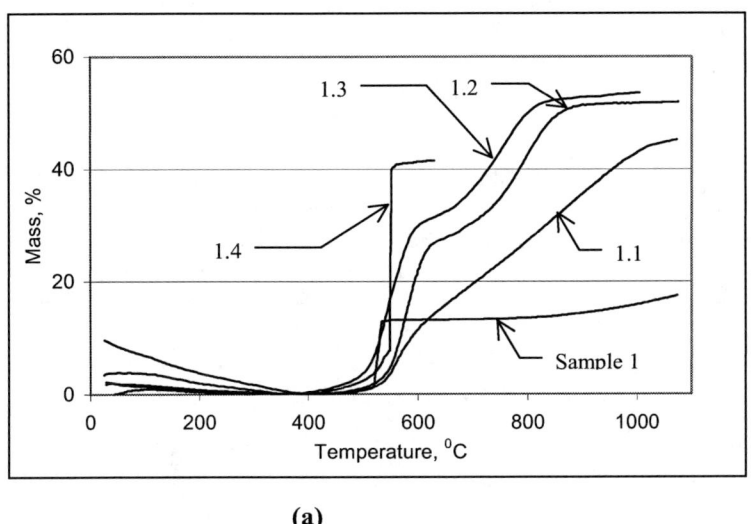

(a)

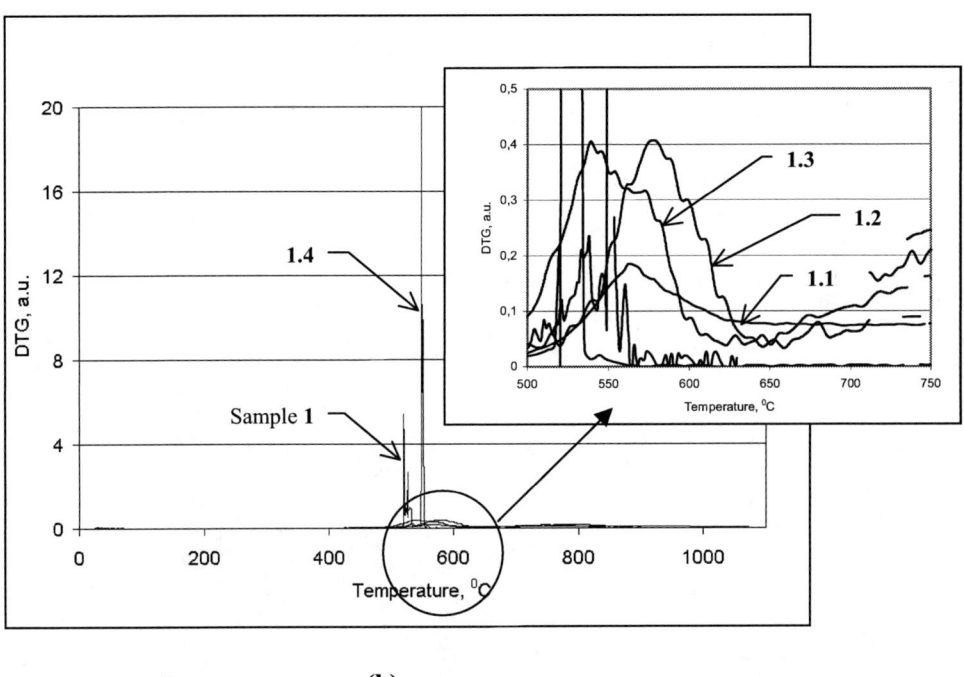

(b)

FIGURE 4. TG (a) and DTG (b) traces of Ba-doped nano-aluminum showing the differences in the oxidation rate of Al+7.5mass% Ba: sample 1 –*as-received*; 1.1 - aged for 3 months; 1.2 - aged for 4 months; 1.3 - aged for 4 months, heating rate is 2^0min^{-1}; 1.4 - aged for 4 months with Pt-foil

Combustion Behavior

The burning rate at initial pressure of 4MPa was measured in a nitrogen atmosphere by two independent techniques, *i.e.,* analyses of the pressure history, and digitized video-images.

Pressure histories are presented in Figure 5. They show that the internal pressure for samples with Ba-doped nano-aluminum is built up four times faster than for samples with nano-aluminum, stabilized with benzine and silicon rubber; and about ten times faster than that of the samples with conventional micron-sized aluminum powder. Fig. 7 shows the averaged U values for investigated samples. The use of Ba-doped nano-aluminum (samples 1.1. and 1.2) instead of micron-sized metal (sample 4) results in a burning rate increasing from 3,7 mm/s to 42,0 mm/s. The use of the benzine (sample 2) and silicon rubber (sample 3) stabilized nanopowders leads to a burning rate increase of about two times, as shown in Figure 6.

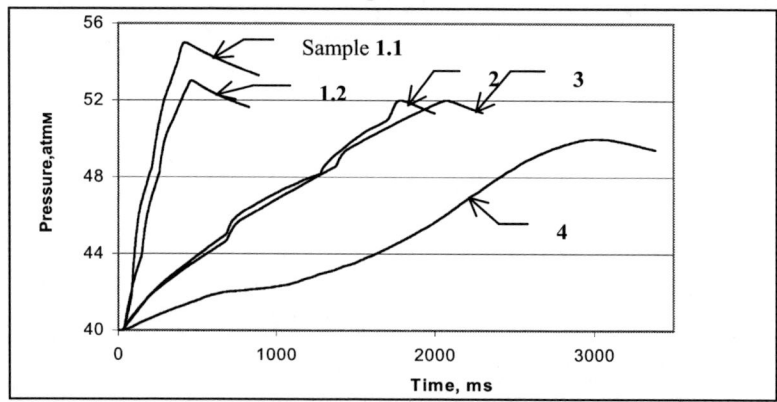

FIGURE 5. Pressure history for the investigated compositions AP/Al with the different Al powders: sample 1.1 – Al+7.5% Ba, aged for 3 months; 1.2 - Al+7.5% Ba, aged for 4 months; 2 – Al+1.5% benzine; 3 – Al+1.5% silicon rubber; 4 – micron-sized Al. The initial pressure of nitrogen is 4MPa.

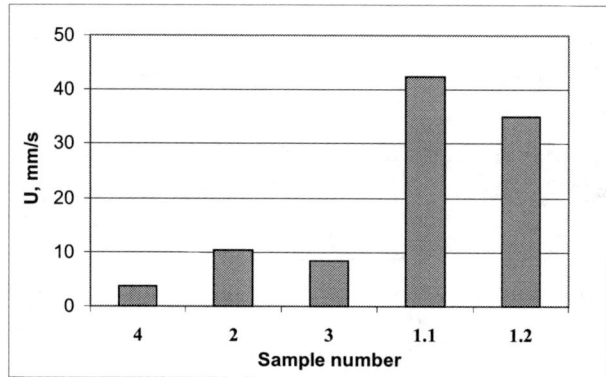

FIGURE 6. Burning rate of stoichiometric compositions of 76%AP/ 24%Al with the different Al powders: sample 1.1 – Al+7.5% Ba, aged for 3 months; 1.2 - Al+7.5% Ba, aged for 4 months; 2 – Al+1.5% benzine; 3 – Al+1.5% silicon rubber; 4 – micron-sized Al. The initial pressure of nitrogen is 4 MPa.

CONCLUSIONS

The present experiments provide the first comparison between the thermal behavior and combustion of different types of plasma-synthesized nano-aluminum and conventional micron-sized aluminum particles. Ba-doped nano-aluminum was found to have an average particle size of 43 nm, and an ideally spherical particle shape. Experiments show that the thermal behavior and combustion parameters of nano-aluminum are strongly dependent on the compounds being used during the particle fabrication process. Additionally, the considerable changes in the chemical activity of Ba-doped nano-aluminum were found during the powder storage, resulting in the oxide layer degradation.

Out of the studied series, the most suitable selected aluminum is Ba-doped powder, which is aged for 4 months, because of its high oxidation rate in air under heating, its high conversion degree in the temperature range 20-1100^{0}C, and its considerable burning rate enhancement for the samples with ammonium perchlorate. Moreover, the process of Ba-doped nano-aluminum could be catalyzed by platinum metal: at temperature of 540^{0}C a dramatic one-step oxidation of 60% of the active metal was found

ACKNOWLEDGMENTS

Financial support of the Russian Foundation of Basic Research (RFFI grant #04-03-32273a) is gratefully acknowledged.

REFERENCES

1. G. Pavlovetz, *Scientific and Technological Basics of Production and Use of Ultra-Sized Metal Powders for the High-Energetic Compositions*, Moscow: Section of Application Problems, Prezidium of Russian Academy of Science, 1999, pp. 80.
2. A. Pivkina, Yu. Frolov, S. Zavyalov, D. Ivanov, J. Schoonman, A. Streletskii, and P. Butyagin, *Proceedings of Thirty-First International Pyrotechnics Seminar*, Fort Collins, Colorado July 11-16, 2004, pp. 285.
3. M. M. Mench, K. K. Kuo, C. L. Yeh, and Y. C. Lu , *Combust. Sci. Technol.* **135** 269 (1998).
4. C. E. Aumann, G. L. Skofronick, and J. A. Martin, *J. Vac. Sci. Technol., B* **13** 1178 (1995).
5. Y. Champion, and J. Bigot, *Nanostructured Materials* **10** 1097 (1998).
6. A. Pivkina, P. Ulyanova, Yu. Frolov, S. Zavyalov, and J. Schoonman, *Propellants, Explosives, Pyrotechnics* **29** 39 (2004).
7. H. C. Burger and P. H. Cittert , *Z. Phys.* **66** 210 (1930).
8. A. G. Goursat, G. Vernet, J. F.Rimpert, J. Foulard, T. Darle, and J. Bigot, *Metal powder manufacture starting with a molten material*, Air Liquide SA pour l'Etude et l'Exploitation des Procedes Georges Claude, France 1984, p. 9.
9. G. V. Ivanov, M. I. Lerner, and F. Tepper, *Adv. Powder Metall. Pat. Meter.* **4** 15/55 (1996).
10. ASTM C1274-00, Standard Test Method for Advanced Ceramic Specific Surface Area by Physical Adsorption, *01/01/2000,* American Society for Testing Materials, Conshohocken, PA, USA.
11. M. Sandstrom, B. Jorgensen, B. Smith, J. Mang, and S. F. Son*, Proceedings of Thirty-First International Pyrotechnics Seminar*, Fort Collins, Colorado July 11-16, 2004, pp. 241.
12. R. Franchy, *Surf. Sci. Reports* **38** 195 (2000).
13. X. Phung, J. Groza, E.A. Stach, L.N. Williams and S.B.Ritchey, *Mater. Sci. and Eng. A* **359** 261 (2003).
14. I.Vol'nov, *Peroxide Compounds of Alkali-Earth Metals* (in Russian), Moscow: Nauka, 1983, pp.136.

Investigation of diaminodinitroethylene (DADNE) thermal decomposition

I.V. Chemagina, V.P. Filin, B.G. Loboiko, M.B. Kazakova,
Yu.A. Shakhtorin, V.M.Lagutina, N.P. Taibinov, N.V. Garmasheva,
A.V. Alekseev

Federal State Unitary Enterprise "Russian Federal Nuclear Center – Zababakhin All-Russia Research Institute of Technical Physics, Snezhinsk, Russia

Abstract. Diaminodinitroethylene (DADNE or FOX-7) is a low-sensitive high explosive. The curve of the differential-thermal analysis (DTA) of DADNE has two obvious endothermal peaks and two exothermal peaks. DTA data permitted to suppose that DADNE has several polymorph modifications, and transitions between them are observed at temperatures ~+115 ^{O}C and ~+170 ^{O}C. Two exothermal peaks (at +220° C and +270 ^{O}C, correspondingly) on DTA curves correspond to the processes taking place in DADNE at temperatures over +210 ^{O}C. One possible reason for appearance of these two peaks is: DADNE has polymorph modifications with substantially different thermal stability.

Attempts were made to obtain metastable (under normal conditions) polymorph modifications of DADNE. To attain this, DADNE was thermally treated in vacuum. Thermoanalysis and IR-spectroscopy demonstrated the residue on the plate to be the substance we called DADNE-T.

DADNE-T was recrystallized to study specific features of its molecular structure. Data of thermal analysis, IR-spectroscopy given for initial DADNE, DADNE-T and recrystallized DADNE-T. Performed investigations allow assumption that DADNE–T is a polymorph modification, which is stable within ~ 20°C to ~ 270°C temperatures.

INTRODUCTION

Diaminodinitroethylene (DADNE or FOX-7) is a low-sensitive high explosive and, just like TATB, has a graphite-like structure of crystals, functional groups (-NO₂ и -H₂N), and a double bond C=C. Specifics in DADNE (FOX-7) behavior during heating are of interest for many researchers [1,2, 3].

DADNE behavior under thermal impacts was earlier investigated using instruments of company "Perkin Elmer": the thermogravimetric analyzer TGA-7NT and the differential-thermal analyzer DTA-7 [4]. The curve of the differential-thermal analysis (DTA) of DADNE has two obvious endothermal peaks (~110°C and ~170°C) and two exothermal peaks (~210°C and ~275°C, heating rate is 2 degree/min). DTA- and TGA-curves for DADNE are given in figure1.

CP849, *Zababakhin Scientific Talks - 2005,*
edited by E. N. Avrorin and V. A. Simonenko
© 2006 American Institute of Physics 0-7354-0345-7/06/$23.00

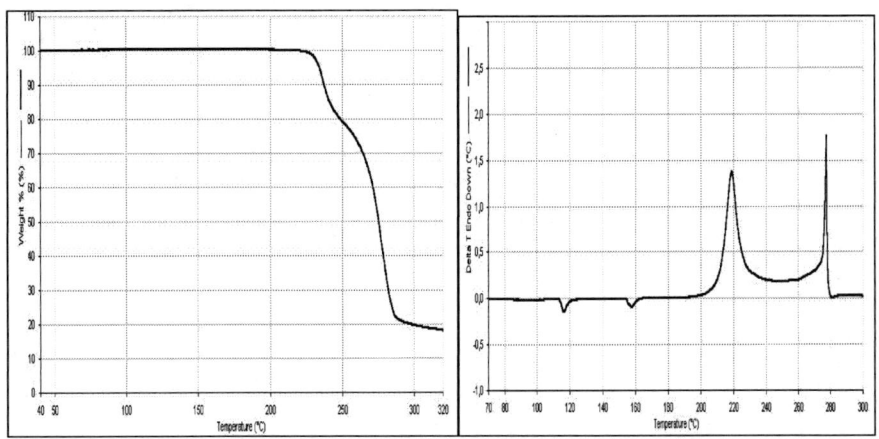

FIGURE 1. DTA-curve and TGA-curve for DADNE

The multi-stage differential thermal analysis, when a sample was repeatedly heated up to 190°C and then cooled, excluded assumption about DADNE melting under these temperatures and allowed assumption that DADNE has at least three polymorph modifications, which transit into each other when DADNE is heated and cooled (figure 2).

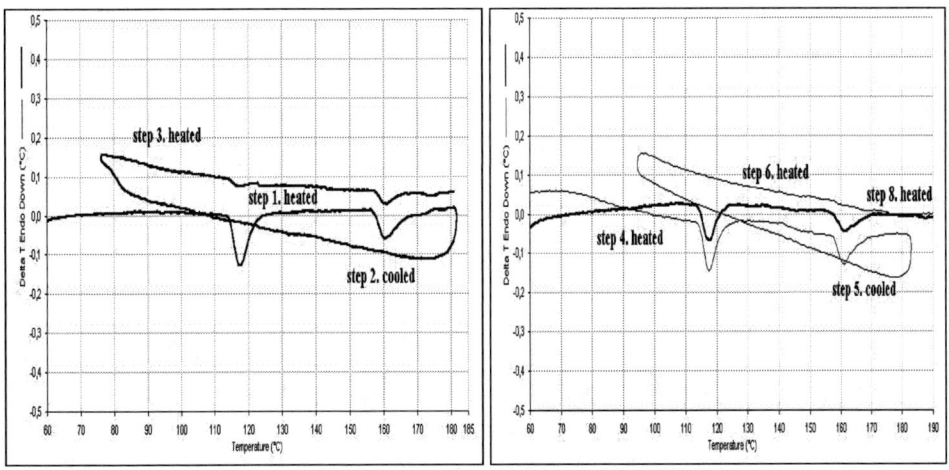

Figure 2. Polymorph transitions of DADNE

Paper [4] demonstrated that when exothermal processes terminate at ~ 210°C ÷ 240°C (after the first peak on the DTA curve), DADNE undergoes irreversible processes resulting in the formation of a new explosive substance, i.e. DADNE – T. The DTA-curve for DADNE – T was discovered to have one exothermal peak, which coincides with the second exothermal peak of DADNE (~ 275°C) in temperature. So, the TGA-curve for DADNE - T is one stage of decomposition (figure 3).

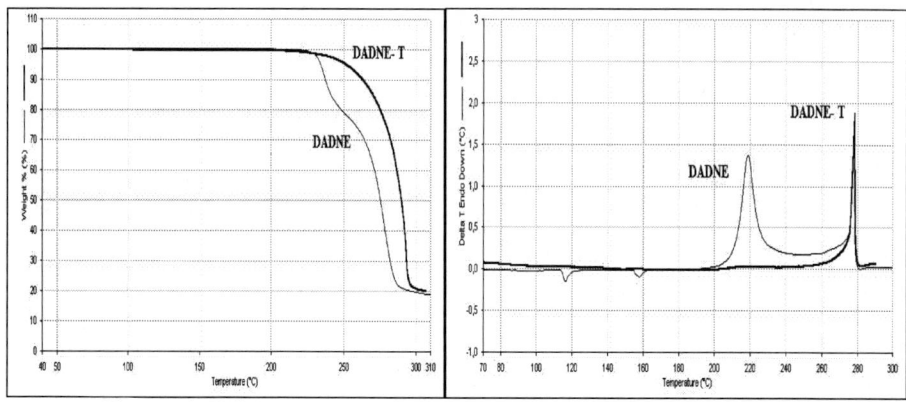

Figure 3. TGA- and DTA-curves for initial and thermally treated

Crystals having IR-spectra and DTA-curves similar to those of DADNE were obtained as a result of DADNE - T recrystallization from water, acetone, and alcohol.

The DTA-curve for DADNE – T after recrystallization from acetone is compared with DTA – curves for DADNE and DADNE - T in figure 4.

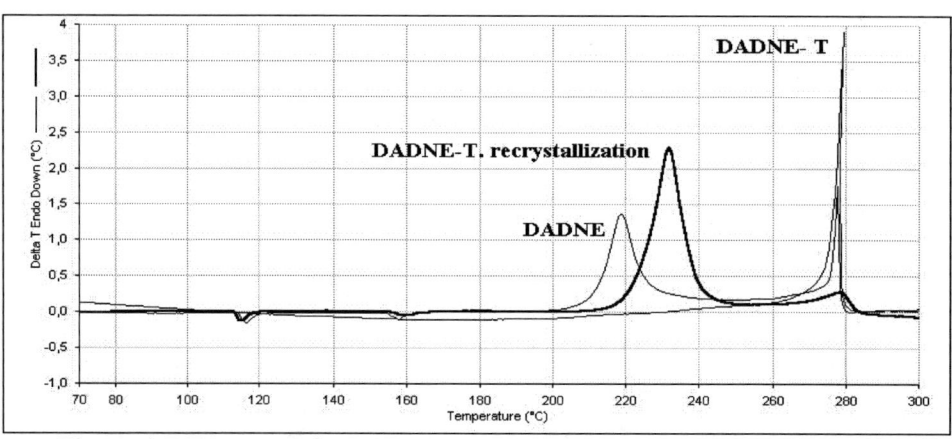

Figure 4. DTA-curves for DADNE, DADNE -T and recrystallized DADNE – T

One more polymorph modification of DADNE was assumed to exist:
– this polymorph modification is formed under high temperatures (~ 210°C ÷215°C);
– initial DADNE is a mixture of polymorph modifications having different thermal stability: the first peak – a less thermally stable modification undergoes decomposition, the second peak – more thermally stable modification.

In any case, specificity of this polymorph modification consists in the fact that it goes in parallel with the exothermal decomposition of a part of DADNE. DADNE -T

176

is brown (in contrast to initial DADNE and recrystallized DADNE –T having bright-yellow color) and contains admixtures.

EXPERIMENTS

The method of sublimation in vacuum during heating was used to get pure DADNE -T.

Heating rate, automatic regulation, and control of temperature under thermal impacts were specified with a help of special instrumentation based on the microprocess regulator of temperature «Miniterm 300.31».

Experimental setup is given in figure 5.

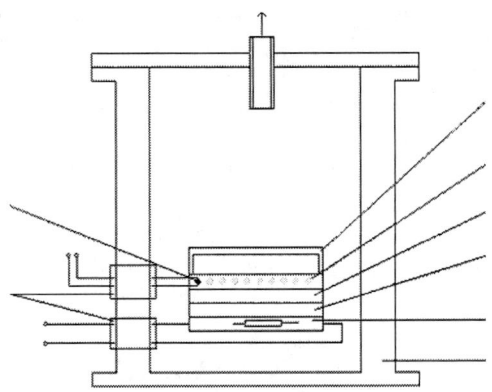

Figure 5. 1 – weighed portion of DADNE; 2 – object plate; 3 – plate having a skirting; 4 – steel support; 5 – heating device; 6 – chromel-Copel thermocouple; 7 – container; 8 – hermetic connection

A thin layer of the ~0.5g weighed portion of DADNE (1) was applied on the object plate (2) with 100x100 mm in size. This weighed portion was covered by a polished stainless-steel plate having a skirting (3). Skirting height was 3÷5mm. The plate with a «cover» was placed on a steel support (4) and further on a heating device (5), which uniformly heated the sample. Temperature of the weighed portion was continuously measured with the help of the chromel-Copel thermocouple (6). This experimental assembly was placed in an explosion-proof container (7). Prior to heating, air was pumped from the container by a vacuum pump down to $8x10^{-2}$ mercury.

During experiments, the sample was heated at the rate of $2÷3^0$C/min up to 180^0C and then at the rate of 1^0C/min up to 200^0C. The exposure temperature was kept to be T=200^0C within 30 minutes. Then, heating was stopped.

After natural cooling, the assembly was taken from the container. The upper plate had a deposit, which was a powder of yellow color.

The thermogravimetric and differential-thermal analyses were used to study the resultant substance. The DTA – and TGA – curves for this substance absolutely coincided with the curves for DADNE -T.

177

In addition, recrystallization of this substance from water, acetone, and alcohol resulted in the formation of crystals whose IR – spectra and DTA curves were similar to those of DADNE.

So, it is possible to say confidently that the substance of yellow color, which was obtained through vacuum sublimation during heating, represents purified small crystals of DADNE -T.

Experiments confirmed feasibility of the high-temperature modification of DADNE cleaned up from admixtures.

CONCLUSIONS

Investigation results:
- if heated above +210^{0}C, DADNE undergoes two-stage decomposition: the product of the first stage of DADNE decomposition is an explosive substance (DADNE - T);
- DADNE has several (at least four) polymorph modifications, transitions between which are reversible;
- DADNE– T is a polymorph modification, which is stable within temperatures ~+20°C to ~+270°C;
- During heating, the vacuum sublimation method can be used to obtain the high-temperature modification of DADNE cleaned up from admixtures.

The pure powder of DADNE -T is planned to be produced for further investigations.

REFERENCES

1. H.Qstmar, A.Langlet, N.Wingbord, U.Wellmar and U.Bemm
1. FOX-7 – a New Explosive With Low Sensitivity and High Performance// 11 Symp. On Det (Int.), pp.807-812, 31 aug – 1 sep, 1998, Snowmass, Colorado, USA.
2. V.P.Sinditskii, A.L.Levshenkov, V.Y.Egorshev, v.V.Serushkin
3. Study on Combustion and Thermal Decomposition of 1,1- Diamino-2,2- dinitroethylene (FOX-7)
4. U.Ticmanis, M.Kaiser, G.Pantel, I.Fuhr, U.Teipel
5. Kinetics and Chemistry of Thermal Decomposition of FOX-7
6. 4. I.V. Chemagina, V.P. Filin, B.G. Loboiko, N.V. Garmasheva, M.B. Kazakova, N.P. Taibinov, Yu.A. Shakhtorin, Investigation Of Diaminodinitroethylene (DADNE), Proceedings of the Zababakhin Conference, Snezhinsk, September 2003.

Development of Explosive Systems for Investigating Multiple Impacts of Solid Projectiles with Moderate Velocity

E.E. Lin, S.I. Bodrenko, V.V. Burtsev, P.S. Bushmelev, M.L. Vasil'ev,
M.A. Vlasova, V.V. Domnichev, S.K. Zhabitskii, V.N. Lobanov,
V.Yu. Mel'tsas, G.F. Portnyagina, S.V. Prokhorov, A.L. Stadnik,
Z.V. Tanakov, Yu.V. Yanilkin

Russian Federal Nuclear Center – VNIIEF, 607190, Sarov, Nizhni Novgorod Region, Russia

Abstract. We present the results of development of explosive systems for investigating multiple action of solid projectiles upon an obstacle at a moderate impact velocity on the order of 1 km/s. Two types of systems are considered, namely: 1) evacuating shock tube with plane high explosive (HE), 2) ballistic gun with distributed shot of plastic HE. In the first system an acceleration of steel spheres are realized in flux of expanding explosive products. The second system provides the acceleration of a "soft" plastic block with steel spheres having certain mutual orientation. The experiments with these explosive systems testify to higher effectiveness of projectiles penetration into obstacles at multiple impact with a moderate velocity.

Keywords: explosive systems, multiple impact, solid projectiles, moderate velocity.

PACS: 82.40Py; 83.50.-v

INTRODUCTION

The series of works [1-6] is concerned to a problem of the joint action of macroscopic solid projectiles upon obstacles. In [1], on the basis of inequalities, which tie together an initial velocity and a size of solid projectile with a critical deformation velocity at whose exceeding a shear strengthening and a disconnect embrittlement occur, it has been stated a hypothesis for heightened effectiveness of penetration into obstacles for flux of projectiles with relatively large sizes when significant compression and heating in shock wave do not occur. Experimental results [2, 5, 6], which have included the data of detailed metallographic examination of saved targets (duralumin obstacles) and projectiles (steel spheres), evidence the correctness of hypothesis [1] as a whole: we have registered more effective penetration of groups of steel spheres with diameter $d = 5$ mm and velocity $U \sim 1$ km/s in the course of the influence upon duralumin obstacles with relative thickness $H/d = 3$ and $H/d = 20$, than that observed in a single impact.

CP849, *Zababakhin Scientific Talks - 2005,*
edited by E. N. Avrorin and V. A. Simonenko
© 2006 American Institute of Physics 0-7354-0345-7/06/$23.00

For the subsequent study of the nature of the joint action of solid projectiles upon obstacles it is necessary to develop explosive systems, which provide a possibility to vary the impact parameters, namely, mass, velocity, form, number and mutual orientation of projectiles, the absence of impacts synchronism, combination of obstacle and projectile materials, ratio of their sizes and so on. In this paper we present the results of development of two explosive systems for investigating multiple impact of solid projectiles upon the obstacle at moderate impact velocity $U \sim 1$ km/s, namely, 1) explosive driven shock tube, in which steel spheres are accelerated by expanding explosive products, 2) ballistic gun, which provides an acceleration of "soft" plastic block carrying steel spheres with certain orientation from one to another.

EXPERIMENTAL METHODS AND RESULTS
EXPLOSIVE DRIVEN SHOCK TUBE

The scheme of steel spheres acceleration in an explosive driven shock tube has been presented in [2]. Two-dimensional numerical calculations show that under initiation of a plane explosive charge in a central point it is possible to accelerate a group of steel spheres within the shock tube up to velocities $U = 1.2 - 2.4$ km/s depending on sphere diameter $d = 2 - 5$ mm and charge thickness $\Delta = 4 - 9$ cm. The calculated distance-time diagram of spheres movement and the diagram of their velocity growth with the coordinate along a tube axis in the case of acceleration of 25 spheres with $d = 5$ mm at charge thickness $\Delta = 4$ cm are adduced in figures 1a), 1b). Initially, spheres were placed in a single plane at a distance of 160 cm from explosive charge. The inner diameter of a shock tube was 9 cm, the sizes of region occupied by spheres were 4×4 cm, a distance between the centers of adjacent spheres was equal to 1 cm (see figure2).

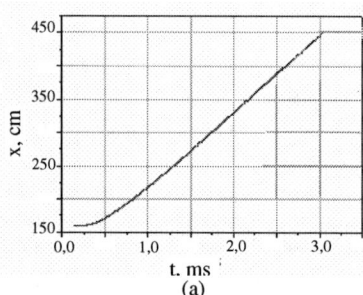
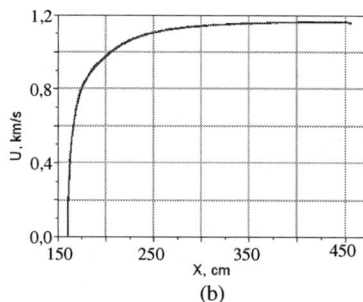

(a) (b)

FIGURE 1. Calculated curves for spheres motion within shock tube ($d = 5$ mm, $\Delta = 4$ cm): a) distance-time diagram; b) velocity growth with distance.

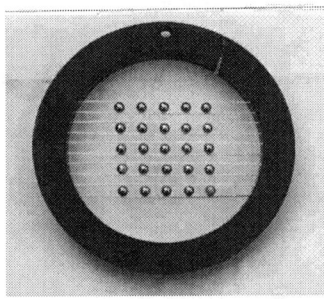

FIGURE 2. The block of steel spheres ($d = 5$ mm).

In experiments we have recorded steel spheres within a shock tube with the help of impulsive X-ray device [7]. The X-ray images of the spheres with $d = 5$ mm before the collision with duralumin obstacle with the thickness $H = 45$ mm are presented on figure 3. A compact leading group consisted of nine spheres is viewed near a target at the instant $t = 2.74$ ms after the explosion.

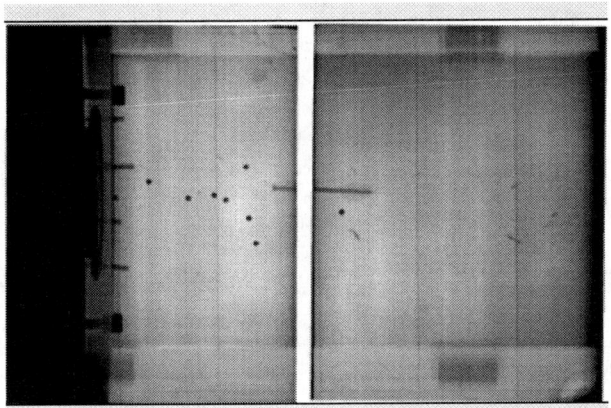

FIGURE 3. X-ray images for leading group of steel spheres$t = 2.74$ ms after the explosion).

A comparison of X-ray data with calculated curves of spheres movement shows that the velocity of nine leading spheres in the immediate vicinity of the target exceeds the calculated value $U_{calc} = 1.18$ km/s for twenty five spheres approximately by 10 percent and was determined at $U = 1.3$ km/s. The time interval between impacts of two spheres nearest to the target was estimated at $\Delta t \approx 20$ μs, the time interval between impacts of the first sphere and the last sphere was estimated at $\Delta t \approx 160$ μs. At these conditions of multiple impact of nine steel spheres with $d = 5$ mm upon "thick" duralumin obstacle the relative depths of spheres penetration lie in the range of $z/d = 2.8 - 4.1$. This result corresponds to the data [5], where under analogous conditions at the velocity and time intervals of multiple impact the relative penetration depths of a group of four steel spheres with $d = 5$ mm into duralumin obstacle with thickness $H = 100$ mm were $z/d = 3.6$. At single impact with the same velocity the relative penetration depth for steel sphere into duralumin obstacle was $z/d = 2.2-2.3$ [5, 8]. Therefore, the joint action of nine steel spheres upon duralumin obstacle under the impact velocity near 1.3 km/s leads to the increase of penetration depth by average factor of 1.6. Thus, heightened penetration effectiveness of solid projectiles into an obstacle at multiple impact with a moderate velocity has been confirmed.

BALLISTIC GUN

A shortcoming of the scheme of spheres acceleration within a shock tube is irregular regime of action, namely, heterogeneity of projectiles distribution in the space and the absence of impacts synchronism. For the study of influence of these parameters on the character of projectiles penetration into obstacle we use an explosive ballistic gun [9] with a calibre of 30 mm which provides acceleration of "soft" plastic block with steel spheres having pre-given orientation from one to another. The block-projectile is a cylinder from polyethylene (the density is $\rho = 0.95 \cdot 10^3$ kg/m^3) with diameter $D = 29$ mm and length of 46 mm. A steel plate is attached at the rear of a cylinder. There are hollows with steel spheres in front of a cylinder. The projectile is accelerated towards duralumin target in a smooth manner by explosion products of a plastic explosive charge. Since both the density and acoustic hardness of polyethylene are smaller than that of duralumin, therefore, the action of significantly hard steel spheres on an obstacle can be marked out evidently in the background of the action of a soft plastic block. The present construction makes it possible to provide both initial depth and mutual orientation of spheres arrangement in a block. Thus, we can vary both the time interval between impacts and the space distribution of projectiles.

The experiments have been carried out with polyethylene projectiles, in which either three steel spheres with diameter 5 mm were placed in the hollows having identical depths of 5 mm (figure 4a) or one steel sphere with the same diameter was placed at the same depth (figure 4b). Thus, in the test with three steel spheres the multiple impact was simultaneous. On the plane of impact the center of each sphere coincides with the apex of equilateral triangle with the side of 12 mm. The calculated velocity at the gun cut was 1480 m/s. The appearance of a fore-part of duralumin target with the thickness $H = 75$ mm after multiple impact is presented in figure 5. The depth of a shell-hole created by polyethylene projectile was near 5 mm. Distances between craters created by steel spheres were 15-17 mm. The depths of spheres penetration into obstacle from the bottom of a shell-hole were $z = 16$ mm, 17 mm, and 18 mm. Under impact of a single sphere in the same polyethylene projectile having the same velocity the penetration depth from the bottom of a shell-hole was 13 mm. Therefore, the higher effectiveness of projectiles penetration under simultaneous multiple impact is observed.

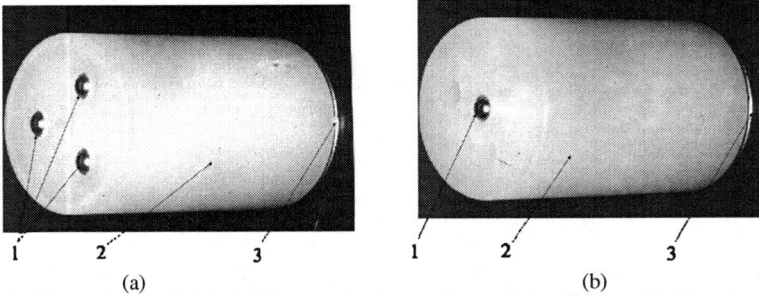

(a)   (b)

FIGURE 4. The soft plastic block with steel spheres (1 – steel spheres with diameter of 5 mm, 2 - polyethylene block, 3 – steel plate).

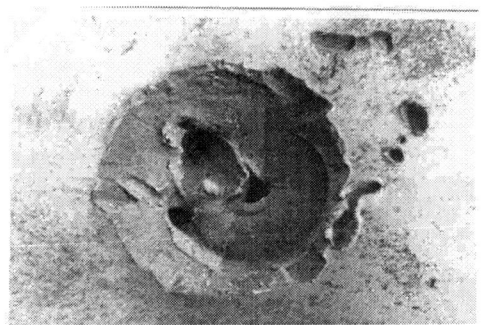

FIGURE 5. The appearance of fore-part of duralumin obstacle after multiple impact of three steel spheres.

NUMERICAL SIMULATION

A simulation of multiple collisions of steel spheres with "thick" obstacle was carried out with the help of the method elaborated in [10]. Two variances were considered, namely, 1) simultaneous impact modeled by introducing absolutely hard walls parallel to the axis of impact, 2) nonsimultaneous impact simulated by setting free boundaries at a certain distance from the axis of impact. Following to [6], the loss of strength along the axis of impact in the obstacle material in a channel with a cross size equal to sphere diameter was modeled by setting a yield stress equal to zero. Corresponding dependences of penetration depth on the time are given in figure 6. The penetration depth under the presence of absolutely hard walls is smaller than that under the presence of free boundaries. Therefore, one can see that nonsimultaneous (delayed) impact of many spheres is more effective than simultaneous impact.

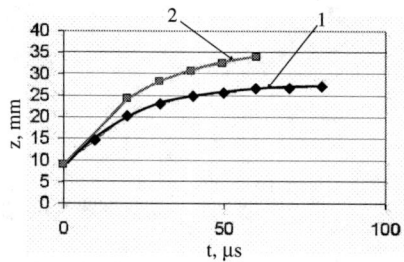

FIGURE 6. Calculated curves of steel spheres penetration into duralumin obstacle (U = 1.3 km/s, d = 5 mm, H = 100 mm): 1 - calculation under given absolutely hard walls parallel to the axis of impact, 2 – calculation under given free boundaries at a certain distance from the axis of impacts

CONCLUSION

The experiments with developing explosive systems testify to higher effectiveness of projectiles penetration into obstacles at multiple impact with a moderate velocity. But the issue of the phenomenon nature requires more careful study.

The authors are grateful to R. S. Lee and E. V. Roos (Lawrence Livermore National Laboratory, USA) for helpful discussion of the issue under consideration.

REFERENCES

1. S. S. Grigoryan, *Dokl. Akad. Nauk SSSR*, **292**, 1319-1323 (1987)
2. E. E Lin., V. Yu. Mel'tsas, S.A. Novikov, E.N. Pashchenko, A.V. Sirenko, B.P. Tikhomirov, "Investigation of Directed Group Acceleration of Solid Fragments by Expanding Explosive Products" in *Proc. of 16th Int. Ballistics Symposium and Exhibition*, San Francisco, CA, September 23-27 1996, Vol. 2., pp. 651-663.
3. I. E. Khorev, S. A, Zelepugin, A. A, Konyaev, *et. al.*, *Dokl. Akad. Nauk* **369**, 481-485 (1999) [Dokl. Phys. **44**, 818-822, (1999).
4. I. E. Khorev, V. K. Yakushev, S. A. Zelepugin, *et. al.*, *Dokl. Akad. Nauk* **389**, 197-202, (2003) [Dokl. Phys. 48, 146-151, (2003).
5. E. E. Lin, S. K. Zhabitskii, V. Yu. Mel'tsas, *et. al.*, Pis'ma Zh. Tekh. Fiz. **31**, 6-12 (2005) [Tech. Phys. Lett. **31**, 48-50 (2005)].
6. E. E. Lin, A. L. Mikhailov, Z. V. Tanakov, *et. al.*, "Determination of Obstacles Destruction Character at Multiple Impact of Solid Projectiles with Moderate Velocity" in Proceedings of International Conference "VII Khariton's Topical Scientific Readings", edited by A.L. Mikhailov, Sarov, 14 March-18 March 2005, pp. 541-546.
7. M.L. Vasiliev, A.V. Gladtsinov, A.V. Tabachkovskii, "Methodical Support of tests of Fragmentation –High-Explosive Ammunition" in *Collection Book «Modern Methods of Designing and Verification of Missile and orDnanceAmmunition»*, edited by R. I. Il'kaev, *et. al.*, Sarov, Russian Federal Nuclear Center – VNIIEF, pp. 276-278.
8. L. A. Merzhievskii and V. M. Titov, *Combust., Expl., Shock Waves*, **23**, 589-604 (1987).
9. V. I. Skokov, E. E. Lin, V. A. Medvedkin, and S. A. Novikov, *Fizikai Goreniya I Vzryva*, **34** (3), 105-106 (1998)[*Combust., Expl., and Shock Waves*, **34**, 346-347 (1998).
10. A. L. Stadnik, V. I. Tarasov, and Yu. V. Yanilkin, *Vopr.At. Nauki Tekh., Ser.: Mat. Model. Fiz. Protsesov*, **3**, 52-60 (1995).

Features of Transitional Regimes for Hydrocarbon Combustion in Closed Volumes and in Opened Clouds

E.E. Lin, Z.V. Tanakov

Russian Federal Nuclear Center – VNIIEF, 607190, Sarov, Nizhni Novgorod Region, Russia

Abstract. We present brief review and analysis of experimental results concerned to simulation of processes both in power-plants and in open-air surface space, when burning hydrocarbons gaseous mixtures. Combustion regimes in closed volumes are considered for acetylene mixtures $C_2H_2 + mO_2 + nN_2$, $C_2H_2 + mN_2O + nN_2$ in tubes with relative length $L/d = 4 - 60$. Combustion of opened fuel-air clouds under regime of their collisions is considered for propane-butane, when dispersing in atmosphere from several closely located reservoirs with liquefied gas.

Keywords: hydrocarbons, combustion, closed volumes, opened clouds.
PACS: 82.40Py.

INTRODUCTION

Invariable interest to investigating combustion regimes for gaseous mixtures based on hydrocarbons in the closed and opened energetic systems, in the first place, is connected with the safety problem when using them to solve scientific and applied tasks [1-8]. The study of the processes of burning to detonation transition requires particular attention because of their greatest danger from the point of view of destroying actions on the objects in environment.

In this work we present the brief report of experimental results concerned to combustion processes both in the closed tubes with gas mixtures based on acetylene and in colliding open fuel-air clouds based on propane-butane. The choice of combustible components in the closed tubes was caused due to the fact that decomposition of the highest hydrocarbons during burnout can proceed through the stage of the acetylene formation [9]. The growing attention to propane is connected with the great number of the wrecks in the industry and on the transport, which were caused by generating fuel-air clouds.

EXPERIMENTAL METHODS AND RESULTS COMBUSTION REGIMES OF ACETYLENE MIXTURES IN CLOSED TUBES

Combustion regimes of acetylene mixtures $C_2H_2 + mO_2 + nN_2$ (first type) and $C_2H_2 + mN_2O + nN_2$ (second type) were studied in horizontally placed stainless steel tubes with closed ends when varying the mole concentrations of components and the initial pressures p_0 in the wide ranges of [7, 8]. Relative length of closed channel was varied

CP849, *Zababakhin Scientific Talks - 2005*,
edited by E. N. Avrorin and V. A. Simonenko
© 2006 American Institute of Physics 0-7354-0345-7/06/$23.00

in the range $L/d = 4 - 60$ (the tube length $L = 1 - 6$ m, the tube diameter $d = 0.07 \div 0.25$ m). The combustion process was initiated near the tube end using the high voltage spark discharge, the explosion of thin wire, the explosion of electric detonator (ED), the explosion of plasticized high explosive (HE) sheet. The combustion regime was determined by the process rate and by the character of pressure change in time at the opposite tube-end and on the side wall of the tube. The pressure was recorded with the help of either piezoelectric or potentiometric pressure gauges. In the detonationless burning regime the gauges readings allowed to determine the maximum pressure p_{max} of combustion products in the tube volume, the total combustion time τ, equal to the time of achieving p_{max}, and the mean combustion rate v, determined as L/τ ratio (figure 1). The measurement of the circulation period of acoustic pressure wave between the closed tube-ends makes it possible to evaluate the sound velocity c in combustion products. The burning to detonation transition regime was determined from the fact of shock wave (SW) formation and from its propagation velocity D_1, which was fixed by gauges on the side wall of the tube. The registration of the amplitude and the velocity of circulating SW (see figure 2) makes it possible to determine the character of energy release in closed volume. In a number of tests a high speed framing photoregistration of combustion products luminescence was carried out through a transparent optical window in the side wall of the tube and through the transparent tube-end (figure 3).

The experimentally determined regions of components concentrations, which correspond to burning, burning to detonation transition, and detonation when initiated by the high voltage electric spark and thin wire explosion, were given in [7, 8]. Upon burning of the mixtures of the first type with concentration of acetylene $\alpha = 4 - 7$ % initiated by electric spark in the tubes with $L/d = 20$, the measured combustion parameters were varied in the following ranges: $p_{max}/p_0 = 3.4 - 6.8$, $v = 30 - 100$ m/s, $c = 660 - 1100$ m/s. In the regime of burning to detonation transition the velocity of SW near the tube-end was $D_1 = 600 - 900$ m/s, in the regime of normal detonation it was $D_1 = 1300 - 2000$ m/s. The thin wire explosion initiation of the mixtures of the second type with $\alpha = 1 - 3.2$ % in tubes of $L/d = 4$ and 10 at $p_0 = 0.5 - 1.1$ MPa led to detonationless combustion with the mean rate $v = 0.23 - 1.00$ m/s. The ratio p_{max}/p_0 was $6.4 - 12.9$. An increase of α up to 8% led to growth of v up to 50 - 100 m/s. The sound velocity in combustion products was $c = 800 - 1000$ m/s. The regime of burning to detonation transition in the tube with $L/d = 10$ at $\alpha > 8$ % was characterized by formation of SW, propagating with the mean velocity $D_1 = 300 - 500$ m/s. Upon SW reflection from the tube-end the behavior of pressure behind the reflected SW front was essentially non-stationary. In a number of tests the pressure amplitude upon reflections of circulating SW were approximately equal to the amplitude of the first reflection (figure 2). Very likely, the mentioned feature of transitional regime of combustion was connected with the mixture burnout within reflected SW, which circulating between the tube-ends.

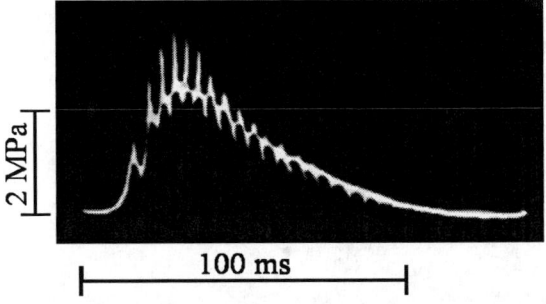

FIGURE 1. The oscillogram of the pressure on the tube-end under detonationless burning (the $C_2H_2+2.5O_2+16.5N_2$ mixture, $p_0 = 0.5$ MPa, $L = 2$m, $d = 0.1$m, initiation by the electric spark).

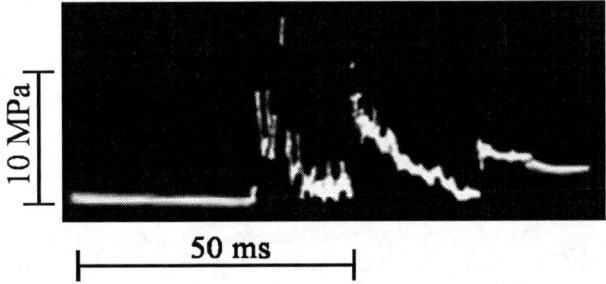

FIGURE 2. The oscillogram of the pressure on the tube-end under burning to detonation transition (the mixture $C_2H_2+1,27N_2O$, $p_0=0.1$MPa, $L=1$m, $d =0.1$m, initiation by the explosion of thin nichrome wire).

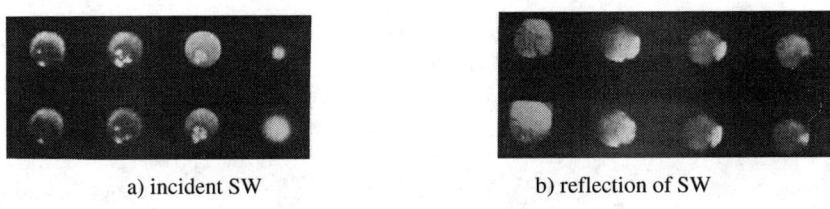

a) incident SW   b) reflection of SW

Intervals between the images is $\Delta t = 16\ \mu s$

FIGURE 3. High-speed framing of combustion process through the transparent tube-end (the mixture $C_2H_2+60N_2O$, $p_0=0.1$MPa, $L=1$m, $d =0.1$m, initiation by explosion of ED).

The processes that were initiated by explosions of ED in investigated mixtures, had a pronounced shock wave character. In addition, in mixturea with acetylene content, close to the normal detonation concentration limits, in different tests with the similar initial conditions we observed different combustion regimes, namely, either the normal detonation, or the low-velocity quasi-detonation process, propagating with the mean velocity $D_1 = 400 - 700$ m/s (see figure 4). In the $C_2H_2 + 60N_2O + 2CO_2$ mixture the explosion of ED created the low-velocity process with $D_1 = 700$ m/s, the explosion of HE sheet created the normal detonation with $D_1 = 1800$ m/s. It is typical for low-velocity process that the energy release occurs near tube-wall similar to "spin" (see figure 3). The velocity of reflected SW, circulating between tube-ends, was approximately equal to the velocity D_1 of incident SW. Therefore, one can see that this combustion regime is connected with burnout of the mixture during circulation of SW.

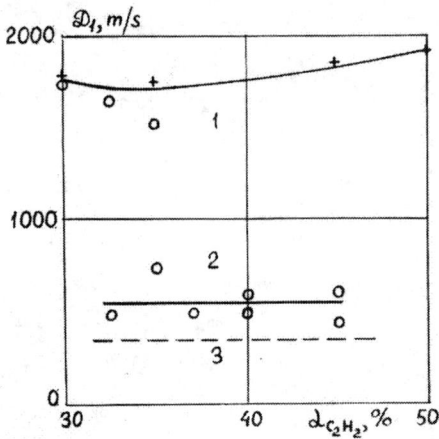

FIGURE 4. Dependence of detonation velocity on concentration of acetylene in the mixture C_2H_2 + air: 1 – normal detonation, 2 – low-velocity quasi-detonation process, 3 –velocity of sound in the initial mixture. Crosses are the data of [10], the circles are the data of the present work (p_0=0.1 MPa, L = 6 m, d = 0.1 m, initiation by explosion of ED).

THE BURNING OF COLLIDING PROPANE-AIR CLOUDS

We created the opened fuel-air clouds by means of explosive dispersing of liquefied propane-butane from several reservoirs, which were placed on the distances providing average concentration $\alpha \approx 2 - 9$ % of fuel in the hydrocarbon-air mixture, appropriated to the limits of ignition. The mixture combustion was initiated by explosive injecting of aluminum powder into extreme cloud. High-speed filming shows that the liquefied propane-butane scatters in the air with the mean velocity U = 30-40 m/s. The sizes of created clouds before their burning were varied in the range of $L \approx 5 \div 20$ m. The high-speed video framing shows that mixture ignition occurs in a set of the local centers on the cloud surface. During the stretch of time near 0.1 s these centers flow together in uniform burning spot with the characteristic size of 1 m. Subsequently, this spot increases and propagates on the cloud surface either as continuous field of luminescence on the all cloud height, or as separate spiral-like flame brush (figure 5). The latter indicates that in our experiments occurs the process of spontaneous combustion, which was described earlier in [5, 6].

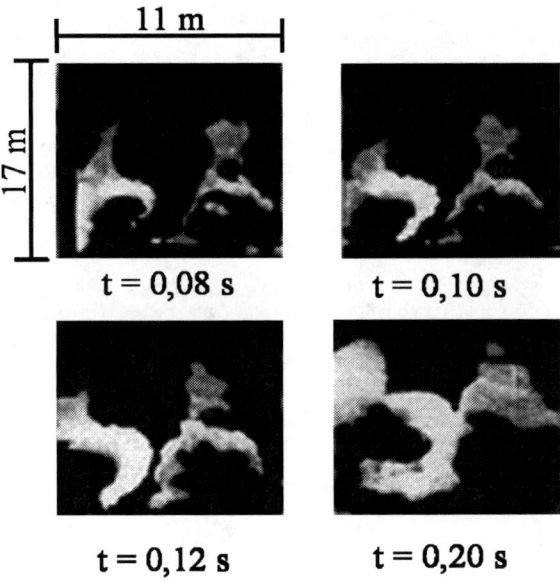

11 m

17 m

t = 0,08 s   t = 0,10 s

t = 0,12 s   t = 0,20 s

FIGURE 5. High-speed filming of the flame brush.

It has been stated experimentally that under creation of the chain of colliding clouds the visible velocity of flame is more than that under creation of the single cloud. One can assume that a turbulization of the propane-air mixture is most probable reason for the observed increase of the flame velocity. To estimate the intensity of auto-turbulization of the flame in the opened clouds we determine the Reynolds number of the flow of reacting mixture as

$$\text{Re} = \frac{\rho \cdot U_f \cdot l}{\eta \varepsilon}. \tag{1}$$

Here ρ is the mean density of gas mixture, η is the dynamic viscosity of the mixture, U_f is the visible velocity oif flame, ε is the degree of expansion of combustion products, and l is characteristic spatial scale of the process (parameter, which is measured with the help of combustion filming). If we use the well-known data for ρ, η, ε togther with the measured data for U_f and for l (for example, in figure 5 $U_f \approx 100$ m/s and the cross size of flame brush is $l = 5$ m) we can obtain from the expression (1) that the Reynolds number of flame in the created colliding clouds on the order of magnitude is equal to $\text{Re} \approx (2\text{-}3) \cdot 10^6$.

CONCLUSION

The main feature of transitional regimes for hydrocarbons burning in the closed tubes and in the opened clouds is localized release of the heat energy within the flame brushes. Evidently, it is connected with turbulent nature of the combustion processes.

ACKNOWLEGMENTS

We are grateful to G. A. Kirillov and A. L. Mikhailov for support of this work. We are indebied to A. I. Funtikov, V. F. Gerasimenko, S. K. Zhabitskii, S. A. Novikov and S. V. Prokhorov for collaboration.

REFERENCES

1. W.E. Baker, P. A. Cox, P. S. Westine, J. J. Kulesz, R. A. Strehlow, *Explosion Hazards and Evaluation*, Elsevier Scientific Publishing Company, Amsterdam - Oxford – New York, 1983.
2. M.A. Nettlreton, Gaseous Detonations: *Their Nature, Effects and Control*, Chapman and Hall Ltd., London, 1987.
3. V. S. Marshall, *Major Chemical Hazards*, Ellis Horwood Ltd., Chichester, 1987.
4. B.E. Gel'fand, G.M. Makhviladze, V. B. Novozhilov, et. al., *Fizika Goreniya I Vzryva* **28 (2)**, 75-81, (1992).
5. S.M. Frolov, B. E. Gel'fand, S. A. Tsyganov, *Fizika Goreniya I Vzryva* **28 (5)**, 13-17, (1992).
6. A.M. Khokhlov, E. S. Oran, Combustion and Flame **108**, 503-517 (1997).
7. V.F. Gerasimenko and É.É. Lin, *Fizika Goreniya I Vzryva* 25, 38-40, (1989).
8. É.É. Lin., A. I. Funtikov, *Fizika Goreniya i Vzryva* 28, 43-46, (1992).
9. P.A. Tesner, *Fizika Goreniya i Vzryva* **15 (2)**, 3-9, (1979).
10. B.A. Ivanov, *The Physics of Acetylene Explosion*, Khimiya, Moscow, 1969.

Dependence of TATB Specific Electric Conductivity on Pressure and Time Behind Shock Wave Front

M.M. Gorshkov, K.F. Grebyonkin, V.T. Zaikin,
V.M. Slobodenjukov, O.V.Tkachev, N.G. Bagavetdinov

Federal State Unitary Enterprise Russian Federal Nuclear Centre –Zababakhin All-Russia Research Institute of Technical Physics» (Zababakhin FSUE RFNC-VNIITF), Snezhinsk, Russia

Abstract. The dependence of specific conductivity of the pure TATB on pressure and time ($\sigma(P,t)$) behind shock wave front (SWF) was determined by measuring resistance of thin specimen(Δ=0.75mm) using four-wire circuit. The specimen positioned in parallel with SWF was placed in the medium, which retains its insulating properties behind SWF. It is shown that the rate of rise for $\sigma(P,t)$ depends on amplitude of the shock wave being introduced into the specimen, but the determining factor is not the pressure of TATB dynamic loading, but the value of specific internal thermal energy being introduced.

Keywords: shock, compression, unexploded HE, TATB, electric conductivity.
PASC: 82.40.Fp.

In the previous work [1] TATB was analyzed on addition of 10 % plasticizing agent, which possesses insulating properties (the specimens to be tested had the initial porosity of ≤ 0.3 %).

On further analysis of the obtained results, it has been speculated that small addition of the plasticizing agent significantly effects the trend of the dependence $\sigma(p,t)$ and its amplitude values.

To test the validity of this suggestion the analogous set of experiments was performed with the specimens made of pure TATB. But unfortunately, in manufacturing the specimens made of pure TATB we have succeeded in attaining the initial density ρ_{00}=1.865 g/cm^3, i.e., the specimens had porosity of ≈ 2.5 %. Thus in the experiments possible is the emergence of pore conductivity, that is, conductivity behind the shock wave front of the air contained in possibly open pores of the specimen, and that can introduce an error in measurement results σ.

This set of experiments was staged in the same manner as in the previous work [1], where they are described in detail. The majority of experiments were performed following the procedure with a "stick" (25*4*0.75 mm), the experiment No 177 was conducted following the procedure with a "disk" ($\varnothing$60*0.75 mm).

For the insulating medium that surrounds the specimen the previously tested mixture (70 % corundum + 30% paraffin) was used, which shock adiabat is more stiff than that of porous unexploded TATB, but this is insignificant within experiment

CP849, *Zababakhin Scientific Talks - 2005*,
edited by E. N. Avrorin and V. A. Simonenko
© 2006 American Institute of Physics 0-7354-0345-7/06/$23.00

accuracies: the calculations have shown that pressure in the second shock wave is increased by ~3%.   Table 1 presents the results of one-dimensional calculations under the program "Volna" [2] for each experiment of this set. Calculations were performed with the equation of state in the form with constants n and h [3].

Table 1.Discribtion of experiment

Test No	Characteristics of loading device		Design state parameters in TATB P (GPa)/E_t (kJ/g)		Geometry of experiment	
	Flyer material	Impact velocity [km/s]	E_t/P_{sv}	$E_{t\ cir}/P_{cir}$	$\Delta t[\mu s]$	L[mm]
149	Cu	1.0	0.179/6.7	0.1634.87	2.1	10.0
146	Cu	1.23	0.282/8.88	0.2556.4	1.95	9.55
142	Cu	1.48	0.422/11.5	0.380/8.2	1.72	8.0
139	Cu	1.78	0.631/15	0.563/10.7	1.55	8.65
144	Al	2.55	0.814/17.8	0.724/12.7	1.5	10.0
154	Al	3.15	1.255/24	16.9/1.107	2.13	10.55
174					1.50	9.7
176	Al	3.15	0.394/23.5	0.350/16.2	1.60	17.45
177	Cu	1.48	0.92/5.75	1.00/6.85	–	

The following notations are given in the table:
P_{sv}, E_t - the values of pressure and internal thermal energy in shock waves incoming to the TATB specimen.
P_{cir}, $E_{t.cir}$ - pressure and thermal energy in the specimen after arrival of the wave reflected from organic glass (rarefaction wave, in experiment No 177 – compression wave);
Δt – circulation time of transmitted wave and reflected wave between the TATB specimen and organic-glass panel;
L – distance between electrodes.

Table 2 Parameters of equation of state

i.i.	MATERIAL	ρ_{00} g/sm^3	ρ_0 g/sm^3	C_0 km/s	N	h
1	Copper	8.93	8.93	3.978	3.263	2.4
2	Aluminum	2.71	2.71	5.524	3.539	2.70
3	Corundum-paraffin mixture	1.965	1.965	2.336	7.43	1.915
4	TATB	1.865 1.2	1.91	2.202	8.16	1.78
5	Foam plastic	0.5	1.05	2.03	4.4	4.35
6	Organic glass	1.186	1.186	2.583	4.75	1.98
7	Paraffin	0.904	0.904	2.7	5.03	2.90

n- exponent of power dependence of cold pressure on compression;
h- ultimate compression.

Numerical values of parameters of the equations of state for the materials used in the experiments (and calculations) are given in Table 2.

In Table 1 the values of an increment of internal thermal energy for HE (E_t) are given without regard for energy liberation during HE decomposition (i.e., for unexploded HE).

In the course of experiments measured is a change in voltage drop (U) on electrodes applied to the TATB specimen. Re-processed oscillograms, in time dependence of specific conductivity ($\sigma(t)$) for each experiment are presented in Fig. 1. For pictorial representation the same oscillograms are given in Fig. 2 in a single scale. The moment of actuation of the reference pin placed on the boundary being correspondent to the upper surface of the "stick" specimen, that is, the moment of shock wave transmitting over the "stick" specimen is taken to be the "0" time.

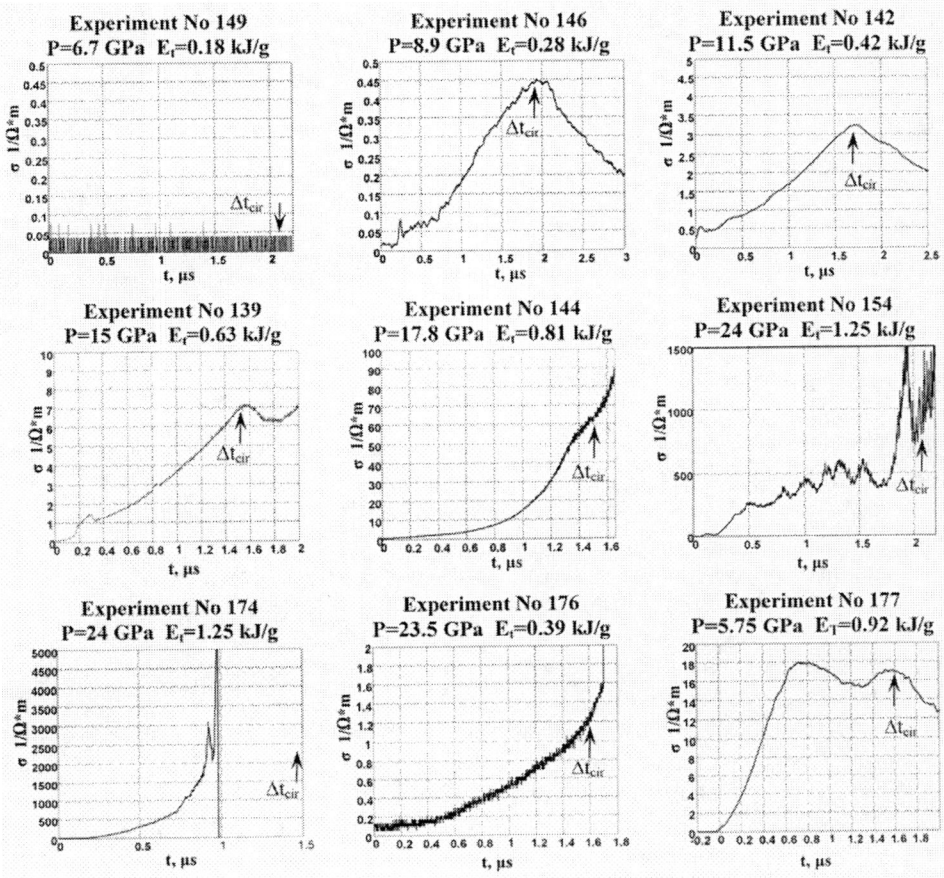

Figure. 1 Oscillograms of experiments.

If pore conductivity takes place (if it is considerable), then interelectrode conductivity would increase as a shock wave is distributed over the "stick" specimen (~ 0.1 μs), and then remain constant because pressure in the specimen is held constant within (1.5÷2) μs. This is not the case and conductivity continues to increase within the registration period.

At t=0 some conductivity is registered and if to take it as pore one, it does not exceed ~ 0.5 1/Ω*m even at P ~ 24 GPa, as it can be seen from Fig. 2 .

In the experiment No 177, when specimen porosity is high k=p_0/ρ_{00}≈1.6, at the moment t=0 σ≤0.5 1/Ω*m and continues to increase within Δt=0.65 μs, i.e., in this experiment pore conductivity also doesn't play a significant role.

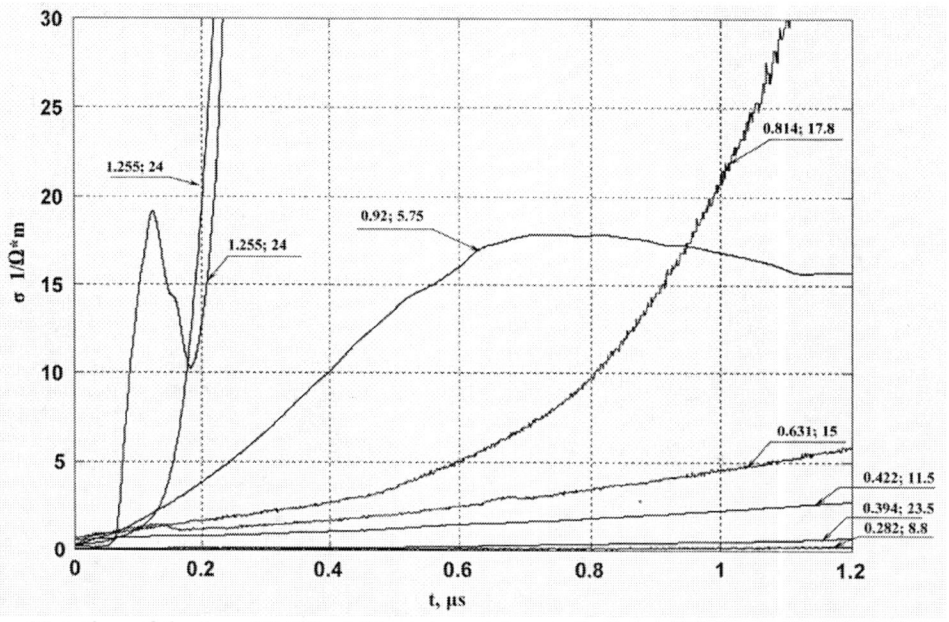

Notation of the type "0.631; 15" – internal thermal energy (kJ/g); pressure (GPa)
Figure. 2 – TATB electric conductivity, assembled diagram.

The experiments of this work show that the dependence σ(t) is varied as shock wave amplitude introduced into the specimen rises.
-   At P=6.6 GPa conductivity remains zero within the whole period of registration.
-   At pressures equal to 8.9 GPa and 11.5 GPa σ(t) rises (and faster at P=11.5 GPa); as the rarefaction wave comes to the specimen conductivity begins to drop.
-   Of interest is the experiment at P=15 GPa, in which conductivity firstly drops, as the rarefaction wave comes, and then it again begins to rise.
-   In the experiment at P=17.8 GPa a drop of conductivity is not observable as the rarefaction wave comes – only the inflection point emerges on the curve σ(t).
-   At pressure P≈24 GPa electric conductivity rapidly rises and
o   within (1÷1.5) μs it reaches the value ≥ 500 1/ Ω*m.

194

To clarify the issue what factor, i.e. pressure or heating, mostly effects the $\sigma(t)$ at dynamic loading of HE, two experiments were performed with varying the level of heating the specimen behind shock wave front. In the experiment No 176, using the interchange of layers made of foamed plastic and cooper the "stick" specimen made of TATB was loaded by a number of shock waves ($P_1/E_{t1}=8.3/0.271$, $P_2/E_{t2}=16/0.348$, $P_3/E_{t3}=23.5/0.394$, GPa/(kJ/g)) up to pressure $P\approx23.5$ GPa, i.e., pressure reached in the experiment No 174 by single-compression wave but, as it follows from calculations, it was heated significantly lower (~ 3 times). The fixed $\sigma(t)$ have an intermediate value between $\sigma(t)$ obtained in the experiments No146 (8.8 GPa) and No 142 (11.5 GPa).

In the experiment No 177 $\sigma(t)$ was determined when the specimen of decreased initial density ($\rho_{00}\approx1.2 \pm 0.15$) g/cm^3 was loaded. Pressure attained in the transmitted wave in the experiment P=5.7 GPa is lower than in the experiment No149 (P = 6.7 GPa) when conductivity remains zero. The design value of heating in the experiment No 177 corresponds to heating of the specimen with ρ_{00}=1.865 g/cm^3 in a single wave by amplitude P $\approx$ 20 GPa, and fixed $\sigma(t)$ at t≤0.6 μs corresponds to this heating (Fig. 2).

CONCLUSIONS

The experimental and calculation data in the performed set of experiments indicate that the determining factor of the development of TATB electric conductivity at dynamic loading is its heating.

The potential pore conductivity in the performed experiments is not of considerable importance.

ACKNOWLEDGMENT

In closing the authors express their sincere gratitude to N.M. Matveev for the development of the technology and assembling of all measuring units.

REFERENCES

1. M.M. Gorshkov, K.F. Grebyonkin, V.T. Zaikin, V.M. Slobodenjukov, O.V. Tkachev, I.R. Shakirov, "Electric Conductivity of Shock-Compressed Unexploded TATB Based HE". The VII-th Zababakhin Scientific Talks, RFNC-VNIITF, September 8 – 12, 2003, Snezhinsk.
2. V.F.Kuropatenko, G.V.Kovalenko, et al. VOLNA Code Package and Non Homogeneous Difference Method for Calculation of Compressible Matter Motion. Rus. J. Questions of Nuclear Science and Engineering (VANT). Series "Techniques and Programs of Computational Solution of Mathematical Physics Problems". 1989, Issue 2, P. 9-25 (in Russian).
3. E.I. Zababakhin. "Some Aspects of Explosion Gas-Dynamics", Snezninsk, 1997.

Two Stages of The Energy Release of Explosive Decomposition of Heavy Metals Azides

E.D. Aluker[*], G.M. Belokurov[*], A.G. Krechetov[*], B.P. Aduev[†],
B.G. Loboyko[**] and V.P. Filin[**]

[*]Kemerovo State University, Krasnaya st. 6, Kemerovo, 650043, Russia
[†]Kemerovo Branch of Institute of Solid State Chemistry and Mechanochemistry, Sovetsky av. 18,
Kemerovo, 650099, Russia
[**]RFNC-VNIITF, P.O. Box 245, Snezhinsk, region Chelyabinsk, 456770, Russia

Abstract. Methods of high time-resolution spectroscopy were used to determine two stages of explosive luminescence in solid energetic materials: the pre-explosive stage and the stage of processes occurring in explosion products. It was experimentally established that in silver azides the heating of a sample at the end of the pre-explosive stage does not exceed ~ 300 K; while at the second stage the temperature of the explosion products reaches ~ 3000 K. As it follows from the data obtained the major portion ($\sim 90\%$) of the energy released during the explosion in AgN_3 is associated with the processes occurring in the explosion products.

Keywords: stages, energy release, explosive decomposition, energetic materials.
PACS: 47.40.-x; 82.33.Vx

TWO STAGES OF EXPLOSIVE DECOMPOSITION

Under pulse initiation two separated in time stages are clearly distinguished in explosive luminescence of energetic materials, i.e. luminescence accompanying their explosive decomposition [1, 2]. This regularity is illustrated by the shown in Figure 1 data for silver azide and PETN.

At the first stage the continuous spectrum luminescence is observed (the two first peaks on the kinetic curves in Fig. 1a), at the second stage there is line spectrum luminescence (the third peak on the kinetic curves Fig. 1a). The first peak amplitude in Fig. 1a depends heavily on the initiating pulse energy, while the second and third peaks' amplitudes do not depend on the initiation energy.

There are no doubts about the nature of the luminescence at the second stage. It's explosion products luminescence, which is attributed to low-temperature plasma generated by the explosion. This conclusion is proved by the fact that the observed lines correspond to the elements of the explosives [3, 2].

The case of the luminescence at the first stage is more uncertain (peaks 1 and 2 in Fig. 1a). The luminescence has continuous spectrum, in principle, it may correspond either to condensed phase luminescence (of the sample)[1, 2], or glow of dense plasma generated by the explosion [4]. In the second case the transition of continuous spectrum into discrete-line spectrum (the transition from the first stage to the second) would simply correspond to the decrease in plasma density due to its dispersion.

CP849, *Zababakhin Scientific Talks - 2005,*
edited by E. N. Avrorin and V. A. Simonenko
© 2006 American Institute of Physics 0-7354-0345-7/06/$23.00

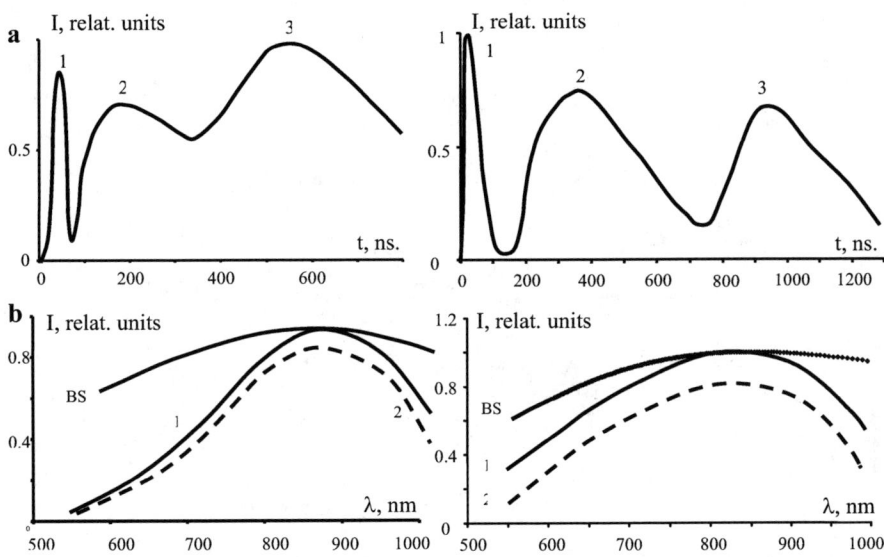

AgN₃　　　　　　　　　　　　　　**PETN**

FIGURE 1. Explosive luminescence of silver azide (AgN₃) and PETN crystals (electron pulse initiation – 250 keV, 20 ns, 2000 A/cm²).
a – luminescence kinetics at λ = 810 nm for silver azides or at λ = 780 nm for PETN; b – radiation spectra at peaks 1 and 2 at times 20 ns and 200 ns for silver azide or 20 ns and 400 ns for PETN; BS – the Planck's spectrum at T = 3500 K.

To choose between these two possibilities the comparison between the observed continuous spectrum and the absolute black-body spectrum should be made. There are considerable differences in the spectra (Fig. 1b), attesting to a non-thermal character of the observed glow and definitely relating the glow at the first stage to the luminescence of the sample before its disintegration during the explosive decomposition, i.e. to pre-explosive luminescence [5].

The similarity of glow spectra of the first and second peaks is a supplementary evidence of the non-thermal character of the luminescence at the first stage. Within the limits of apparatus time resolution the duration of the first peak coincides with the initiating pulse duration (this peak is attributed to the sample radioluminescence [8]). Simple estimate shows that under the energies of the initiating pulse used the radiation heating cannot exceed ~ 10 K. Hence, the glow in the first peak can not be thermal radiation, and since the glow spectrum of the second peak coincides with the first peak glow of the second peak is not thermal radiation either, i.e. pre-explosive luminescence.

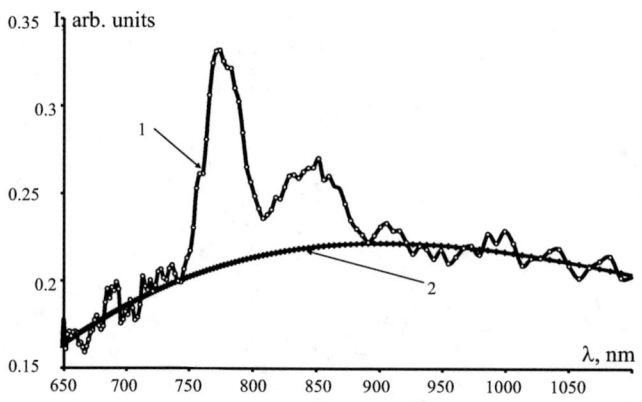

FIGURE 2. Spectrum of explosion products radiation of the silver azide pressed sample with soot additives at time t = 10 μs (electron pulse initiation, 250 keV, 5 ns, 200 A/cm^2).
1 – experimental curve, 2 – approximation of the continuous constituent of the spectra by the Planck's formula at T = 3200 K.

HEATING AT THE PRE-EXPLOSIVE STAGE AND AT THE EXPLOSIVE PRODUCTS DISPERSION STAGE

An assumption that the main source of energy release in HMAs explosions is the exothermic reaction $2N_3 \rightarrow 3N_2$ was made as early as mid-50s. However, even in the study in question [6] it was pointed out, that certain fundamental obstacle arise when the exothermic reaction takes place in crystal lattice. These restrictions are removed on the lattice surface or in gas phase. In application to the topic of the given article, the matter concerns the heating resulting from the exothermic reaction at the stage of explosive products dispersion (peak 3 in Fig. 1a).

For the quantitative estimate of the temperature of the explosion products the following method was used. A small amount of soot was added into silver azide tablets during pressing. Routine spectral-kinetic tests of the explosive luminescence under pulse initiation by the high-current electron accelerator were made on the samples. In the sample with soot impregnations almost no luminescence was observed at the pre-explosive stage. This is associated with the fact that pre-explosive luminescence is reabsorbed by the soot particles (the sample is opaque!).

The picture emerging at the stage of explosive products luminescence is more informative and of greater interest to us (Figure 2). The line spectrum of plasma luminescence is substituted by the broadband luminescence of the red-hot soot particles with plasma lines standing out against its background. The broadband luminescence within the intensive luminescence range t = 10 μs is well described by the Planck formula if T = 3200 K (Fig. 2).

Since the above stage corresponds to the plasma extension usually resulting in its cooling [7], the obtained outcome indicates significant energy release not only compensating the plasma cooling but providing its heating as well.

198

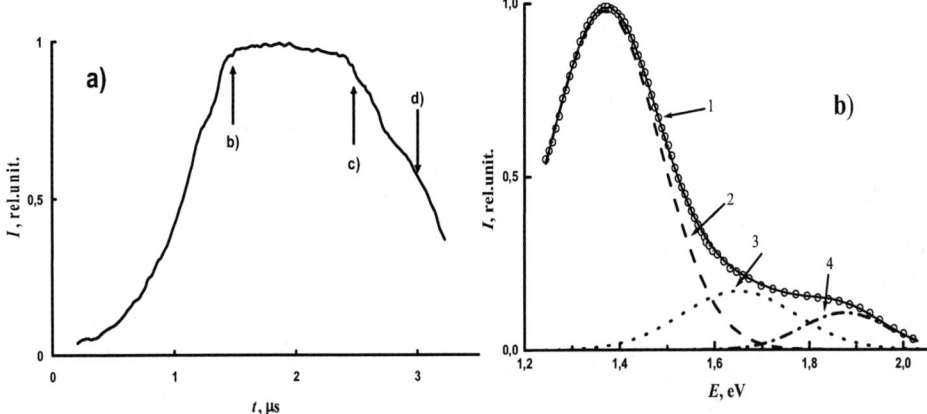

FIGURE 3. Spectral-kinetic characteristics of pre-explosive luminescence of silver azide crystals (initial temperature of the sample is 80 K, electron pulse initiation, 250 keV, 5 ns, 200 A/cm^2). a – luminescence kinetics at hv = 1.4 eV; b – luminescence spectra at 1.5 μs: 1 – registered spectrum; 2 – spectrum obtained by eliminating bands 3 and 4 from the spectrum 1; 3,4 – shortwave shoulder resolution into Gaussians with maxima 1.65 eV and 1.87 eV (maxima position and Gaussian half-width are obtained from [8]).

It is even more interesting that at times t ~ 3 μs corresponding to the final pre-explosion stage we didn't manage to register any thermal glow. Considering the spectral characteristics and apparatus sensitivity this means that in this segment the temperature is no higher than 1500–2000 K.

To give a more definite answer to the question of the heating degree at the pre-explosion stage we studied spectral-kinetic characteristics of pre-explosive luminescence under the initiation of 80 K (Figure 3).

In this case at the short-wavelength slope a distinct shoulder, consisting of two overlapping bands 1.65 and 1.87 eV (Fig. 3b) is observed. The nature of these bands is discussed in [8]. We are mostly interested in the temperature dependence of these bands' intensity. Under heating the quenching of 1.65 eV band starts at 80 K, and the quenching of 1.87 eV band starts at 120 K, and at 300 K these bands are almost quenched. Thus, a comparatively low heating of the sample at the pre-explosion stage may eradicate these bands from the pre-explosive luminescence spectrum. The very effect was observed in the given experiment. By 3 μs both bands are quenched and don't emerge in the pre-explosive luminescence spectrum, i.e. the sample is heated up to 200–300 K.

The comparison of the data in Fig. 2 and in Fig. 3 shows that the sample heating during the explosive decomposition reaction has the following dynamics: 200–300 K at the pre-explosive stage (an intact sample), 1000–1500 K at the initial stage of products dispersion (dispergated sample particles and, possibly, dense low-temperature plasma), 1500–2000 K at the final dispersion stage (plasma with line spectrum).

199

CONCLUSIONS

The problem of the thermal and chain mechanism of HMAs explosion has been discussed a lot these years [9, 10].

The results of this study permit to look at this discussion from a different angle. Indeed as it is shown in Figure 1 there are two distinct stages in HMAs explosive decomposition: pre-explosive stage and the stage of the explosion product dispersion. At the pre-explosive stage the actual processes are of chain nature and provide a smaller portion of energy-release (no more than 10–20%), which is connected with the nonradiative trapping and the recombination of electrons and holes, multiplying by chain mechanism. The objective of this stage is to provide the primary heating (up to 500–600 K) and the dispergation of the model, that means to set conditions for the second stage of the product dispersion. At this stage the major energy ($\sim$ 90%) connected with the exothermic reaction $2N_3 \rightarrow 3N_2$ is released. The possibility of its realization is more preferable on the surface or in the gas phase than in a solid state [6]. At this stage the explosion process has undoubtedly a thermal nature. So in HMAs the first stage of the explosion decomposition is ruled by the chain mechanism and the second (the dispersion stage) is ruled by the thermal mechanism.

This is why the question about the chain or the thermal nature of explosion is not correct. The correct question is: which mechanism of the explosive decomposition rules the given stage and what input this stage has to the total explosion energy.

Since the two stages of the explosion decomposition are observed in the secondary explosives (Fig. 1) we may presuppose that this question may be essential for these systems too.

The research was supported by RFBR.

REFERENCES

1. B. P. Aduev, G. M. Belokurov, S. S. Grechin, and E. V. Tupitsin, *Techn. Phys. Letters* **30**, 774–775 (2004).
2. B. P. Aduev, G. M. Belokurov, S. S. Grechin, and E. V. Tupitsin, *Techn. Phys. Letters* **30**, 660–662 (2004).
3. B. P. Aduev, E. D. Aluker, A. G. Krechetov, and A. Y. Mitrofanov, *Physica Status Solidi (b)* **207**, 535–540 (1998).
4. V. I. Korepanov, V. M. Lisitsyn, V. I. Oleshko, and V. P. Tsypilev, *Techn. Phys. Letters* **29**, 669–671 (2003).
5. B. P. Aduev, E. D. Aluker, G. M. Belokurov, et al., *J. Experim. Theor. Physics* **89**, 906–915 (1999).
6. F. P. Bowden, and A. D. Yoffe, *Fast Reaction in Solids*, London: Butterworths Scientific Publications, 1958.
7. D. A. Frank-Kamenetsky, *Lectures by Plasma Physics*, Moskow: Nauka, 1998, in russian.
8. E. D. Aluker, B. P. Aduev, A. G. Krechetov, et al., *Combust., Explos., and Shock Waves* **41**, 467–473 (2005).
9. M. M. Kuklja, B. P. Aduev, E. D. Aluker, et al., *J. Appl. Physics* **89**, 4156–4166 (2001).
10. V. I. Korepanov, V. M. Lisitsyn, V. I. Oleshko, and V. P. Tsypilev, *Combust., Explos., and Shock Waves* **40**, 612–614 (2004).

On the Physical Mechanism of Electro-conductivity of Detonation Products of Condensed High Explosive

M.M. Gorshkov, K.F. Grebenkin, A.L. Zherebtsov

Federal State Unitary Enterprise "Russian Federal Nuclear Center —Zababakhin All–Russia Research Institute of Technical Physics", 456770, Snezhinsk, Chelyabinsk region, Russia

Abstract. Interpretation of the recent measurement of electro conductivity kinetic in shock-compressed TATB-based explosive is suggested. From this point of view the models of the slow energy release due to carbon clusters growth in explosion products are examined. Similar experiments with explosives containing no or low amount of the solid carbon are proposed to conduct.

Keywords: detonation, TATB, electro conductivity, carbon cluster.
PACS: 47.40.Rs, 72.20.-i

INTRODUCTION

The physical mechanism of electro conductivity of explosion products (EP) of condensed high explosive (HE) has been discussing for many years. Several candidate processes were considered (see overview [1]), but to present day there is no commonly accepted physical model of this effect. Recent measurements of the plasticized TATB (PST) EP electro conductivity [2] gave cause to return to this problem, again. Results of these experiments appeared to be rather unexpected. Analysis of these results is represented in this work. Also, the physical mechanism, which could explain the results of [2] and known experimental data on electro conductivity of other explosives EP, is discussed.

STATEMENT AND RESULTS OF THE EXPERIMENTS.

The experimental technique was described in detail in [2]; here we just note two essential differences in the experiments setting from all previous measurements of EP electro conductivity. First, a very thin HE sample of 0.75 mm in thickness was placed into the buffer non-conducting medium having same dynamic compressibility as that of the sample. Secondly, the sample was loaded by a step pressure pulse, and the pressure was kept at constant level during ~1 – 1.5 μs.

Dependences of specific electro conductivity of plasticized TATB on time, obtained in two experiments, where the shock pressure was close to that of the stationary detonation, are given in the fig. 1. Time is counted off from the moment the shock wave came out of the sample; this moment was registered by means of special contact sensor.

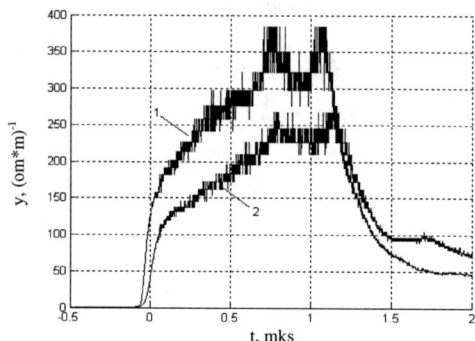

FIGURE 1. Dependence of specific electro conductivity on time in the experiments where the pressure at the shock wave front were equal to 34.3 GPa (line 1) and 28.6 GPa (line 2); t=0 corresponds to the moment of the shock coming out of the sample.

Two conductivity growth stages are distinctly observed in the oscillograms. The duration of the first of them is about 0.1μs; that is close to the time of chemical reaction when detonation of plasticized TATB [3]. Hence, the first stage may be interpreted as electro conductivity growth in the process of chemical transformation of HE into EP. During the second stage, lasting 0.5 – 1.0 μs, the conductivity growth was continued, but with a rate being several times lesser than that at the first stage, and then, beginning with the moment of unloading wave arrival to the sample, the decrease of the conductivity occurred. At the second stage the EP specific electro conductivity increases approximately twice.

The second (slow) stage of the conductivity growth was not observed earlier. The possible reason of its appearance, namely what physical process may result in times of 0.5 - 1.0 μs, which exceeding by order the chemical reaction time at such conditions, and how this process may be related with electro conductivity, will be discussed in the next section.

DISCUSSION OF THE RESULTS

The above results may be explained under assumption that observe EP conductivity is a result of thermal electron emission from graphite nanoparticles [4]. Also, this mechanism enables to explain the experimental ratio of EP conductivities of TNT, TNT-RDX, PETN and PST, which correlate rather well with the graphite content in the EP.

The signs of the graphite thermal emission conductivity mechanism are: first, observed in the experiments obvious correlation of conductivity with graphite concentration in EP, and, second, the exponential dependence of conductivity on the EP temperature

$$\sigma \sim n_e \sim \exp\left(-\frac{E_a}{RT}\right), \tag{1}$$

where E_a – is the conductivity activation energy, estimated in [4] as about 2 ev.

Probably, the last dependence becomes apparent in the experiments as the second (slow) stage of the EP conductivity growth, if some slow energy release process resulting in temperature growth beyond the chemical reaction takes place in EP. This process is known; it is growth of the carbon nanoparticles in EP resulting in the corresponding energy release.

Assuming the exponential dependence of EP electro conductivity on temperature, it is possible to evaluate the EP energy release kinetics from the results of the electro conductivity measurements [2]. The idea of such estimation is rather simple; the energy release, related with the slow kinetics, will result in the EP temperature growth beyond the chemical reaction zone, and this will result in the corresponding growth of electro conductivity observed in the experiments.

Several physical models of kinetics of the energy release caused by the carbon clusters growth were proposed in literature. Thus, in [5] an exponential dependence of the energy release with time was considered

$$f_s(t) \simeq 1 - \exp\left(-\frac{t}{\tau_s}\right), \tag{2}$$

where the parameter was evaluated as $\tau_s = 75$ ns, and $f_s(t)$ is a fraction of carbon condensation total energy released by the given moment t.

Another model equation for the slow kinetics was proposed in [6]

$$f_s(t) = 1 - \left(1 + \frac{t}{\tau_s}\right)^{-\frac{1}{n-1}}. \tag{3}$$

Assuming the graphite thermal emission conductivity mechanism, one may estimate the temperature growth during the slow kinetics stage from the measured EP conductivity dependence on time:

$$T(t) \approx \left[\frac{1}{T_f} - \frac{R \cdot \ln\left(\frac{\sigma(t)}{\sigma_f}\right)}{E_a}\right]^{-1}, \tag{4}$$

where T_f and σ_f – temperature and conductivity by a moment of completion of the fast stage of the energy release. The result is given in the fig. 2 along with the corresponding estimations obtained with slow kinetics models [5,6]. One can see that EP temperature $T(t)$ evaluated from the experimental data under the assumption that the conductivity activation energy is about 1 ev, agrees with the calculation result obtained from the model (3).

Thus, both the value of the conductivity growth during the second stage and its typical time may be explained, considering that generation of conductivity electrons happens according with graphite thermal electron emission mechanism, and the conductivity electrons concentration is increased as a result of the EP temperature growth during the slow stage of the energy release related with the carbon nanoparticles coagulation in EP.

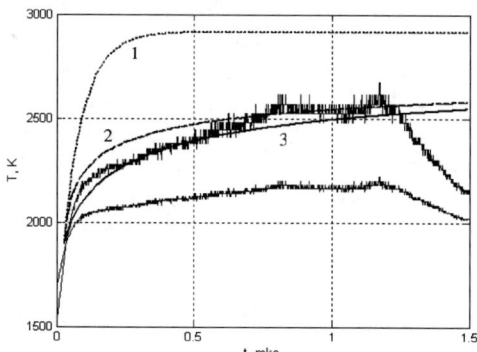

FIGURE 2. Estimated EP temperature – time dependence for two values of the conductivity activation energy: 2 ev (lower line) and 1 ev (upper line). Estimation results by different models are also represented: 1 - model (2) with ϕ_s=75 ns, 2 - model (3) with ϕ_s=18 ns и n=5, 3 - model (3) with ϕ_s=75 ns and n=4. It is supposed that 40% of total PST caloricity was released during the slow stage.

By the way, this interpretation results in a conclusion that may be examined in the experiment: it is expected that the second conductivity growth stage will be lacking in some HE having oxygen balance is near zero or positive, for example in PENT or ammonium nitrate.

Basing on the thermal electron EP conductivity mechanism one may easily explain the fact that, in contrast to [2], the second conductivity growth stage was not observed in other EP conductivity measurements, including those performed with HE having high carbon content. The probable reason is that in experiments [2], after passing the shock wave, EP were kept at high pressure during ~1μs while, in all previous experiments, an unloading taken place immediately after the shock coming to the sample. Thus, the conductivity decrease due to unloading prevailed over the effect of the slow energy release.

CONCLUSION

It has been shown that the second, relatively slow, stage of EP PST conductivity growth kinetics may be interpreted as a result of the EP temperature growth caused by the energy release in the process of carbon nanoparticles growth. The similar conclusion about the late stage of the electro conductivity kinetic in detonation HE has been made in [7].

Total PST caloricity released during the slow stage is estimated as 40%, and this value agrees with the results presented in [8] where, basing on analysis of the thin plate pushing experiments, it was estimated as 30%.

To examine the suggested interpretation of the EP conductivity measurement results it is desirable to carry out similar measurements in statement [2] for another HE, having a little amount of solid of carbon in EP or lacking it at all, for example, for PETN or ammonium nitrate. It is expected that the slow conductivity growth stage will be lacking in these HE.

Development of a detailed physical model of the thermal electron emission of the graphite nanoparticles in EP medium, especially an ab initio calculation of the EP conductivity activation energy and comparison it with the estimation $E_a \approx 1\,ev$, obtained in this work, is of great interest.

REFERENCES

1. S.D. Gilev. Proc. of Int. Conf. VI Zababakhin Scientific Reading. Snezhinsk. Russia. 2001. http://www.vniitf.ru/rig/index/konfer.htm.
2. M.M. Gorshkov, K.F. Grebenkin, V.T. Zaikin e.a. Proc. of Int. Conf. VII Zababakhin Scientific Reading. Snezhinsk. Russia. 2003. http://www.vniitf.ru/rig/index/konfer.htm.
3. B.G. Loboiko, S.N. Lyubyatinsky. Fizika Gorenia i Vzriva (in Rus). 2000, V. 36, N. 6, p. 45.
4. F.P. Ershov. Fizika Gorenia i Vzriva (in Rus). 1975, V.11, N. 6, p. 938.
5. W.L. Seitz, H.L. Stacy, R. Engelke e.a. Proc. IX Sympos. on detonation. Portland. USA. 1989. p. 657.
6. J.A. Viecelli, F.H. Ree. J. Appl. Phys. 1999, V. 86, N. 1, p. 237.
7. D.G. Tasker, R.J. Lee. Proc. 9th Symposium (International) on Detonation. 1989, p. 396.
8. C.M. Tarver, J.W. Kury, R.D. Breithaupt. J. Appl. Phys. 1997. V. 82(8), p. 3771.

Radiowave technique for studying the gas-dynamic characteristics and safety performances of the explosive compositions

A.V. Lebedev, Yu.A. Belenovskii, A.V. Vershinin,
V.A. Pestrechikhin, V.V. Shaposhnikov

RFNC-VNII Technical Physics named after E.I.Zababakhin, Snezhinsk, Russia

Abstract. The paper presents the results of application of the 8-mm range radiointerferometer for studying EC combustion processes and also their shock wave and mechanical response. The features and perspectives of using the radiointerferometer for these purposes are discussed.

The radiowave technique for studying a shock wave sensitivity and the combustion processes of the explosive components (EC) has been used at the RFNC-VNIITF for a long time [1,2]. Recently, this technique is widely used for the analyzing the EC mechanical sensitivity too. The radiointerferometry technique, up-to-date recording devices and corresponding techniques allow the analyzing of the fast processes to be more developed. The results of using the 8 mm radiointerferometer МБИМ-08 for analyzing the processes of EC combustion and also their shock wave and mechanical sensitivity are also presented in this report. The features and perspectives of such using have been considered.

INVESTIGATE THE SHOCK WAVE SENSITIVITY OF THE EXPLOSIVE COMPONENTS

The experimental unit used for the analysis of a shock wave sensitivity of the explosive components is given in Figure 1.

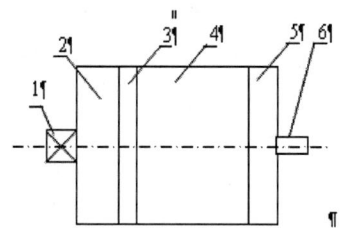

1 Electrical detonator;
2 Initiator from the explosive A ($\varnothing60\times20$);
3 Inert spacer (plexiglas) ($\varnothing60\times$Hmm);
4 Investigated sample from the explosive B ($\varnothing60\times50$mm);
5 Investigated sample from the explosive B ($\varnothing60\times10$mm);
6 Polyethylene waveguide ($\varnothing6$ mm).

FIGURE 1. Schematic diagram of the experiment conduction on shock wave sensitivity determination by inert barrier method.

CP849, *Zababakhin Scientific Talks - 2005*,
edited by E. N. Avrorin and V. A. Simonenko
© 2006 American Institute of Physics 0-7354-0345-7/06/$23.00

The presence or absence of the explosive transformation in the investigated samples of the explosive components (4) and (5) are recorded. A shock wave generated as a result of initiator detonation (2) passes through the investigated specimens. The initiator is separated from the investigated specimens by the plexiglas spacer (3) of different thickness (H, mm).

The determination criterion of the fact of presence or absence of the explosive transformation is the velocity values of passing of shock or detonation waves through the investigated specimens. Thus, the velocity measurements of passage of shock or detonation waves along the investigated specimens are necessary. The measurement accuracy of these velocities using the interferometer depends extensively on permittivity (ε) of the investigated samples. Taking into account that the investigation of shock wave sensitivity is conducted for compositions with different densities including prethermalized EC, it is necessary to determine the permittivity- density relationship. To estimate the range with the variable ε the experiments on permittivity determination were carried out according to the second of above-mentioned methods. In the experiments we have used the standard EC samples based on ТАТБ 60mm in diameter and 50 mm height of two different densities. The Table 1 shows the values of permittivity and wave -length of probe radiation in the sample corresponding to the different densities.

TABLE 1. Results of determination of ε.

Density ρ, g/cm^3	Permittivity ε	Wave length of probe radiation in the sample λ_ε, mm
1,845	4,255	3,930
1,910	4,556	3,798

The experiments confirmed the expected relationship between density and permittivity of the sample and therefore, wavelength of probe radiation. As shown in Table 1 an increase of sample density of 3,5% results in an increase of ε of 7,1%, that results in turn an contraction of $\lambda\varepsilon$ of 3,4%. Therefore, neglected ε for these samples during the velocity measurement of the explosive transformation can introduce an error 3,4%. Nevertheless, we can conclude the fact of presence or lack of an explosive transformation in the investigated samples.

During the investigation of shock wave sensitivity of EC the three processes could be realized: detonation occurrence at a depth, absence of detonation and, so called, low velocity process.

Figure 2 shows the oscillogram and corresponding D-x diagram for the case when detonation is at a depth. The characteristic points of the process are marked on the oscillogram. So point A corresponds the entry of a shock wave into the sample. Due to low reflection power of the shock front the signal amplitude is low, so the location of the point A is more difficult. Furthermore, due to its instability the typical signal distortions are observed that results in spread in velocity values during the further data processing. At the moment of detonation front generation (point B) the amplitude of the recorded signal increases step-wise that is explained by the high reflection power of the detonation front. The further exponential amplitude increase is connected with the approximation of the reflection boundary to the probe waveguide, and consequently with decreasing of probe radiation damping in the sample. The time of detonation front output in the back surface of EC (point C) is also easy determinated according to the specific signal curvature. The processing of the oscillogram presented in Figure 2 was made by means of discrete method (according to the signal minimum and maximum). The processing was made from the «end» of the signal, since in the result of distortions of the initial area it is dimplier to localize the point C than the point A, resulting in reduction of determination error of the point coordinate of shock wave entering into the sample. The processing result is presented as a D-x diagram for the shock wave passing through the sample.

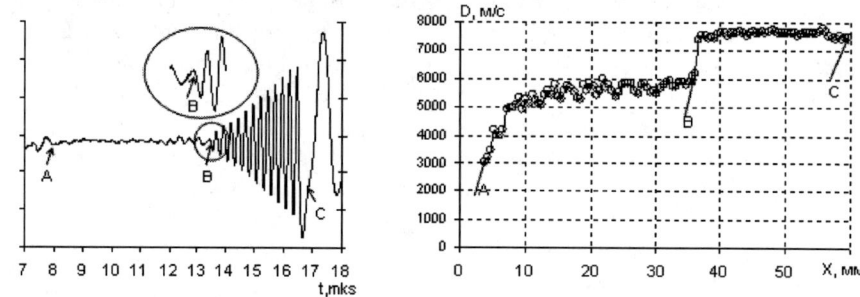

A – shock wave entry into the investigated sample, B –detonation occurrence;
C – meeting point of detonation front of EC boundary joining to the waveguide.

FIGURE 2. Detonation occurrence at a depth.

Figure 3 shows the oscillogram and the diagram D-x for the case when detonation is absent. In this test it was a low amplitude signal comparable with the amplitude of part AB of Figure 2 and the shock wave velocity decreases slowly.

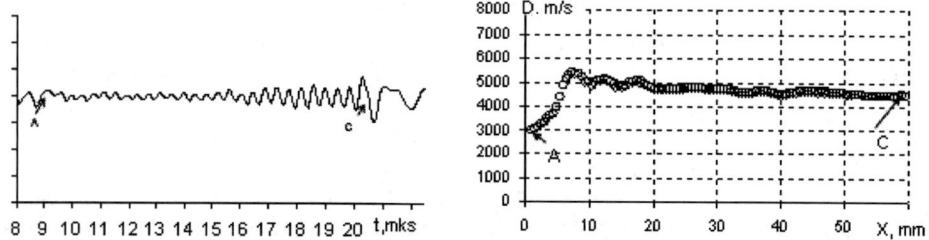

A –SW entrance into the investigated sample, C –SW output from the investigated sample.

FIGURE 3. The case of no detonation.

Figure 4 shows the oscillogram and the diagram D-x for the low velocity process.

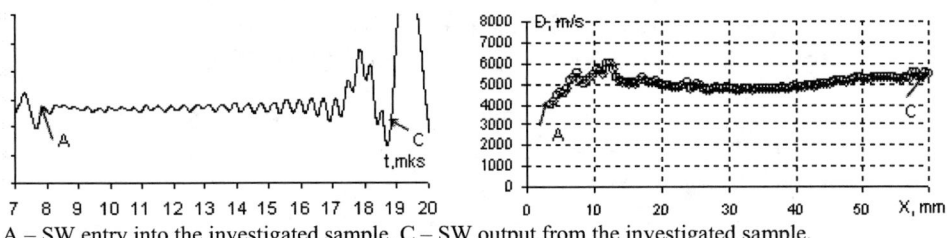

A – SW entry into the investigated sample, C – SW output from the investigated sample.

FIGURE 4. Low velocity process.

Let us consider the peculiarities of oscillograms and diagrams D-x (Figures 2-4). As shown from the diagrams D-x the velocity values of shock wave propagation are given not from the coordinate origin and have a significant spread up to its transformation time into the detonation wave. It is caused by a low reflection power of the low shock wave front, as a result of which the signal amplitude is low. Simultaneously, due to the front instability the typical signal distortions are observed, which together with the low signal amplitude result in a spread of velocity values at the initial part or even loss of information at first 2-4 mm.

According to the oscillogram form (Figure 2) and, namely, step-wise increase of the signal amplitude we can conclude that detonation occurs at a depth, furthermore, we can make approximately estimation of the depth (according to the quantity of signal cycles between points B and C). The D-x plotting (Figure 8) allows the detonation depth to be determined more exactly.

The oscillograms for the case of detonation absence and low velocity process (Figures 3,4) are similar and only to the form it is difficult to make an express evaluation, what the process takes place. Hence, the diagrams D-x help us to see the process dynamics, from the analysis of which we can conclude that the test relates to one of two cases.

So, from the given in Figure 3 diagram we can see that the shock velocity is initially increased up to ~5300 m/c at a depth of 8 mm then it is slowly decreased up to ~4200 m/c at a depth of 60 mm. Thus, we can conclude that the shock rate attenuation takes place and this case is true for the detonation absence.

The given in Figure 4 the diagram D-x shows the part of initial velocity increase up to ~6000 m/c at a depth of 12 mm and then its decrease up to ~4700 m/c and then the part of slowly velocity increase, which value reaches ~5500 m/c to the end of the second sample. Obviously, that in this case there is no wave transformation, but there is no wave attenuation process too. Thus, we can conclude that there is a self-sustained low velocity process.

Taking into account the high information density of the radiointerferometer methods in comparison with an electrocontact one it is possible to calculate the following parameters: the depth of detonation initiation, the shock rate attenuation.

INVESTIGATE THE MECHANICAL SENSITIVITY
OF THE EXPLOSIVE COMPONENTS.

We use now so-called "piston methods" to analyze the responsibility of EC samples to the mechanical effects. Figure 5 shows the experimental unit used for the determination of piston flying up velocity.

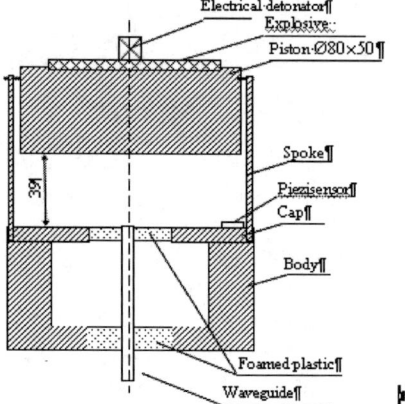

FIGURE 5. Experimental unit.

The typical oscillogram from the experiments on determination of the piston flying up velocity is given in Figure 6. We can see a part corresponding to the piston arrival to the point A. The signal after the point A corresponds to the piston jump from the unit. The difference of the end movement time of the piston from the piezosensor impulse is explained with the thickness of the latter (1,5 mm). The processed oscillogram as a diagram V-x is presented in the same Figure. Unfortunately, the 8 mm radiointerferometer does not allow the acceleration part to be recorded, since its size is significantly shorter than the wavelength.

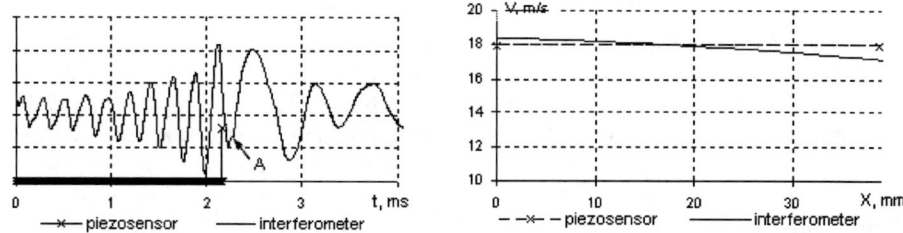

FIGURE 6. Oscillogram and diagram V-x for the case of the piston flying up.

To calculate the velocity of spoke penetration into the EC investigated material we have used the similar unit given in Figure 5, but in contrast to it, with the installed spoke and an EC sample. The waveguide is installed to the lower end of the investigated EC. With such experiment we knowingly deprive ourselves the information concerning piston flying up obtaining instead the dynamics of spoke penetration into the sample. For the preliminary experiments we have chosen the composition based on trotyl due to its sufficiently low responsibility to the mechanical effects concerning to other composition that allows the spoke to be inserted deeply. The Figure 7 shows the experiment and the diagram V-x.

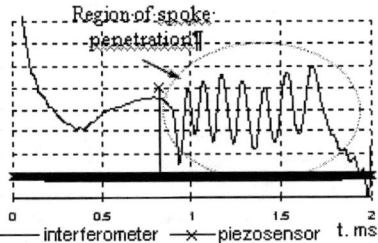 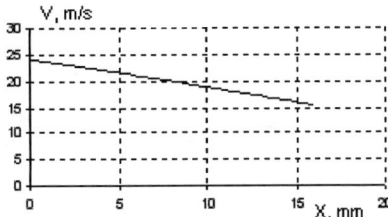

FIGURE 7. Oscillogram and diagram V-x for velocity of spoke penetration into the sample.

In this experiment the explosive transformation in the investigated sample was not recorded. The spoke penetrates the sample up to 16 mm that corresponds to the value measured in the experimental unit after the experiment completed. Using the 8 mm radiointerferometer for ocsillogram extreme processing we can plot the diagrams V-x with 1 mm discreteness that did not allow the end time of the spoke movement in the sample to be exactly identified, therefore the velocity on the base of 16 mm differs from zero.

The use of the radiointerferometer methods for the analysis of the mechanical responsibility makes the experiment more informative.

INVESTIGATE THE COMBUSTION PROCESSES OF THE EXPLOSIVE COMPONENTS.

To study the combustion behavior of the EC samples in the closed volume (manometric bomb) we have used the measuring and computing complex allowing the combustion rate to be measured both discrete (as time register we have used thermocouples) and continuously (using the radiointerferometer). The bomb pressure is measured simultaneously with the combustion rate measurement, and also with the temperature measurement in four points of the experimental unit. A pressure in the safety container is measured using the two-channel inductive transducer ИД-2И. Figure 8 shows experiment interferogram for combustion rate of HMX sample (∅20×20) and the processing result.

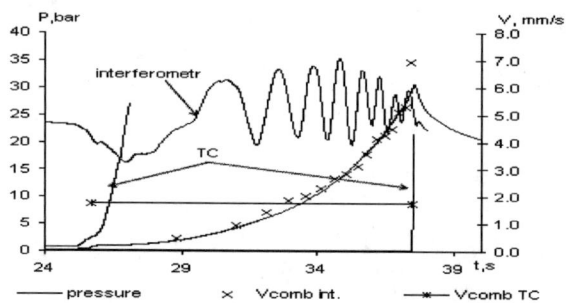

FIGURE 8. Determination of HMX combustion rate.

Furthermore, we can see in this figure the pressure profile and the thermocouples signals arranged on the ends of the investigated sample. From the analysis of Figure 8 we can see that using the interferometer we can consider the combustion condition more exactly and also the changes of combustion rate with time.

CONCLUSION

From above mentioned we could conclude that the radiowave methods is the effective means for investigation of the gas dynamic characteristics and safety factor of the explosive compositions. The main indisputable advantage of this method is the absence of the disturbances induced in the investigated process and the recording continuity. At the same time the experiments have a high information density.

REFERENCES

1. Batalov C.V, Filin V.P., Shaposhnikov V.V. Radiowave method for investigation of physical effects and chemical transformation in the heterogeneous explosives under shock wave effect. Physics of combustion and explosive , №6, Moscow, 1991.
2. Lebedev A.V, Loboyko B.G., Filin V.P., Shaposhnikov V.V. Radiowave method for measurement of combustion rate of the explosive materials in the sealed volume. Chemical physics, v.17, №9, Moscow, 1998.

Application of Energy-Saturated Complex Perchlorates

M.A. Ilyushin, I.V. Tselinsky, I.V. Bachurina, L.O. Novoselova,

E.N. Konyushenko, A.S. Kozlov, Yu.A. Gruzdev

Saint-Petersburg State Institute of Technology(Technical University),
Moskovsky,26, Saint-Petersburg, 190013, Russia

Abstract. We have synthesized several of new energetic complex perchlorates According to the data of derivatographic analysis, the complexes are thermally stable compounds. One of the promising ways of regulating ballistic characteristics of highly filled energetic formulations is the introduction of modifiers of burning. Complex compounds of d-metals are effective modifiers of the process of burning. Ballistic tests were made in a constant pressure bomb and demonstrated that the prepared compounds are promising modifiers of the burn rate and can be used in the developed formulations of rocket propellants.
Keywords: complex perchlorates, 1,5-pentamethylenetetrazole, modifiers of burning, derivatographic analysis, drop hammer test, ballistic test.

PACS: 65.69+i;61.66Fn

INTRODUCTION

The report is devoted to coordination compounds in the series of perchlorates of d-metals containing substituted tetrazole as ligand. Considerable attention to this class of energetic materials is caused by their application in safe detonators and rocket propellants.

APPLICATION OF COMPLEXES

We have carried out the synthesis of new energetic d-metal complex perchlorates with 1,5-pentamethylenetetrazole (PMT) as ligand.

1,5-Pentamethylenetetrazole (PMT)

CP849, *Zababakhin Scientific Talks - 2005,*
edited by E. N. Avrorin and V. A. Simonenko
© 2006 American Institute of Physics 0-7354-0345-7/06/$23.00

Preparation of metal complexes was performed made according to the following reactions:

For Ni(II) complexes

$$Ni(HCOO)_2 + n(CH_2)_5CN_4 + 2HClO_4 \rightarrow [Ni((CH_2)_5CN_4)_n](ClO_4)_2 + 2HCOOH$$

$n = 2, 4, 6;$

For Cu(II) complexes

$$Cu(CH_3COO)_2 + n(CH_2)_5CN_4 + 2HClO_4 \rightarrow [Cu((CH_2)_5CN_4)_n](ClO_4)_2 + 2CH_3COOH$$

$n = 2, 4$

For Co(III) complex

$$[Co(NH_3)_5H_2O]^{3+}(ClO_4^-)_3 + PMT \rightarrow \{Co(NH_3)_5[PMT]\}^{3+}(ClO_4^-)_3 + H_2O$$

The following compounds were prepared and investigated: $Ni(PMT)_6(ClO_4)_2$ (1), $Ni(PMT)_4(ClO_4)_2$ (2), $Cu(PMT)_4(ClO_4)_2$ (3), $Co(NH_3)_5(PMT)(ClO_4)_3$ (4). Composition and structure of the complexes **1-4** were supported by the results of elemental analysis; IR-, UV-, and NMR(H^1)-spectroscopy. The results of studies of complex perchlorates thermal properties by means of derivatographic method are presented in Tables 1 and 2.

TABLE 1. Derivatographic analysis of complexes **1-3**

№	Products of thermal decomposition	$\Delta T\ ^0C$	Effect	Residue %	
				Found	Calculated
1	$[Ni((CH_2)_5CN_4)_6](ClO_4)_2$	50-200		100	
		200-295	Exo	45	47,5
	$Ni((CH_2)_5CN_4))_3(ClO_4)$	295-460	Exo	17,5	18,1
	$Ni((CH_2)_5CN_4))_2(ClO_4)$	460→		37,5	40,0
2	$[Ni((CH_2)_5CN_4)_4](ClO_4)_2$	50-180	Endo	100	-
	$Ni (CH_2)_5CN_4)(ClO_4)_2$	180-320 (max 280)	Exo	47,5	51,1
		320-440 (max390)	Exo	22,5	
	$NiClO_4, NiO$	440→		25	
3	$[Cu((CH_2)_5CN_4)_4] (ClO_4)_2$	30-175	Endo	100	-
	$Cu ((CH_2)_5CN_4))_3 (ClO_4)$	175-240	Exo	29,3	29,6
	$Cu((CH_2)_5CN_4))_2$	240-305	Exo	29,3	29,6
	CuC_xN_y	350→		40,9	

TABLE 2. Derivatographic analysis of complex 4

Products of thermal decomposition	$\Delta T\,^0C$	Residue %	
		Found	Calculated
$[Co(NH_3)_5(C_6H_{10}N_4)](ClO_4)_3$	60-245 (max 243)	100	
$Co(C_6H_{10}N_4)(ClO_4)_3$	245-360 (max 275, 295, 320)	85.0	85.3
$1/2Co_2O_3$	360-500	14.0	14.3

The data listed in Tables 1 and 2 shows that complexes **1-4** are decomposed in several steps. Decomposition of the complexes begins when temperature rises above 200 ^{0}C. Removal of PMT is the first step of decomposition of perchlorate complexes **1-3.** During the first stage the process of inactivation and removal of NH$_3$ molecules from complex **4** inner sphere is observed. Subsequently the thermal decomposition of the tetrazole ligand occurs with the involvement of the perchlorate anion into oxidation reactions at the late stages of the complex destruction. According to the data of derivatographic analysis, complexes are thermally stable compounds.

The drop hammer test of impact sensitivity by GOST 4545-88 demonstrated sensitivities to impact of perchlorates **1, 2, 4** are similar to that of PETN. Sensitivity to impact of copper complex **3** was at the level of tetryl. Complex 4 has a short distance of deflagration-to-detonation transition and can initiate charge of RDX in blasting cup No 8 (See Table 3) [1].

TABLE 3. Explosive characteristics of perchlorate complex **4.**

Density ρ, g/sm^3	Oxygen coefficient, α	Sensitivity to impact, drop hammer K-44-II, 2 kg/25 cm, % of explosion	Minimal charge for RDX, g detonator №8	Rate of detonation D, mm*μs^{-1}, at maximum density (calculated)*
1,82	0,55	16	~0,4	~0,4

The impact sensitivity of PETN is 12% under similar conditions. *calculated by method of A.A. Kotomin [2].

Consequently, complex **4** may serve as an initiating material with the rate of detonation close to that of TNT and sensitivity to mechanical action like other modern high explosives. Perchlorate complexes **1-3** do not initiate HE.

One of promising the ways of regulating ballistic characteristics of highly filled energetic formulations is the introduction of burn modifiers into their composition. Complex compounds of d-metals are effective modifiers of the burning process [3]. Energetic copolymer of 2-methyl-5-vinyltetrazole (PVMT) with methacrylic acid (TC–38-403-208-88) was used as a high-molecular base for a model formulation of solid rocket propellant.

Properties of PVMT are given in Table 4.

Poly-2-methyl-5-vinyltetrazole

TABLE 4. Physical and chemical characteristics of Poly-2-methyl-5-vinyltetrazole

N	Characteristics	Value
1	Intrinsic viscosity in DMF, $10^{-3} m^3$/kg	0.64
2	Mass fraction of the residual monomer, %	0.09
3	Mass fraction of ashes, %	0.16
4	Mass fraction of moisture, %	1.9
5	Weight average molecular weight	68400
6	Number average molecular weight	27700
7	Polydispersity coefficient	2.5
8	Density, g/cm^3 [4]	1.28
9	Enthalpy of formation, kJ/kg [4]	1255
10	Temperature of the onset of decomposition, ^{0}C	105
11	Temperature of the onset of intense decomposition, ^{0}C	197

Liquid aliphatic azide 1,5-diazido-3-nitro-3-azapentane (5) was used as active plastificator of the polymer base (TC–13–01–76)[3]. Properties of **5** are presented in Table 5.

$$N_3-(CH_2)_2-N-(CH_2)_2-N_3$$
$$|$$
$$NO_2$$

1,5-Diazido-3-nitro-3-azapentane

TABLE 5. Properties of 1,5-diazido-3-nitro-3-azapentane

MM	Density, g/cm^3	Enthalpy of formation $\Delta H_{f,}$ kJ/kg
200	1,344	3025

Model formulations had the following composition: ammonium perchlorate of bimodal granulometric composition:grade «C» (<50 μm) - 32% of grade «A» under TC-84-942-82 ($S_{specific}$ = 12000 sm^2/g) - 32%, fuel - aluminum powder of grade ACD - 6 ($S_{specific}$ = 6000 cm^2/g) - 6%, binder - 28%, additions of the complexes were introduced into formulations in amount of 2%. Ballistic tests were made in a constant pressure bomb. Results of ballistic tests are presented in Table 6.

TABLE 6 Effect of additions of perchlorate complexes **1-4** on the burning of model propellant

Index of formulation	Order of deflagration U=A·P^v		U$_{10}$, mm/s	U$_4$, mm/s
	A	v		
Basic formulation	11,1	0,71	47,2	26,6
[Ni((CH$_2$)$_5$CN$_4$)$_6$)](ClO$_4$)$_2$	12,5	0,62	52,1	29,5
[Ni((CH$_2$)$_5$CN$_4$)$_4$)](ClO$_4$)$_2$	12,9	0,62	50	28,2
[Cu((CH$_2$)$_5$CN$_4$)$_4$)](ClO$_4$)$_2$	16,1	0,56	58,4	35,0
[Co(NH$_3$)$_5$((CH$_2$)$_5$CN$_4$)](ClO$_4$)$_3$	10,5	0,60	41,8	24,1

U$_{10}$- rate of burning under 10 MPa pressure;

U$_4$- rate of burning under 4 MPa pressure;

The results obtained permit to make the following conclusions:

The pressure exponent o burn rate law in the pressure of additions decreases from 0,71 to 0,62-0,56 and this is an indirect confirmation of the fact that these additions affect the burning processes in the condensed phase,

Complex perchlorates **1-3** increase the burning rate of a model propellant; the highest catalytic effect is observed for a complex based on copper - 35,5%.

Therefore, the prepared energetic complexes are promising modifiers of the burning rate and can be used in development of solid rocket propellant formulations.

ACKNOWLEDGMENTS

This work was performed under financial support of RF Ministry of Education and Science via scientific and technical program "Development of scientific potential of higher school" to Subprogram II "Applied studies and developments of priority directions of science and engineering", Grant of the Authorities of Saint-Petersburg 41/05 and State Contract № 02.444.11.7051(RI-111.0/001/028)

REFERENCES

1. M.A.Ilyushin, A.Yu. Zhilin, I.V.Tselinsky, Yu.A. Nikitina, A.S.Kozlov and I.V Shugalei. *Russian Journal of Applied Chemistry.* **78**. No.2. 195-199 (2005) (in Russian).
2. M.A.Ilyushin, A.V.Smirnov, A.A.Kotomin and I.V Tselinsky. *Hanneng Cailiao=Energetic materials.* **2**. No 1. 16-20.(1994).
3. A.S. Dudyrev, I.V. Tselinsky and M.A. Ilyushin "Technology of energetic materials and products on their base" in *Chemical technologies,* edited by P.D. Sarkisov. Ministry of Education of Russia. Russian University of Chemical Technology named after D.I.Mendeleev. Moscow. 2003. pp. 403-488. (in Russian)
4. D.B. Lempert, G.N. Nechiporenko and S.I. Soglasnova *Chemical Physic.* **23** No 5. 75-81.(2004) (in Russian)

217

SECTION 3

DENSE PLASMA PHENOMENA AND INTENSIVE ELECTROMAGNETIC PROCESSES

Contribution of neutron reactions in hybrid targets of inertial heavy ion fusion (HIF)

V.S. Imshennik [*, **], V.T. Zhukov [**]

* A.I.Alikhanov Institute for Theoretical and Experimental Physics
** Keldysh Institute of Applied Mathematics, Russian Academy of Sciences

ABSTRACT

Recently some simple estimations (Koskarev, Sharkov) were made   for capability of achievement of critical conditions in an uranium shell of HIF energetic target, and afterwards it was proved use of an uranium shell (pusher) for substantial expansion of energy-release in a such hybrid target. The mentioned justification is included accounting of neutron-induced fission in the pusher. This accounting is formulated as generalization for cylindrical geometry of the well-known Axiezer-Pomeranchuk solution.

A corresponding analytical solution of one-speed Payerls equation allows sufficiently accurately to compute the critical parameters of the uranium pusher in hydrodynamic model of compression and fusion of HIF target (taking into account of development of chain nuclear fission reaction under critical condition achievement). Nevertheless the implemented computations show that the most essential effect is forced nuclear fission of uranium under the influence of thermonuclear neutrons generated by fusion of deuterium- tritium fuel in the central region of the target. In these computations we use a simple analytical description of forced nuclear fission of uranium by thermonuclear neutrons. The critical conditions are not achieved in the considered (not optimized) hybrid targets but they are close to accomplishment in the investigated shock-free compression regime. This regime of compression is the most adequate one for hybrid HIF targets.

The obtained results allow us to make conclusion of advisability of further development of energetic hybrid HIF targets particularly their optimization and utilization of natural uranium as pusher materials.

1. INTRODUCTION

Development of hybrid targets is quite natural in the frame of controlled nuclear fusion HIF, in which the high-current heavy ion beams execute a role of an original energy input for compression and heating of target materials. As it is well known [1] the construction of such targets includes at least two shells from normal density high-Z metals. A paraxial capsule filled by cryogenic deuterium-tritium (DT) fuel is enclosed into the interior shell . This shell is called the pusher and/or the radiation shield as it provides both fuel compression and a barrier for volumetric loss of energy by radiation. The ion beam deposit their energy in the region placed outside the pusher and bounded from outside by the second metallic shell called the tamper. It

CP849, *Zababakhin Scientific Talks - 2005*,
edited by E. N. Avrorin and V. A. Simonenko
© 2006 American Institute of Physics 0-7354-0345-7/06/$23.00

restrains expansion of the energy deposit region. The pusher and the tamper are made from heavy metals, lead and gold for example. For transition to a hybrid (fusion-fission) target it is evidently necessary to make barely the pusher from a fissible materials , for example from pure isotope of ^{235}U. In [2] it was firstly given some estimations of capability to obtain chain reaction of nuclear fission in the conditions of HIF targets.

On Figure 1 we show the scheme of the HIF target [3] presented by ITHEP on the international conference HIF 2002. This target exposes the two-stage energy input: 1) on stage of compression and 2) on stage of fuel ignition. The parameters of two ion beams used on these stages are presented on Figure 1. Let us remark the target HIF 2002 (conventional name) has axial-symmetric construction but the compression stage ion beam rotated sufficiently fast is braked in a metal shell (for instance porous gold). On Figure 1 it is schematically shown a detonation wave of fuel ignition in which energy-release exceeds energy input in G >100 times (without taking thermonuclear neutrons into account) that gives needed energetic effect.

Let us emphasize that cylindrical symmetry is natural one for HIF targets in contrast to laser induced inertial synthesis targets because of known properties of high-current accelerators (drivers), sources of heavy ion beams. Therefore in this paper we restrict ourselves consideration of axial-symmetric targets assumed certainly only direct action targets as it is presented on Figure 1. Actually the hybrid HIF targets possess only axial symmetry because of both non-uniform energy input along of the target axis and finite length of cylinder target. That is way it is possible to take into consideration the scheme of hybrid HIF target where in comparison with the target on Figure 1 the pusher is made from uranium ^{235}U and some other non-important modifications of the target construction are done.

In [4] the initial parameters of the target and the results of computations for numerical models are given and it is shown significant growth of the relative coefficient of energy-release in the hybrid target with U-pusher in comparison with the traditional target with Au-pusher.

We introduce multiplication coefficients K_{totl}, K_U, K_{DT}, described relative energy-release in the target, U-pusher and DT fuel respectively by the following formulas:

$$K_{totl} = K_U + K_{DT}, \quad K_U = \mathcal{E}_{U \exp l} / \mathcal{E}_{input},$$

$$K_{DT} = \mathcal{E}_{DT \exp l} / \mathcal{E}_{input}, \quad \mathcal{E}_{expl} = \left\{ [1 + 4(1-\delta)] K_{DT} + K_U \right\} \mathcal{E}_{input},$$

(1)

here $\mathcal{E}_{U \exp l}$ is the total energy-release in the U-pusher, $\mathcal{E}_{DT \exp l}$ is the total energy-release in DT fuel, and in addition $\mathcal{E}_{expl}$ is the genuine total energy-release where the parameter δ means a rate of absorbed thermonuclear neutrons and finds easily as $\delta = 2.13 \cdot 10^{-2} (K_U / K_{DT})$. The expression $\mathcal{E}_{expl}$ from (1) can be rewritten by extraction from it the uranium part $K_U^* = K_U - 4[1 - (1-\delta)] K_{DT}$ of the energy of thermonuclear neutrons absorbed by the pusher. We imply that $K_{DT}^* = 5 K_{DT}$ as it is true without the uranium pusher and so $\varepsilon_{expl} = K^*_{DT} + K^*_U$. The expression for K^*_U can be simplified by substitution in it the expression for δ :

$$K^*_U = K_U - 4\delta K_{DT} = 0.915 K_U .$$ (1a)

Finally we write the total energy-release for the hybrid target with DT fuel and an uranium pusher as

$$\varepsilon_{expl} = [5K_{DT} + 0.915 K_U] \varepsilon_{input} ,$$ (1b)

where K_{DT} and K_U are the partial rate of energy-release in the fuel and the pusher respectively. To run ahead we say that the target ignition is implemented when $\varepsilon_{input} \gtrsim 10$ MJ and level of uranium burnup is a few percents (see Table 1 in Section 4). We emphasize that all quantitative values of energy-release as well as energy-input are computed for the cylindrical target of one centimeter length that it is much more typical radial sizes.

2. DETERMINATION OF THE CRITICAL PARAMETERS OF THE URANIUM PUSHER

The critical parameters which we are interested in might be find sufficiently accurately by direct solution of well-known integral Paierls equation written down for neutron concentration in the cylindrical symmetry case. This problem is considered under the simplifying conditions: 1) homogeneity of the pusher materials;2) constancy of all neutron cross-sections (f, r, s – type of nuclear reactions); 3) isotropy of neutron elastic scattering (s); 4) one –speed approach (neutron velocity is $v_n = const$ and is identified with maximum of nuclear fission spectrum). The problem is solved in mentioned geometry by the longstanding method [5]. Here we give only final interpolation formula between the exact solutions of two limit cases of diffusion and kinetics, see in details [6]. So for the critical radius R_{cr} of the uranium rod we have the seeking solution:

$$\frac{1}{\beta R_{cr}} = \begin{cases} 0.720\xi + 0.368\xi^2, & 0 \le \xi \le 0.433 \\ 1.35 + 1.199(\xi^2 - 1), & 0.433 \le \xi \le 1.00, \end{cases}$$ (2)

where the dimensionless parameter $\xi = \sqrt{1 - \alpha/\beta}$ is expressed by atomic constants of the pusher matter:

$$\alpha = N(\sigma_f + \sigma_r + \sigma_s), \quad \beta = N(v\sigma_f + \sigma_s), \quad N = \frac{\rho}{A_0 m_0}$$.(3)

Here σ_f , σ_r и σ_s are the cross-sections of nuclear reactions: fission (f), radiation absorption (r) and elastic scattering (s), N is concentration of material atoms (m_0 is atomic mass unit), v is the coefficient of neutron multiplication per nuclear fission reaction. For the material under interest, ^{235}U with normal density $\rho = \rho_0 = 18.7 \, g/cm^3$ and taking known atomic constants into account we find from (3) $R^{bar}_{cr} = 5.37$ cm, that differs from more rigorous determination [6] in a few percents. Note that in (2) for

$\xi \ll 1$ diffusion approach is correct but for $\xi \to 1$ kinetics limit is true. In the case of ^{235}U the parameter $\xi = 0.502$, placed in the middle point of the interpolation between mentioned approaches, see [6].

It seems the obtained solution (2) is correct for the simplest configuration of a pure cylindrical rod and does not have direct connection with considered more complicated case of the uranium shell. But it take place practical coincidence the critical radiuses of the rod and shell under condition that the averaged density $\overline{\rho}_{env}$ of the shell is equal to the averaged density ρ_{bar} of the rod [6]:

$$\rho_{bar} = \overline{\rho}_{env} = \frac{R^2 - R'^2}{R^2}\rho_{env}, \quad \rho_{bar0} = \overline{\rho}_{env0} = \frac{R_0^2 - R_0'^2}{R_0^2}\rho_0, \qquad (4)$$

where R и R´ is outer and interior radius of the shell respectively, symbols with zero subscript are related to the initial state when $\rho_{env0} = \rho_0$. Here we imply all suppositions mentioned in the beginning of this section in particular homogeneity of the pusher matter. The critical radius value R_{cr} is defined by the outer shell radius when the interior radius R´ varies in the limits $R \geq R' \geq 0$). It is clear the critical radius is defined by the formula (for given matter):

$$\rho_{bar} \quad R_{cr} = const = \rho_0 R_{cr}^{bar},$$

from which with (4) we have the following dependencies of the seeking values of the averaged density values of the shell included the initial state:

$$R_{cr} = \left(\frac{\rho_0}{\overline{\rho}_{env}}\right) \cdot R_{cr}^{bar}, \quad R_{cr0} = \left(\frac{\rho_0}{\overline{\rho}_{env0}}\right) \cdot R_{cr}^{bar}, \quad \left(\frac{R_{cr}}{R_{cr0}}\right) = \left(\frac{\overline{\rho}_{env0}}{\overline{\rho}_{env}}\right). \qquad (5)$$

Under definition (4) of the averaged densities the expressions (5) give us possibility to find the critical radius in the process of hydrodynamic compression of the target. In this process under condition of cylindrical symmetry the conservation law of linear mass of the pusher holds:

$$\overline{\rho}_{env}R^2 = \overline{\rho}_{env0}R_0^2, \qquad (6)$$

therefore the third relation in (5) one can easily express through the variable shell radius $R = R(t)$:

$$\left(\frac{R_{cr}}{R_{cr0}}\right) = \left(\frac{R}{R_0}\right)^2, \qquad (7)$$

from which it follows comparatively faster reduction of the critical radius R_{cr} in the process of the pusher compression when the inequality $R_{cr} < R$ holds (in the initial state, generally speaking, it takes place the strict inequality: $R_{cr0} \gg R_0$). It is easily to see from (7) the critical state arises under radial compression in hundred times taking into attention typical sizes of HIF targets [1], [6].

Followed by [5] one can now extract from solution (2) an important generalization in the form of non-stationary solution of Paierls equation for supercritical condition (R > R$_{cr}$). Let us consider the time dependent neutron concentration in the form :

$$n(r,t) = n(r,t_{cr}) \exp\left(\int_{t_{cr}}^{t} \lambda(t')dt'\right),\qquad(8)$$

in the cylindrical symmetry case (r is cylindrical radius) for t ≥ t$_{cr}$, where t$_{cr}$ is time of occurrence of supercriticality. Substituted (8) into original equation one can check correctness of this exponential solution if time-varying function λ defines as:

$$\lambda = \left[\beta\left(1-\xi_t^2\right)-\alpha\right]v_n\qquad(9)$$

where value ξ_t is the root of the algebraic system of equations (2)   under substitution: R$_{cr}$ → R, a ξ → ξ_t (R > R$_{cr}$) in it. The parameters α and β are given in (3), v_n is a neutron velocity. Recall we consider as before one-speed approach. In the solution (8) the exponent index is an integral over time until a moment t$_{fin}$ of loss of supercritical condition. Introduction of the integral in (8) is justified in the usual case of large domination of neutron velocity v_n under a typical velocity of hydrodynamic target compression [6]. In this sense the solution (8) might be considered as quasi-stationary (with respect to variable hydrodynamic characteristics). The solution (8)) for supercritical condition defines by the initial neutron concentration n(r,t$_{cr}$.). It is important to emphasize the chain self-sustaining fission reaction gives significant impact in energy-release of the hybrid HIF target only in the case of sufficiently high values   of   this   concentration   n(r,t$_{cr}$).   One   can   show   by   estimations   that $n(r,t_{cr}) \geq 10^{16}\,\mathrm{cm}^{-3}$ [6]. The mentioned value can be provided by powerful outlet of thermonuclear neutrons from preceding of thermonuclear fuel fusion in the paraxial fuel region of the target. Before achievement of critical condition in the pusher the burst of neutrons has already created good energy-release in the pusher thanks to nuclear reactions of forced uranium fission.

3. FORCED NUCLEAR FISSION IN THE PUSHER

One can write rather obvious formula for the thermonuclear neutron concentration within the pusher with cylindrical radiuses $R'(t) \leq r \leq R(t)$ using simplification of «filamentary» neutron source (coinciding with the target axis), and neglecting by impact of scattered neutrons and secondary neutrons:

$$n(r,t) = \frac{Q(t)}{2\pi v_{th}\, r} \int_0^{\infty} \frac{d\zeta}{1+\zeta^2} \cdot \exp\left[-\sqrt{1+\zeta^2}\cdot\Phi(r,t)\right],\qquad(10)$$

where the function Φ(r,t) has the following explicit view:

$$\Phi(r,t) = \frac{1}{m_0}\left[\frac{\sigma_{totl}^{(1)}}{A^{(1)}}\int_0^{R'(t)} \rho(r',t)dr' + \frac{\sigma_{totl}^{(2)}}{A^{(2)}}\int_{R'}^{r} \rho(r',t)dr'\right],\qquad(11)$$

225

its physical sense is «optical» radial width, as long as $\sigma_{totl}^{(1)}$ and $\sigma_{totl}^{(2)}$ are the total cross-sections in the fuel and the uranium pusher respectively for the thermonuclear neutrons of velocity $\mathcal{V}_{th}$ ($A^{(1)}$ и $A^{(2)}$ are the atomic weights of DT fuel and uranium). Further, the power Q(t) of «filamentary» source expresses easily through the energy-release density $\varepsilon_{th}(r', t)$:

$$Q(t) = \frac{2\pi}{\varepsilon_{0th}} \cdot \int_0^{R'(t)} \varepsilon_{th}(r',t) \cdot \rho(r',t) \cdot r' dr', \qquad (12)$$

where ε_{0th} is local energy-release per act of deuterium-tritium fusion reaction, dimension of Q(t) is $cm^{-1}s^{-1}$. In (12) it takes evidently the total power that concentrates in «filamentary» source (10).

Finally the energy-release density ε_{fth} of the considered forced fission reaction of uranium expresses through the neutron concentration n_{th}, given by (10) - (12):

$$\varepsilon_{fth}(r,t) = \frac{1}{A^{(2)} m_0} n_{th}(r,t) \sigma_{fth} \, \mathcal{V}_{th} \varepsilon_0 , \qquad (13)$$

where σ_{fth} is the fission cross-section of thermonuclear neutrons, ε_0 is energy-release per a fission act. All atomic constants in (10) - (13) (v_{th}, ε_{0th}, $A^{(1)}$, $\sigma_{totl}^{(1)}$, $\sigma_{totl}^{(2)}$, ε_0, $A^{(2)}$, σ_{fth}) are well-known and here we do not give their values. Let us remark only the functions R(t) и R´(t) and the integrand functions $\varepsilon_{th}(r',t)$, $\rho(r',t)$ are found out from hydrodynamic computation, see section below. It makes sense to stress that given simplified description of uranium forced fission effect in the pusher underestimates most likely the neutron concentration and the corresponding energy-release due to the main assumption of lack of impact of scattered neutrons and secondary neutrons. Their role can be significant especially when the pusher drives near the critical state but it needs incomparably more complicated mathematical tool then it is used for derivation of the expressions (10) - (13).

Before achieving the critical state the energy-release in the pusher estimates by value ε_{fth} from(13). For the supercritical pusher we add to this value the energy-release of chain fission that expresses similarly to ε_{fth} through the neutron concentration increasing in time according to the non-stationary solution (8):

$$\varepsilon_f(r,t) = \frac{1}{A^{(2)} m_0} \cdot n(r,t) \sigma_f \, v_n \varepsilon_0, \qquad (14)$$

where besides $n(r,t)$ from (8) the fission cross-section σ_f and the neutron velocity v_n appear. In the frame of our one-speed approach it is natural to choose them with averaged velocity of neutrons over fission spectrum of uranium. Let us remark that at the transition of the pusher in the supercritical regime, $t = t_{cr}$, the ratio $\varepsilon_f / \varepsilon_{fth} = (\sigma_f v_n)/(\sigma_{fth} v_{th}) \sim 1$; strictly speaking this ratio is a little less than 1, but it is acceptable for the considered simplified description of fission reaction in the pusher matter.

4. HYDRODYNAMICS SIMULATION OF COMPRESSION AND HEATING OF HYBRID HIF TARGETS AND GENERAL RESULTS

In this work such simulation is carried out by the 1D (cylindrical symmetry) one-temperature (equality of ion, electron and photon temperature) ideal hydrodynamics with the simplest material equation of state. Nevertheless this model includes the full kinetics of thermonuclear reactions in deuterium-tritium fuel (four reactions for three nucleuses , D, T и ^3He), as well as the above presented model of nuclear fission in the uranium pusher. The aim of the computations is to present the conception of hybrid HIF targets under a complicated combination of thermonuclear reactions (fusion) and nuclear fission in the typical hydrodynamic processes for HIF targets.

As a typical target we take four-shell target depicted on Figure 1, its parameters - radial sizes of shells and their densities - are given on Figure 2. The shell masses (per centimeter of target length) are: $m_1 = 0.00628$; $m_2 = 0.2575$; $m_3 = 2.0347$ и $m_4 = 2.8283$, [g] . In [4] one can find a few variations of this construction as well as its detailed elaboration which do not play a principal role in the presented conception.

As stated above to achieve sufficiently high compression in radial direction (this is important for enhancement of neutron reaction impact) the energy-input in the lead (*Pb*) shell is specified uniformly according to its mass but its time distribution is profiled by special rule. Such profiling provides sufficiently long shock-free regime of target compression. The rule of energy-input profiling is proposed in [7] and implemented in the presented computations. Figure 3 shows plots of power of energy-input per unit mass $Q_{input}(t)$ (TW/g) and the total energy-input in the lead (*Pb*) shell

$$\varepsilon_{input}(t) = m_3 \int_0^t Q(t')dt' \text{ [MJ]}$$ for given parameters of the total energy-input $\varepsilon_{input}^{totl}$ and

the maximal power of energy-input Q_{max} (we consider two typical variants: 12 MJ and 500 TW /g , 21 MJ and 800 TW /g). There is a moment of time t^* when for all

$t > t^*$ $\varepsilon_{input}(t) = \varepsilon_{input}^{totl}$. Before this time the regime of shock-free compression is

interrupted (see Figure 3) a «forepart» that, firstly, satisfies practically a requirement dictated by physics of HIF driver and bounded the total power of heavy ion beam by the value 1000 TW [8] and, secondly, initiates finally a shock wave for ignition of thermonuclear fuel. It is appropriate that a correct physical description of fuel ignition process make necessary consideration of additional factors, particularly non-equilibrium state of radiation processes in deuterium-tritium plasma - at least one can consider three-temperature approach [9] instead of one-temperature one.

Table 1 presents a few general results of hydrodynamic computations (these results are analogous to ones from [4]). Definition of dimensionless multiplication

coefficients K_{totl} и K_U are given in (1). Table 2 besides the total energy-inputs $\varepsilon_{input}^{totl}$

gives maximal power of the energy-input Q_{max}, accordingly above-mentioned shock-free regime. As in [4] here we compare the traditional target with the golden pusher (first column of Table 1 with designation Au) and the hybrid target with the uranium pusher. The corresponding results are given in the next two columns. We can see the multiplication coefficient K_{totl} for the hybrid target increases almost in three times in comparison with the corresponding background variant and impact in this total energy-release growth gives also growth of thermonuclear energy-release. Let us recall Table 1 gives only partial energy-release due to definition of K_{totl}. From (1) one can easily obtain the total energy-release growth taking into account thermonuclear neutrons outgoing from the target outside (apart from they small portion δ absorbed by the uranium pusher). In accordance with (1) this total energy-release is $\varepsilon_{expl}/\varepsilon_{input}^{totl} = K$ and equal to 135.2, 103.8 and 82.4 for the corresponding energy-input of 12, 16, 21 MJ from Table1. Using the representation (1b) for the total energy–release one can find that the partial multiplication coefficients for the fuel are 105, 80 and 65 respectively. These values ought namely be compared with the values $5 \cdot K_{totl}$ of the traditional variants , i.e. with 90, 70 and 55. Thus for the hybrid target the energy-release in the DT fuel increases approximately at 15% due to stimulation of DT fuel fusion. The effect is conditioned by the uranium pusher implosion that it is sequent of forced fission reactions with their powerful energy-release. In the whole for each variant the total energy-release increases about in 1.5 times (taking into account outgoing neutrons).

In addition Table 1 shows the uranium burnup rates (about 2%). Let us remark that energy-release in the pusher has non-uniform distribution with peak near the pusher interior side in according to (10)-(13). Such peak makes for developing implosion, although this effect is quantitatively not too large. Finally we can mark on satisfactory coincides of the presented results with the results from [4]. These results together show that in the case of the hybrid targets dependence of efficiency from input data (characterizing energy-input) is very insignificant. And what is more the highest coefficients K_{totl} are obtained near threshold of thermonuclear fuel fusion, i.e. near $\varepsilon_{input}^{totl} \cong 10$ MJ ($Q_{max} = 500$ TW/g) accordingly Table 1. The most economical variant corresponds to $\varepsilon_{input}^{totl} \cong 12$ MJ ($Q_{max} = 500$ TW/g).

Table 1. Multiplication coefficients K_{totl} and K_U for the traditional and hybrid targets

$\varepsilon_{input}^{totl}$, MJ ; Q_{max} , TW/g	Au	Uranium 235		
	K_{totl}	K_{totl}	K_U	
12 ; 500	18	54	33 (burnup: 2.29%)	
16 ; 500	14	42	26 (2.40%)	
21 ; 800	11	32	19 (2.28%)	

5. THE DETAILED RESULTS OF HYBRID HIF TARGET HYDRODYNAMIC SIMULATION

For the variant presented in the first row of Table 1 we give below a few detailed results of implemented hydrodynamic computations. On Figure 4 (upper frame) we show r-t diagrams of both interior and outer radiuses of the uranium pusher on the whole time interval. The plots corresponding to the interval of significant compression are presented on Figure 4 (down frame) on a large scale. This frame shows the most interesting hydrodynamic processes in the pusher and DT fuel. These processes have pointed extremal characteristics of averaged densities of DT и and U layers. Even for the pusher the value of compression is very large achieved extremal value ~10^3 times. Both lagrangean trajectories R and R' practically merge into one on the left frame long before extremum point. However on the right frame they distinguish well and show at $t = 1031.6$ ns the explosion of the pusher with typical phenomenon of above mentioned implosion. On Figure 4 it is also given the plot of critical radius R_{cr}, computed by (4) и (5). This plot is to compare with the plot of the outer pusher radius R. One can see that in each time undercritical condition $R_{cr} > R$, holds and in the minimum of the plot R_{cr}, $t = 1031.3$ ns, the ratio $R_{cr} / R \cong 6$. It corresponds approximately to maximal convergence of the considered plots. One ought to have into consideration that on the initial stage, $t = 0$, such ratio is about 280 as from (4) and (5) and initial data from Figure 3 we find $\overline{\rho}_{env0} = 1.86$ g cm^{-3}, and $R_{cr0} = 58.0$ cm. The relation (7) is performed. In addition on Figure 4 it is given the coefficients K_U и K_{DT} (as in Table 1) and the total energy-release $\varepsilon_{expl} = 1622$ MJ , see formula (1b) .

The comparison of the hybrid target with the uranium pusher and the traditional target with golden pusher is given on Figure 5. On this Figure we show r-t diagrams both shells obtained in the hydrodynamic computations of both above-mentioned targets. To see the differences distinctly we use more larger scale than one on the down frame of Figure 4. The dashed lines are outer and inner radiuses (R and R') of the Au-pusher, they are marked by symbol « x » and « o » respectively. The solid lines are corresponding radiuses of the U-pusher, they are marked by symbol « ◊ » and « □ » respectively . At the moment of the maximal compression the Au-pusher drives differently than the U-pusher. Differences begin at t $\cong$ 1031.1 ns, when forced fission reactions have heated significantly the U-pusher. As far as this heating acts the minimal interior radius $R' \cong 0.25 \cdot 10^{-2}$ cm of the U-pusher becomes distinctly less than the minimal interior radius $R' \cong 0.35 \cdot 10^{-2}$ cm thanks to an implosion phenomena of the U-pusher.

The acute expressed moment $t = 1031.6$ ns of implosion coincides practically with thermonuclear explosion (maximum of energy-release power) given on Figure 6 where it is also shown the plot of average neutron concentration in the U-pusher. We can see that the maximal power corresponds to maximal gradients of functions $\varepsilon_{U\,expl}$ (t) and $\varepsilon_{DT\,expl}$ (t) (exactly at $t = 1031.6$ ns), as well as maximum of average neutron concentration. The latter achieves value 10^{23} cm^{-3} and significantly exceeds the above-mentioned estimation of original concentration (10^{16} cm$^{-3)}$ that needed for efficient chain fission reaction. Finally we pay attention on extremely short-term thermonuclear fuel explosion roughly speaking   it is equal to 0.1 nsec. We remark that excess of energy-release $\varepsilon_{U\,expl}$ (t) over $\varepsilon_{DT\,expl}$ (t) (Figure 6) occurs only for partial energy-releases in the pusher and the fuel whereas the ratio of the full energy-releases taking outgoing thermonuclear neutrons into account (see (1) and (1b)) is inverse in these computations.

6. CONCLUSION

In this paper the method of description of neutron-induced fission in the uranium pusher is developed. It provides needed foundation for hydrodynamic computations of hybrid HIF targets with large energy-release. The considered hybrid targets differ mainly from pre-existing energetic HIF targets by exploiting the shock-free regime [7]. This one is intended for achievement of so very high density of matter in the pusher and the fuel that energy powers of thermonuclear fusion and neutron-induced fission are compared. We ought remark   the total energy-input, typical time distribution of power and its maximal value are the same that it is established for energetic HIF targets from point of view of heavy ion driver physics. However the presented consideration is restricted a typical cylindrical symmetry geometry, one-temperature equation of state. The implemented analytical constructions and results of hydrodynamic computations allow us to state that the use of hybrid HIF targets provide: 1) essential growth of energetic output of plant; 2) advance of efficiency of thermonuclear fusion of deuterium-tritium fuel due to the pusher implosion; 3) significant increase of fuel compression and generation of self-sustaining ignition process in the system «pusher-fuel»; 4) approaching of critical state of the uranium pusher in which besides of forced fission of uranium appears chain self-sustaining fission.

Critical conditions are nod achieved in the considered (no optimized) HIF targets but they are very close to attainment in the investigated shock-free compression regime. Let us remark that impact of chain self-sustaining fission might be increased by using other fissible matter as well as by optimization of combination of nuclear fusion and neutron fission in hybrid targets.

The obtained estimations of this work corresponds with previous paper [4] and confirms by sufficiently complete formulation [10], in which description of neutron reactions and equation of state were improved. It n naturally appear that in the approach of this work the investigated effects were generally speaking underestimated.

ACKNOWLEDGEMENTS

The authors would like to thanks gratefully V.I. Subbotin who stimulates this work and suggest an idea to use natural uranium in hybrid HIF targets, A.V. Zabrodin and G.V. Dolgoleva for useful discussions, L.G. Bass and O.V. Nikolaeva for helpful impact by independent verification and computation of critical parameters of uranium rod and shell, B. Yu. Sharkov, D.G. Koskarev, M.D. Churazov, M.V. Maslennikov, Yu.N. Orlov for interest and attention to this work.

REFERENCES

1. Basko M.M., Imshennik V.S., Churazov M.D., Overview of Directly Driven HIF Targets. Particle Accelerators, 37-38, 505-512 (1992).
2. D.G. Koskarev, B. Yu. Sharkov. Nuclear fission under inertial confinement. Letters in JETF. 75(7), 2002, p. 371-373.
3. M.M. Basko, C. Yu. Guskov, C.L. Nedoseev, M.D. Churazov. HIF targets. In: Nuclear fusion under inertial confinement. Edit by B. Yu. Sharkov. Moscow, Fizmatlit, 2005, p.53-56
4. N.N. Alekseev, M.M. Basko, E. A. Zabrodina, V.S. Imshennik, D.G. Koskarev, M.D. Churazov, B. Yu. Sharkov, G.V. Dolgoleva, V.T. Zhukov, A.V. Zabrodin, M.V. Maslennikov, Yu.N. Orlov, V.I. Subbotin. Development of energetic plant of fusion and fission on basis of microtargets of direct action and heavy ion driver. Atomic energy, 97(3), 2004, p. 200-209
5. A.I. Axiezer, I.Ya. Pomeranchuk. Some questions of nucleus theory. Moscow-Leningrad, Gos. Izd.tech.-theor. Lit, 1950
6. V.S. Imshennik. Analytical method of computation of chain reaction fission for cylindrical uranium rod and estimation of energy-release of hybrid HIF targets. Nuclear physics. 2006, to appear
7. G.V. Dolgoleva, A.V. Zabrodin. Energy cumulation in layer structure systems and implementation of shock-free compression. Moscow, Fizmatlit, 2004
8. D.G. Koskarev. Heavy ion driver of megawatt power level. Preprint ITEP, 03-06, 2006
9. A.V. Zabrodin, G.P. Prokopov. Method of numerical simulation of 2D non-stationary flows of heatcontucting gas in three-temperature model. VANT. Mathematical modeling of physical processes. 3 (1998).
10. G.V. Dolgoleva, A.V. Zabrodin. Computational constructing a combined target conjoining implementation of fusion and fission under heavy ion driver. XXXIII Int. Conf. on Plasma Physics and Controlled Fusion, Zvenigorod , 13 – 17 February, 2006

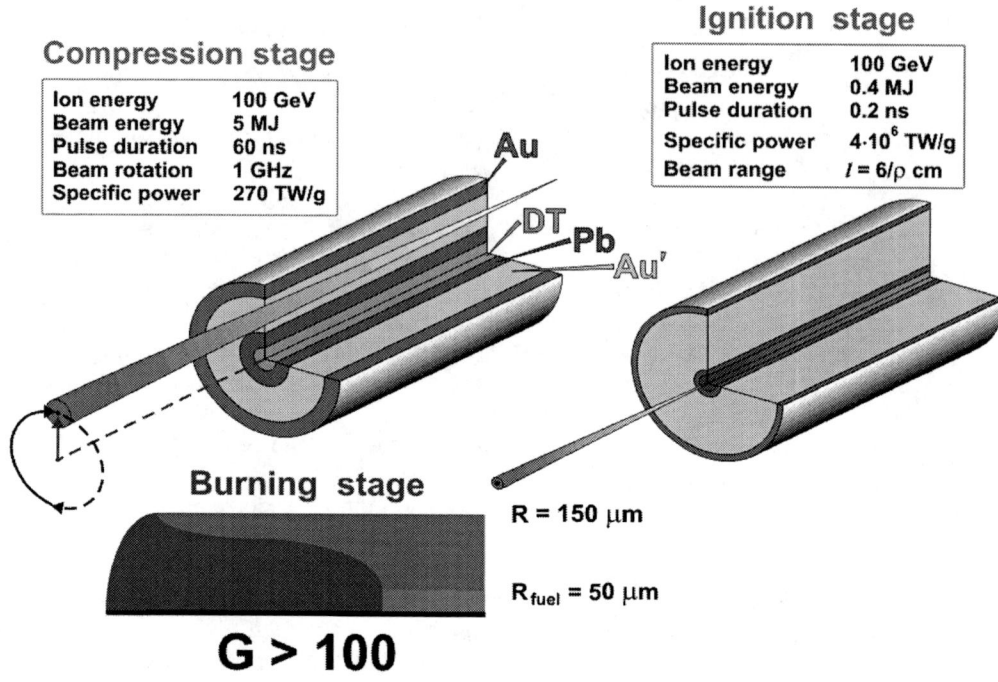

Density :	0.05	20	6	20	g/cm^3
	DT fuel	Pusher U (Au)	Tamper Pb	Au	
r : 0	0. 20	0.21	0.39	0.444 cm	

Figure 2. The initial parameters of the hybrid and traditional targets in hydrodynamic computations:
shell materials, density of materials and radial sizes of shells

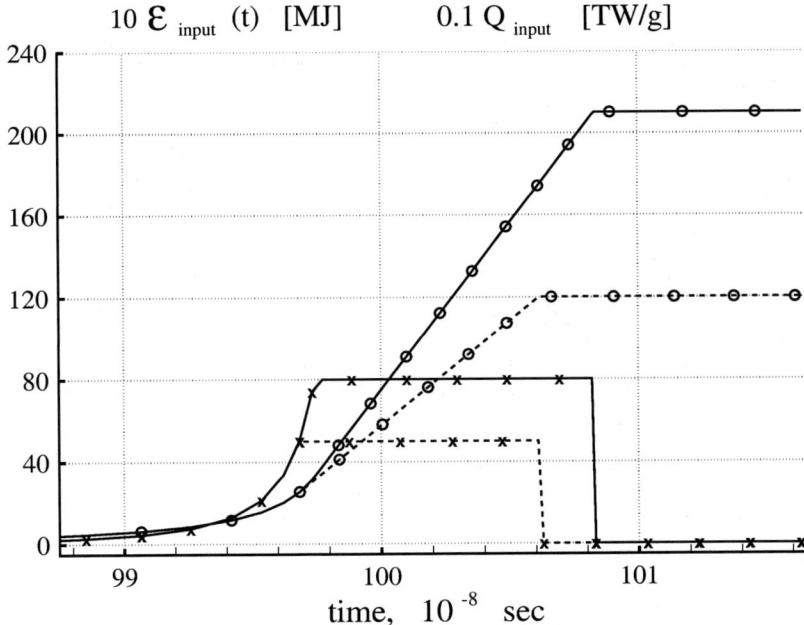

Figure 3. The power energy-input $Q_{input}(t)$ (multiplied by 0.1 and depicted by « x ») and the total energy-input $\varepsilon_{input}(t)$ (multiplied by 10 depicted by « o ») for two variants of values Q_{max} and $\varepsilon_{input}^{totl}$ (500 TW/g and 12 MJ - dashed lines , 800 TW /g and 21 MJ - solid lines) .

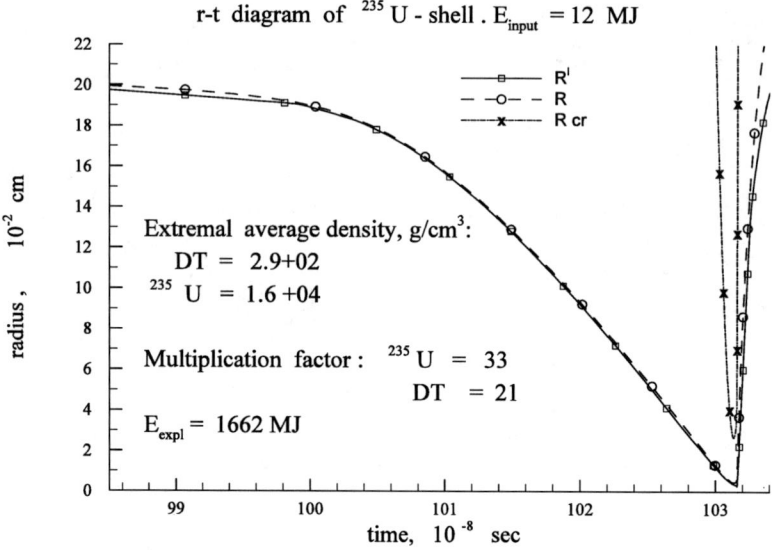

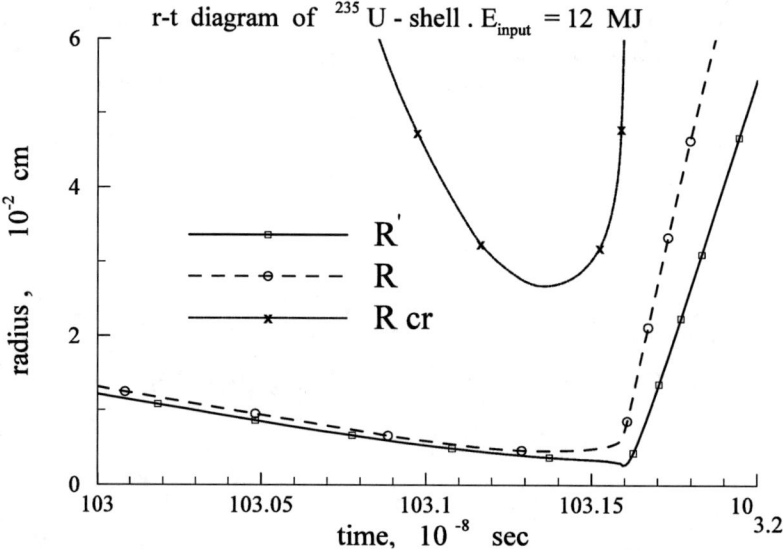

Figure 4. The results of hydrodynamic computations of the hybrid target (see Figure 1) on application of 12 MJ energy input. R-t-diagrams of outer radius (R) and interior one (R′) of the U-pusher and its critical radius (R_{cr}). The down frame is a fragment of the upper frame.

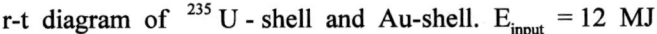

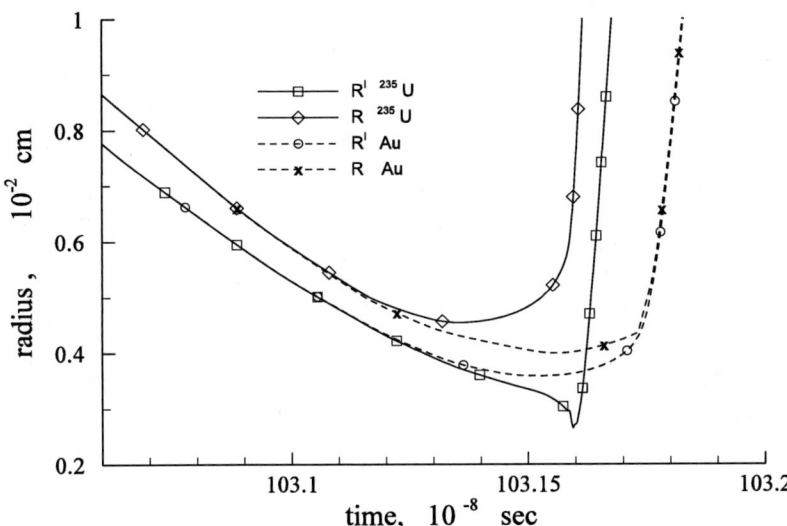

Figure 5. Comparison of r-t-diagrams for hydrodynamic computations of the hybrid target (see Figure 4 caption) and the traditional target with the golden pusher. The scales are reduced to show time interval of maximal compression.

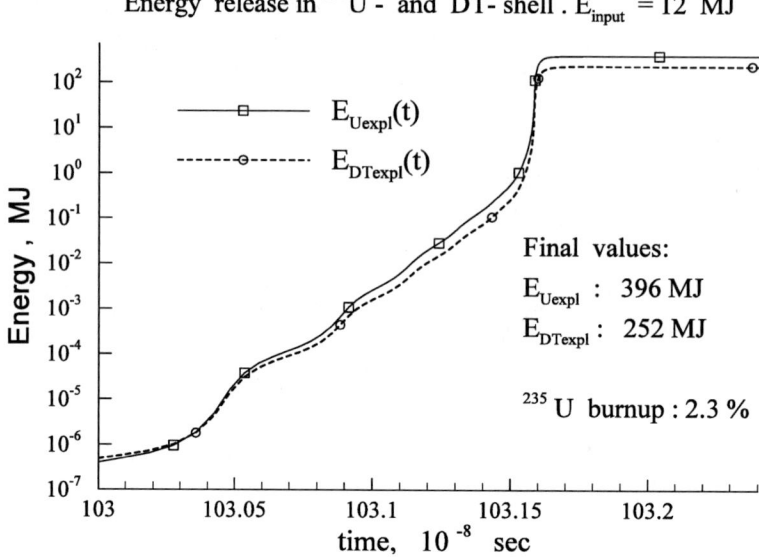

Energy release in ^{235}U - and DT- shell . E_{input} = 12 MJ

Energy , MJ

E$_{Uexpl}$(t)

E$_{DTexpl}$(t)

Final values:

E_{Uexpl} :   396 MJ

E_{DTexpl} :   252 MJ

^{235}U burnup : 2.3 %

time, 10^{-8} sec

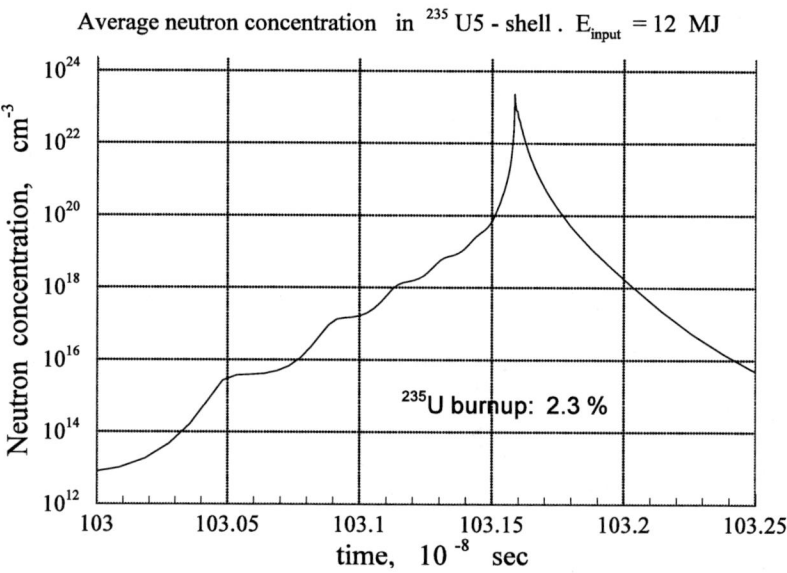

Average neutron concentration   in ^{235}U5 - shell . E_{input} = 12 MJ

Neutron concentration,   cm^{-3}

^{235}U burnup: 2.3 %

time, 10^{-8} sec

Figure 6. Plots of the partial energy-release (upper) in the pusher and the fuel and the radial average values of the thermonuclear neutron concentration (down) in the U-pusher . The hydrodynamic data are given in Figure 5.

High-Intensity Laser Pulse Interaction With Solids of Variable Density

V.S. Belyaev[1], A.P. Matafonov[1], V.I. Vinogradov[1], S.G. Garanin[2],
N.G. Borisenko[3], A.M. Chekmarev[3], N.N. Demchenko[3], A.I. Gromov[3],
S.Yu. Gus'kov[3], A.M. Khalenkov[3], Yu.A. Merkul'ev[3], V.B. Rozanov[3],
V.P. Andrianov[4], G.N. Ignat'ev[4], Yu.E. Markushkin[5] and V.G. Pimenov[6]

[1]Central Research Institute of Machine Building, Korolev, Russia
[2]Institute of Laser Physical Research of Russian Federal Nuclear Center, Sarov, Russia
[3]Lebedev Physical Institute of Russian Academy of Sciences, Moscow, Russia
[4]Research Institute of Pulsed Technique, Moscow, Russia
[5]Bochvar Research Institute of Inorganic Materials, Moscow, Russia
[6]Zelinsky Institute of Organic Chemistry RAS, Moscow, Russia

Abstract. Experimental studies are presented on composition, density and structure influence onto the neutron yield resulting from deuterium-containing target interaction with laser light of 1018 W/cm2. Experimental data is compared with theoretical and simulation results.

Keywords: neutron yield, laser experiment, deuterated beryllium
PACS: 52.57.Bc, 52.38.Ph, 52.70.Nc

1. INTRODUCTION

The dependence of neutron generation from targets containing deuterium on pulse energy, intensity, contrast (prepulse) is studied in many laboratories with super-high intensity lasers. Solid density CD_2, C_8D_8, C_2D_4 of density ~ 1 g/cc and jets of deuterated clusters D_2, CD_4 with diameters less than 0.01 micron [1] were mostly used as targets for such laser shots.

In our experiments we have studied the effect of chemical composition of the target (the comparison of plates of $(CD_2)_n$ and BeD_2), target density (the comparison of solid polyethylene 1g/cc, powder of meshed density 0.1g/cc, and polyethylene foam 0.01-0.04 g/cc), and target structure (foam compared to films 20- and 100-micron thick separated by 40-micron spacing) on the neutron yield at the constant laser irradiation conditions. Laser pulse is characterized by the following: the beam energy 10 J, the wavelength is 1.055 micron, pulse duration 1.5 ps, contrast by intensity $K > 10^4$. The focusing system allows to concentrate not less than 40% of laser energy in the focal spot of 15 micron, resulting in intensity on the target surface of up to 10^{18} W/cm^2.

CP849, *Zababakhin Scientific Talks - 2005*,
edited by E. N. Avrorin and V. A. Simonenko
© 2006 American Institute of Physics 0-7354-0345-7/06/$23.00

2. EXPERIMENT

The scheme of experiment is given in Figure1. The laser beam is focused onto the target surface by the off-axis parabolic mirror of a focal distance 20 cm. The focused beam struck the solid target surface at the angle of 40° in the vacuum chamber BK of diameter 30 cm and of height 50 cm. The residual gas pressure inside the chamber was no more than 10^{-3} Torr.

The hard X-ray registration was made on a basis of 4 scintillation detectors standing 0.3 m, 3 m, 3.5 m, and 4.1 m away from a target. Lead filters of 1 to 13.5 cm thick were placed in front of the detectors. Three neutron detectors D5 – D7 were used for neutron registration. Helium counter D5 was fixed at 16 cm from the rear side of the target, counter D6 was placed from 1 to 10.4 m in front of the target. Activation Indium detector D7 was placed 22 cm away from the target.

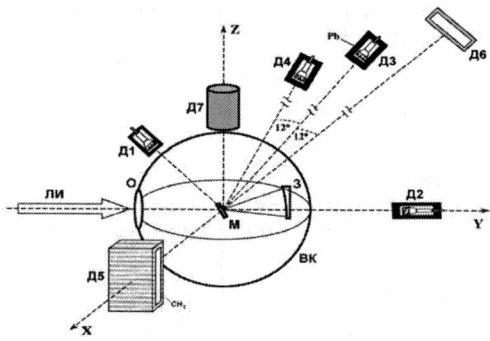

FIGURE 1. Experiental scheme. M – target; BK – vacuum chamber; O – vacuum-chamber window; 3 – off-axis parabolic mirror; laser radiation from the left; D1 – D4 – scintillation detectors of γ-radiation; D5, D6 – helium-counter neutron detectors; D7 – activation detector of neutrons.

TARGETS

In our experiments on picosecond lasers the plates or films from deuterated polymers were used (see Figure. 2). In Fig. 2a the target from CD_2 after 10 different shots is presented.

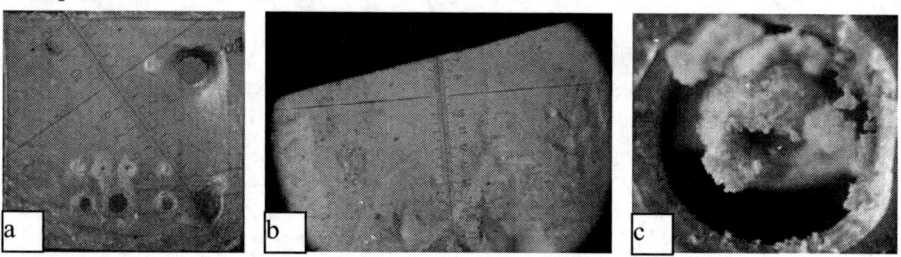

FIGURE 2. a) Target – CD_2 plate with thickness 350 μm after 10 shots; b) the surface $(CD_2)_n$ foam before the shot; c) the same target after shot.

The lower row of craters is obtained at intensities of 10^{18} W/cm^2 and the upper row – at $5 \cdot 10^{15}$ W/cm^2. It is obvious, that the shape and dimensions of the crates are different with almost same intensity of laser pulse. Probably such result was obtained due to inaccurate focusing (caustic length is about 10 μm) and properties of the target itself. For better focusing one should obtain the targets with minor roughness of the surface. The small pore $(CD_2)_n$ foam layers with densities 10, 20, 30 and 40 mg/cc (see Fig 2b and 2c) were used to investigate the influence of density and structure on the neutron yield in strong e.m. fields [4]. Earlier in the references we failed to find the data concerning the fabrication of foam layers from $(CD_2)_n$ with densities lower than 50 mg/cc. The mentioned above foams and also the layers of deuterated epoxy resin with densities as low as 4 mg/cc were synthesized by V.G. Pimenov (IOC RAS).

The thick (0.7- 0.9 mm) plats of extruded BeD_2 (see Figure 3) were also used as targets. The density and dopant concentration fluctuations were measured by microradiograph methods at maximum energies of 2.5 keV. The major achievement in development of fabrication methods of ultra-thin (0.1-0.2 μm) films from BeD_2 was made by N.A. Chirin (Bochvar Institute of Inorganic Materials).

The development of such deuterated material allows us to propose the new target designs – models of ultra low-density foams, cavities filled with "BeD_2 gas" which could be irradiated in vacuum without utilizing protecting windows and so on.

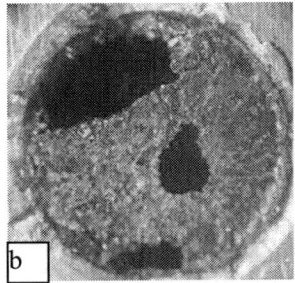

FIGURE 3. a) BeD2 disk with thickness 0.7 mm before shot; b) the same target after 3 several shots.

RESULTS

Neutron yield was measured experimentally for all types of target, while the best results were obtained with double-layered target. Fig.2 exemplifies pulses registered by the neutron detectors D5 and D6 from the deuterated polyethylene of the density 0.01 g/cc (Fig. 2b,c - foam) and 1 g/cc (Fig. 2a - full-density polymer). Figure 5 represents neutron yield measurements for various targets at 10^{18} W/cm^3 . From Fig. 5 it is seen that neutron yield drops with the target density, but even at 0.01 g/cc it is as high as 10^4 neutrons inside 4π steradian for a single pulse. Maximum γ-quant energy also drops from 2 MeV for $\rho = 1$ g/cc to 0.5 MeV for $\rho = 0.01$ g/cc. The neutron yield Y_n calculated per 1 J laser energy varied from $2.9 \cdot 10^4$ to $3.5 \cdot 10^4$, and the absorbed laser energy efficiency, δ_a , was 0.046-0.083.

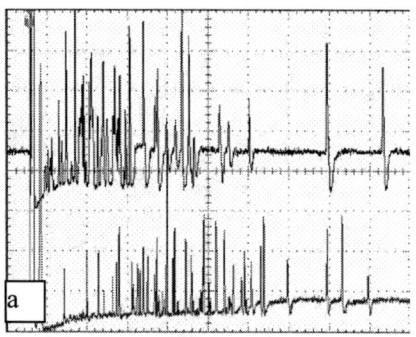 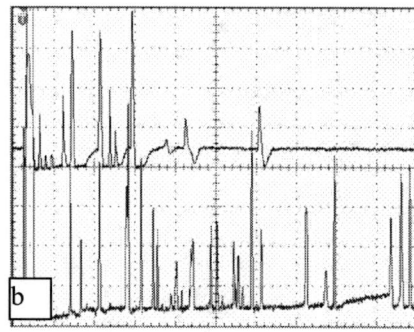

FIGURE 4. Pulse oscillograms from helium-counter neutron detectors D5 and D6 (the upper signal from the D6 analogous detector exit, placed 1 m apart from the target, the lower signal is from the D5 analogous detector exit 16 cmapart from the target): a – a target of deuterated polyethylene foam 0.01 g/cc, b - a target of deuterated polyethylene of 1 g/cc solid density.

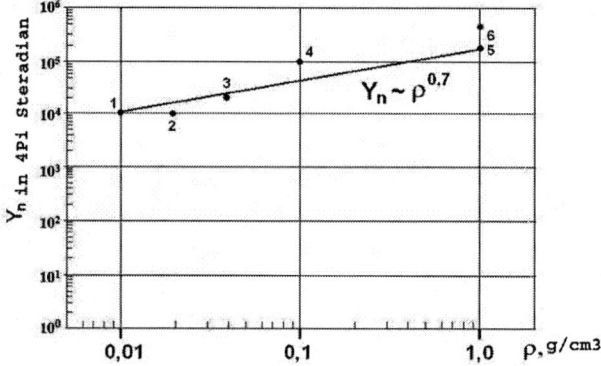

FIGURE 5. Target dependent neutron yield: 1 – foam polyethylene; $\rho = 0.01$ g/cc, $Y_n1 = 10^4$; 2 - foam polyethylene, $\rho = 0.02$ g/cc $Y_n2 = 10^4$; 3 - foam polyethylene, $\rho = 0.04$ g/cc $Y_n3 = 2\times10^4$; 4 - powder, $\rho = 0.1$ g/cc $Y_n4 = 10^5$; 5 - (CD2)n and BeD2 , $\rho = 1$ g/cc $Y_n5 = 2\times10^5$; 6 – 2-layered (CD2)n target. $Y_n6 = 4\times10^5$.

The dependence of Y_n on q_{ASE} was weak. Low values of δ_a are due to a stimulated scattering of a main pulse in a plasma produced by the pre-pulses.At pulse energy of 4 J the calculated neutron yield proves to be close to the experimental value. The effect of the many-fold neutron yield enhancement, at using of a two-layer target, was considered in [2,3]. It was connected with an extra heating of ions at collision of the plasma flows produced by the fast-electron heating of the target layers. It is natural that the neutron yield drops with the target average density. If the density is homogeneous, then the reaction rate drops proportionally to ρ^2, and the plasma volume increases, $\sim\rho$. Under equal temperatures, Y_n drops proportionally to ρ. As the foam-like target is being irradiated the density remains nonhomogeneous for a long time. This explains a weaker dependence of Y_n on ρ.

3. CONCLUSIONS

Experimental studies of deuterium containing target composition, density and structure effects on the neutron yield under laser irradiation of 10^{18} W/cm^2 are carried out.

For the first time the neutron yield Yn $= 10^4 - 2 \cdot 10^4$ is measured from the deuterated polyethylene foam target of the density of $0.01 \div 0.04$ g/cm^2 .

ACKNOWLEDGEMENTS

The work was partly supported by the International Science and Technology Center, projects #2155, #2917. The work was also partly supported by RFBR grants ## 05-02-16551a, 05-08-18167

REFERENCES

1. K.W. Madison, P.K. Patel, T. Ditmire et al., Phys. Rev. A 70 053201 (2004).
2. N.N.Demchenko, V.B.Rozanov, ECLIM 2002, Proc. SPIE, 5228, 427 (2003).
3. N.N.Demchenko, S.Yu.Gus'kov, V.B.Rozanov, Proc. of 28th Europ. Conf. on Laser Interact. with Matter,Roma, 2004, Italy, September, 6th-10th, p. 181.
4. Belyaev V.S., Vinogradov V.I., Kurilov A.S., Matafonov A.P., Andrianov V.P. et al., JETP, 2004, v. 125, #6, p. 1-7.

Intensive (up to 10^{15} W/cm^2) Laser Light Absorption and Energy Transfer in Subcritical Media with or Without High-Z Dopants

N.G. Borisenko[1], I.V. Akimova[1], A.I. Gromov[1], A.M. Khalenkov[1], V.N. Kondrashov[2], J. Limpouch[3], Yu.A. Merkuliev[1], V.G. Pimenov[4]

[1]*P.N. Lebedev Physical Institute of RAS, Leninskyi Pr-t. 53, 117924 Moscow, Russia*
[2]*Troitsk Institute of Innovation and Thermonuclear Research, 142190 Troitsk, Russia*
[3]*Institute of Physics, AS CR, Na Slovance 2, 18221 Prague 8, Czech Republic*
[4]*Zelinsky Institute of Organic Chemistry, 49 Leninskiy Pr-t. 119991, Moscow, Russia*

Abstract. Fabrication methods for low-density fine-structure (cell size <1 μm) 3-D networks of cellulose triacetate (TAC) are developed. Target densities ranged 4-20 mg/cc, similar polymer structures were produced both with no load and with high-Z cluster dopant with concentration up to 30%. Foams of varying density down to 0.25 plasma critical density at the third harmonic of iodine laser wavelength are supplied for laser shots. Experiments with underdense foam targets with and without clusters irradiated on the PALS laser facility are analyzed preliminary, showing strong influence of target structure on process of laser light absorption. Heat and radiation transport in such targets are considered.

Keywords: low-density foams, laser experiments, energy transfer.
PACS: 52.38.-r; 52.50.Jm; 52.57.Bc

1. INTRODUCTION

Doping of target materials with high-Z elements for improving the resulting plasma characteristics and for diagnostic purposes is widely used in laser-plasma experiments. The admixed material can appear in the target either in atomic state or in the form of clusters – association of thousands of atoms. The resulting targets will have the clearly visible fluctuations of density and/or structure on microscopic scales while staying homogeneous on macroscopic scales [1]. We call such targets microheterogeneous targets. In the present work we discuss the recently developed cellulose triacetate (TAC) targets of regular web-like structure and preliminary data illustrating laser interaction with such undercritical low-density and cluster foam targets obtained on the Prague Asterix Laser Facility (PALS) [2].

2. AEROGEL LAYERS AS SMALL-SCALE (REGULAR) 3-D NETWORKS

Most of low-density polymer materials are produced by gel structure formation from the solution with the subsequent drying. Introducing high (>10%) concentration

CP849, *Zababakhin Scientific Talks - 2005,*
edited by E. N. Avrorin and V. A. Simonenko
© 2006 American Institute of Physics 0-7354-0345-7/06/$23.00

of clusters influences the gel formation process essentially. It proves possible only with high polymer concentration (2-3%) solutions. Such structures could be undercritical only for second or third harmonics of Nd-laser and on the whole for wavelengths less then 0.5 µm. Copper clusters were preferred because of their low chemical activity, though the surface still needed to be passivated before processing. The well-known radiation constants necessary for energy transport simulations were of additional advantage in favor of copper. With cellulose triacetate densities of 20 mg/cc and a copper loading of 35%, the gelling process drops rapidly, so the enhanced densities of 3% polymer concentration become obligatory which result in foam densities of 40-50 mg/cc. To research effects of structure, density and dopant on the laser radiation absorption and energy transport processes in structured plasma the targets of TAC, agar and trimethylol propane triacrylate (TMPTA) either with or without clusters (with densities as low as 4.5 mg/cc or 9.1 mg/cc for TAC respectively), have been fabricated. Doping the low-density material with high-Z clusters led to certain structure changes and roughening of the aerogel structure. Scanning electron micrographs of the TAC target surface microstructure are shown in Figure. 1

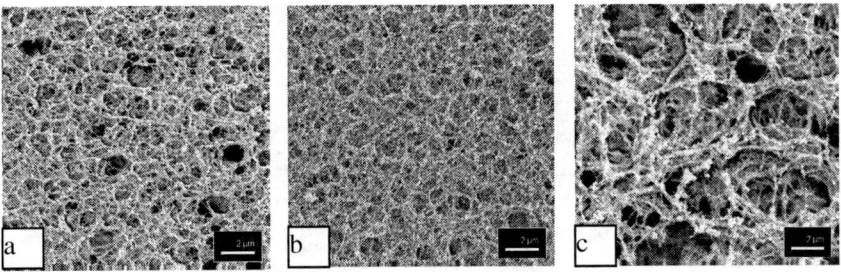

FIGURE 1. SEM images of cellulose triacetate (TAC) foams. a) 4.5 mg/cc; b) 9.1 mg/cc; c) TAC doped with 9.9% Cu clusters, average density 9.1 mg/cc. Scale of 2 µm is provided on each frame.

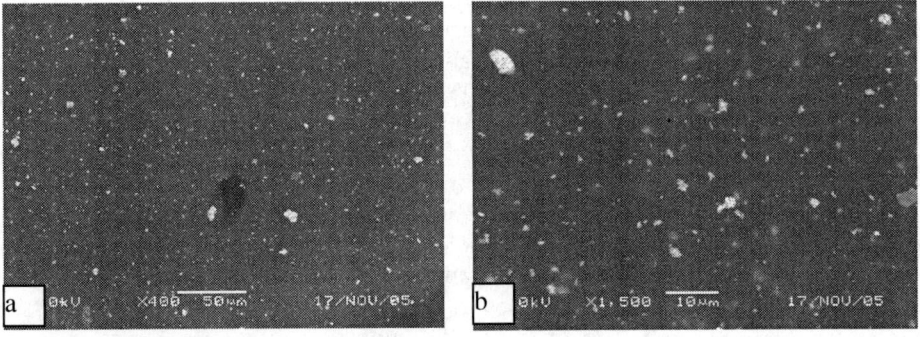

FIGURE 2. SEM images of cellulose triacetate (TAC) foam of density 10 mg/cc with Cu nanoparticles 15% by wt. a) scale 50 µm; b) scale 10 µm.

The characterization process of low-density polymer networks with high-Z adding requires much more parameters to be measured compared to solid targets. Also the interpretation of the results of the experiment and adequate numerical simulations are impossible without clear knowledge of target parameters, both macroscopic and microscopic. The microscopic parameters includes the average pore size, the average diameter of structure fiber, pore size distribution and density fluctuation, in presence of high-Z nanoparticles the particle size distribution and fluctuations of particles concentrations, etc. As the characteristic scales of mentioned parameters are less than 1 μm, the only high-resolution method to obtain the structure is SEM (scanning electron microscopy). The example of space distribution of Cu nanoparticles is presented at Figure 2, the sample is TAC with 10 mg/cc density and 15% of Cu by weigh. The black spot in the middle of the Fig. 2 is due to the foam destruction by electron beam. The data was obtained in "low-vacuum" mode of scanning microscope without evaporation of conducted material on the surface of the foam.

3. TARGET DEPENDENT EXPERIMENTAL DATA

The primary experimental data consists of spatially resolved streak records of the x-ray emission from the lateral side of the target and records of time-resolved optical self-emission from the rear side of the Al foil, presented in the Figure. 3. The x-ray and optical frames for each shot are combined and oriented along the horizontal time axis. Time increases from left to the right, the two-nanosecond scale given by the bars is the same for optical and x-streak records and also refers to the width of laser pulse, presented in the upper left corner of the each frame. Also, the pulse duration in appropriate scale is presented at the upper left corner of the frame. The dashed horizontal line in the x-streak image corresponds to the initial foam target surface, the solid horizontal line corresponds to the Al foil surface. Dashed vertical lines correspond to the times of the pulse maximum and strong optical emission from the rear side of the Al foil. The time fiducial mark is presented in the lower left corner and corresponds to the maximum of the laser pulse intensity. Laser light is incident from the top for x-ray streak and from the left for the optical streak image. We consider the fast heating wave reaching the Al-foil surface at the moment when its luminescence starts. It is indicated by the very left small arrow, the slope yields the fast velocity. The second arrow in the middle position is the tangent to the outside contour of the coronal plasma emitting region. This indicates the second ($V_{X\text{-ray}}$) velocity considered for the visible x-ray emission front propagation in the beginning of the laser-foam interaction. It is not constant in time. Slowing down of the X-ray front is withdrawn from the curvature of the lower boundary for the emitting region. The third arrow (dotted in Fig 2a) is drawn as tangent to the brightest region witnessing the hottest front movement. It may be called the slow "hot" wave (V_{hot}). Strong optical emission from the rear side of Al-foil appears no earlier than this hydrothermal wave provides energy transfer through the foam and foil. The above mentioned transfer mechanisms are indicated for three targets at 3 ω (TAC 4.5 mg/cc, TAC 9.1 mg/cc, and TAC 9.1 mg/cc with Cu clusters of 9.9% wt.) and are presented in Fig. 2a, b, c. The slowing down of the corresponding velocities with density growth (still remaining undercritical!) and changes with clusters added are illustrated qualitatively. The time period between X-ray radiation (of E>1.4keV) cessation and strong optical self-emission of the Al rear side onset is density and dopant additive dependent for a given laser frequency as

244

could be clearly seen from the frames. Both higher density and high-Z clusters present in an aerogel enhanced the time delay of bright optical self-emission. The velocities of the X-ray emission front and hot region propagation front (in 10^8 cm/s) are indicated for each type of target.

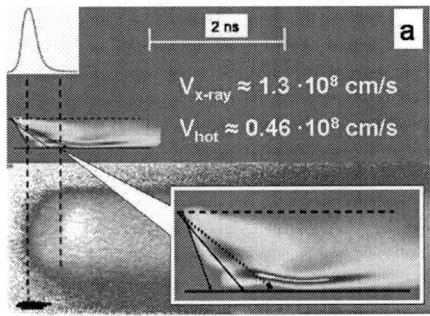

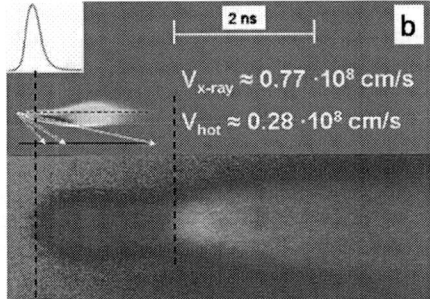 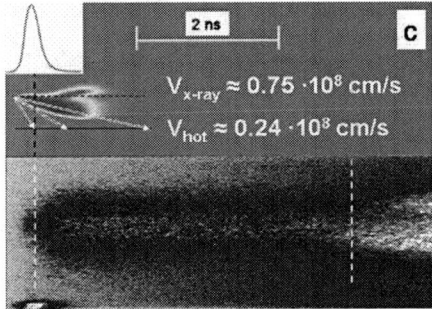

FIGURE 3. X-ray front propagation (upper in each frame) and time delay of the bright optical emission appearing in the optical streak record (lower in each frame) for 3 ω (λ=0.439 μm) shots. a) TAC (cellulose triacetate) 4.5 mg/cc b) TAC 9.1 mg/cc c) TAC with 10% Cu, average density 9.1 mg/cc. Time increased from left to right, time scale (2 ns) is given by the bars.

4. DISCUSSION

High-Z additives in the foam fundamentally change the near-critical plasma behavior. In spite of almost equal mass density and electron density the homogenization process slows down and heating wave arrival to the rear side of *Al*-foil happens later than in the foam without clusters. The present plasma experiments with undercritical polymer targets showed that the laser light is absorbed near the boundary of the low-density layer. A "fast wave" corresponds to the initial heating and homogenization of the low-density material by hard X-rays. As the foam transforms into a homogeneous plasma, the energy transfer process slows down. Further it results in the *Al* ablating into an underdense region thus causing a diminished radiation transfer rate in a highly ionized plasma.

The "X-ray emission front" corresponds to the propagation of soft x-rays in partly homogenized plasma and also causes the intensive Al ablation. The hot front velocity is a slower "hydrothermal" velocity of mass and heat transfer [3], and is responsible for the second, considerably brighter maximum in the Al emission. Different transfer process interplay could be responsible for the outward foam boundary shape. In experiments with clusters there exists an essential time-delay in the optical signal appearing from the rear side of the Al-foil on the foam. This might be connected, firstly, with the radiation cooling of the rarified plasma, and, secondly, with strong flux of the hard X-ray radiation from "foam-born" plasma onto the surface of the aluminium. The latter transfers into an Al-plasma flux encountering the main plasma stream that also cools hot plasma.

5. CONCLUSION

Materials and targets were developed and used for laser experiments studying foam smoothing and transport. Densities as low as $0.25N_{cr}$ were realized for some laser targets, with absolute density for optically transparent 3-D networks being as low as 4.5 mg/cc. High-uniform open cells structures of 1-3 µm cell sizes were demonstrated. Layers of foam with high-Z loadings were realized on the target surface with spread of nanoparticle dispersion up to 30% wt. The introducing of clusters in the target structure strongly affects the process of laser absorption and energy transfer.

ACKNOWLEDGMENTS

Part of the work was supported by the INTAS project # 2001-0572. The authors gratefully acknowledge the heads and staff of PALS laser facility for organizing and performing the experiments, N. N. Demchenko, S. Yu. Gus'kov, O. Renner for guidance and useful discussions, teams of B.Rus and D. Batani and all other participants of the third year of INTAS project # 2001-0572.

REFERENCES

1. N.G. Borisenko, Yu. A. Merkuliev and A.I. Gromov J. Moscow Phys. Soc., v.4, #3, (1994), 47-273

2. K. Jungwirth, A Cejnarova, L. Juna, B. Kralikova, et al. Phys.Plasmas 8, (2001) 2495–2501.

3. A.E. Burov, S. Yu. Guskov, V.B. Rozanov, I.N.Buurdonskiy, V.V.Gavrilov, A.Yu. Goltsov, E.V. Zhuzhukalo, N.G. Kovalskiy, M.I. Pergament, and V.M.Petrakov, Journal of Experimental and Theoretical Physics, 84(3), (1997), 497.

ELECTRIC ENERGY TRANSMISSION
BY MEANS OF LASER RADIATION

O. N. Krokhin

P. N. Lebedev Physical Institute, Russian Academy of Sciences
Leninskii Pr. 53, Moscow 119991, Russia
E-mail: krokhin@sci.lebedev.ru

The possibility of energy transmission with the usage of laser light as an energy carrier is discussed. Now it became more realistic due to appearance of optical fibers capable to transmit quite powerful light. The principle and the scheme for light–electricity conversion are considered. The physics of this approach consists in the optical saturation of the electron interband transitions when Fermi quasilevels for electrons in the conduction and the valence bands are split on the value of quantum energy. The saturation effect and carrier separation will produce an electromotive force if the conducting and valence electrons have a possibility to move in the same direction. To realize this the gates have to be applied which is either p − n -junctions or heterostructure barriers. This device reminds of the injection laser operating in the inverted regime.

The realization of highly efficient powerful lasers, for example, semiconductor diode lasers with a power and an efficiency up to 20 W and 60%, respectively, gives us the possibility of analyzing the problem of the construction of a transmission line for the distant transport of electric energy by means of its conversion into the monochromatic coherent laser light and the subsequent reconversion of the latter into electricity (see Fig. 1).

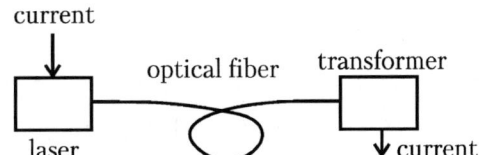

Fig. 1: Principal scheme of an optical fiber line for the electric energy

transmission.

In these scheme, the laser light–electricity converter is the only unknown element. At present, the photoelectrical transformer (a unit of the solar battery) is widely used but transformers of this type are designed for conversion of the wide-thermal-spectrum light of very low intensity. For example, the solar flux at the Earth surface is only a quarter–one half of a watt per square centimeter but the efficiency of its transformation using the photoelectric transformer is sufficiently high being about 30%.

CP849, *Zababakhin Scientific Talks - 2005,*
edited by E. N. Avrorin and V. A. Simonenko
© 2006 American Institute of Physics 0-7354-0345-7/06/$23.00

In the scheme proposed we consider the transformation of coherent light into electric energy that leads to different physics and requires another design of the unit. On the one hand, the coherent light is monochromatic that allows us to fit the laser radiation frequency to the optimal value of the absorption by variation of the frequency of the emitted laser radiation or by variation of the forbidden gap energy. On the other hand, one can use the spatial coherence of the radiation transmitted. Indeed, the line that consists of the laser, optical fiber, and converter forms a long cavity. It means that the electromagnetic energy transmitted has to exist in the form of the specific modes. In other words, the radiation generated by the laser can be focused in the region of the same volume as the emitting one (see Fig. 2). This is the main feature differing such a converter from the conventional photoelectric cell. In the case under consideration we have a very high concentration of the electromagnetic energy in the active volume of the converter of the order of that in the generating volume of the semiconductor laser. In terms of radiation intensity this value is of a few MW/cm^2.

With so high intensity the rate of optical transitions due to absorption of the monochromatic radiation exceeds the rate of relaxation processes and provides the saturation of population between levels of the valance and the conduction bands [1] (see Fig. 3).

In this case the distance between Fermi quasilevels for electrons in the conduction band and the valance band that characterizes the nonequilibrium of these states tends to $\hbar\omega_0$, the energy of the quantum. The absorption coefficient k tends to zero, and the specific energy absorbed is equal to the specific energy $\hbar\omega_0 R$ associated to the relaxation process, where R is the number of recombination transitions in the unit volume per the unit time interval. Thus

$$k(\omega_0) = \alpha(\omega_0)\left(f_v - f_c\right) \approx \alpha(\omega_0)e^{-\frac{E_v - \mu_v}{T}}\left(1 - e^{-\frac{\hbar\omega_0 - \mu_c - \mu_v}{T}}\right), \qquad (1)$$

where indexes c and v refer to the conduction and valence band, respectively, f is the Fermi distribution function $f = \left(\exp\dfrac{E - \mu}{T} + 1\right)^{-1}$, E is the energy level, μ is the Fermi level (quasilevel), $\alpha(\omega_0)$ is the absorption coefficient in the nonsaturated (conventional) case, and $E_c - E_v = \hbar\omega_0$ (see Fig. 3).

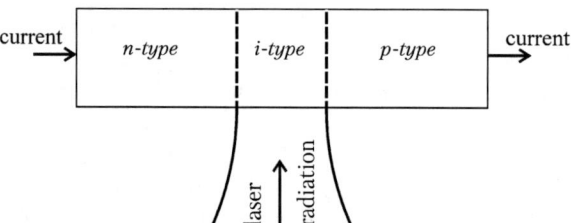

Fig. 2: Structure of the converter of optical radiation into electric current. The radiation is directed to the facet of a $p-i-n$ structure.

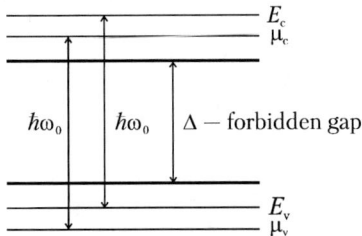

Fig. 3: Illustration of the saturation effect. Optical transition takes place between levels E_v and E_c. The distance between them is equal to the quantum energy $\hbar\omega_0$. The distance between Fermi quasilevels $\mu_c - \mu_v$ tends to $\hbar\omega_0$ from below with an increase in the light intensity.

In the case of strong saturation it follows from Eq. (1) that

$$k(\omega_0) \cong \alpha(\omega_0)e^{-\frac{E_v-\mu_v}{T}} \frac{\hbar\omega_0 - \mu_c + \mu_v}{T}, \tag{2}$$

The number of quanta absorbed in the unit volume per the unit time interval is determined from the following equation for the flux density of quanta:

$$\frac{\partial I}{\partial x} = kI . \tag{3}$$

Equating I to the recombination rate R, one obtains

$$\hbar\omega_0 - (\mu_c - \mu_v) = \frac{RT}{I\alpha(\omega_0)\exp\left(\dfrac{E_v - \mu_v}{T}\right)}. \tag{4}$$

As can be seen from (4) the saturation effect produces an electromotive force if we are able to separate the electrons and holes created. The latter is achievable using $p-n$ or $p-i-n$ structures with p- and n-parts of the semiconductor highly doped close to the degenerate state. Such a structure provides a unidirectional current. Even more effective way of electron–hole separation is connected with the usage of heterostructures. Both these schemes are shown in Fig. 4.

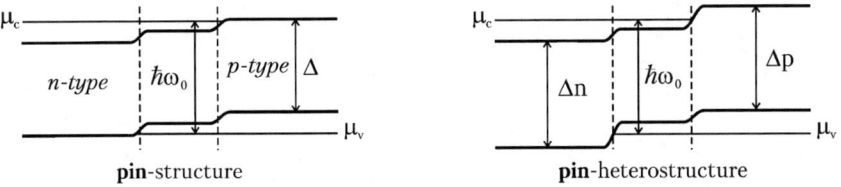

Fig. 4: Separation of electric charges at the contacts of the region of carrier generation and the strongly doped layers of n- and p-types. μ_c and μ_v are Fermi quasilevels in the conduction and valance bands, respectively.

Being included into the electric circuit the converter introduces an additional load in the active region of electron and hole generation. This load is equal to the number of electrons that leave the active volume, i.e., to the total flux of particles (electrons) from the unit volume of the active region, J/eV, where J is the electric current, e is the electron charge, and V is the active region's volume. This means the following replacement in expression (4):

$$R \rightarrow R + J/eV \approx J/eV. \tag{5}$$

The latter estimate reflects high efficiency of the converter where the internal relaxation is small as compared to the load.

The basis of the converter operation is quite similar to that of the semiconductor diode laser. One can say that the converter operates as an inverter with respect to a laser and can expect practically the same efficiency of conversion and amount of the converted energy up to 60% and 20 W, respectively [2]. These characteristics are also supplemented by rather high energetic capacities and low losses of optical fiber lines of 100 W and 0.1 Db/km, respectively [3].

Concluding I would like to specify possible applications of this type of energy transmission lines. They can be especially suitable in the case where a metallic wire line is hardly applied, for example, inside high-voltage systems or in energy transport through conductive media.

References

1. Basov, N.G., & Krokhin, O.N., Transforming the high-power monochromatic radiation into electric current. *Fiz. Tverd. Tela* **5** (1963) pp. 2384–2386.
2. Bugge, F., 12 W continuous-wave diode lasers at 1120 nm with InGaAs quantum walls. *Appl. Phys. Lett.* **79** (2001) p. 1965.
3. Broderick, N.G., Large mode area fibers for high power applications. *Opt. Fiber Tech.* **5** (1999) p. 185.

1D-Simulation of Thermonuclear Target Compression and Burning for Laser Facility NIF and LMJ

R.Zh. Valiev, M.N. Chizhkov, N.G. Karlyhanov, O.V. Lusganova,
V.A. Lykov, D.S. Netsvetayev, M.S. Timakova

Federal State Unitary Enterprise "Russian Federal Nuclear Center — Zababakhin All–Russia Research Institute of Technical Physics", 456770, Snezhinsk, Chelyabinsk region, Russia

Abstract. The high-power laser facilities NIF [1] and LMJ [2] with the pulse energy as high as 2 MJ are being created in the USA and France. The basic cryogenic indirect-drive targets for thermonuclear ignition on these facilities are a spherical shell from polystyrene doped with oxygen and bromine. (CH+5%O+0,25%Br), whose inner surface is covered with DT-ice layer [1, 3]. The central region of targets is filled with DT-gas. The targets for NIF [1] and LMJ [3] have different external radii (1,11 and 1,215 mm, correspondingly), masses of DT-fuel (210 и 310 µg), X-ray radiation temperature dependences in time. The thermonuclear yield from the NIF target calculated with LASNEX code is 15 MJ [1], the yield from the LMJ target calculated with FCI1 code is 25.4 MJ [3]. In RFNC-VNIITF calculations of compression and burning of basic NIF and LMJ targets were performed by using of the 1D ERA code [4] in the spectral diffusion approximation for radiation transfer. We used tabulated opacity calculated by the mean ion model [5]. Thermonuclear yield calculated with ERA code is about 18 MJ for the NIF target and nearly 23 MJ for the LMJ target. Calculated yields are in good agreement with published results. Performed calculations justified the possibility to simulate ICF targets in RFNC-VNIITF. In paper are also presented analysis results of target sensitivity to opacity and X-ray temperature variations.

Keywords: Indirect-drive targets; Thermonuclear ignition.
PACS: 52.57.Bc

INTRODUCTION

The high-power laser facilities NIF [1] and LMJ [3] with the pulse energy as high as 2 MJ are being created in the USA and France. The basic cryogenic indirect-drive targets for thermonuclear ignition on these facilities are a spherical shell from polystyrene doped with oxygen and bromine. (CH+5%O+0,25%Br), whose inner surface is covered with DT-ice layer [1, 3]. The central region of targets is filled with DT-gas. Capsules are shown in Figure 1.

CP849, *Zababakhin Scientific Talks - 2005*,
edited by E. N. Avrorin and V. A. Simonenko
© 2006 American Institute of Physics 0-7354-0345-7/06/$23.00

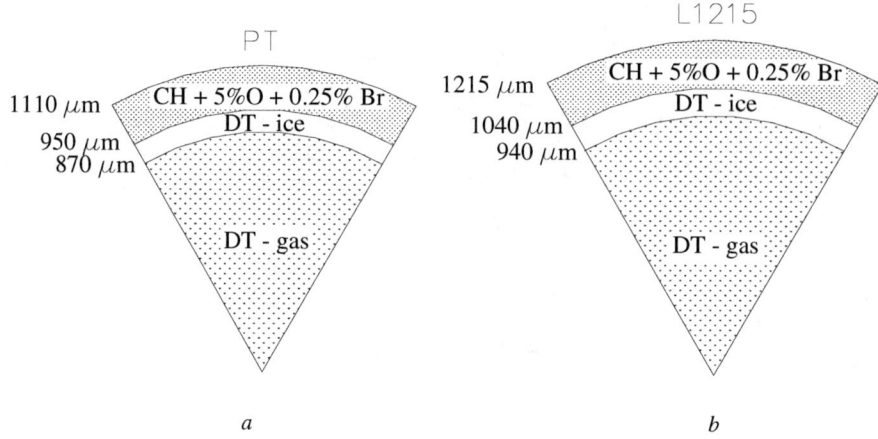

a   *b*

FIGURE 1. Cryogenic targets for NIF (PT [1]) and LMJ (L1215 [3]).

The targets for NIF [1] and LMJ [3] have different external radii (1,11 and 1,215 mm, correspondingly), masses of DT-fuel (210 и 310 µg). The time dependences of radiation temperature on the target surface for PT and L1215 are presented in Figure 2.

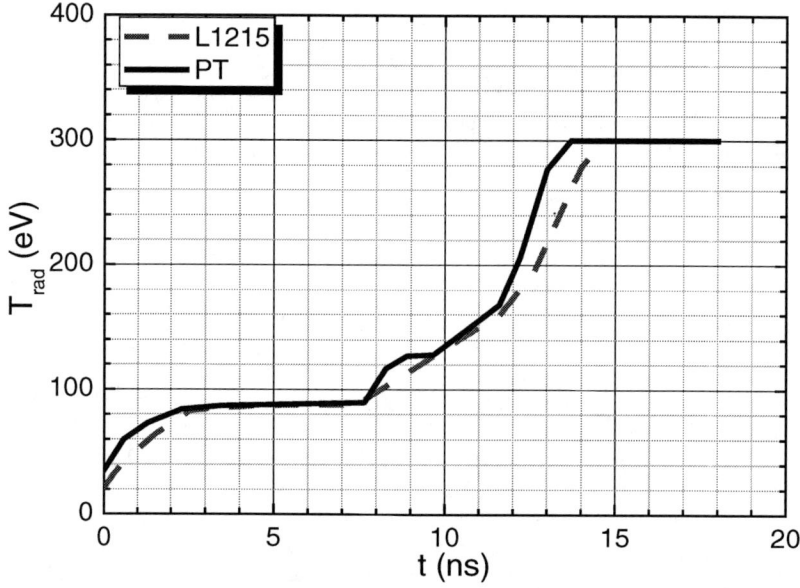

FIGURE 2. Time dependences of radiation temperature on the target surface for PT [1] and L1215 [3].

Calculations of compression and burning of basic targets for NIF and LMJ were also performed in RFNC-VNIITF.

1D-SIMULATION OF COMPRESSION AND BURNING OF TARGETS

Calculations of compression and burning of basic NIF and LMJ targets were performed by using of the 1D ERA code [4] in the spectral diffusion approximation for radiation transfer. We used tabulated opacity calculated by the mean ion model [5].

Results of calculations on the 1D ERA code are compared with results presented in [1, 3] in Table 1.

TABLE 1. Results of Calculations of Compression and Burning of Targets PT and L1215 on the LASNEX Code [1], FCl1 Code [3] and ERA Code

Target	Code	$\langle\rho_{DT}\rangle_{max}$, g/cm^3	$\langle T_{DT}\rangle_{max}$, keV	E_{fusion}, MJ	α	V_{imp}, km/s	μ %	E_{abs}, kJ
PT	ERA	393	64	18.1	0.26	410	86	156
	LASNEX	–	–	15	0.21	410	–	153
L1215	ERA	313	47	23.5	0.22	390	85	201
	FCl1	–	–	25.4	0.24	380	–	172

Following symbols are used in Table 1:

$\langle\rho_{DT}\rangle_{max}$ – peak DT-fuel density;

$\langle T_{DT}\rangle_{max}$ – peak ion temperature;

E_{fusion} – fusion energy yield;

α – DT-fuel burning-out;

V_{imp} – implosion velocity;

μ – ablated fuel mass;

E_{abs} – absorbed energy;

In Figure 3, profile of ion temperature at the moment of intensive burning is shown. Calculated with the ERA code characteristics of compression and burning of targets are in good agreement with data presented in [1, 3].

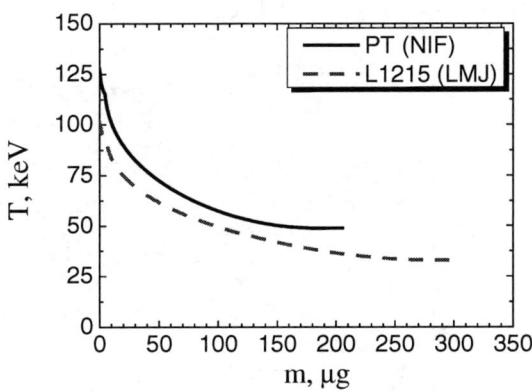

FIGURE 3. Profile of ion temperature at the moment of intensive burning.

As shown in Figure 2 ion temperature is maximal at the centre of target and decreases with increasing distance from centre. Ion temperatures in DT-fuel are higher for PT than for L1215.

In Figure 4, profile of fuel density at the moment of maximal compression is shown.

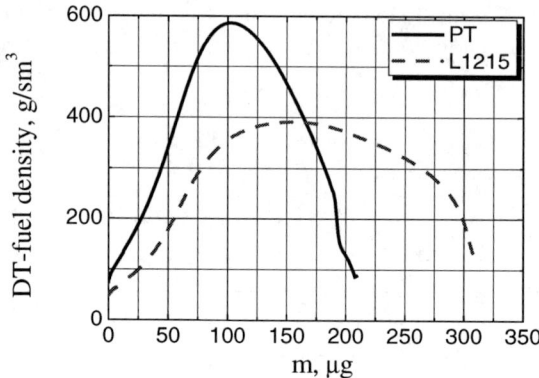

FIGURE 4. Profile of fuel density at the moment of maximal compression.

Density of DT-fuel is higher for PT than for L1215.

TARGETS SENSITIVITY TO VARIATIONS OF TIME DEPENDENCE RADIATION TEMPERATURE ON THE TARGET SURFACE, OPACITY AND DT-REACTION RATE

Calculations on the ERA code were performed to determine targets sensitivity to variations of some physical parameters. Tabulated opacity was multiplied on coefficient K_L to determine targets sensitivity to variations of opacity. Dependence of fusion energy yield on coefficient K_L is presented in Fig. 5. Fusion energy yield is normalized to nominal fusion energy yield at $K_L = 1$.

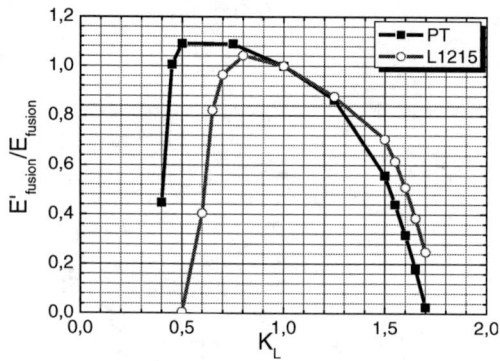

FIGURE 5. Targets sensitivity to variations of opacity. Opacity was changed by multiplying on coefficient K_L. Fusion energy yield is normalized to nominal fusion energy yield at $K_L = 1$.

Figure 5 shows that opacity variation by 60% results to decreasing two times of fusion energy yield from the target PT. Fusion energy yield from the target L1215 decreases two times by opacity decreasing by 40% or by opacity increasing by 60%. So the target PT is less sensitive to variations of opacity.

Time dependence of radiation temperature on the target surface was multiplied on coefficient K_T to determine targets sensitivity to variations of time dependence of radiation temperature. Dependence of fusion energy yield on K_T is presented in Fig. 6. Fusion energy yield is normalized to nominal fusion energy yield at $K_T = 1$.

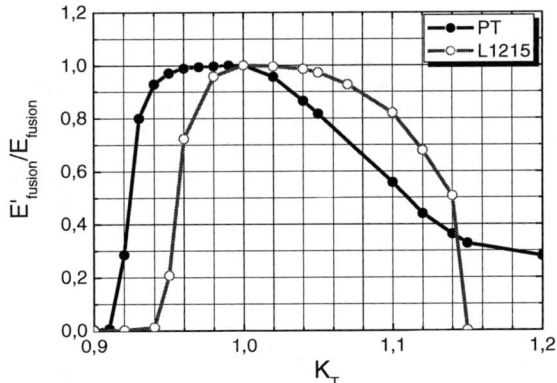

FIGURE 6. Targets sensitivity to variations of time dependence of radiation temperature on the target surface. Time dependence of radiation temperature was changed by multiplying on coefficient K_T. Fusion energy yield is normalized to nominal fusion energy yield at $K_T = 1$.

Figure 6 shows that the target PT is less sensitive to variations of time dependence of radiation temperature on the target surface than the target L1215.

DT-reaction rate was multiplied on coefficient K_R to determine targets sensitivity to variations of DT-reaction rate. Dependence of fusion energy yield on K_R is presented in Fig. 6. Fusion energy yield is normalized to nominal fusion energy yield at $K_R = 1$.

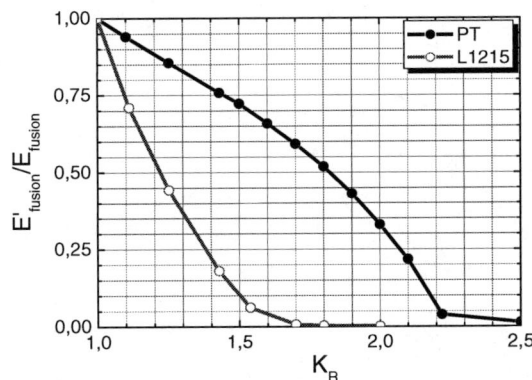

FIGURE 7. Targets sensitivity to variations of DT-reaction rate. DT-reaction rate was changed by multiplying on coefficient K_R. Fusion energy yield is normalized to nominal fusion energy yield at $K_R = 1$.

Figure 7 shows that DT-reaction rate decreasing 1.7-1.8 times results to decreasing two times of fusion energy yield from the target PT. Fusion energy yield from the target L1215 decreases in 2 times by DT-reaction rate decreasing 1.3 times. So the target PT is less sensitive to variations of DT-reaction rate.

CONCLUSION

1D-simulations of compression and burning of basic targets PT and L1215 for NIF and LMJ were performed by using the ERA code. Calculated fusion energy yield differs from published results less than 20%. Calculations were performed to compare targets sensitivity to variation of some physical parameters: opacity, time dependence of radiation temperature on the target surface and DT-reaction rate. Analysis of obtained results has showed that the target PT (LLNL, USA) is less sensitive to variation of physical parameters than the target L1215 (CAE, France). In this paper, we did not consider such important issues as influences of mixing, compression asymmetry, ice roughness and ablator roughness. These problems will be the subject of our further investigations.

REFERENCES

1. J.Lindl et al. Phys. Plasmas 11(2), pp.339-491, 2004.
2. C.Cavailler et al. Plasma Phys. and Control. Fusion, 46, pp. B135-B141, 2004.
3. P.-A.Holstein et al. C. R. Acad. Sci. Paris, t. 1, Serie IV, p. 693-704, 2000.
4. N.A.Barysheva et al. Computational mathematics and mathematical physics (In Russian), 22(2), pp.401-410, 1982.
5. A.F.Nikiforov et al. Quantum-statistical models for high-temperature plasmas (In Russian). Moscow: FIZMATLIT, 2000.

Some Features of Taking into Account the LPM Effect for Research into Radioemission of Cascades from Particles of Super-High Energy in Dense Environments

P.I. Golubnichiy, S.V. Dzyobark, A.D. Filonenko

Department of Physics, East-Ukrainian national university by Vladimir Dahl.
bl. Molodezhny, 20a, 91034, Lugansk, Ukraine

Abstract. In the article we present results of calculations of influence of LPM effect on spectrum-frequency characteristics of radio impulse accompanying cascade process, formed owing to interaction of high-energy particles with matter. Opportunity of detecting neutrino of ultrahigh energy by radio-method in a range of frequencies about units of MHz is discussed.

INTRODUCTION

It is known, that detection of cosmic rays of high energy, i.e. definition of energy of an initial particle, direction of its movement, structure etc. is carried out with the help of traditional methods of registration of ionizing radiation. It is basically ionization and scintillation counters. In modern installations (detectors) designed for research of particles with energy $W_0 > 10^{19}\ eV$, the amount of them can reach thousand of units and even more. Together with necessary for work of the detector power and signal communications this system of counters is placed on an area of 10-20 km^2.

Cosmic rays with energy $W_0 > 10^{20}\ eV$ a little bit change direction of their movement in galactic and intergalactic magnetic fields. It allows to carry out search of sources of the charged particles and their mechanism of acceleration up to such high energy. However the amount of such events up to now is not enough for definite conclusions. In all the history of research of CR the number of the registered particles with energy $W_0 > 10^{20}\ eV$ does not exceed several tens.

It is obvious, that this statistics is not enough for certain conclusions (especially concerning particles with higher energy), and for continuation of research it is necessary to increase the working area of detectors. It is known, that intensity of a flux of cosmic rays is defined by the expression $J_v(> W_0) = A \cdot W_0^{-(\gamma-1)}$, where $\gamma \approx 2.5$ is the parameter of power spectrum dependent on the range of energy. According to this expression, expansion of power border by one order (without change of frequency of events) demands increase of the area of the detector approximately by two orders. In

CP849, *Zababakhin Scientific Talks - 2005,*
edited by E. N. Avrorin and V. A. Simonenko
© 2006 American Institute of Physics 0-7354-0345-7/06/$23.00

the range of energy $W_0 > 10^{20}$ eV the intensity of flux of cosmic rays is such, that on an area of one square kilometer we have in average one particle in one century.

The traditional method of detection of particles of high energy is hardly capable to overcome the boundary of 10^{21} eV in the foreseeable future. Such detectors should occupy the huge areas ($> 10^6$ sq.km.) and for the realization would demand a big material inputs. Research into finding-out of opportunities of application of radio-physical methods for detecting cosmic rays of ultrahigh energy (methods of radiodetection) therefore are actual.

1. PRINCIPLES OF RADIODETECTION OF COSMIC RAYS

For the first time the idea of radiodetection was published approximately forty years back by Askarian. According to him the electron-photon cascade in a medium causes Cherenkov coherent radiation of excessive electrons. Registration of a rather high level of signal with a narrow diagram of orientation in this case is expected. But such approach somewhat reduces efficiency of use of the method as frequency of registered events owing to the narrow diagram of radiation will decrease. Besides, Landau-Pomeranchuk-Migdal effect (LPM) also results in reduction of the level of useful signal (owing to reduction of excessive charge of the cascade). As an example of one of the working detectors using the given mechanism, one might mention FORTE experiment (Fast On-orbit Recording of Transient Events) [1].

Another opportunity of radio-detection of cosmic rays is in consideration of low-frequency mechanisms of radio-emission from the general excessive charge of the cascade (from now on the talk will be about them). In this case the signal should have the wide diagram of orientation similar to a half-wave dipole. It considerably allows to increase frequency of registration of events in comparison with Cherenkov mechanism. Laboratory of physics of cosmic rays of ultrahigh energy of East-Ukrainian national university by Vladimir Dahl developed methods of radiodetection of particles with energy up to 10^{23} eV. In one of them for the working area of the detector it was planned to use the surface of the Moon [2].

For confirmation of the theory the authors offered two methods. One of them is radio-astronomical. It is now being realized together with Radio-astronomical institute of NAS of Ukraine with the help of decameter telescope of UTR - 2 (Kharkov). Another method is connected to measurements in Antarctica. It is shown, that the cascade from a particle with energy higher than 10^{22} eV reaches the surface of ice (3-4 km. above sea level) of Antarctica long before its maximum [3]. Therefore the basic share of a cascade shower will be on the depth of 5-10 meters under the surface of ice. The electromagnetic pulse accompanying this process, having passed 200-300 km to ionosphere will be reflected from it. The aerial placed on the continent or near it, will register this signal.

Estimations allow to speak that for a level of a useful signal > 25 mkV/m·MHz for energy of an initial particle $> 10^{22}$ eV at Ukrainian Antarctic station "Academician Vernadskiy " the working area of the "detector" of the continent will be about 7×10^6 km^2. The amount of expected particles in such range will be from 1 for 8 a day for γ=2.7 up to 3 a day at γ=2.6 (the flux of the particles is proportional to $W^{-(\gamma-1)}$, where W is energy of an initial particle in GeV).

Besides, there is an opportunity of registration of particles with energy above 10^{19} eV by transitive radiation of a wide atmospheric shower split in magnetic field of the Earth. For a signal > 30 mkV/m·MHz an estimation of amount of expected particles is 1 for 7 days at a parameter γ=2.7 up to 2 a day at γ=2.6.

Both methods could be realized on the station "Academician Vernadsky " due to its particular arrangement. In seasonal research of 2005 the methods were tested in Antarctica and preliminary results are have been received.

2. CALCULATION OF RADIO SIGNAL FROM THE CASCADE TAKING INTO ACCOUNT INFLUENCE OF LPM-EFFECT

We put forward a hypothesis about background contribution to the general number of the facts of registration by interaction of neutrino of ultrahigh energy with Antarctica's ice. However the mechanism of development of the cascade in this case differs from the previous one. As it has been already told, wide atmospheric shower (WAS) reaches the surface of the continent shortly before its maximum. In the maximum the basic share is made by particles with energy, already insufficient for change of the cascade curve essentially owing to LPM-effect (it results in increase of radiating unit of length with the growth of energy; calculations show, that the longitudinal size of the cascade in a dense environment due to this effect can increase by tens and even hundreds of times). For start of the shower directly in a dense environment (interaction of neutrino with ice) one needs to take into account LPM effect for in this case characteristics of a radio pulse depend not only on number of charged particles in the cascade, but on the kind of cascade curve as a whole too. Let's emphasize, that we speak in this case about the mechanism of radioemission of the general excessive charge of the cascade appearing owing to annihilation of positrons and involving to the avalanche of Compton-and δ-electrons.

Spectrum function $\left|\vec{E}_{\omega}\right|$ for the bremstrahlung mechanism obviously depends on cascade function $N(t)$:

$$\left|\vec{E}_{\omega}\right| = \frac{\omega \cdot \sin\Theta}{4\pi\varepsilon_0 c^2 R_0} \left|\int_{-\infty}^{\infty} \vec{j}(t) \cdot e^{i\omega t(1-\sqrt{\varepsilon}\cos\Theta)} \cdot dt\right|, \tag{1}$$

where $\vec{j}(t) = \vec{j}(\vec{r}_0(t)) = Q \cdot \vec{V} \cdot \delta(\vec{r} - \vec{r}_0(t)) \approx 0.25 \cdot e \cdot N(t) \cdot c \cdot \delta(\vec{r} - \vec{r}_0(t)) \cdot \vec{e}_{\vec{v}}$ and precision of the last one will determine the quality of calculation of the expected signal.

LPM effect is necessary for taking into account of particles with energy $\varepsilon > \varepsilon_{LPM}$, where $\varepsilon_{LPM} = 61.5 \cdot X_0$ TeV, X_0 - radiating length in cm. In the work [1] it is offered to use the following approximation which gives accuracy up to 20 % for definition of lengthening of radiating length: $X_{LPM} = \sqrt{\varepsilon/\varepsilon_{LPM}} \cdot X_0$. But practical use of this function is complicated, as it demands breaking up of the cascade to separate parts, for every one of which it is necessary to take into account their own ε, $d\varepsilon/dX$, X_{LPM}. This will result in enlarging the volume of calculations.

259

For simplification of calculation of integral (1) one needs to set up the cascade function in a simple analytical form. We used the results of numerical calculation of cascades using LPM effect allowing to estimate fluctuations of updated Monte Carlo method of [4] for water and have received such approximation for function $N(t, W_0)$ [5].

The use of the received result has allowed us to obtain estimations of spectral decomposition $\left|\vec{E}_\omega\right|$ for a signal of radio pulse of the cascade in ice in a range of energy of the initial particles 10^{18}-10^{24} eV, and to determine dependence of $\left|\vec{E}_\omega\right|$ on a degree between a direction on the observer and the axis of the avalanche Θ.

The account of LPM effect for consideration of some questions connected with radioemission of cosmic showers, results in displacement of spectral maximums of radio signal intensity to the aside of smaller frequencies and to reduction of level of the signal.

Results show that for higher frequencies the diagram is more sharply directed, and in its vicinities prevail $\left|\vec{E}_\omega\right|$ with smaller frequencies. The effect of enlargement of the diagram of orientation along with reduction of level of the signal at reduction of frequency, can appear powerful for development of radiomethods of detecting of deeply penetrating particles of ultrahigh energy.

3. APPLICATION OF THE RESULTS AND CONCLUSIONS

The received results were used by us for estimation of contribution of acts of neutrino interaction with ice to frequency of registration in the above mentioned experiment with the Ukrainian Antarctic station by definition of a flux of cosmic rays of ultrahigh energy. By the results the frequency of neutrino registration at minimum by one order exceeds theoretically predicted frequency of registration on the signal reflected from ionosphere for cosmic rays [6]. Thus the power spectrum of detected neutrinos is above 10^{23} eV.

The decision of the question of definition of the nature of detected signal (from neutrino to a space particle) is possible by the spectral analysis of the latter in a wide range of frequencies. Besides, the mechanism of radioemission offered for cosmic rays uses a wide diagram of orientation for frequency 3MHz and calculated diagram for this frequency from the "neutrino" cascade is rather narrow (enlargement is 4^0-5^0). This fact gives one more opportunity of separation of useful signal for conducting duplicating experiments with the stations placed on a certain basis.

REFERENCES

1. Nikolai G. Lehtinen, Peter W. Gorham et. al. FORTE satellite constraints on ultra-high energy cosmic particle fluxes, arXiv:astro-ph/0309656, v.2, October 4, 2003.
2. P.P. Golubnichy, and A.D. Filonenko. Detection of space rays of super-high energy using the Moon artificial satellite, Space science and technology, v. 5, edition 4, 1999, pp. 87-92.
3. A.D. Filonenko. Radioemission of cascade showers and detection of space rays of super-high energy, Lugansk, 2002.
4. L.G. Dedenko, I.M. Zheleznyh, and S.G. Kolomatsky. Opportunities of using the LPM effect to

investigate space rays of super-high energy, Izvestiya of U.S.S.R Academy of Science (physics), v. 53. edition 2, 1989, pp. 350-354.

5. P.P. Golubnichy, and S.V. Dz'yobak. Properties of radiopulse induced by neutrino of super-high energy in dense medium taking into account the LPM effect, Visnik SNU, edition 10 (92), 2005, pp. 25-32.

6. P.P. Golunichy, and S.V. Dz'yobak. Effect of interaction of super-high energy neutrinos on record rate in the experiment to determine super-high energy space rays at Ukrainian Antarctic Station "Academician Vernadsky", Visnik SNU, edition 10 (92), 2005, pp. 32-35.

Effect of the Thermal Instabilities on Electrical Explosion of Thin Metal Wires

V.I. Oreshkin[a], R.B. Baksht[a], N.A. Ratakhin[a], A.Yu. Labetsky[a],
A.G. Rousskikh[a], A.V. Shishlov[a], P.R. Levashov[b], K.V. Khishchenko[b],
I.V. Glazyrin[c], I. Beilis[d]

[a] High Current Electronics Institute SB RAS, Tomsk, Russia
[b] Institute for High Energy Densities RAS, Moscow, Russia
[c] RFNC- Zababakhin Institute of Technical Physics, Snejinsk, Russia
[d] Tel Aviv University, Tel Aviv, Israel

Abstract. An electrical explosion of thin metal wires at a current rise time of several tens of nanoseconds and at a current density of $\sim 10^8$ A/cm^2 was studied. A two-dimensional magnetohydrodynamic code based on the particle-in-cell method is used to calculate the formation of striations and a low-density plasma corona surrounding the wire.

Recent interest in the electrical explosion of conductors (EEC) in the nanosecond range stems from successful experiments with imploded wire arrays for soft X-ray production on the Z-setup [1]. Experimental studies of the nanosecond EEC (at current densities of $> 10^8$ A/cm^2) [2-4] have disclosed that in the explosion of the conductor in vacuum a low-density plasma corona surrounding a more dense core and strata with a wavelength of several tens of microns are formed. The occurrence of these strata can not be credited to the development of magnetohydrodynamic (MHD) instabilities of the sausage type ($m=0$), since in a nanosecond EEC the times of energy delivery to the conductor are shorter than the characteristic times in which this type of instabilities evolves. In [5] it has been shown that in these modes strata with such a wavelength may result from the development of overheat instabilities. In this work, the conditions for the formation of this type of strata have been investigated by numerical simulation of the electrical explosion of aluminum and tungsten microconductors.

To simulate the explosion of conductors, the MHD program JULIA based on the PIC method was used [5]. This program allows simulating the EEC in the two-dimensional approximation. In this program, the system of MHD equations has the form:

$$\frac{\partial \rho}{\partial t} + \nabla(\rho \mathbf{v}) = 0 \; ; \tag{1}$$

$$\rho \frac{\partial \mathbf{v}}{\partial t} + \rho(\mathbf{v}\nabla)\mathbf{v} = -\nabla p + \frac{1}{c}[\mathbf{j} \times \mathbf{H}] \; ; \tag{2}$$

CP849, Zababakhin Scientific Talks - 2005,
edited by E. N. Avrorin and V. A. Simonenko
© 2006 American Institute of Physics 0-7354-0345-7/06/$23.00

$$\rho \frac{\partial \varepsilon}{\partial t} + \rho(\mathbf{v}\nabla)\varepsilon = -p\nabla\mathbf{v} + \frac{\mathbf{j}^2}{\sigma} - \nabla\mathbf{W} \ ; \tag{3}$$

$$-\frac{1}{c}\frac{\partial \mathbf{H}}{\partial t} = \nabla \times \mathbf{E} \ ; \tag{4}$$

$$\nabla \times \mathbf{H} = \frac{4\pi}{c}\mathbf{j} \ ; \tag{5}$$

$$\mathbf{j} = \sigma(\mathbf{E} + \frac{1}{c}[\mathbf{v}\times\mathbf{H}]), \tag{6}$$

where ρ is the density of a matter, $\mathbf{v}$ is the average-mass velocity, p, ε are the pressure and the internal energy, respectively, $\mathbf{W} = -\kappa\nabla T$ and is the heat flux due to thermal electron conduction, T is the temperature of a matter; $\mathbf{H}$, $\mathbf{E}$ are the magnetic and electric field strengths (the latter is in a fixed coordinate system), $\mathbf{j}$ is the current density, κ, σ are the thermal conductivity and the conductance.

The system of equations (1)–(6) was solved numerically using the following algorithm. Equation of motion (2) was solved for each Lagrangian particle, whereupon, by summing over all particles, the average-mass velocities and the matter density in each cell were determined. In this method, continuity equation (1) is automatically fulfilled, since the particles are assumed to be Lagrangian constants. To solve energy equation (3) and Maxwell equations (4), (5), we employed a fixed Euler grid which was constructed early in the calculation, being unchanged throughout the numerical solution.

In solving the system of equations (1)–(6), the following boundary conditions were prescribed. To impose the boundary conditions on equations of motion (2), one can either specify the velocity or the pressure at the boundary. So, equations (2) were integrated taking $p = 0$ at the free boundary (the vacuum boundary) and the boundary conditions in the center at $r = 0$ to correspond to the condition of symmetry about the axis, i.e., the radial projection of the velocity $v_r = 0$. At both axial boundaries, i.e., at $z = 0$ and $z = Z$, the boundary conditions corresponded to the conditions of symmetry about the axes $z = 0$ and $z = Z$, i.e., the axial projection of the velocity $v_z = 0$.

The boundary conditions for energy equation (3), which involve the heat equation, were imposed by specifying the heat flux at the boundaries. The vector of the heat flux (in out case, due to thermal conduction) consists of two components (radial and axial) and has the form $\mathbf{W} = \left\{-\kappa\frac{\partial T}{\partial r}; 0; -\kappa\frac{\partial T}{\partial z}\right\}$. The heat flux at the boundaries was taken to be everywhere zero that corresponds to the absence of external heat sources and heat sinks.

The boundary conditions for Maxwell equations (4), (5) were the following: at $z = 0$ and $z = Z$ the radial projection of the electric field strength $E_r=0$ that also corresponds to the condition of symmetry about the axes $z = 0$ and $z = Z$. For the magnetic field strength, the boundary conditions were $H_\varphi = 0$ at $r = 0$ and $H_\varphi = 2I(t)/(cR_{max})$ at $r = R_{max}$, where R_{max} is the maximum radius of the Euler grid, $I(t)$ is the current through a wire. This current was calculated in integrating the electric circuit equations.

In the program, wide-range semi-empirical equations of state [6] were employed which took into account the effects of high-temperature melting and evaporation and possible existence of metastable states of the liquid and gas phases. In calculating the electrical characteristics of the metals, we referred to the tables of Al conductivity drawn by M. Desjarlais at SNL, USA [7]. The tables were based on the Lee-Moor model modified with account of experimental data. For tungsten we used the tables of conductivity [8] drawn according to the experimental-calculated procedure described in [9]. A characteristic feature is that with low densities an increase in temperature causes the conductivity of metals in the gas-plasma region to increase, whereas that of metals in the condensed state to decrease. In both procedures of drawing of conductivity tables the temperature dependence is assumed to change at metal densities close to the density at the critical point.

Let us consider the results of simulation of the explosion of Al and W wires in the mode corresponding to the experimental conditions [10]. The system of MHD equations was solved simultaneously with the equations of the LC-circuit of the setup. For both wire materials the initial diameter (the diameter of cold wires) was 15 μm

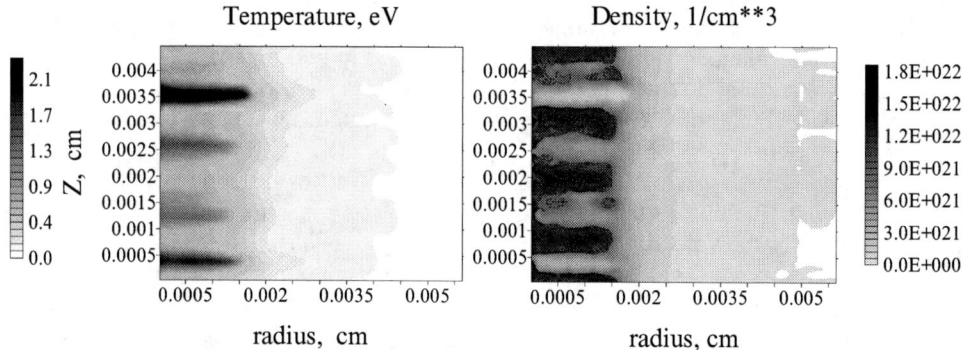

FIGURE 1. Distribution of the thermodynamic parameters of the exploded Al wire (t=28 ns).

and the length was 2 cm. Before simulation, the initial nonuniformity of the metal density was randomly specified (less than 5%). In physical terms, the presence of such nonuniformity can be attributed, e.g., to the granular structure of metal. The characteristic grain size therewith varies widely from 1 to 1000 μm [11]. The total number of grains in the calculations was $3.75 \cdot 10^5$. Before calculation, the particles were in 25 radial cells, i.e., at the initial moment in time 200 particles fell, on average, within each cell. The initial temperature of the matter was equal to room temperature, i.e., to 300 K.

Figure 1 shows the distribution of the parameters of the exploded Al wire immediately after the voltage peaks, when most of the energy has been already delivered to the wire. It can be seen in this figure that on heating a low-density corona involving metal vapors is formed around the dense wire core. Moreover, in the dense wire core a periodic structure has appeared with layers of rather cold and dense matter alternating layers of hot and less dense one, much as in the case observed in [2-4]. In so doing, early in the explosion there are instabilities with small wavelengths and once the energy release on Joule heating is completed the instabilities with small wavelengths are damped (due to dissipations associated with heat transfer through thermal induction) and modes with long wavelengths are left over. In the case under consideration, the wavelength of this kind of instabilities is 10–15 μm.

The strata are due to the development of overheat instabilities, i.e., instabilities attributable to nonuniform heating. The structure of overheat instabilities is determined by the character of the temperature dependence of the matter conductivity. In the case where the matter conductivity increases with increasing temperature, as it takes place in classical plasmas at Spitzer conductivity, the overheat instabilities brings about the formation of current channels. In the case where the matter conductivity decreases with increasing temperature, as occurs in a metal in the condensed state, the overheat instabilities result in layered structures with the layers perpendicular to the current flow.

Note that the development of sausage-type MHD instabilities (mode $m = 0$) is not responsible for the formation of strata, since the magnetic pressure in an explosion

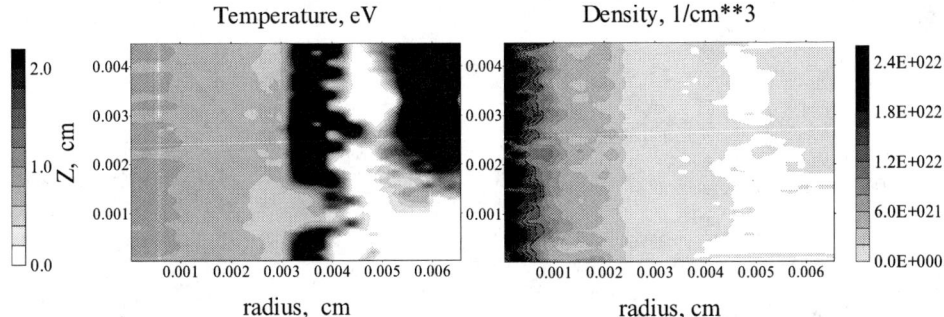

FIGURE 2. Distribution of the thermodynamic parameters of the exploded W wire (t=35 ns).

is small compared to the thermal one. The maximum magnetic pressure is 0.5 kbar, whereas the characteristic thermal pressure, which is equal to the pressure at the critical point, is an order of magnitude higher (in our equations of state the pressure at the critical point is 4.8 kbar). Therefore, the electromagnetic forces can not markedly affect the explosion dynamics and the role of the current through the wire reduces only to Joule heating of the matter. The strata are due to the development of overheat instabilities, i.e., instabilities attributable to nonuniform heating. The structure of overheat instabilities is determined by the character of the temperature dependence of the matter conductivity. In the case where the matter conductivity increases with increasing temperature, as it takes place in classical plasmas at Spitzer conductivity, the overheat instabilities brings about the formation of current channels. In the case where the matter conductivity decreases with increasing temperature, as occurs in a metal in the condensed state, the overheat instabilities result in layered structures with the layers perpendicular to the current flow.

Another situation he is observed in simulation of the electrical explosion of a tungsten wire (Fig.2). A dense core and a hot low-density corona are also formed. The corona intercepts the current and is scattered, with the wire remaining cold, but no strata appear, as opposed to an electrical explosion of the Al wire in the same mode

(Fig.1). Why no strata occur in electrically exploded W conductors. To answer this question, let us compare the time-dependences of the energy delivered to the electrically exploded wires.

Figure 3 shows the time-dependences of the energy delivered to the Al wire. It can be seen in this figure that by the moment in time 28 ns, for which the metal parameter distributions are presented in Fig.1, an energy higher than the energy of sublimation (the energy required for complete evaporation of the matter) is delivered to the wire. In contrast, the energy delivered to the tungsten wire by this moment (Fig.4) is two times lower than the energy of sublimation. Thus, it can be concluded that for strata to occur the energy delivered to a wire must be comparable with the energy of sublimation. It should be noted that MHD calculations may reveal stratifications in tungsten conductors, too. To do this, however, requires an increase in the rate of energy delivery to the wire that can be attained, e.g., by increasing the voltage of the capacitor bank.

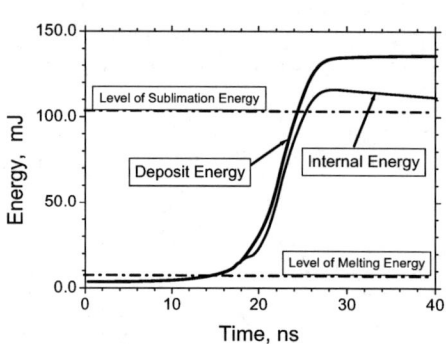

FIGURE 3. Time dependence of the energy delivered to the electrically exploded Al wire and of the internal energy of the metal.

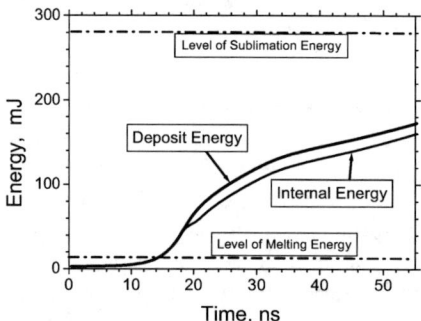

FIGURE 4. Time dependence of the energy delivered to the W wire and of the internal wire energy in electric explosion.

Thus, the MHD calculations show that the stratification due to the development of overheat instabilities manifests itself when the energy delivered to a matter approximates the energy of sublimation for this material. Such an opinion about the correlation between the energy delivered to a conductor and the formation of strata is held by the authors of [12]. They observed stratification using dark-field laser photographing and outlined that strata occur only if in an EEC the energy delivered to the conductor is equal to or higher than the energy of sublimation, otherwise the conductor expands in the form of an unstructured vapor-droplet cloud. This viewpoint on the correlation between the energy delivered to a conductor and the formation of strata agrees with the results of MHD simulation reported above.

The work was supported by RFBR grant No. 05-02-16845 and No. 05-02-08351 by MSTC project No. 2151

REFERENCES

1. Spielman R.B., Deeney C., Chandler G.A. et al. // *Phys. Plasmas*. – 1998. – Vol. 5. – P. 2105.
2. Sinars D.B., Shelkovenko T.A., Pikuz S.A.et al/ // *Phys. Plasma*. – 2000. – Vol. 7. – P. 429-432.
3. Sinars D.B., Hu M., Chandler K.M., Shelkovenko T.A. et al/ // *Phys. Plasmas*. – 2001. – Vol. 8. – P. 216-230.
4. Sarkisov G.S., Rosental S.E., Struve K.W.et al. // Proc. 5 Intern. Conf. on Dense Z-pinches. – Albuquerque, US, 2002. – AIP Conf.Proc. – 2002. – Vol. 651. – P. 213-216.
5. Oreshkin V.I., Baksht R.B., Ratakhin N.A. et al. // *Physics of Plasma*. – 2004. – Vol. 11, No 10. – P. 4771-4776.
6. Fortov V.E., Khishchenko K.V., Levashov P.R., Lomonosov I.V. // *Nucl. Instr. Meth. Phys. Res. A*. – 1998. – Vol. 415, No 3. – P. 604–608.
7. Desjarlais M.P. // *Contrib. Plasma Phys*. – 2001. – Vol. 41, No 2-3. – P. 267.
8. Oreshkin V.I., Baksht R.B., Labetsky A.Yu. et al. // *Zhur. Tekh. Fiz.*, (in Russian) – 2004 - Vol. 74 - No. 7- P.38-43.
9. Bakulin Yu.D., Kuropatenko V.F., Luchinsky A.V. // *Zhut. Tekh. Fiz.* (In Russian) – 1976 - Vol. 20, p. 1963.
10. Rousskikh A.G., Baksht R.B.,.Oreshkin V.I et al // Proc. 5 Intern. Conf. on Dense Z-pinches / Ed. by N. Pereira, J. Davis, C. Deeney. – AIP Conf. Proc. – Vol. 651. – New York: Melville, 2002. – P. 217-220.
11. Metals Handbook. Vol. 9. Fractography and Atlas of Fractograph. – Ohio: Amer. Soc. of Metals, Metals Park, 1974.
12. Sarkisov G.S., Struve K.W., McDaniel D.H. // *Phys. Plasma*. – 2004. – Vol. 11, No 10. – P. 4573-4581.

The dynamics of under surface condensed substance irradiated by intense energy stream

[a]N. B. Volkov, [a]A. Ya. Leyvi, [b]A.E. Mayer, [c]K.A. Talala , [b]A.P. Yalovets

[a]Institute of Electrophysics, Russian Academy of Sciences, Ural Branch, 106 Amundsen Street, Ekaterinburg, 620016, e-mail: nbv@ami.uran.ru
[b] Chelyabinsk State University, Physical Department, 129 Kashirin Brothers Street, Chelyabinsk, 454136,Russia
[c]South-Ural State University, Physical Department, Lenina 76, Chelyabinsk 454080,Russia

Abstract. The new method for simulation of dynamics of target surface irradiated by electron beams. There are two irradiation regimes: precritical and supercritical. We have conducted the research of crater formation for different beam parameters. Also 3D calculations of crater interaction and evolution of craters from initial irregular perturbation are carried out.

Keywords: Rayleigh-Taylor instability, ion, electron beams, microcrater formation, mixing.
PACS: 47.11.-j, 47.20.Ma

INTRODUCTION

Intensive charged particle beams are applied for modification of near-surface layers [1, 2, 3]. A wide range of beam parameters (current density, pulse dutation, particle energy) cause a variety of surface phenomena such as microhardness increase, crystal structure changing, microcrater formation, material droplet detachment, surface structure generation, surface layer mixing.

In case of ion radiation treatment the model of microcrater formation was proposed in Ref. [6]. Irradiated target is considered as three-layer system (plasma – melted phase (liquid) – solid phase) with different mass densities and sharp boundaries between phases. This approach is justified only for ion treatment when there is a sharp boundary of energy release zone. Microcrater is as a result of Richtmyer-Meshkov instability between plasma- liquid (RMI) [7].

Due to multiple fast electron scattering in substance the dose varies smoothly with target width, plasma spread and thermal expansion result in the smooth dependence of mass density ρ on width, which makes impossible to identify the sharp boundaries between the phases and to use the model [6] for the description of microcrater formation driven by electron irradiation.

The first aim of the given article is the development of the method to simulate the nonlinear dynamics of near-surface target layers irradiated by intense energy stream. The second one is the research of near-surface dynamics depending on beam parameters. We have considered clean target surface with micro-prominences or micro-cavities as initial perturbations.

THE NONLINEAR DINAMICS OF NEAR-SURFACE TARGET LAYERS IRRADIATED BY INTESIVE ENERGY STREAM

In case of irradiation by particle beams with power density $\geq 10^6$ W/cm^2 there are two regimes: precririeal and supercritical. The changeover between them has threshold nature.

In precritical regime the target is in condensed state (solid or liquid). It is subject to thermal expansion. According to estimations and numerical simulation RMI is limited by surface tension in precritical regime. And so microcraters do not form.

CP849, *Zababakhin Scientific Talks - 2005*,
edited by E. N. Avrorin and V. A. Simonenko
© 2006 American Institute of Physics 0-7354-0345-7/06/$23.00

In supercritical regime irradiation generates plasma jet which moves with velocity $v_f \geq 10^3$ m/s and acceleration $g_f = \dot{v}_f = 10^9 - 10^{11}$ m/s^2. Both high acceleration and absence of surface tension on plasma surface excite RMI after irradiation and stimulate development of microcraters from initial perturbation on the surface.

As inertia force on free surface is maximal the instability dynamics determines material flow in volume. In consequence of plasma evaporation target layers which are crystallized have density close to one of solid state. Therefore after crystallization the surface relief is determined by perturbation on the contact surface between plasma and liquid which is the surface of maximal mass density gradient.

During irradiation it is appropriate to use linear analysis for dynamics of small initial perturbation as inertia forces induced by material acceleration in near-surface layers delay Rayleigh-Taylor instability and produce gravity waves. After irradiation on the contact surface perturbation amplitude grows driven by RMI. Since by this moment plasma jet has low density it does not influence the contact surface dynamics. The further simulation uses the method [8]. Considering potential flow of incompressible liquid we have derived the equations set which determine nonlinear interfacial dynamics (for example RMI) in 2D and 3D geometries.

The RMI evolution goes with temperature reduction due to heat conductivity. Crystallization of melted material delays perturbation growth limiting microcrater depth. We have estimated the time $t_c = \tau + t_{tc}$ as the upper bound of crystallization time, where τ -the beam pulse duration, $t_{tc} = R^2 / \chi$, χ - thermal diffusivity, R - fast electron range in material. The second factor limiting microcrater depth is the depth of melted layer.

Let us consider the initial (linear) stage of crater formation. Electron beam is directed alone axis Oz and normal to the target surface. If the target surface is an ideally plane the material flow is one-dimensional. The free surface is the plane $z = z_f$, the contact surface is the plane $z = z_c$. The perturbation on the free surface is $a_f(t)\cos(kx)$, on the contact surface - $a_c(t)\cos(kx)$. The melted substance is homogeneous incompressible liquid with $\rho = \rho(z,t)$ and $a_{f,c}k \ll 1$, $(a_{f,c}/\rho)(d\rho/dz) \ll 1$, where k is wavenumber. The target field of displacements determining by surface perturbation and corresponding to incompression condition may be written as :

$$u_z(x,z,t) = \left(b_f(t)e^{-k|z-z_f|} + b_c(t)e^{-k|z-z_c|} \right)\cos(kx),$$

$$u_x(x,z,t) = \left(b_f(t)e^{-k|z-z_f|}sign(z - z_f) + b_c(t)e^{-k|z-z_c|}sign(z - z_c) \right)\sin(kx) \tag{1}$$

where b_f and b_c - system generalized coordinates.

As it follows from (1) amplitudes of perturbation on the free and contact surfaces are

$$a_f = b_f + b_c e^{-(z_c-z_f)}, \quad a_c = b_c + b_f e^{-(z_c-z_f)}. \tag{2}$$

The work done the field on volume element $\delta x \delta z$ to move it from point x,z to point x,z+uz is $\delta A = \delta x \delta z\, g(z,t)$ $(\partial\rho/\partial z)\, u_z^2/2$. As a result of displacement potential energy increment may be derived by integration of (1) on coordinate z:

$$U(bf,bc) = -\alpha_{ff}(k,t)b_f^2 - \alpha_{cc}(k,t)b_c^2 - 2\alpha_{fc}(k,t)b_f b_c. \tag{3}$$

Kinetic energy is:

$$T\left(\dot{b}_f, \dot{b}_c\right) = \frac{1}{2}\int_0^{2\pi/k} dx \int_{z_f}^{\infty} dz\, \rho(z,t)(\dot{u}_x^2(x,z,t) + \dot{u}_z^2(x,z,t)) =$$

$$= \beta_{ff}(k,t)\dot{b}_f^2 + \beta_{cc}(k,t)\dot{b}_c^2 + 2\beta_{fc}(k,t)\dot{b}_f \dot{b}_c. \tag{4}$$

In (3) and (4) $g(z,t)$ is local material acceleration,

$$\beta_{ij}(k,t) = \frac{\lambda}{2}\int_{\xi_{ij}}^{\infty} \rho(z,t)\exp\left(-k\left(|z - z_i| + |z - z_j|\right)\right)dz,$$

$$\alpha_{ij}(k,t) = \frac{\lambda}{4}\int_{z_f}^{\infty} g(z,t)\frac{\partial\rho(z,t)}{\partial z}\exp\left(-k\left(|z - z_i| + |z - z_j|\right)\right)dz,$$

Indexes i and j are equal c or f, $\xi_{ij} = z_f \delta_{ij} + z_c(1 - \delta_{ij})$.

269

From (3) and (4) we have determined Lagrangian and have derived the set of ordinary differential equations for calculation generalized coordinates b_f and b_c:

$$\dot{\psi}_1 = \alpha_{ff}b_f + \alpha_{fc}b_c, \qquad \dot{b}_f = \left(\beta_{cc}\psi_1 - \beta_{fc}\psi_2\right)\left(\beta_{ff}\beta_{cc} - \beta_{fc}^2\right)^{-1},$$
$$\dot{\psi}_2 = \alpha_{fc}b_f + \alpha_{cc}b_c, \qquad \dot{b}_c = \left(\beta_{ff}\psi_2 - \beta_{fc}\psi_1\right)\left(\beta_{ff}\beta_{cc} - \beta_{fc}^2\right)^{-1}. \qquad (5)$$

In case of sharp boundaries the set (5) turns into Taylor equation [7].

The equation set (5) with (2) is numerical integrated during irradiation when the material movement is accelerated and boundary diffusiveness is very important. The computer code BETAIN [9] calculates $\rho(z,t)$ and $g(z,t)$. It is numerical consistent solution of kinetic fast particles equation, one-dimensional continuum mechanics equations for elasto-plastic flows. It takes into account thermal conductivity and wide-range state equation.

NUMERICAL RESULTS

Computed dependences of growth rate on wavenumber (Figure1) have am extremum at k_0 that is firstly determined by distance between the free surface and the contact surface. As perturbation growth on the contact surface is induced by free surface the reduction of growth rate $\dot{a}_c$ at large wavenumbers is connected with strong short-wave damping with width. Long-wave perturbations develop longer than short-wave ones [7] therefore $\dot{a}_c$ decrease at small wavenumbers.

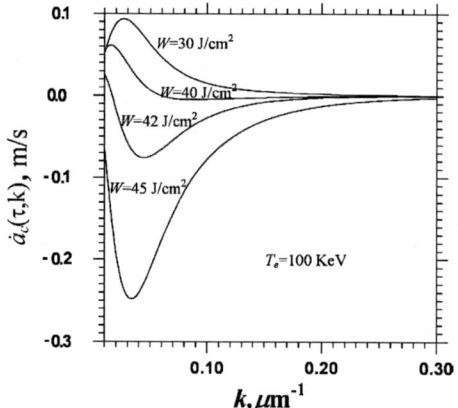

FIGURE 1. Growth rate versus wavenumber at the moment of pulse end.

The microcrater formation has threshold nature as a consequence of the existence of two irradiation regimes: precritical and supercritical.

Having defined initial perturbations on the free target surface we have computed the contact surface relief and velocity field at $t = \tau$ using linear analysis. They are initial conditions for simulation of nonlinear dynamics of the contact surface between plasma-liguid at $t \geq \tau$ computed by the method [8].

Let note the main evolution regularities of initial perturbation – a tubular cavity with depth h_0 $(1 \leq h_0 (\mu m) \leq 4)$ and diameter D_0 $(10 \leq D_0 (\mu m) \leq 200)$ for experiments with eletron irradiation [5]. In initial spectrum the fastest modes have wavenumbers k close to k_0 which determines perturbation dynamics. As a result microcrater tends to have a diameter $D^* = \pi / k_0$. At $D_0 < D^*$ there is stronly increase of diameter and depth $h(t_c)$, at $D_0 > D^*$

diameter and depth change slowly. The microcrater diameter $D(t_c)$ does not depend on h_0 and is defined only by D_0.

The increase of h_0 results in increase of $h(t_c)$ but the value of $h(t_c)$ is limited so that at large h_0 depth $h(t_c)$ becomes stationary value which corresponds to melted layer width $20-27\ \mu m$.

It may come to a conclusion in these irradiations regime cases the largest amount of experimental microcrates should have diameters about $D^* = 100-120\ \mu m$ and crater depths $15-25\ \mu m$ which is adjusted with results [5].

To research crater interaction we have conducted 3D simulation of two crater dynamics for equal and different initial diameters and depths. All results reveal the commom tendency: when intercrater distance is less than D^* they merge into elongated one with size close to D^*. If intercrater distance is about D^* material ejected from one crater delay another crater growth (depth reduces by 10-15 %). If the intercrater distance is larger than D^* craters grow independently.

The scale D^* appears also in crater development for irregular shape perturbation (as rule such craters have round shape). If initial size exceeds D^* an initial crater divides into several ones with diameters close to D^* (Figure 2).

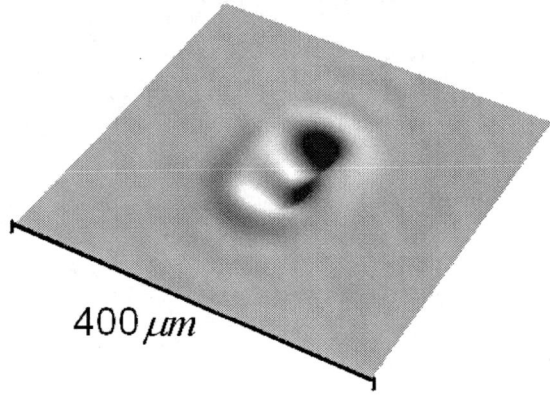

FIGURE 2. Breaking up craters on Fe surface. The initial perturbation was a cavity with sizes 100μm, 50μm, 1μm. Electron pulse duration τ=30μs, electron energy T=120 keV, power density W=35 J/cm².

Let study the dynamics of support film AL/FE in case of electron irradiation (voltage U=120 KeV, current density J=20 A/cm², pulse duration t=9 μs, power density W=19.2 J/cm²). Precritical irradiation regime is characterized by roughness reduction while in supercritical regime film material penetrates into supporting material and results in layer mixing (Figure 3).

CONCLUSION

We have developed the method for simulation of near-surface target layer dynamics in case of intensive energy stream irradiation. After irradiation target material may be in condensed state (precritical irradiation regime) or plasma jet forms and fastly widens (which corresponds to supercritical regime). The changeover between precritical irradiation regime and supercritical one has threshold nature. In precritical regime case perturbation growth is limited by surface tension. In supercritical irradiation case Taylor instability is essential and causes growth of surface perturbations, crater formation, mixing of near-surface layers.

271

ACKNOWLEDGMENTS

The research was supported by RFBR-Ural grant No 04-01-96074 and Chelyabinsk region government; by Ministry of Education of Russian Federation (program «Development of science potential of Higher education» No 45437), the presidium of the Ural branch of Russian Academy of Science under of the integration projects program of the Ural Branch and the Siberian Branch of Russian Academy of Science. We appreciate discussions with V.I. Engel'ko and V.S.Kuznetsov.

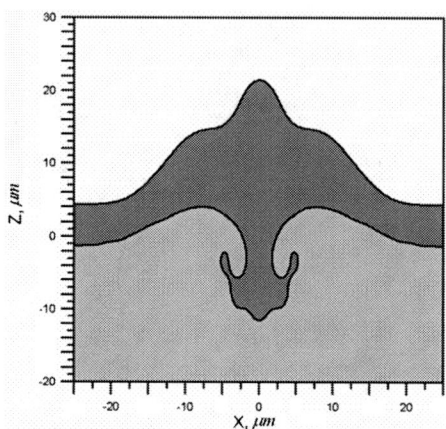

FIGURE 3. Mixing of film and supported material.

REFERENCES

1. Boiko V.I, Valyaev A.N., Pogrebnyak A.D. // UFN. 1999. V. 169. p. 1243 (in Russian)
2. Bluhm H., Engelko V., Mueller G., et. al. // AIP CP650, BEAMS 2002: 14th International Conference on High-Power Particle Beams. Melville, New York, 2002. P. 9.
3. Korotaev A.D., Ovchinnikov S.V., Pochivalov Yu.I., et al. // Surface and Coatings Technology. 1998. V. 105. P. 84.
4. Korotaev A.D., Tyumentsev A.N., Tretjak M.V., et al. // Physics of Metals and Metallography. 2000. V. 89. P. 54.
5. Shulov V.A., Engelko V.I., Kovalev I.V., Mueller G. // Proceedings of the 7th Int. Conf. on Modification of Materials with Particle Beams and Plasma Flows. Tomsk. 2004. P. 289.
6. Volkov N.B., Mayer A.E., Yalovets A.P. // Jour. Techn. Phys, 2002. V. 72. No. 8. p. 34. (in Russian)
7. Richtmyer R.D. // Comm. on Pure and Appl. Math. 1960. V. XII. P. 297;
8. Meshkov E.E. // Izvestiya of Academy of Science of USSR. Mech. of Fluid and Gas. 1969. V. 5. p. 151. (in Russian)
9. Volkov N.B., Mayer A.E., Yalovets A.P. // Jour. Techn. Phys, 2003. V. 73. No. 3. p. 1; (in Russian)
10. Volkov N.B., Mayer A.E., Talala K.A., Yalovets A.P. // Conference - Phys. of extremal substance state – 2004, Chernogolovka, 2004. p. 155. (in Russian)
11. Yalovets A.P., Mayer A.E. // Proceedings of the 6th Int. Conf. on Modification of Materials with Particle Beams and Plasma Flows. Tomsk, 2002. P. 297.

Experimental results on single-turn solenoid powering from helical MCG with current opening switch

A. S. Boriskin, A. Ya. Brodsky, Yu. V. Vlasov, S. N. Golosov,
V. A. Demidov, Ye. M. Dimant, S. A. Kazakov, I. M. Markevtsev,
A. N. Moiseenko, A. P. Romanov, V. D. Selemir, O. M. Tatsenko,
A. V. Filippov, Ye. V. Shapovalov

Russian Federal Nuclear Center – VNIIEF
37 Mira Ave., Sarov, Nizhny Novgorod region, 607188, Russia

Abstract. Experimental results on powering of a single-turn solenoid with a current of ~3 MA with a rise time of ~0.5 µs from a helical magneto-cumulative generator of 100 mm diameter (MCG-100) having explosive current opening switch are presented. To protect the solenoid from destruction with HE-charge explosion products, it was located behind an armored plate. The magnetic field in the solenoid was ~300 T. Experimental results showed possibility of single-turn solenoids powering, used for investigation of samples behavior in ultra-high magnetic fields, from MCG-100, having the opening switch, and possibility of the samples preservation after the experiments.

Introduction

Traditionally capacitor banks are used as power sources to obtain high magnetic fields with the magnetic inductance of ~300 T in single-turn solenoids [1-3]. Obligatory destruction of the solenoid is minimized by application of fast (quarter of the discharge period of about several microseconds) capacitor bank with very low inductance and resistance, discharging to the light turn with optimized wall thickness [2]. The most important advantage of this method is saving the studied samples, located inside of the exploding turn. However, such capacitor banks are stationary and rather expensive, though the information on development of transportable megagauss generator has been published [3].

Experiments on single-turn solenoid powering from magneto-cumulative generator (MCG) were held. In papers [4, 5] bus-type MCG was used for single-turn solenoid powering. Creating the field of ~100 T with MCG the authors managed to save the studied sample from destruction by the detonation products of HE-charge [5]. In paper [4] double-sided bus-type generator was used to get the magnetic field of ~200 T. However, the maximum field was limited by the value of about 250 T due to large (tens microseconds) operation time of the generator, and increase on the inner sizes of the coil and transmitting line under effect of ponderomotive forces [6].

CP849, *Zababakhin Scientific Talks - 2005,*
edited by E. N. Avrorin and V. A. Simonenko

To obtain the magnetic field of more than 300 T and minimize the solenoid destruction under effect of the ponderomotive forces it is necessary to sharpen the MCG current pulse with a current opening switch. In this case the conditions of the solenoid powering will be similar to ones realized at its powering from fast capacitor banks. To protect the solenoid from destruction by the explosive charge detonation products it is proposed to locate the solenoid behind the armored plate, and to transfer the energy from the MCG to the solenoid with low-inductive transmitting line. Use of this device would make it possible to perform search studies in wide range of the power source and the solenoid parameters.

In this paper we present the results of the first experiment on single-turn solenoid powering with the current of ~3 MA and current rise time of ~0.5 μs from the helical magneto-cumulative generator of 100 mm diameter (MCG-100) [7] having explosive current opening switch. To protect the solenoid from destruction with the explosive charge detonation products it was located behind the armored plate. Magnetic field in the solenoid was ~300 T.

Experimental setup

To power the single-turn solenoid with current pulses on the level of 3 MA we developed the device, based on fast helical MCG-100 having explosive current opening switch. To protect the load from the MCG fragments and shock waves it is possible to transport the MCG energy to the solenoid with low-inductive coaxial and bus-type transmitting lines. External view of the MCG-100 and approximate scheme of the experiment are presented in Figure 1.

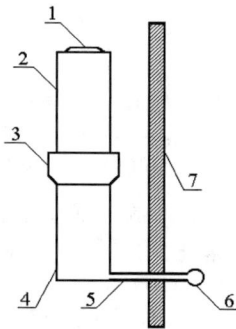

FIGURE 1. External view of the MCG-100 generator and experimental scheme. 1 – electric detonator,   2 – MCG-100, 3 – explosive current opening switch, 4 – coaxial transmitting line, 5 – bus-type transmitting line, 6 – single-turn solenoid, 7 – armored plate

A current derivative in the generator and the solenoid contour are measured by inductive magnetic probes. Another approach to the magnetic inductance measuring is based on application of optical method, based on Faraday effect [8].

Experimental results

In the experiment the single-turn solenoid of 5 mm length, inner diameter of 5 mm and wall thickness of 1.5 mm was powered by helical generator MCG-100 having explosive opening switch with the diameter of 100 mm and the length of 200 mm. HE-charge and ribbed barrier were used in the opening switch to destroy the current conductor. Total inductance of the load, which consists of coaxial and bus-type transmitting lines and the solenoid was ~20 nH.

The MCG-100 generator was powered from the initial current of 22 kA and provided maximum current of 6.8 MA in the final contour of ~20 nH. After MCG current switching with the explosive opening switch the current pulse of 2.4 MA amplitude with the rise time from 0.1 up to 0.9 of its maximum value (τ0.1-0.9), equal to 0.37 μs was obtained in the solenoid. Figure 2 presents characteristic oscillograms of the MCG and the solenoid current derivative and oscillogram of Faraday rotation recorded by the probe.

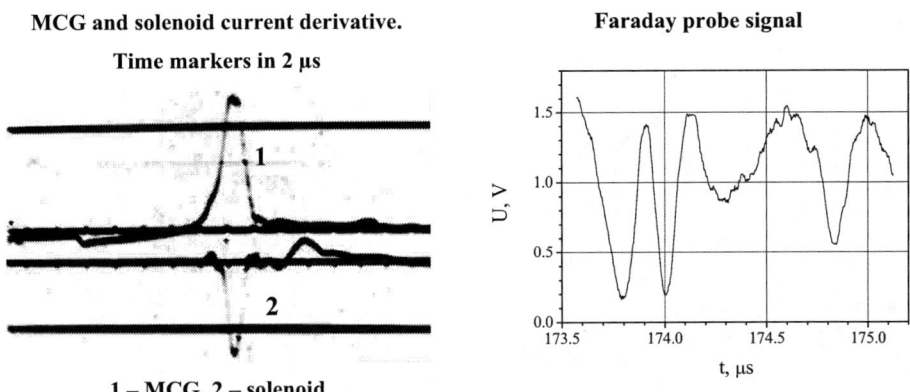

MCG and solenoid current derivative.

Time markers in 2 μs

Faraday probe signal

1 – MCG, 2 – solenoid

FIGURE 2. Oscillogram of the MCG and the solenoid current derivative and Faraday probe signal by which the magnetic field in the solenoid was found

To determine the size of the copper coil at the magnetic field maximum, in the experiment we carried out x-ray imaging of the solenoid. The diagram of x-ray facility elements location, cassette with the film, the solenoid, the MCG and protecting armored plate is presented in Figure 3. Triggering of the x-ray imaging facility was made from the voltage pulse, taken from the inductive probe, located in the current contour of the single-turn solenoid.

Figure 4 presents picture of the solenoid made before the experiment and its x-ray photograph made during the experiment. Fig. 5 presents experimental time dependencies of currents in the breaking contour (If) and in the solenoid (Il), and time dependence of the magnetic induction in the center of the solenoid. Maximum value of the magnetic induction in the solenoid was 290 T, and its rise time up to the maximum was 0.7 μs.

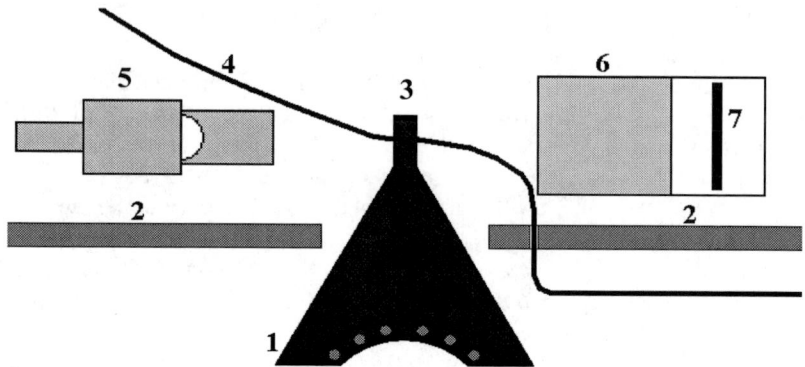

FIGURE 3. Diagram of the elements location in the experiment. 1 – MCG-100, 2 – armored plate, 3 – solenoid, 4 – fiber of the magneto-optical probe, 5 – x-ray source, 6 – foam plastic, 7 –cassette with the film

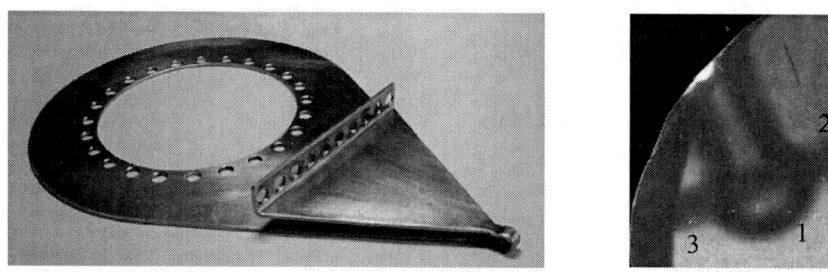

FIGURE 4. The solenoid before and during the experiment. 1 – turn, 2 – Faraday probe protected by ceramic tube with outer diameter of 3.5 mm, 3 - shield

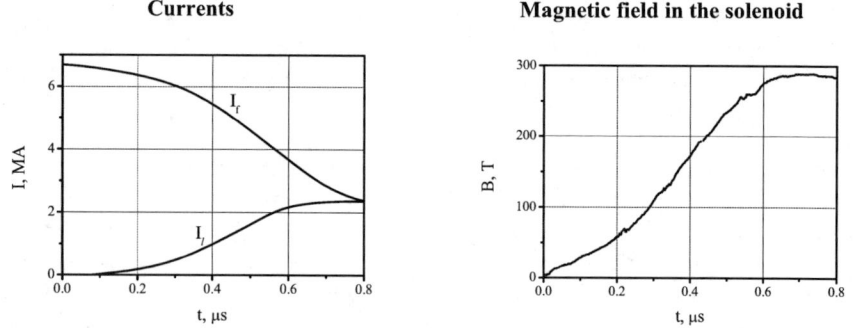

FIGURE 5. Experimental results

Conclusion

Current pulse of 2.4 MA amplitude and characteristic rise time of ~0.4 μs was obtained in the solenoid in the first experiment on the single-turn solenoid powering from the helical generator MCG-100 with explosive current opening switch. The magnetic field in the center of the solenoid was ~300 T. To protect the solenoid from destruction by the explosive charge detonation products the solenoid was located behind the armored plate. Experimental results showed possibility of single-turn solenoids powering with MCG-100 having the current opening switch. In the future this will make it possible to use this device for samples behavior studies in the ultra-high magnetic fields saving the sample after the experiment.

REFERENCES

1. F. Herlach, "Megagauss Fields from Single Turn Coils," in *Ultrahigh Magnetic Fields. Physics. Techniques. Applications*, V.M.Titov and G.A.Shvetsov, Ed., Moscow: Nauka, 1984, pp.58-69.
2. F. Herlach, M. von Ortenberg, "Megagauss Fields in the European Union," *in Megagauss and Megaampere Pulse Technology and Applications*, V.K.Chernyshev, V.D.Selemir, L.N.Plyashkevich, Ed., Sarov: VNIIEF, 1997, pp.171-175.
3. O. Portugall, N. Puhlmann, H.-U. Muller, M. Barczewski, I. Stolpe, M. Thiede, M. von Ortenberg, F. Herlach, "Design and Performance of a Transportable Low-Cost Setup for the Generation and Application of Megagauss Fields," in *Megagauss and Megaampere Pulse Technology and Applications*, V.K.Chernyshev, V.D.Selemir, L.N.Plyashkevich, Ed., Sarov: VNIIEF, 1997, pp.221-225.
4. F. Herlach, H. Knoepfel and R. Luppi, "Magnetic Field and Current Amplification with Non-Cylindrical Explosive Systems," in *Proc. Conf. on Megagauss Magnetic Field Generation and Related Experiments*, Frascati, Italy, Sept. 21-23, 1965, H. Knoepfel and F. Herlach, Ed. Brusselles, EUR 2750.e. Euratom, 1966, pp.287-304.
5. H. Yokoi, Y. Kakudate, S. Usuba, M. Yoshida, S. Fujiwara, "Development of a Small-Size Bellows-Type Megagauss Generator," in *Megagauss and Megaampere Pulse Technology and Applications*, V.K.Chernyshev, V.D.Selemir, L.N.Plyashkevich, Ed., Sarov: VNIIEF, 1997, pp.138-143.
6. "High and Ultra-High Magnetic Fields and their Applications," *Transl. from English, F.Herlach*, Ed., Moscow: Mir, 1988.
7. V. D. Selemir, V. A. Demidov, L. N. Plyashkevich, A. S. Kravchenko, S. A. Kazakov, A. M. Shuvalov, A. S. Boriskin, V. A. Zolotov, G. M. Spirov, and M. M. Kharlamov, "High-Current (30 MA and More) Energy Pulses for Energizing of Inductive and Resistive Loads," in *Megagauss and Megaampere Pulse Technology and Applications*, V. K. Chernyshev, V. D. Selemir, L. N. Plyashkevich, Ed., Sarov: VNIIEF, 1997, pp.241-247.
8. I. Pavlovsky, V. V. Druzhinin, O. M. Tatsenko, N. P. Kolokolchikov, A. I. Bykov, M. I. Dolotenko, "Magneto-optical investigations in ultra-high magnetic fields," in *Ultrahigh Magnetic Fields. Physics. Techniques. Applications*, V.M.Titov and G.A.Shvetsov, Ed., Moscow: Nauka, 1984, pp.130-135.
9. V. K. Chernyshev, V. A. Demidov, V. N. Veselov et al., "Investigation of the Speed Response Dependence of the Explosive Current Opening Switch on Initial Conditions," in *Megagauss Fields and Pulsed Power Systems*, V.M.Titov and G.A.Shvetsov, Ed., N.Y.: Nova Science Publishers, 1990, pp.527-531.

Current Pulse Sharpening
of Multi-Element Disk Generator
with Electric Exploded Opening Switch

A S. Boriskin, Yu. V. Vlasov, S. N. Golosov, V. A. Demidov,
P. V. Duday, S. A. Kazakov, S. V. Kutumov, Y. N. Lashmanov,
I. M. Markevtsev, A. N. Moiseenko, A. P. Romanov, V. D. Selemir,
O. M. Tatsenko, A. V. Filippov, E. V. Shapovalov

Russian Federal Nuclear Center – VNIIEF
37 Mira Ave., Sarov, Nizhny Novgorod region ,607188, Russia

Abstract. Investigation results of current pulse sharpening of multi-element disk generator with
HE-charge diameter of 240 mm with electric exploded opening switch are presented. Both solid
conductor (copper foil), and layer structure metal-dielectric were used in electric-exploded
opening switch. Experimental results with three-element generator showed the possibility to
form the current pulse of ~30 MA with characteristic rise time of ~1 µs in the load of ~1.5 nH. It
is supposed to use ten-element generator with electric exploded opening switch as an energy
source in EMIR complex for soft x-ray radiation generation at multi-wire liner implosion.

INTRODUCTION

Multi-element disk MCG of 240 mm (DMCG-240) diameter [1] are supposed to be
used as one of the types of an energy source of explosive electro-physical complex
EMIR [2] intended for generation of pulsed fluxes of soft x-ray radiation (SXR). To
decrease expenses the development of base modules is carried out with three-element
generators DMCG-240-3. Output characteristics of the generator DMCG-240-3,
obtained in the first experiments, were worse than calculated [3]. Performed two-
dimensional hydrodynamic calculations made it possible to explain some peculiarities
of the generator DMCG-240 operation and put corresponding changes into the design
in order to improve its output characteristics [4]. In this paper we present the results of
the experiment with modified three-element generator DMCG-240-3M. Since
imploding of the liner unit for SXR pulse generation should be performed at
comparatively short time (≤ 1 µs) [5], we have carried out two experiments on the
current pulse sharpening of the DMCG-240-3 generator with electric-exploded
opening switch.

CP849, *Zababakhin Scientific Talks - 2005*,
edited by E. N. Avrorin and V. A. Simonenko
© 2006 American Institute of Physics 0-7354-0345-7/06/$23.00

EXPERIMENT ON INDUCTIVE LOAD

The first experiment with the generator DMCG-240-3M was carried out on the inductive load. General view of the generator is shown in Figure 1.

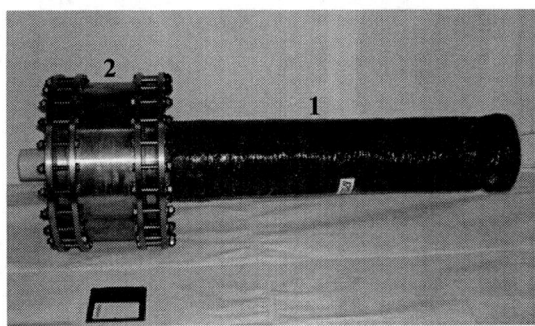

FIGURE 1. General view of DMCG-240-3M equipped with energy preamplifier. 1 – preamplifier, 2 – disk MCG.

In this experiment the preamplifier - a helical MCG of 120 mm diameter (MCG-120) [3], having initial inductance of 22 μH was powered with the current of 26 kA. Initial inductance of the DMCG contour was 27 nH, The powering current of the DMCG was 5.2 MA up to the end of the preamplifier operation. A current pulse of 60 MA amplitude with maximum derivative of ~1.9·10^{13} A/s and characteristic rise time of ~3.3 μs was recorded in the DMCG contour in 19 μs after the beginning of its operation (explosion time of electric detonators, located on the DMCG axis, was selected as a start moment). The DMCG contour inductance in this moment was ~1.5 nH.

Calculated (using a program [1]) and experimental dependencies of the DMCG current, obtained in this experiment, are shown in Figure 2. One can see that the experimental results agree well with the calculation.

SHARPENING OF THE DMCG CURRENT PULSE WITH ELECTRIC-EXPLODED OPENING SWITCH

Electrically exploded foil or wires, located in dielectric (arc-suppressing) medium, is the main unit of electric-exploded current opening switches (EEOS). In contrast to other types of opening switches, EEOS differs with simple of the design, small size and, consequently, with low own inductance. Due to their simplicity and low inductance EEOS are now used at creation of high-power energy sources, based on disk MCG to power liner ponderomotive units [6].

The current pulse of ~35 MA at time of ~1.5 μs [6] was obtained in the DMCG of 400 mm diameter equipped with foil current opening switch. The generator DMCG-240-3M equipped with EEOS was developed to study the possibility to obtain similar pulses from disk MCG of smaller diameter.

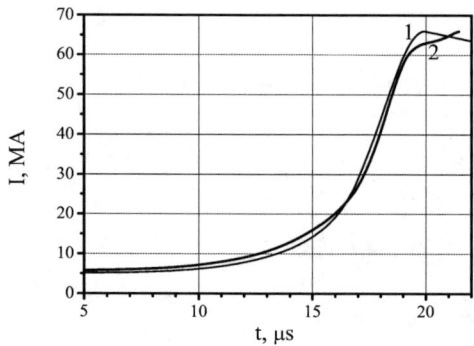

FIGURE 2. Time dependence of the DMCG current. 1 – calculation, 2 – experiment.

In the experiments the DMCG-240-3M was powered with the preamplifier MCG-120. In the first experiment the initial current of the preamplifier was 25 kA. The EEOS represented itself a copper foil of 0.19 mm thickness, located between dielectric cylinders of 2 mm thickness (Figure 3). Initial inductance of the DFCG contour was 27 nH, inductance of rigid section of the DFCG contour, located above the diameter of 240 mm and lower the EEOS conductor (foil) was equal to 1.6 nH. A cavity between an external surface of the EEOS and a reverse current conductor, having the inductance of 1.7 nH (taking into account the inductance of the closing switch), was used as a load. The load was connected at specified time moment using the explosive closing switch.

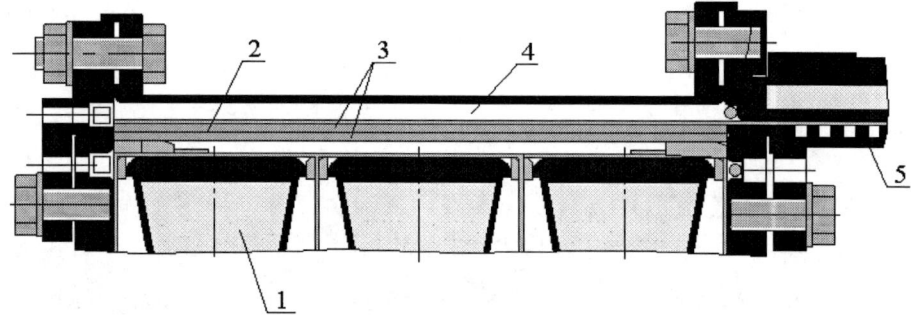

FIGURE 3. Fragment of the sketch of DMCG-240M with EEOS. 1 – DMCG HE-charge, 2 – EEOS conductor, 3 – dielectric cylinders, 4 – load, 5 – explosive closing switch.

DMCG powering current was 4.9 MA. Characteristic oscillograms of the DMCG current derivative and current derivative in the load, obtained in this experiment using the inductive probe, are shown in Figure 4.

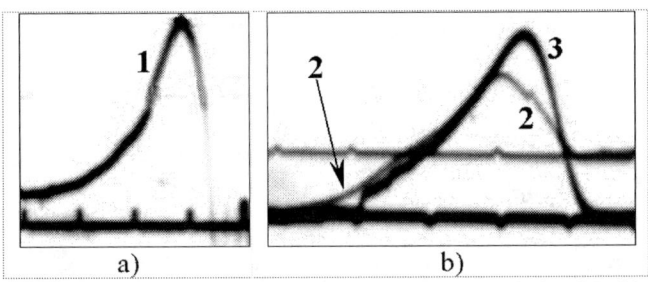

FIGURE 4. Characteristic oscillograms of the DMCG current derivative (a) and the load current derivative (b) in the first experiment. 1 – DMCG current derivative in the process of current amplification, time markers are set in 2 μs; 2 – DMCG current derivative in the process of the current switching; 3 – current derivative in the load. Time markers are set in 0.5 μs.

The current pulse of 40 MA amplitude with maximum rise time of ~1.1·1013 A/s was recorded in the DMCG contour in 18.4 μs after the beginning of its operation. The DMCG current was also measured using an optical method based on Faraday effect [7]. According to the estimations based on hydrodynamic calculations, held earlier [4], the inductance of the DMCG contour in the current maximum was 1.5 nH. The current in the load appeared in 17.8 μs after the beginning of the DMCG operation. A current pulse of 28 MA with maximum derivative of ~3.5·1013 A/s was recorded in the load of ~1.7 nH inductance. The characteristic current rise time (from 0.1 up to 0.9 of its maximum value) was 1.0 μs. Experimental dependencies of the DMCG current and the current in the load on time, obtained in this experiment, are presented in Figure 5.

In the second experiment the EEOS represented itself alternating layers (total 5 layers) of copper foil and dielectric lavsan film [8]. Section area of all foil layers in this experiment is almost equal to section area of the foil in the experiment with the single-layered EEOS. Parameters of the preamplifier, the DMCG and the load, and powering conditions of the generator in the moment of the load switching were the same as in the first experiment.

Powering current of the DMCG up to the moment of preamplifier operation termination was 5.1 MA. A current pulse of 41 MA amplitude with maximum derivative of ~1.1·10^{13} A/s was recorded in the DMCG contour in 18.1 μs after the beginning of its operation. As in the first experiment, the DMCG current was measured with the inductive probes and optical method. Characteristic oscillograms of the DMCG current derivative and the load current derivative, obtained in this experiment using the inductive probes, are shown in Figure 6.

Experimental dependencies of the DMCG current and current in the load on time, obtained in this experiment, are shown in Figure. 7. The current in the load appeared in 17.7 μs after the beginning of the DMCG operation. A current pulse of ~23 MA amplitude with maximum rise time of ~3.2·1013 A/s and characteristic current rise time (from 0.1 up to 0.9 of its maximum value) of ~0.8 μs was recorded in the load of ~1.7 nH. Time dependence of the EEOS resistance is presented in Fig. 8. In the same place for comparison we present time dependence of the resistance, obtained in the experiment with the single-layered EEOS.

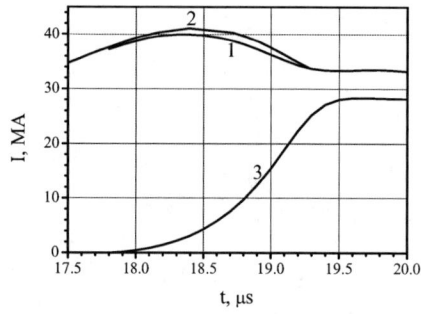

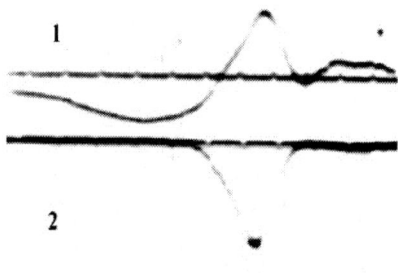

FIGURE 5. Time dependencies of the DMCG current and the load current in the first experiment. 1 – DMCG current based on the inductive probes data; 2 – DMCG current based on Faraday probe data; 3 – current in the load.

FIGURE 6. Characteristic oscillograms of the DMCG current derivative and the load current derivative in the second experiment. 1 – DMCG current derivative, 2 – current derivative in the load. Time markers are set in 0.5 µs.

From Figure 8 one can see that in the experiment with the multi-layered EEOS the resistance is growing faster, but up to smaller values than in the experiment with the single-layered EEOS.

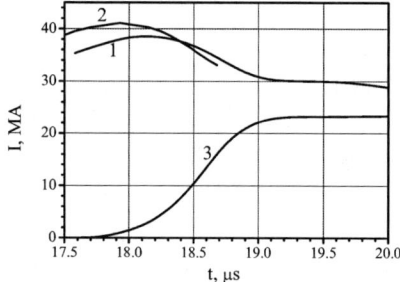

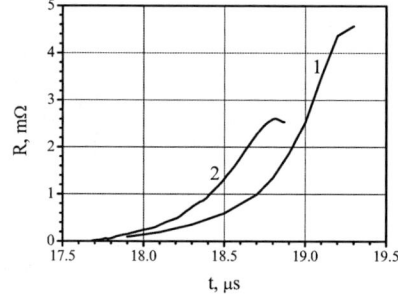

FIGURE 7. Time dependence of the DMCG current and current in the load in the second experiment. 1 – DMCG current based on the inductive probes data; 2 – DMCG current based on Faraday probe data; 3 – current in the load.

FIGURE 8. Time dependence of the EEOS resistance, obtained in two experiments. 1 – experiment with single-layered EEOS; 2 – experiment with multi-layered EEOS.

The value of a specific action integral through the EEOS conductor in the first experiment was ~$1.3 \cdot 10^{17}$ A^{2}s/m^4 in the moment of the current appearance in the load and ~$1.9 \cdot 10^{17}$ A^{2}s/m^4 after termination of the process of the current switching, and the EEOS conductor resistance increased in 250 times. The obtained results agree well with the results presented in [9].

The value of the specific action integral of the current through the EEOS conductor in the second experiment was ~0.9·1017 A2s/m4 in the moment of the current appearance in the load and ~1.4·1017 A2s/m4 after termination of the process of the current switching. The EEOS conductor resistance in this experiment increased in approximately 170 times. This result agree not so well with the results [9]. Evidently, this is associated with the fact that in this experiment the EEOS conductor represented itself a layered structure, that is why the estimations made are averaged on the whole space, occupied by this layered structure.

The results of the experiments performed show that use of the layered structure metal-dielectric instead of solid metal foil in the EEOS construction could provide faster EEOS operation. However, detailed simulation of electric explosion of the layered structure metal-dielectric will require development of one-dimensional MHD-code. Besides, additional investigations on the results of calculation optimization of EEOS construction will require.

CONCLUSION

The results of the experiment with three-element DMCG-240-3M generator on the inductive load are presented. The current of 60 MA was recorded in the DMCG contour in 19 μs after the beginning of its operation. The DMCG contour inductance in this moment was 1.5 nH.

The results of two experiments on the current pulse sharpening of the DMCG-240-3M generator with electric-exploded opening switch showed possibility of current pulse formation of ~30 MA with the characteristic rise time of ~1 μs in the load of ~1.7 nH inductance.

REFERENCES

1. V. A. Demidov, S. A. Kazakov, A. S. Kravchenko, L. N. Plyashkevich, V. D. Selemir, Yu. V. Vlasov, R. M. Garipov, and Ye. V. Shapovalov, "High-Power Energy Sources Based on the FCG Parallel and Series Connection," in *Digest of Technical Papers 11th IEEE International Pulsed Power Conference*, G. Cooperstein and I. Vitkovitsky, Ed., Baltimore, Maryland, 1997, pp.1459-1464.
2. V. D. Selemir, V. A. Demidov, A. V. Ivanovsky, V. F. Yermolovich, V. G. Kornilov, V. I. Chelpanov, S. A. Kazakov, Yu. V. Vlasov, and A. P. Orlov, "Explosive Complex for Generation of Soft X-ray Radiation Pulsed Fluxes," in *12th International Conference on High Power Particle Beams BEAMS'98. Program and Abstracts*, Haifa, Israel, June 7-12, 1998, p.21.
3. V. D. Selemir, V. A. Demidov, S. A. Kazakov, Yu. V. Vlasov, A. P. Romanov, R. M. Garipov, Yu. N. Lashmanov, V. A. Yanenko, and I. K. Fetisov, "Investigation of Disk FCG for Electro-Physical Complex EMIR," in *Digest of Technical Papers 12th IEEE International Pulsed Power Conference*, C. Stallings and H. Kirbie, Ed., Monterey, California, USA, 1999, pp.728-731.
4. S. Boriskin, Yu. V. Vlasov, V. A. Demidov, I. N. Pavlusha, T. A. Toropova, "Calculation of current conducting shells motion of disk magneto-cumulative generator," in *Accepted Abstracts for Contributed Papers the Xth International Conference on Megagauss Magnetic Field Generation and Related Topics*, July 18-23, 2004, Humboldt-University, Berlin, C 78-poster: Russian.
5. Deeney, C. A. Coverdale, M. R. Douglas et al., "Radiative properties of high wire number tungsten arrays with implosion times up to 250 ns," *Phys. Plasmas*, vol.6, pp.3576-3585, No.9, 1999.
6. V. K. Chernyshev, A. M. Buyko, V. N. Koctyukov et al., "Investigation of Electrically Exploded Large Area Foil for Current Switching" In *Megagauss Fields and Pulsed Power Systems*. Ed. V.M.Titov and G.A.Shvctsov. N.Y.: Nova Science Publishers. 1990. P.465-470.

Calculation of Extreme Characteristics of an Explosive Plasma Opening Switch to Break Currents of 50-100 MA

A.S. Pikar, P.V. Korolev, A.S. Russkov

Russian Federal Nuclear Center – VNIIEF
37 Mira Ave., Sarov, Nizhny Novgorod region, 607188, Russia

Abstract. The paper is devoted to evaluative calculation of explosive plasma opening switch circuit to break 50-100 MA currents from RANCHERO type generator to form profiles current pulse for isentropic compression of materials. The calculation was carried out using empirical formulas to estimate duration of formed voltage pulse, shape of fore and back fronts of current pulse and an amplitude of generated voltage obtained with analysis of experimental data on investigation of opening switches having commutated energy from several kilojoules up to 10 MJ. Module system, consisting of 8 parallel explosive plasma opening switches, located between two current conducting disks, is considered. Main advantage of this scheme is that megavolt voltage is applied perpendicular to an insulator surface, and high-voltage system of axial initiation is used for the opening switch. In this case energy supply to the load is performed with conical electrodes, separated by solid state insulators.

INTRODUCTION

An explosive pulsed system for 20 megabar experiments on isentropic compression [1] was considered in the joint paper of authors from Los Alamos and Laurence Livermore National Laboratories. This system is based on application of explosive magneto-cumulative generators for 50-100 MA currents obtaining, shortening of the current pulse duration up to 0.5-1 µs using explosive opening switches and obtaining a magnetic pressure of sub-megabar range at the obtained current passage through the plain current guide.

The authors of the proposed system underline that the current achieved in the load depends on the voltage pulse amplitude of the opening switch and how the load is connected to it. The opening switches with "flux conservation" are more efficient, but the load has smaller inductance if the opening switch is connected in parallel to it. Though it is necessary to achieve higher voltage, the general circuit becomes more preferable.

In this paper we consider the circuit in which it is possible, in some extent, eliminate this drawback. With its characteristics it is approaching to the circuit "with flux conservation". This is achieved with using multi-module design of explosive plasma opening switches, split by the insulators. This circuit makes it possible to transport the energy to the load with the cone line with solid state or vacuum insulation. The variant of 8-modules explosive plasma opening switch with the vacuum magnetically insulated line is shown in Figure 1.

CP849, *Zababakhin Scientific Talks - 2005,*
edited by E. N. Avrorin and V. A. Simonenko

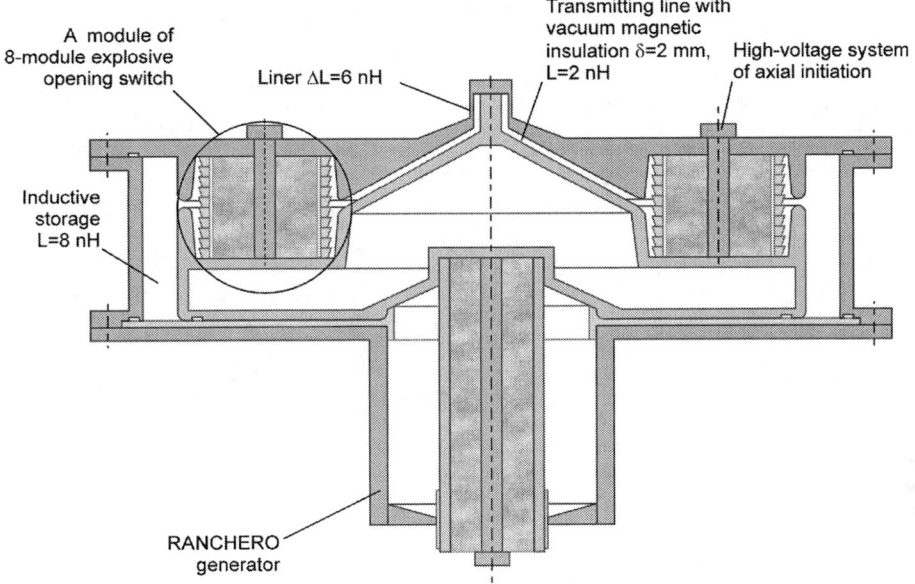

FIGURE 1. Schematic drawing of the multi-module opening switch.

An explosive plasma opening switch [2], in contrast to explosive foil opening switches [3, 4], has higher specific characteristics (specific dissipation power per surface unit is 2÷3 times higher). The experimentally reached strength of the electric fields at megampere currents is 30÷40 kV/cm. This makes it possible to decrease total inductance of the opening switch at the expense of smaller length. Besides, varying the shape of the insulator surface, where the plasma channel is destroyed, one can change the voltage pulse shape, providing the required regime of shock-free compression. Use of a vacuum line with the magnetic insulation makes it possible to decrease the inductance of the current supply to the load, thus increasing the scheme efficiency.

THEORY OF THE EXPLOSIVE PLASMA
OPENING SWITCH CALCULATION

A theoretical model and a calculation method of boundary characteristics of the explosive plasma opening switch was considered by the authors in paper [5]. It was experimentally shown that if the inductive storage pumping time exceeds the current contour breaking time in not more than 100 times, the main energy release in the plasma channel takes place at the breaking moment. At small linear current density up to 10 MA/m the opening switch resistance is limited by the detonation products conductivity. At higher current densities the power of the opening switch is limited by the HE-charge power, since in this regime the pressure in the channel could exceed the pressure in the shock wave at the expense of the Joule heating, and the channel will expand.

285

The specific power of the plasma opening switch is limited by the conductivity of the detonation products on the one hand, on the other hand it can not exceed the specific power of the HE-charge substance. These limitations could be described graphically in coordinates E(H), where areas for two opening switches with different HE-charges views are presented in Fig. 2.

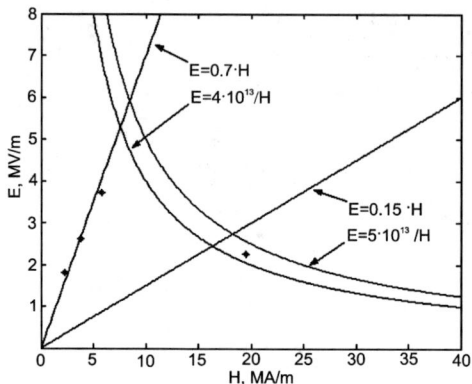

FIGURE 2. The magnetic field strength dependence of electric field strength.

The first one has the resistance of 0.7 Ω and HE-charge of $4 \cdot 10^{13}$ W/m^2 specific power, the second one has the resistance of 0.15 Ω but more powerful HE-charge of $5 \cdot 10^{13}$ W/m^2. In the same diagram we pointed experimental dots for the plasma opening switch with cylindrical HE-charge TH50/50 $\varnothing$300x$\varnothing$150x180, with the inductive storage powering from the capacitor battery, coil MC-generator [6] and helical generator C-320 [7].

One can see that use of more powerful HE-charge does not obligatory lead to increase of the opening switch power. The power of the opening switch is a complicated function of time and the magnetic field value. It depends on overall sizes of the opening switch, current, storage inductance, load inductance and resistance. It is possible to use the notion of Ohmic resistance for the opening switches with the current density higher than 10 MA/m, but with large reservations, and in case when the storage and the load contour parameters change weakly. In a general case, to calculate the field strength and the breaking time of the current contour, it is necessary to solve an integral equation:

$$H(t) = H_0 - \frac{l}{L_0 h} \cdot \int_0^t E(H,\tau)d\tau \qquad (1)$$

where L_0 - given inductance of the storage; l and h - length and width of the plasma channel. $E(H,t)$ - is determined either experimentally or with numerical methods for the setup geometry of the HE-charge.

DESIGN SCHEME
OF THE MULTI-MODULE OPENING SWITCH

Let us consider the design of the explosive plasma opening switch for 50÷100 MA currents breaking. In Fig. 2 we can estimate that at current density of 20 MA/m the power of the opening switch will be limited by the HE-charge power and, for switching of 40 MJ energy at time of 1-2 μs , for example, it will be necessary to use the opening switch with the HE-charge surface area of 1 m². The opening switches of smaller diameter have the larger area at one and the same mass. That is why it is preferable to use the multi-module scheme.

The plasma opening switch, made in the form of 8 parallel modules, has both advantages and disadvantages. The technology of small opening switches fabrication is simpler, and operating efficiency of the separate module could be checked with small generators. The drawbacks are heterogeneity of the current flow on the plasma channel surface, but this problem is partially solved by location of the opening switch module in the current conducting cylinder-screen and approach of high-voltage and grounding electrodes to the distance of 2÷3 mm. Mutual inductance of the plasma channel and the cylinder-screen provides homogeneous current flow. The same phenomenon also decreases the inductance of the plasma opening switch. Having the thickness of the insulator of 1 cm, the vacuum gap between the insulator and the screen of 6 mm, outer diameter of the plasma channel of 150 mm, length of 200 mm, the inductance of one module is equal to 8 nH. So, the inductance of 8 parallel modules will be 1 nH.

Existence of vacuum between the electrodes makes it possible to use the vacuum magnetic self-insulation, strength of the electric field of which could be estimated with the formula $E \approx \sqrt{\mu_0 / \varepsilon_0} \cdot H$. At current in the storage of 80 MA at maximum radius of the high-voltage electrode of 0.45 m, we have $E \approx \dfrac{8 \cdot 10^6}{2\pi \cdot 0.45} \cdot 377 \approx 1000$ MV/m ≈ 1 MV/mm. For the variant with the film solid state insulation this value will be, at least, five time less.

Up to the moment of the voltage maximum the storage current will decrease up to 40-60 MA, and the current in the load increases up to 20÷25 MA. So, for the insulation of 600 kV it is enough to provide the vacuum gap of δ=2 mm. For the conical electrodes with the gap, decreasing from 3 mm on the diameter of 0.9 mm, up to 1 mm on the diameter of 1 cm, the inductance will be approximately 2 nH.

Using the results of MHD calculations of the inter-electrodes gap change from paper [1] and solving the integral equation (1) we can calculate the operating scheme of 8 module opening switch with the generator of Ranchero type. Equivalent electric circuit at the breaking moment is presented in Fig. 3.

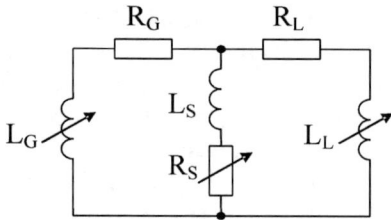

FIGURE 3. Equivalent electric circuit at the breaking moment. $R_G=10^{-4}$ Ом; $R_L=10^{-4}$ Ом; $L_G=10$нГн$-2\cdot10^{-3}\cdot t/\tau$; $L_L=2$нГн$+6\cdot10^{-3}\cdot t^2/2\tau$.

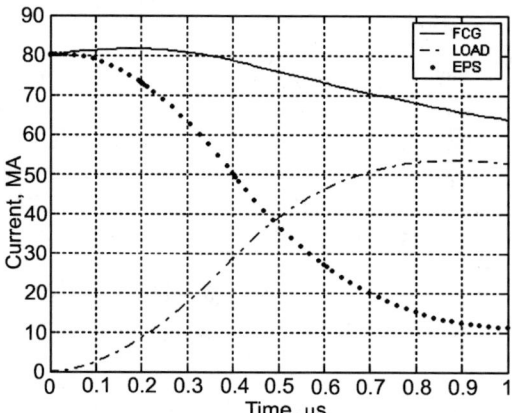

FIGURE 4. Current in the generator (FCG), lid (LOAD), and explosive plasma opening switch (EPS).

Fig. 4 shows the current in the generator, the current in the load and the current in the explosive plasma opening switch, corresponding to the beginning of the breaking. An expected voltage on the opening switch has the rectangular shape with the fore front duration of the voltage pulse of 0.4 μs, length on the basement of 1.5 μs and maximum amplitude of 600 kV.

For the liner located on the diameter of 1 cm at the current of 50 MA the magnetic field value will be 17 MOe, the pressure of 20 Mbar correspond to this.

Let is note that the presented values are the upper estimate and the real values will be lower.

CONCLUSION

The presented in the report calculation results of 8 module explosive plasma opening switch show that at current contour breaking at the current of 80 MA from the RANCHERO type generator it is possible to obtain the currents on the level of 50 MA in the dynamic load with the inductance, changing from 2 nH up to 8 nH during 1 μs. The calculated value of the magnetic field is 17 MOe using the liner, with the energy supply on magnetically insulated line, at a load on the diameter of 1 cm.

REFERENCES

1. J.H. Goforth, et al., Design of high explosive pulsed power systems for 20 MB isentropic compression experiments, in *Proc. Ninth International Conference on Megagauss Magnetic Field Generation and Related Topics*, 2004, Sarov, VNIIEF, p.137.
2. A.I. Pavlovskii, et al., MC-Generator Current Contour Break By Explosiv-Driven Plasma Switch, in *Proc. Fifth International Conference on Megagauss Magnetic Field Generation and Related Topics*, 1989, Novosibirsk, RUSSIA, p.503.
3. V.D. Selemir, et al., High-current (30 MA and more) energy pulses to power inductive and active loads. in *Megagauss and Megaampere Pulse Technology and Applications*. Ed. V.K. Chernyshev, V.D. Selemir and L.N. Plyashkevich. Sarov: VNIIEF. 1997. p.241-247.
4. V.K. Chernyshev, et al., Commutation of electric circuits with megaampere current by a small amount of HE in confined volumes, in *Proc. Ninth International Conference on Megagauss Magnetic Field Generation and Related Topics*, 2004, Sarov, VNIIEF, p.308.
5. A.S.Pikar, V.E.Gurin, P.V.Korolev, A.S.Russkov Effect of detonation products on electric characteristics of explosive plasma opening switch, in *Proceedings of International Conference V Khariton Readings, "Substances, materials and constructions at intense dynamic effects"*, Sarov, March 17-21, 2003. –Sarov: VNIIEF, 2003, p.522-525.
6. Magneto-cumulative generator of current. A.Ya. Koshelev, V.S.Fomenko, V.I.Chizhov –a.s.321190, priority 20.02.70. *Bulletin Discoveries. Inventions.* / CNIIPI. –M., 1974. -№ 33
7. A.I. Pavlovskii, et al., A Multiwire Helical Magnetic Cumulation Generator in *Megagauss Physics and Technology* ed. by P.J. Turchy. N.Y.-L.: Plenum Press. 1980. p.585-593.

Autonomous Magneto-Cumulative Energy Source Based on Permanent Magnets

V.E. Gurin, A.S. Boriskin, Yu.V. Vlasov, S.N. Golosov, E.M. Dimant,
V.A. Demidov, S.A. Kazakov, N.R. Kazakova, M.V. Klimashov,
P.V. Korolev, S.V. Kutumov, A.S. Pikar, A.P. Romanov,
E.V. Shapovalov, Yu.M. Shibitov, E.I. Shchetnikov, V.A. Yanenko

Russian Federal Nuclear Center – VNIIEF
37 Mira Ave., Sarov, Nizhny Novgorod region, 607188, Russia

Abstract. In the paper we describe a design of autonomous source of electromagnetic energy with initial field, created by permanent barium oxide magnets and present results of its tests. The energy source consists of magnetic system with initial energy of 1 J, helical magneto-cumulative generator with helix diameter of 50 mm, amplifying the field energy in 130 times, and transforming unit for a load matching. At peak power on helical generator of 66 MW, the voltage of 16 kV was obtained at inductive load of 40 µH.

Weight of the energy autonomous source is 5 kg, HE-charge mass is 100 g.

Several variants of magnetic systems were tested. Comparison of the magneto-cumulative generator operation, powered from permanent magnets and powered from capacitor bank was carried out. Optimizing the magnetic system with numerical methods the authors managed to get almost identical distribution of the fields in the generators. Finally, the generator based on permanent magnets operates and provides similar output characteristics as the generator powered from the capacitor bank.

In contrast to other magneto-cumulative energy sources powered, for example, from an accumulator or piezo-generator, this source, based on permanent magnets, has higher specific energy characteristics, does not require any switching elements and always ready for operation.

I. INTRODUCTION

Efforts to create the magneto-cumulative generator with acceptable parameters, where the initial magnetic field is created with a system made of permanent magnets, belong to the 60-s. Thus, the first 2-strips MC-generator [1] was presented at the first conference on ultra-high magnetic fields in 1965 in Italy. In this generator the magnetic circuit consisted of four magnets and two gaps, limited by the central tube with HE-charge and two stripes, located diametrically, turning to the cylinder.

Further, [2] a design of helical MC-generator, where the magnetic field was created by the permanent magnet, located over the helix, and created the magnetic flux of 1.3 mWb in the first section of the helix, was proposed. The initial energy in the contour was 0.07 J. Small value of the energy is associated with the fact that at the magnets location above the winding the significant part of the flux is distributed outside of the area, covered with the winding.

CP849, *Zababakhin Scientific Talks - 2005*,
edited by E. N. Avrorin and V. A. Simonenko
© 2006 American Institute of Physics 0-7354-0345-7/06/$23.00

This paper is devoted to development and investigation of the helical magneto-cumulative generator EMG-50 with the magnetic system, based on ferrite of barium magnet. General view of this generator assembled with a magnetic system and a matching transformer is presented in Figure 1.

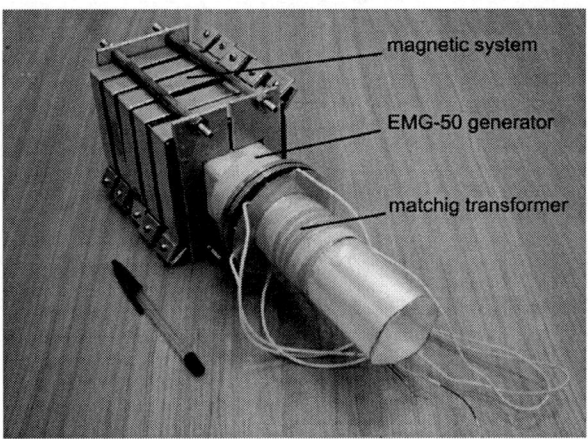

FIGURE 1. General view of the helical generator with the magnetic system.

II. CALCULATION AND OPTIMIZATION OF THE MAGNETIC SYSTEM

During the research we have developed three variants of the magnetic system design for the helical magneto-cumulative generator EMG-50. In the first two experiments we used the magnetic systems with the initial field energy of 0.25 and 0.7 J [3], correspondingly. The sketch of the generator with the magnetic system of 0.25 J energy and the magnetic field distribution in it are presented in Figure 2. Figure 3 shows the sketch of the generator with the magnetic system of 0.7 J energy and the magnetic field distribution in it. One can see that the magnetic system was located outside of the generator itself, covering the first sections of the multi-sectional generator and creating the initial magnetic field in them.

In the third variant (Figure 4) the magnetic system consisted of a set of permanent ferrite of barium magnets. Selection of the magnets configuration and optimization of the shape and amplitude of the magnetic field were carried out with 2D simulation.

Numerically we solved an equation system for axial component of the vector magnetic potential A_ϕ:

$$\nabla \times \left(\mu^{-1} \nabla \times A_{\phi i} - \overline{M} \right) = 0$$

where μ — magnetic penetrability, $\overline{M}$ — magnetization vector.

Residual magnetization of the ferrite of barium magnets was 0.3 T. Since the vector of the magnetic inductance of the initial field in the generator working volume should be directed along the generator axis, the axial component of the magnetic inductance B_z was the main calculation value. With the same way we also calculated the magnetic energy and the magnetic flux in the working volume of the generator. A space between the helix and the liner of the generator considered to be the working volume.

Calculated distribution B_z under the first three sections is presented in Figure 5a. From the diagram one can see that we managed to get homogeneous distribution of the initial magnetic field in the first two sections of the generator. Thus, the magnetic field became more approximate to the initial field in the generator powered from the capacitor bank. The magnetic induction in the first two sections is directed along the axis, and the axial component of the magnetic field inductance in the third section is still high.

It is important that the field increased greatly in the first section, where the most turns of the generator helix are located. So, the magnetic flux, captured by the generator, is higher.

Calculated energy of the initial magnetic field in the working volume was 1.6 J.

The magnetic flux, captured by the generator helix, was calculated with the following formula:

$$\Phi = 2\pi \int_0^{\ell_1} \frac{R}{h} A_\phi(R,\ell)d\ell - 2\pi r \int_0^{\ell_1} \frac{r}{h} A_\phi(r,\ell)d\ell = 23 \text{ mWb},$$

where A_ϕ — axial component of the magnetic potential; R — inner diameter of the helix; r — outer diameter of the liner; h — pitch; ℓ_1 — length of the generator.

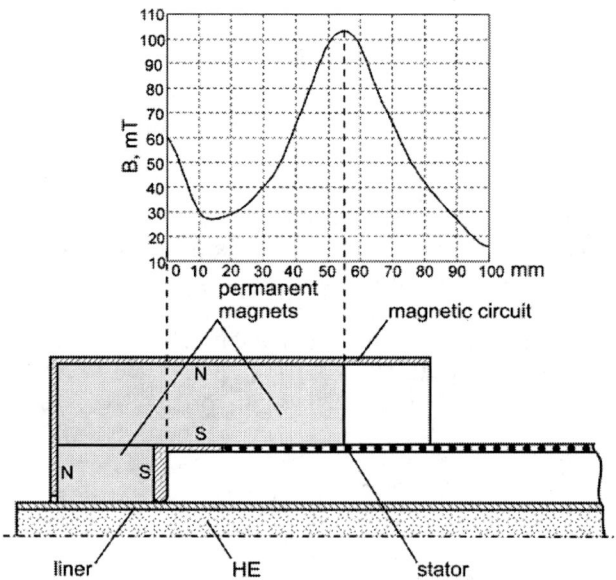

FIGURE 2. The sketch of the generator and distribution of the longitudinal component of the magnetic field inductance for the magnetic system of 0.25 J energy.

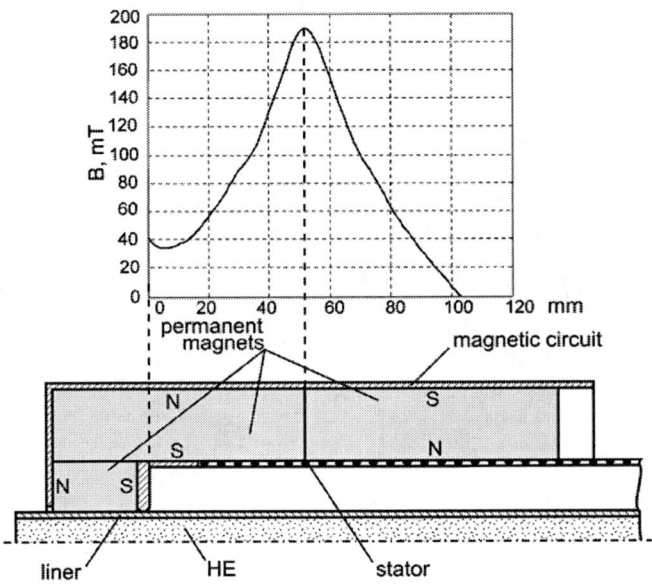

FIGURE 3. The sketch of the generator and distribution of the longitudinal component of the magnetic field inductance for the magnetic system of 0.7 J energy.

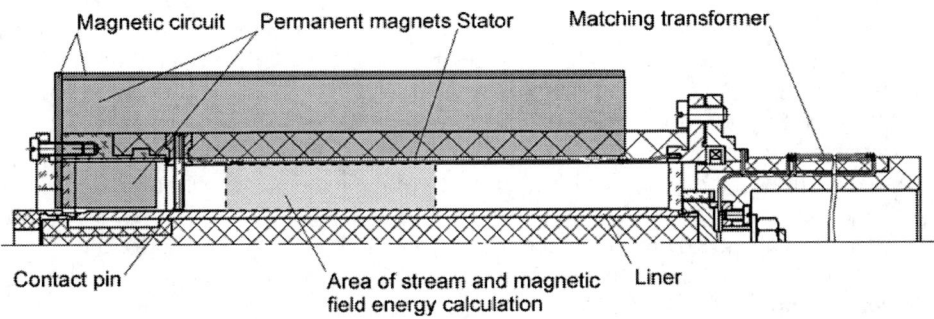

FIGURE 4. EMG-50 generator scheme assembled with the magnetic system of the third variant.

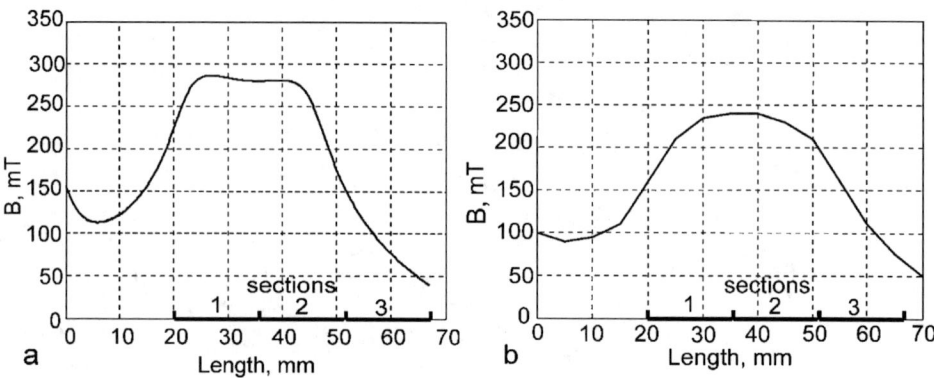

FIGURE 5. Distribution of the axial component of the magnetic inductance B_z under the helix for the magnets with the residual magnetization of 0.3 T: a) calculated; b) measured.

III. DESIGN OF THE MAGNETIC SYSTEM

In the result of the calculations performed we developed the system of permanent magnets that consists of four plain ferrite of barium elements of 84x64x24 mm, connected with each other. They form an internal volume of a rectangular section of 54x54 mm size. A ring magnet, covering the liner and connecting the contact rod from the side of the initiation system, and the first several sections of EMG -50, are installed coaxially in the volume. Analogous poles of the ring magnet and plain elements are directed to the internal volume. To concentrate the magnetic field we used magnetic conductor, assembled of four plates of 1.5 mm from magnetically soft material, located on external surface of the plain elements. The longitudinal magnetic field is created in spare cavity of the internal volume of rectangular section at abovementioned orientation of the poles of the plain elements and the ring magnet. The first section of the EMG are located in part of the magnetic field.

The permanent magnets unit is a separate device and it is assembled separately from the helical generator. In this case there is a possibility to control the initial parameters before the experiment. The measured curve of the axial component of the magnetic induction distribution under the first three sections of the generator for the third variant is presented in Figure 5b. A special case, made of non-magnetic material, is used to keep the magnet unit in working position. Connection of the magnetic unit and the helical EMG-50 is carried out after equipment of the generator with HE-charge and assembly of the helical EMG before the experiment.

A measured energy of the magnetic field for the third variant of the magnetic system in the volume of rectangular shape was 2.4 J, in the EMG-50 compression cavity volume —1.2 J.

IV. EXPERIMENTS

The helical generator EMG-50 with an inductive load L=40 µH, connected through the matching transformer, was tested at explosive experiment with this system. Initial inductance of the generator was 160 μH. The generated energy was 120 J. In the load we obtained the magnetic energy of 40 J at a current of 1.45 kA. Maximum voltage on the load was ~16 kV.

For comparison we carried out the experiments with the generator powering from the capacitor bank. The results coincide at initial energy of the generator on the level of 1 J both on the generated energy and the shape of the current pulse (Figure 7).

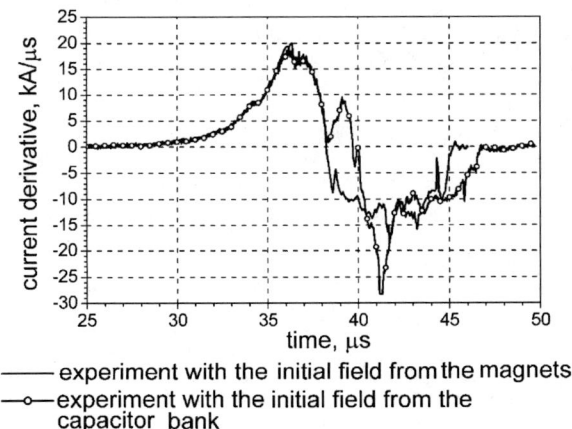

———— experiment with the initial field from the magnets
—o— experiment with the initial field from the capacitor bank

FIGURE 6. Diagrams of the current derivative in the primary winding of the transformer.

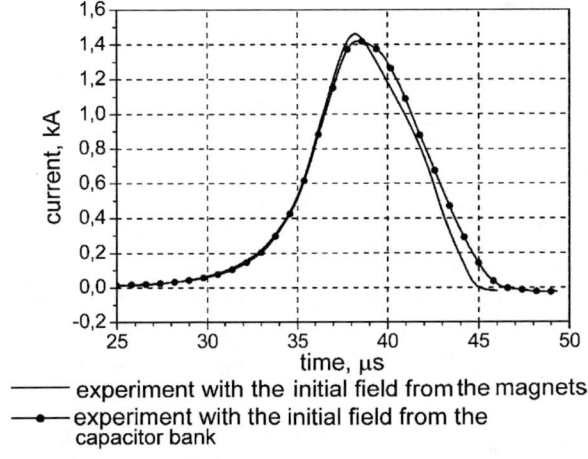

———— experiment with the initial field from the magnets
—•— experiment with the initial field from the capacitor bank

FIGURE 7. Diagrams of the current in the inductive load of 40 μH.

Difference of the initial energy on the level of 0.2 J is explained by not exact coincidence of the magnetic field distribution on the volume for the generator with the current and for the magnetic system.

V. CONCLUSION

The source of the electromagnetic energy of 66 MW power with the initial magnetic field, created by the system of ferrite of barium magnets, is developed. The source is autonomous, simple in operation, and does not require switching elements for its operation. The mass of the source does not exceed 5 kg at overall size of 100x100x200 m.

It could be used both as independent source of pulsed electromagnetic energy, and as a component of a cascade system of magneto-cumulative generators to obtain the finite energy of the megajoule range.

REFERENCES

1. V.E. Gurin, R.Z. Lyudaev, A.I. Pavlovskii, L.N. Pljashkevich, E.N. Smirnov. MC-2 Generator With Permanent Magnet (brief information). Kurchatov Nuclear Energy Institute, Moscow, USSR in *Megagauss Magnetic Field Generation by Explosives and Related Experiments*, EUR 2750.e.eds. H. Knopfel and F. Herlach. Brussels, 1966, P.517.
2. B.A.Bojko, V.E.Gurin, R.Z.Lyudaev and Pavlovskii Autonomous Cascade MC-System With Constant Magnets in *Megagauss Magnetic Field Generation and Pulsed Power Applications* Nova Science Publishers, NY, 1994. PP.467-473.
3. V.E.Gurin, A.B.Boriskin, V.A.Demidov, S.N.Golosov, M.V.Klimashov, A.S.Pikar, E.V.Shapovalov Helical magneto-cumulative generator powered from permanent magnets in *Book of abstracts of "VII Zababakhin scientific readings"* – Snezhinsk: RFNC-VNIITF publishing house, 2003, p.106.

Effects of Current Guides Destruction at Ultra-fast Acceleration of Macrobodies

V.N.Kataev, A.S.Boriskin, S.N.Golosov, V.A.Demidov, M.V.Klimashov, P.V.Korolev, G.F.Makartsev, A.S.Pikar, A.S.Russkov, E.V.Shapovalov, Yu.M.Shibitov

Russian Federal Nuclear Center – VNIIEF
37 Mira Ave., Sarov, Nizhny Novgorod region, 607188, Russia

Abstract. The paper is devoted to discussion of current guides destruction effects in different accelerators: thermal-electric and electro-magnetic rail accelerator at macrobodies acceleration value of 10^8-10^9 m/s^2.

Experimental results with thermal-electric accelerators powering from megajoule capacitor battery and helical magneto-cumulative generator MCG-100 at currents up to 3.5 MA are analyzed.

The process of rails destruction at railgun at pressure magnetic field excess over the limit of metal fluidity is presented. Methods of efficiency coefficient increase of capacitive storage energy transmission to kinetic energy of accelerating body are discussed.

INTRODUCTION

Experiments on acceleration of macrobodies in electromagnetic devices are carried out for a long time [1]. However, achievement of high velocities comes across some difficulties. Their analysis and review of perspectives of this direction are presented in a brilliant paper [2].

One of the obstacle limiting obtaining hypersonic velocities of the macrobodies in the electric projectile devices is a deformation and destruction of the current guides parts and insulators of the devices. In this paper we try to estimate the effect of the current guides and insulators deformation on the process of the macrobodies acceleration in the devices like railgun and electro-thermal gun.

RAILGUN

Several types of effects on the parts exist in the railguns.

A. Deformation of rails under effect of a magnetic field:

$$P = \frac{\mu_0 H^2}{2} \approx \frac{\mu_0 I^2}{2a}$$

where P — pressure, I — current, a — rails width, $\mu_0 = 4\pi \cdot 10^{-7}$ Гн/м.

CP849, *Zababakhin Scientific Talks - 2005*,
edited by E. N. Avrorin and V. A. Simonenko

B. Erosion of the rails surface under effect of the electric current. As one can see in Fig. 2 the deformation and erosion take place at the very beginning of the macrobody acceleration. Effect of the magnetic field and erosion decreases significantly at macrobody velocity increase.

In the presented experiment the device was powered from the capacitor bank of $C=6 \cdot 10^{-3}$ F capacity at voltage $U=10^4$ V. The maximum current was $I_{max} = 1.1 \cdot 10^6$ A, maximum acceleration $2.4 \cdot 10^8$ m/s^2, the projectile mass $m=3$g.

C. A plasma pressure to the insulators surface at a macrobody acceleration, made of dielectric. The pressure to insulators surface causes distribution of the Gook wave along the surface with the velocity:

$$c_g = \sqrt{\frac{G}{\rho}}$$

where c_g - velocity of the flexural wave, G – Gook module of the insulator material, and ρ - its density. If for the time $\Delta t \approx l \sqrt{\frac{\rho}{G}}$, where l – projectile length, the projectile do not get the velocity $v \geq c_g$, the flexural wave will significantly increase the gaps between the projectile and the insulators, and probability of the inter-rails gap breakdown before the projectile will increase.

ELECTRO-THERMAL GUN

A barrel deformation in the discharge area and erosion of the barrel walls take place in the electro-thermal gun. But the erosion causes lesser effect than in the railgun.

Figures 3-6 shows the barrel deformation at different pressures in a discharge cavity. In the experiments we used the capacitor bank of $C=6$ mF capacity at voltage $U=20$ kV.

Figure 3 shows the barrel after the experiment. A pressure of $P_{max}=9.9 \cdot 10^8$ Pa and acceleration $a_{max}=1.85 \cdot 10^7$ м/с^2 was achieved in the discharge cavity at current $I_{max}=1.2$ MA. In these experiments the contour inductance was 370 nH. The projectile mass was 17 g.

Figure 4: $P_{max}=6.8 \cdot 10^9$ Pa and acceleration $a_{max}=1.37 \cdot 10^8$ m/s^2, at current $I_{max}=1.42$ MA.

Figure 5: $P_{max}=6.6 \cdot 10^9$ Pa and acceleration $a_{max}=1.2 \cdot 10^8$ m/s^2, at current $I_{max}=1.81$ MA.

An experiment on the macrobodies acceleration in electro-thermal gun, powered from a magneto-cumulative generator MCG-100 [3] was carried out. The following data: $P_{max}=4.22 \cdot 10^{10}$ Pa and acceleration $a_{max}=1.06 \cdot 10^9$ m/s^2 at current $I_{max}=3.5$ MA were obtained in the experiment. The projectile was made of steel and its mass was $m=24.85$ g. Figure 6 shows a section of the barrel with erosion traces.

It is evident that the pressure in the discharge area could reach significant values. For illustration let us estimate the energy recovered in the discharge area in the experiment (see Figure 5).

At careful examination of the deformed barrel one can see that minimum blow walls thickness of is ≈ 5 mm. Assuming that the blow walls thickness does not change after cracks appearance, let us estimate the maximum radius of the cavity before the cracks appearance. It is equal to ≈ 20 mm. Let us assume also that the blow size along the barrel does not change in time. In this case the center of gravity of the geometrical figure, appearing from the generatrix of the inner cavity surface and the generatrix of the barrel channel (Figure 1) lies on the distance of ≈ 13 mm from the barrel axis. The area of this figure is ≈ 4 cm^2, and the tore volume, obtained by rotation of this figure around the barrel axis is ≈ 38 cm^3. Maximum pressure, obtained in the experiment was $\approx 6.6 \cdot 10^9$ Pa. Assuming that in the first approximation the pressure grows linearly, let us consider $\overline{P} = 3.3 \cdot 10^9$ Pa. In this case the deformation energy of the barrel was:

$$E_{def} = \overline{P} \cdot \Delta V = 3.3 \cdot 10^9 \cdot 3.8 \cdot 10^{-5} = 1.25 \cdot 10^5 \text{ J.}$$

At the pressure of $3.3 \cdot 10^9$ Pa the mass velocity in an iron $u \approx 160$ m/s (steel barrel) and the considered cavity appeared at $\approx 62^{nd}$ μs. The projectile at this time moment was on the distance of 160 mm from the initial position. The volume of this section of the barrel is ≈ 50 cm^3. If we consider the gas to be ideal, the density of the thermal energy will be: $\dfrac{\partial E_T}{\partial v} = \dfrac{3}{2} \cdot \overline{P}$.

The gas contain $E_T = 4.3 \cdot 10^5$ J. A kinetic energy of projectile at the moment of time 62 μs $E_{kin} \approx 2 \cdot 10^5$ J The energy recovered in the discharge area is: $E_{def} + E_T + E_{kin} \approx 7.6 \cdot 10^5$ J. For comparison, the stored energy is $E = \dfrac{cu^2}{2} = 1.2 \cdot 10^6$ J.

One can see that electro-thermal gun consume significant portion of the energy of the electric contour. If we profile the current pulse in a way that the projectile could compensate some portion of deformation at its motion, it will be possible to transfer some deformation energy to the kinetic energy of the projectile.

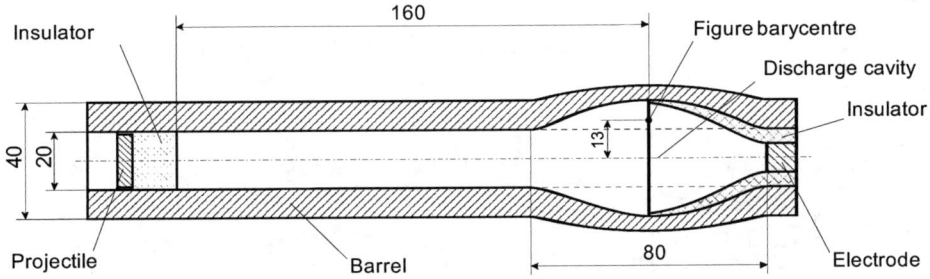

FIGURE 1. Scheme of the barrel deformation

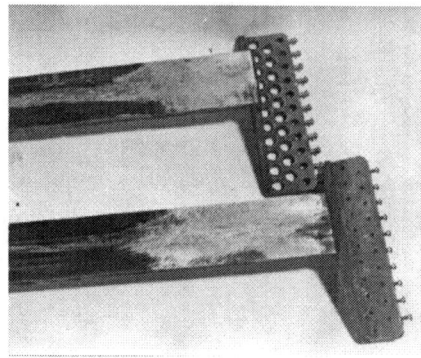

FIGURE 2. Rails view after the experiment.

FIGURE 3. The barrel after the experiment. The pressure P_{max}=9.9·10⁸ Pa and acceleration a_{max}=1.85·10⁷ m/s² was achieved in the discharge cavity.

FIGURE 4. The barrel after the experiment. The pressure P_{max}=6.8·10⁹ Pa and acceleration a_{max}=1.37·10⁸ m/s² was achieved in the discharge cavity.

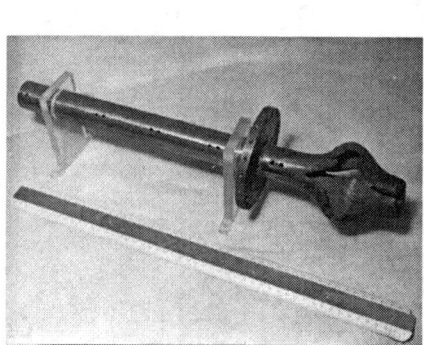

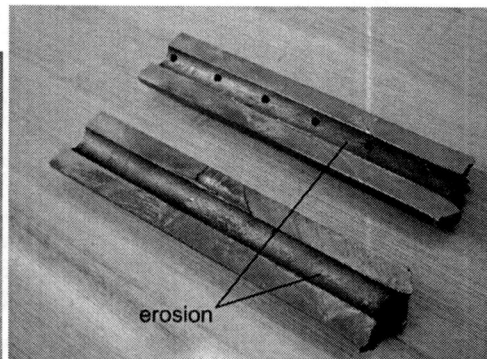

FIGURE 5. The barrel after the experiment. The pressure P_{max}=6.6·10⁹ Pa and acceleration a_{max}=1.2·10⁸ m/s2 was achieved in the discharge cavity.

FIGURE 6. The barrel after the experiment. The pressure P_{max}=4.2·10¹⁰ Pa and acceleration a_{max}=1.6·10⁹ m/s² at current I_{max}=3.5 MA was achieved in the discharge cavity.

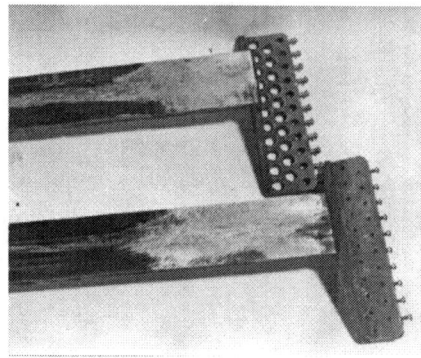

FIGURE 2. Rails view after the experiment.

FIGURE 3. The barrel after the experiment. The pressure $P_{max}=9.9\cdot10^8$ Pa and acceleration $a_{max}=1.85\cdot10^7$ m/s² was achieved in the discharge cavity.

FIGURE 4. The barrel after the experiment. The pressure $P_{max}=6.8\cdot10^9$ Pa and acceleration $a_{max}=1.37\cdot10^8$ m/s² was achieved in the discharge cavity.

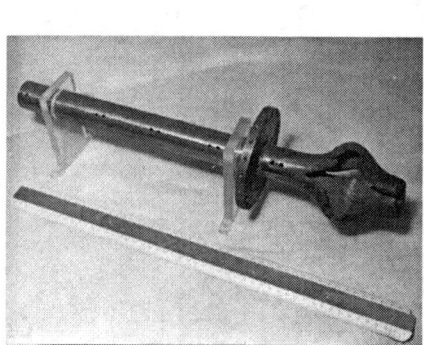

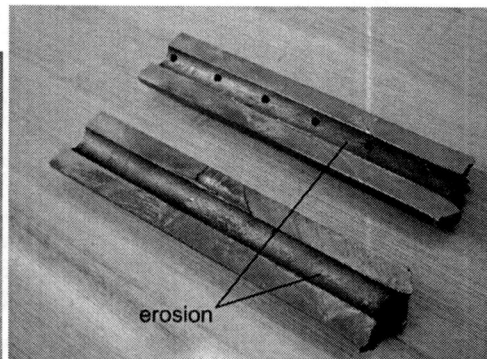

FIGURE 5. The barrel after the experiment. The pressure $P_{max}=6.6\cdot10^9$ Pa and acceleration $a_{max}=1.2\cdot10^8$ m/s2 was achieved in the discharge cavity.

FIGURE 6. The barrel after the experiment. The pressure $P_{max}=4.2\cdot10^{10}$ Pa and acceleration $a_{max}=1.6\cdot10^9$ m/s² at current $I_{max}=3.5$ MA was achieved in the discharge cavity.

REFERENCES

1. S.C. Rashleigh, R.A. Marshall, Electromagnetic acceleration of microparticles to high velocities, *J. Appl. Phys.*, 1978, v.49, p.2540-2542.
2. G.A. Shvetsov, A.G. Anisimov, Yu.L. Bashkatov, S.V. Stankevich Rail electromagnetic accelerators of solid bodies. Achievements. Problems. Perspectives. in *Hydrodynamics of energy high densities. Proceedings of international workshop*, Novosibirsk, Russia. 2004, p.282-304.
3. V.D.Selemir, V.A.Demidov, L.N.Plyashkevich, A.S.Kravchenko, S.A.Kazakov, A.M.Shuvalov, A.S.Boriskin, V.A.Zolotov, T.M.Spirov, M.M.Kharlamov, High-current energy pulses to power inductive and active loads, in *Megagauss and megampere pulsed technology and applications.* V. 1, 1997, p.248-254.

SECTION 4

HYDRODYNAMIC INSTABILITY
AND TURBULENCE

Wave Formation in Symmetric Collision
of Metal Plates

V.V. Pai, Ya.L. Lukyanov, G.E. Kuzmin, I.V. Yakovlev

Lavrentyev Institute of Hydrodynamics, Novosibirsk 630090, Russia

Abstract. Collision of metal plates is treated in the frames of model of a viscous incompressible liquid. In such terms a wave formation process corresponds to an initial stage of occurrence of turbulence in accordance to L.D.Landau's scenario. To verify conformity between the theoretic description and the real process, a series of experiments on symmetric collision of aluminum plates is carried out with identical collision angles and different velocities of a collision point. It is established that with great magnitudes of the Reynolds number R and consequently, with the big velocities of a collision point, excitation of wave formation has soft type, with smaller magnitudes of R there is an area of metastability where the mode of excitation has rigid type, and with even smaller magnitudes of R any influence does not lead to wave formation in conformity with Landau's theory.

Keywords: Symmetric collision, wave formation, explosive welding, turbulence, Reynolds number.
PACS: 62.50.+p; 47.27.Cn.

Tens of works, a brief list of which can be found in [1], are devoted to studying of process of wave formation in explosive welding of metals. The analysis of the published works, which are devoted to wave formation, is extremely complicated because authors in their experiments used samples of different materials, of the various form, and in different conditions of collision. In this work we don't concern a task of a prediction of quantitative wave characteristics, but we focus our efforts on an explanation of the main qualitative phenomena at wave formation.

With this purpose we discuss one series of the experiments performed in the most simple configuration. Results of these experiments represent clearly the basic ideas of a wave formation mechanism which we put forward.

We investigate symmetric collision of two identical aluminum plates under an angle φ with a collision point velocity V_c ; the experimental setup is shown in Fig. 1. With this setup a series of experiments with a constant collision angle $\varphi = 45° \pm 1°$, and with a collision point velocity V_c changing from one experiment to another was carried out. The change of magnitude of V_c with constant φ, was provided with use of explosives of different composition and by the selection of a proper initial angle of installation of plates φ_0 (Fig. 1). Measurements of collision parameters were carried out by a rheostat method [2].

CP849, *Zababakhin Scientific Talks - 2005*,
edited by E. N. Avrorin and V. A. Simonenko
© 2006 American Institute of Physics 0-7354-0345-7/06/$23.00

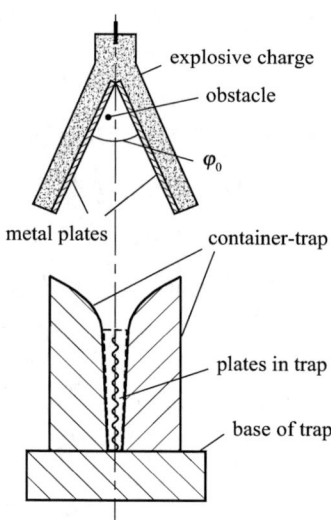

explosive charge

obstacle

φ_0

metal plates

container-trap

plates in trap

base of trap

FIGURE 1. Experimental setup.

In some experiments into initially symmetric flow an obstacle was inserted to break the symmetry of flow. This obstacle in the form of a wire 3 (Fig. 1) was placed in parallel to surfaces of plates in a gap between them.

After collision the samples are preserved in a special container-trap. Then character of a contact boundary (smooth or wavy, with waves, which grow or fade) does not change. Naturally, numerical characteristics of waves, that is, their length and amplitude are distorted in these circumstances, but, as marked above, these characteristics are not a subject of given research. Conditions of collisions and results of experiments are listed in Table 1.

Let us show, that transition from smooth to wavy flow can be described within the frame of the model of incompressible viscous fluid. Then the beginning of wave formation is the manifestation of instability of stationary fluid flow at certain magnitudes of the Reynolds number R and such a behavior can be explained in terms of Landau's theory of arising of such instability [3, 4].

TABLE 1. Conditions and results of experiments.

№	V_c, km/c	Obstacle	Waves
1	3.20±0.03	no	yes
2	1.32±0.03	no	no
3	1.32±0.03	yes	yes
4	0.90±0.03	no	no
5	0.90±0.03	yes	no

It follows from these works that at small exceeding of the critical Reynolds number $R > R_{cr}$, the laminar flow existing at $R < R_{cr}$ becomes unstable and there appears periodically changing additive of a kind

$$\mathbf{V} = A(t)\mathbf{f}(x,y,z) \tag{1}$$

to the flow velocity, where $A(t) = \text{const} \cdot \exp(\gamma t) \cdot \exp(-i\omega t)$ at small amplitudes.

With growth of amplitude (but when it still remains small) the differential equation for $|A|^2$ averaged on the time interval τ, such that $1/\gamma \ll \tau \ll 2\pi/\omega$, is

$$\frac{d|A|^2}{dt} = 2\gamma|A|^2 + \alpha|A|^4 + \beta|A|^6 + \cdots, \tag{2}$$

where α is the Landau constant $\gamma = C(R - R_{cr})$, $C > 0$. Then at $\gamma > 0$ the instability arises with a soft mode of excitation, i.e. any small perturbation grows with time. This case takes place in experiment № 1 (see Table 1). Waves arise at the most accurate ensuring of collision symmetry. In this case $V_c > V_{cr}$, where V_{cr} is the critical velocity and hence $R > R_{cr}$. On contact boundary of plates after collision (Fig. 2) transition from smooth to wavy flow with growing amplitude is seen. Authors of work [5] are closest to such understanding of the mechanism of wave formation. However, we shall see below, that the full scenario of formation of waves is essentially more complicated.

Taking, as in [4], only the three first members of decomposition in the equation (2) and equating the right side of (2) to zero, it is possible to determine the established amplitude of fluctuations of velocity.

For the further analysis it is necessary to know whether the constant α is positive or negative. Let us show, that in our experiments the case is realized with $\alpha > 0$. Really, at $\alpha > 0$, as it is underlined in [3, 4], there should be an area $R_{cr}^* < R < R_{cr}$ (R_{cr}^* is the second critical Reynolds number) and hence $V_{cr}^* < V_c < V_{cr}$ (V_{cr}^* is the velocity corresponding to R_{cr}^*) where the mode of excitation is rigid, and smooth flow is

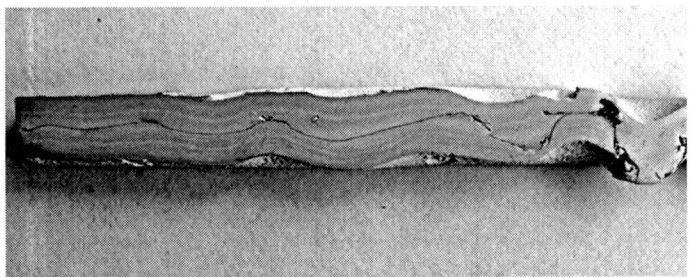

FIGURE 2. Zone of collision of plates in conditions of a soft mode of excitation of instability.

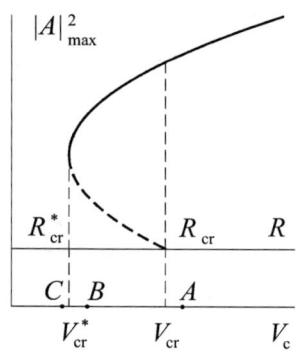

FIGURE 3. The qualitative form of dependence of $|A|^2_{max}$ on the Reynolds number R and corresponding values of collision point velocity V_c.

metastable. It means that at exceeding by amplitude of some threshold the right part of the equation (2) becomes positive, and the arising fluctuations do not fade.

The magnitude of threshold initial disturbance depends on closeness of the Reynolds number to the first or second critical value. If R is close to R_{cr}, then a rather weak initial disturbance is sufficient for initiation of instability and development of waves, but, if R is close to R^*_{cr}, then a disturbance should be big enough for occurrence of waves. These statements are well illustrated by Fig. 3, which shows a square of amplitude of the steady-state fluctuations in dependence on the Reynolds number in accordance with [4]; the dashed curve here represents a branch of the instable solution. If with the given Reynolds number the amplitude of disturbance turns out greater than its value on the instable solution branch, the amplitude of fluctuations will sharply increase up to value on the stable solution branch (the continuous line in Fig. 3), otherwise original disturbance will fade. For the first time the importance of character of excitation of wave formation at explosive welding is indicated in [6]. The authors of [6] concluded that wave formation is self-oscillatory process with rigid excitation. However, as is clear from the preceding, this conclusion is correct only in an interval of values of the Reynolds number $R^*_{cr} < R < R_{cr}$.

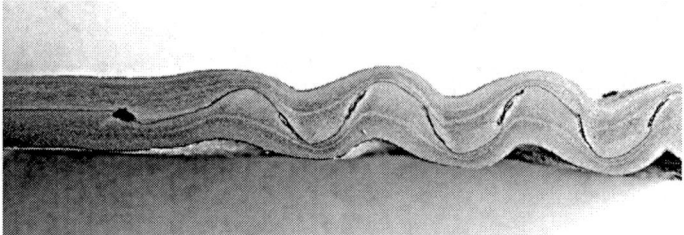

FIGURE 4. Zone of collision of plates in conditions of a rigid mode of excitation of instability.

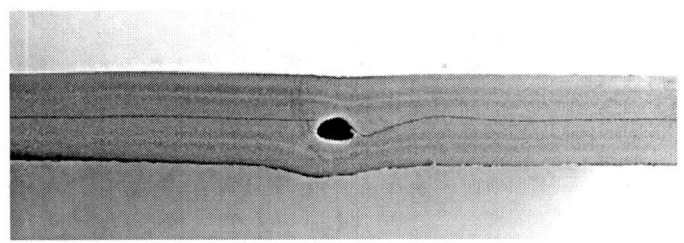

FIGURE 5. Zone of collision of plates in conditions of the small Reynolds number ($R < R_{cr}^*$) with attenuation of initial disturbance.

Let us consider experiments № 2 and № 3 (see Table 1). In these experiments $V_{cr}^* < V_c < V_{cr}$. The distinction between them consists only that in experiment № 3 between plates the obstacle 3 (Fig. 1) breaking symmetry of collision was established. Figure 4 shows that non-fading wave formation begins immediately behind the obstacle. In contrast to this, in experiment № 2 (without an obstacle) the flow has laminar character.

Having reduced else the Reynolds number (experiments № 4 and 5 in Table 1), we have smooth flow in experiment № 4, whereas in experiment № 5 (see Fig. 5) initial disturbance fades immediately behind the obstacle, in exact agreement with the theory for the case $R < R_{cr}^*$. In this experiment, obstacles of the different sizes were established, however the result remained unchanged. In Fig. 3, points A, B, C designate, accordingly, the experimental conditions of experiments № 1, № 3, № 5 of the Table 1.

Values of R_{cr}, R_{cr}^*, α, β depend on a collision angle and can be determined experimentally. In addition we notice that $\beta < 0$ in the considered approximation and the steady-state amplitude of fluctuation is given by

$$|A|^2 = -\frac{\alpha}{2\beta} + \sqrt{\frac{\alpha^2}{4\beta^2} - \frac{2\gamma}{\beta}}.$$

From the last it is evident that a wave amplitude depends weakly on a collision point velocity V_c if $V_c \approx V_{cr}$ (because in this case $\gamma \approx 0$) that quite corresponds to various experimental data. The further development of wave formation process at increasing velocity of the collision point, according to the theory, should lead to complication of the collision boundary shape from near-sinusoidal one, for example, to the shape very distorted by vortex zones, because, besides fluctuations with frequency ω, with growth of the Reynolds number there arise fluctuations with other frequencies what is in full conformity with experimental data.

It is essential, that the scenario of instability development by Landau [3, 4], generally speaking, does not depend on flow geometry. Therefore all the preceding conclusions concern equally to asymmetrical plate collisions and to axisymmetric flows, in particular, to the jet flow at explosive loading of metal conic funnel-shaped sample. With increase in an angle at the top of a cone, with the conserve specimen throw speed, the collision point velocity decreases, and, hence, the Reynolds number decreases too, and metal flow becomes laminar. With decrease of this angle, beginning from its certain value when R becomes equal R_{cr}^*, flow becomes metastable, more and more sensitive to imperfections in explosive and in metal as well as to smoothness of a cone surface. At last, at angles less than critical magnitude, when the condition $R > R_{cr}$ is achieved, instability develops at the most strict control of initial symmetry of flow. Certainly, all the above reasoning make sense only provided that flow remains subsonic in coordinate system related to the collision point, otherwise compressibility of a material becomes essential, and the conclusions made above, will lose force.

Thus, it is established that Landau's theory of turbulence onset explains the key quality features of the initial stage of wave formation at oblique collision of metal plates.

ACKNOWLEDGMENTS

The work is maintained by the grant of the Russian Foundation for Basic Research № 05-01-00398.

REFERENCES

1. *Volnoobrazovanye pri kocykh coudareniakh* (Sost., perev. and red. by I. V. Yakovlev, G. E. Kuzmin, V. V. Pai), Izd. Inst. discr. math. and inform., Novosibirsk, 2000, 222 pp.
2. Kuzmin G. E., Pai V. V., Yakovlev I. V., *Eksperimentalno-analiticheskie metody v zadachakh dinamicheskogo nagrughenya materialov*, Izd. Sib. Otd. Ross. Akad. Nauk, Novosibirsk, 2002, 312 pp.
3. Landau L. D., *Dokl. Akad. Nauk*, **44**, No. 8, 339-342(1944).
4. Landau L. D., Lifshits E. M., *Gidrodinamika*, Izd. Nauka, Moskva, 1986, 736 pp.
5. Godunov S. K., Deribas A. A., Kozin N. S., *J. Prikl. Mekh. Tekhn. Fiz.*, No. 3, 63-72(1971).
6. Cowan G. R., Bergmann O. R., Holtzman A. H., *Metallurgical Transactions*, **2**, No. 11, 3145–3155(1971).

Evolution of Diffusion Layer of Two Gases Mixing at Interaction with Compression Waves

E.G.A. Ruev*, A.V. Fedorov[¶], V.M. Fomin[¶]

*Novosibirsk State University of Architecture and Civil Engineering,
630008 Novosibirsk, Russia
[¶]Institute of Theoretical and Applied Mechanics, Siberian Branch, Russian Academy of Sciences,
630090 Novosibirsk, Russia

Abstract. A mathematical model of mechanics of a two-velocity and two-temperature mixture of gases is developed. Based on this model, the evolution of the mixing layer of two gases of different densities, which are accelerated by a compression wave, is considered by methods of numerical simulation. The problem of interaction of this layer with the compression wave, the heavy medium being accelerated by the light medium, is solved numerically. Problems of instability development in a sine-perturbed mixing layer accelerated by a compression wave are resolved numerically in a two-dimensional unsteady formulation. The calculated width of the mixing region is in reasonable agreement with experimental data.

Keywords: mixing layer, Rayleigh - Taylor instability, two-velocity two-temperature gas dynamics of mixtures.
PACS: 52.35.Py, 51.10+y.

INTRODUCTION

Accelerated motion of the mixing layer separating two gases of different densities is accompanied by development of instability arising if the light medium accelerates the heavy medium or if the heavy medium is located above the light medium in the gravity field. This type of instability is called the Rayleigh - Taylor instability. The mixing layer is traditionally considered as a density discontinuity surface, i.e., as a contact discontinuity. Numerous papers on numerical simulation of Rayleigh-Taylor instability development (see, e.g., [1-3]) based on the Euler equations did not take into account the influence of mutual penetration of the gases. It is known that replacement of a stepwise density profile on the contact discontinuity by a continuous distribution in a certain layer of finite width can reduce the growth rate of disturbances at the initial stage of instability development. This was noted, e.g., in [4, 5] where the disturbance-amplitude growth was studied theoretically and in experimental works [6, 7]. Theoretical studies on the basis of gas dynamics equations were mainly qualitative, and the mixing layer was simulated by a layer of variable density. Therefore, it is of interest to consider this problem on the basis of equations of a two-velocity two-temperature mixture of gases where each component has its own velocity and temperature. This approach offers a description of both the processes of mutual penetration of the gases and interaction of the mixing layer with the compression wave. A semi-empirical model of turbulent mixing of a multispecies medium with

CP849, *Zababakhin Scientific Talks - 2005*,
edited by E. N. Avrorin and V. A. Simonenko

each species having its own velocity was constructed in [1, 2]. This model implies an instantaneous onset of turbulent mixing. Processes at the initial stages of mixing are considered below on the basis of equations of two-velocity two-temperature gas dynamics of mixtures. In [8, 9], this approach was used to study interaction of the mixing layer with the shock wave.

FORMULATION OF THE PROBLEM

We study the evolution of a transitional layer separating two pure gases with different densities by using the model of a two-dimensional unsteady flow of a two-velocity two-temperature mixture accelerated by a compression wave propagating from the light to the heavy gas. The formulation of the problem is illustrated in Fig. 1. The compression wave propagates from gas 2 to gas 1. At the time $t = 0$, the compression wave arrives at the boundary of the layer. The shape of the mixing-layer centerline is described by the equation $x_0(y) = a_0(1 - \cos(\theta y))$, where a_0 is the disturbance amplitude, $\theta = 2\pi/\lambda$ is the wavenumber, λ is the length of the disturbance wave, L is the total width of the mixing layer, and $\delta(y)$ is the width of the layer in the cross section y. At the initial time, $\delta_0(y) = const$.

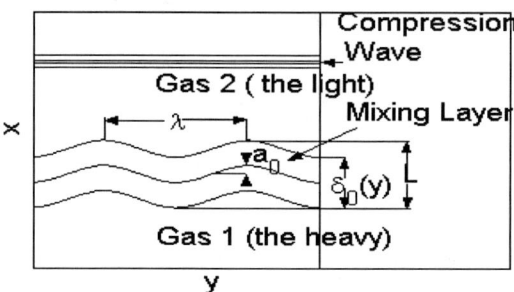

FIGURE 1. Initial configuration of the mixing layer.

The Rayleigh - Taylor instability developed in an accelerated transitional layer was experimentally investigated in [7]. Acceleration was imparted to the contact region by a compression wave formed ahead of the flame front. An oxygen - hydrogen mixture (with a molar weight of 18.5) was used as gas 2. The parameters of the mixture in the layer are described by the following equations of two-velocity two-temperature gas dynamics of mixtures [10]:

$$\partial \rho_i / \partial t + div(\rho_i \vec{u}_i) = 0, \quad \rho_i \partial \vec{u}_i / \partial t + \rho_i (\vec{u}_i \cdot \nabla) \vec{u}_i + \nabla p_i = K(\vec{u}_j - \vec{u}_i),$$

$$\rho_i \partial e_i / \partial t + \rho_i \vec{u}_i \nabla e_i + p_i div(\vec{u}_i) = \beta_i K(\vec{u}_j - \vec{u}_i)^2 + q(T_j - T_i),$$

$$p_i = kn_i T_i, \quad e_i = kT_i/(\gamma_i - 1), \rho_i = m_i n_i, \ i, j, = 1, 2\,(i \neq j).$$

Here ρ_i is the density, u_i and v_i are the velocity components, e_i is the internal energy, p_i is the pressure, T_i is the temperature, m_i is the molecular weight, n_i is the number density of the i th type, x and y are the Cartesian coordinates, t is the time, k is the Boltzmann constant, $K = 16\rho_1\rho_2\Omega_{12}^{(1,1)}/(3(m_1+m_2))$, $\Omega_{12}^{(1,1)}$ is the collision integral, $\beta_i = m_iT_i/(m_1T_1+m_2T_2)$, $q = 3m_1K/(m_1+m_2)$, $c_{iv} = k/(m_i(\gamma_i-1))$, and γ_i is the ratio of specific heats. The interaction potential of hard spheres is described by the relation

$$K = \frac{16}{3}\frac{\rho_1\rho_2}{m_1m_2}\sqrt{\frac{k\pi}{2}}\sqrt{\frac{T_1}{m_1}+\frac{T_2}{m_2}}\sigma_{12}, \sigma_{12} = \frac{(\sigma_1+\sigma_2)}{2}$$

(σ_i is the diameter of the molecule of the i th gas).

For low (or zero) values of concentration of the j th gas, we use the Euler equations for a pure i th gas and determine the parameters of the other gas from the relations

$$\frac{\partial n_j}{\partial t} + \frac{\partial n_j u_j}{\partial x} + \frac{\partial n_j v_j}{\partial y} = 0, u_j = u_i, v_j = v_i, T_j = T_i.$$

The transition to the heavy gas is performed if the molar concentration of the light gas is $x_j = n_j/(n_1+n_2) < 0.1\%$; the transition to the light gas is performed if the mass concentration of the heavy gas is $\alpha_j = \rho_j/(\rho_1+\rho_2) < 0.1\%$.

For a description of the initial transitional layer formation we have obtained an asymptotic solution of the original mathematical model, which gives a fair agreement with experimental data on the formation of the initial diffusion layer [9].

NUMERICAL RESULTS

Let us consider acceleration of a disturbed mixing layer by a compression wave passing from the light gas (oxygen-hydrogen mixture) into a heavy gas (argon or xenon) in terms of Fig. 1. The initial distribution of parameters in the layer was determined by an asymptotic solution obtained in the one-dimensional approximation [9]. The parameters in the compression wave are determined by relations for the centered compression wave. The calculations were performed in a rectangular domain $[x_n, x_k; 0, \lambda/2]$. The upper and lower boundaries (x_n, x_k) were subjected to the condition of zero derivatives. To eliminate the influence of the boundary conditions on waves reaching these boundaries, the computational domain was extended. The condition of symmetry was imposed on the side boundaries.

Figures 2 show the isolines of molar concentration of the light gas 2 (a), the vector field of the dimensionless mean mass velocity of the mixture (b), and the changes in the mixing-layer width (c). These results are plotted for different times after the beginning of compression-wave interaction with the layer. Figures 2b also show the boundaries of the mixing layer (with molar concentration of 5%) and the middle of the layer (with molar concentration of 50%). The compression wave passed from gas 2 into argon. The disturbance wave length was $\lambda = 15$ mm, its amplitude was $a_0 = 3$ mm, and the initial width of the layer was $\delta_0 = 9$ mm. The width $\delta(y)$ was found as the distance between the points of the maximum depth of penetration of gas 1 into gas 2 and of gas 2 into gas 1 (molar concentrations $x_2 = 0.95$ and $x_2 = 0.05$).

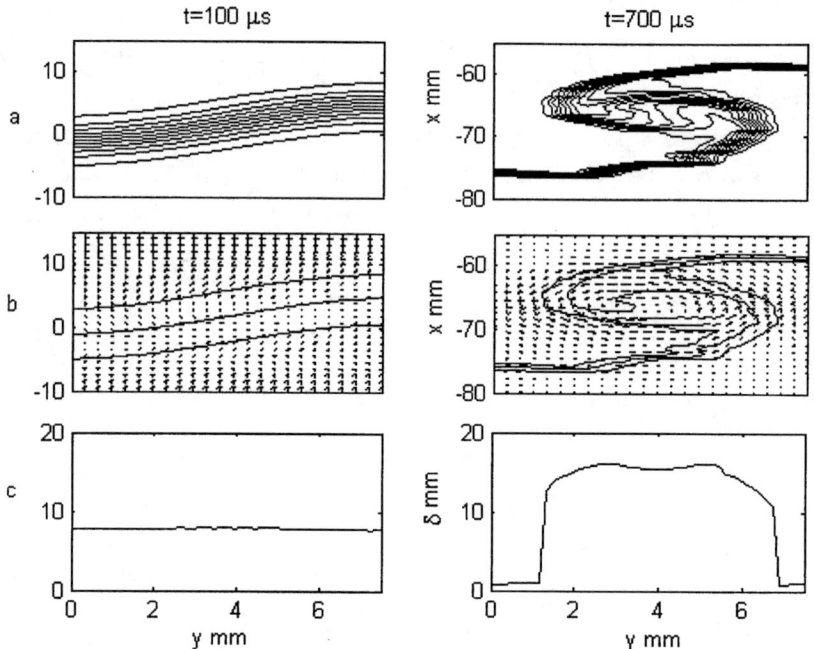

FIGURE 2. Evolution of the mixing layer at $t = 100$ s and $t = 700$ s: isolines of molar concentration (a), field of the dimensionless mean mass velocity of the mixture (b), and width of the mixing layer (c).

Compression-wave interaction with the layer first leads to uniform compression of the layer (the initial width of the layer was 9 mm). Simultaneously, different parts of the layer move with different velocities because the layer is curved, and a vortex with its center in the layer starts forming. The action of the vortex results in an increase in the layer width approximately in the region where $y = \lambda/4$. Compression persists in the valley ($y = 0$) and on the crest of the disturbance wave ($y = \lambda/2$). As a result, a heavy gas jet is formed. Further evolution of this process leads to origination of a mushroom-type structure, which, in turn, leads to reduction of the growth rate of disturbances in the streamwise direction and to intense growth of the layer in the transverse direction. In the course of time, the neighboring jets start interacting with each other, and a region of turbulent mixing appears.

Figure 3a shows the mixing-layer width normalized to the half-length of the disturbance wave at the time $t = 400$ μs after beginning of interaction with the compression wave for different values of the wave length, its amplitude, and initial width of the mixing layer. Gas 1 is argon (curves 1-4) and xenon (curve 5). The calculations were performed for the following values of parameters: $\Delta = 15$ mm; $\lambda = 15$ mm, $a_0 = 3$ mm, and $\delta_0 = 9$ mm (1); $\lambda = 30$ mm, $a_0 = 3$ mm, and $\delta_0 = 9$ mm (2); $\lambda = 15$ mm, $a_0 = 1$ mm, and $\delta_0 = 9$ mm (3); $\lambda = 15$ mm, $a_0 = 3$ mm, and $\delta_0 = 1$ mm (4); $\lambda = 15$ mm, $a_0 = 3$ mm, and $\delta_0 = 9$ mm. The presence of jumps in the curves indicates the presence of the mushroom-type structure. It is seen that an increase in the disturbance wave length leads to later origination of the mushroom-type structure and to slower growth of the width (curves 1 and 2). A similar behavior is observed for lower initial disturbance amplitudes (curves 1 and 3). The width of the mixing layer in the valley and on the wave crest is higher. At the same time, a decrease in the initial diffusion width of the mixing layer leads to earlier origination of the mushroom-type structure and to faster growth of the layer width (curves 1 and 4). The calculations of interaction of the compression wave passing from gas 2 into xenon (molecular weight of 131.2) are also plotted here (curve 5). A higher ratio of molecular weights results in more intense growth of the layer width.

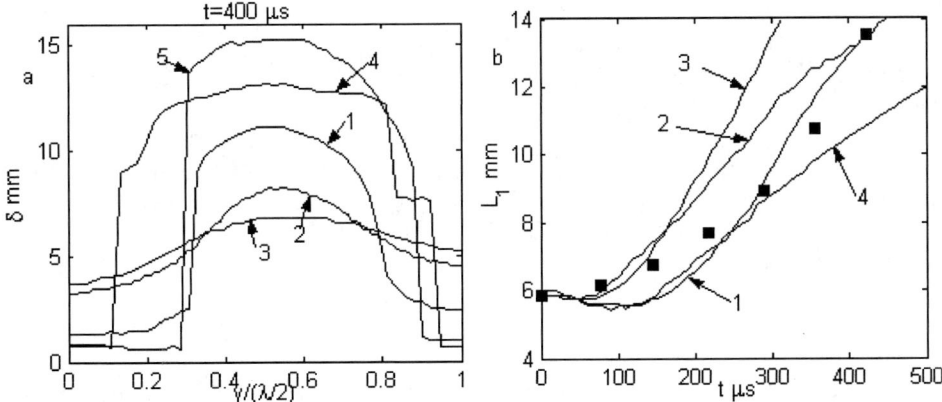

FIGURE 3. 3a is the variation of the mixing-layer width at the wave half-length, 3b is the growth of disturbance amplitude.

Figure 3b shows the depth of penetration of the heavy gas into the light gas $L_1(t)$ (disturbance amplitude), which was determined as the distance between the maximum and minimum values of x where the molar concentration of the heavy gas exceeds 5%. At the initial time, $L_1 = 2a_0$. The calculations were performed for $\lambda = 15$ mm, $a_0 = 3$ mm; curves 1, 2, 4 refer to the case where argon was used as gas 1 and curve 3 refers to the case where gas 1 was xenon; the remaining parameters were $\delta_0 = 9$ mm and $\Delta = 15$ mm (1); $\delta_0 = 1$ mm and $\Delta = 15$ mm (2); $\delta_0 = 7$ mm and $\Delta = 15$ mm (3); $\delta_0 = 9$ mm and $\Delta = 7$ mm (4). Δ is the compression-wave width. A decrease in the initial width of the mixing layer leads to faster growth of disturbances at the initial stage (curves 1 and 2). Similarly, an increase in the ratio of molecular weights results in faster growth of disturbances (curves 1 and 3). A decrease in the compression-wave width leads to slower growth of disturbances (curves 1 and 4). Figure 3b also shown the experimental point obtained in [7]. Both the calculations and the experiments revealed that the disturbance-amplitude growth can be divided into two stages: slow growth of the disturbance after a small decrease and rapid growth whose rate further decreases because of formation of the mushroom-type structure. The duration of the first stage decreases with increasing Atwood number and with decreasing initial width of the mixing layer.

315

CONCLUSION

The mathematical model of a two-velocity two-temperature mixture of gases [10] is applied in the present work to describe the processes in interaction of compression waves with the region of mixing of two gases.

The problem of development of the Rayleigh - Taylor instability with acceleration of the mixing layer sinusoidally disturbed by the compression wave is examined. It is shown that allowance for the initial finite width of the mixing layer leads to a lower growth rate of the dimensionless width of the layer.

It is found that vortices with the center in the mixing layer arise on the jet boundaries; each component in these vortices is characterized by its own velocity. Origination of such vortex structures leads to formation of a mushroom-type structure and to intense expansion of the layer in the transverse direction, which, in turn, results in interaction of the neighboring jets with each other.

The mathematical model is verified on the basis of the measured growth of the mixing-layer disturbance amplitude.

REFERENCES

1. Youngs, D.L., Numerical simulation of turbulent mixing by Rayleigh - Taylor Instability, Physica D. 12 (1984), p.32-44.
2. Youngs, D.L., Numerical simulation of mixing by Rayleigh - Taylor and Richtmyer-Meshkov instabilities. Laser and Particle Beams, 12(4) (1994), p. 725-750.
3. Nikishin, V.V., Tishkin, V.F., Zmitrenko, N.V., Lebo, I.G., Rozanov, V.B., Favorsky, A.P., Numerical simulations of nonlinear and transitional stages of Richtmyer-Meshkov and Rayleigh-Taylor instabilities, Proceedings of the sixth international workshop on the physics of compressible turbulent mixing, (1997), p. 382-387.
4. Chandrasekhar, S., Hydrodynamics and hydromagnetic stability, Oxford University, 1961.
5. Chakraborty, B.B., Rayleigh - Taylor instability of heavy fluid, Phys. Fluids 18(8), (1975), p. 1066-1067.
6. Duff, R. E., Harlow, F.H., Hirt, C.W., Effects of diffusion on interface instability between gases, Phys.Fluids 5(4), (1962) , p. 417-425.
7. Zaitsev, S.G., Lebo, I.G., Rozanov, V.B., Titov, S.N., Chebotareva, E.I, Hydrodynamic instability of the contact of gas media moving with acceleration, Izv. Ross. Akad. Nauk, Mekh. Zhidk. Gasa 6, (1991), p. 15-21.
8. Ruev, G.A., Fedorov, A.V., Fomin, V.M., Evolution of the diffusion mixing layer of two gases upon interaction with shock waves, J. Appl. Mech. Tech. Phys. 45(3), (2004), p. 328-334.
9. Ruev, G.A., Fedorov, A.V., Fomin, V.M. Development of the Richtmyer-Meshkov instability upon interaction of diffusion mixing layer of two gases with shock waves, J. Appl. Mech. Tech. Phys. 46(3), (2005), p. 307-314.
10. Kiselev, S.P., Ruev, G. A., Trunev, A. P., et al., Shock-Wave Processes in Two-Component and Two-Phase Media [in Russian], Nauka, Novosibirsk ,1992.

On Possible Availability of the Cumulative Effect Under Shock Loading of the Curved Surface of Condensed Matter

S.M. Bakhrakh, I.Yu. Bezrukova, A.D. Kovaleva, S.S. Kosarim,
O.V. Ol'khov

RFNC-VNIIEF, 37 Mir Avenue, Sarov, 607190, Nizhny Novgorod region,

Abstract. The paper presents the results of theoretical-and-computational studies of the processes associated with a plane shock wave arrival at the curved free surface (FS) of some condensed matter. It is shown that instability in this case may develop as jet shaping. Main laws describing the development and growth of jet-shaped perturbations depending on the initial conditions and some rheological properties (flow properties) of materials have been obtained.

Keywords: Richtmyer-Meshkov instability, free surface, jet shaping, strength.
PACS: 60.20.Fe, 62.50.+p, 47.20.-k

INTRODUCTION

In case of classic hydrodynamic Richtmyer-Meshkov (R-M) instability development with a shock wave (SW) passing across a point of contact discontinuity from a "heavy" material to a "light" material the interface material acquires a velocity increment, which is opposite in its phase to the initial perturbation profile [1]. The interface perturbation growth is inertial and determined by the initial conditions specified for the problem. The perturbation growth velocity during the linear phase can be estimated using the following Richtmyer expression [2]:

$$\delta \upsilon = Aak \cdot \Delta \upsilon_{CII} \tag{1}$$

where A and a are Atwood number and the amplitude of surface perturbations after a shock wave has passed, respectively; k is the wave number; $\Delta \upsilon_{CII}$ is the interface velocity increment under shock loading with no perturbations.

When SW arrives at the curved free surface (FS) of a condensed matter, implementation of a distinct scenario of perturbation growth is possible. A rarefaction wave (RW) goes away from the point of SW contact with the curved interface of the material. The flow on compression and rarefaction shocks changes its direction to the opposite one. The velocity component directed to the initial perturbation troughs occurs in the material behind the shock wave front. In this case a local change of the angle of FS inclination to a separate perturbation axis of symmetry is insignificant and doesn't change the perturbation sign. The resultant transversal velocity component

CP849, *Zababakhin Scientific Talks - 2005,*
edited by E. N. Avrorin and V. A. Simonenko
© 2006 American Institute of Physics 0-7354-0345-7/06/$23.00

leads to cumulation of flow on the axis and may cause jet shaping. In solids microcumulative phenomena could be observed during SW arrival at a rough surface [3], such phenomena were also found during special experiments [4].

1. STUDYING THE CUMULATIVE BEHAVIOR (JET SHAPING) OF INSTABILITY

Consider a 2D interface of some material shown in Fig.1. Assume, for simplicity, that we have saw-tooth perturbations. Assume also that the material contacts with vacuum and this corresponds to the limit value of the material density ratio on the both sides of the material interface, $\eta \equiv \rho_0 / \rho_b \gg 1$. It's not difficult to extend the results below to the case of a finite value of ratio η.

Let's ignore the multiple-wave configuration of flow near the free surface and consider only the incident wave and the rarefaction wave moving away from the point of intersection of this wave front with the surface.

Assume that the shock wave is weak enough, so the material heating behind the wave front can be neglected. The material behavior is described by the equation:

$$P = A(\delta^n - 1), \quad \delta = \rho / \rho_0. \tag{2}$$

The time-independent pattern of the flow in the coordinate system associated to the point of intersection of the shock wave with surface (point O) is given in Fig.1 In this coordinate system the flow of velocity $U_0 = D / \cos\alpha$ (D is a given velocity of the shock wave) is surging towards the shock wave (0-1). Arrows show lines of flow.

The incident shock wave is weak, so its velocity is almost equal to sound velocity. The rarefaction wave coming from the point of intersection of the incident wave with the surface has the same velocity. The rarefaction wave width can be neglected and replaced by its thin discontinuity. The angle of incidence to the free surface ϑ, for the incident wave coincides with the angle of reflection off the initial position of boundary for the rarefaction wave, because the shock and rarefaction waves have the same velocities. The angles of flow turns at discontinuities should also be coincident in view of the assumptions above.

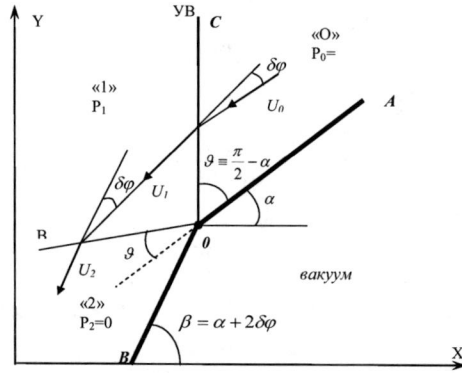

FIGURE 1. SW arrival at the angled surface in the coordinate system associated to the point of SW/surface intersection.

Write Hugoniot relations at discontinuities. On shock wave 0-1 and for discontinuity 1-2 they are:

$$\rho_0 U_0 \cos\alpha = \rho_1 U_1 \cos(\alpha + \delta\varphi),$$ (3)

$$\rho_0 U_0^2 \cos^2\alpha = P + \rho_1 U_1^2 \cos^2(\alpha + \delta\varphi),$$ (4)

$$U_0 \sin\alpha = U_1 \sin(\alpha + \delta\varphi).$$ (5)

$$\rho_1 U_1 \cos(\alpha - \delta\varphi) = \rho_0 U_2 \cos(\alpha - 2\delta\varphi),$$ (6)

$$\rho_0 U_2^2 \cos^2(\alpha - 2\delta\varphi) = P + \rho_1 U_1^2 \cos^2(\alpha - \delta\varphi),$$ (7)

$$U_1 \sin(\alpha - \delta\varphi) = U_2 \sin(\alpha - 2\delta\varphi).$$ (8)

With the shock wave assumed to be weak, compression behind the wave front can be represented as

$$\sigma = \frac{\rho_1}{\rho_0} \approx 1 + \varepsilon, \ \varepsilon \ll 1.$$ (9)

The expressions (3) - (9) allow us to determine the flow parameters in region «2». From the equations (3), (5) we have:

$$tg(\alpha + \delta\varphi) = \sigma \cdot tg\alpha.$$ (10)

With regard to smallness of the flow inclination angle on the wave front, $\delta\varphi$, one can obtain the following from the expression (10):

$$\delta\varphi \approx (\sigma - 1) \cos\alpha \sin\alpha,$$ (11)

hence, angle $\delta\varphi$ has the same order of smallness, as the material compression on the SW front. The following relation determines the angle, at which the flow surges towards the hollow axis:

$$\beta = \alpha + 2\delta\varphi \approx \alpha + 2(\sigma - 1) \cos\alpha \sin\alpha.$$ (12)

For the flow velocity in region «2», we obtain, up to values of the first order of smallness of ε, that

$$U_2 \approx D / \cos\alpha.$$ (13)

Turning to immobile coordinate system write the velocity components for material in region "2":

$$\begin{cases} W_x = D - U_2 \cos(\alpha + 2\delta\varphi) \approx 2D \sin^2\alpha \cdot \varepsilon = 2D\varepsilon \cdot \sin^2\alpha, \\ W_y = D\, tg\alpha - U_2 \sin(\alpha + 2\delta\varphi) \approx -D\varepsilon \cdot \sin 2\alpha, \end{cases}$$ (14)

$$W = \sqrt{W_x^2 + W_y^2} = 2D\varepsilon \cdot \sin\alpha.$$ (15)

Let's obtain an approximate expression for velocity of the shaped jet. The shaped jet, to be formed, requires a compression shock to be generated at the point of impact of the flow coming onto the hollow axis of symmetry and going away up-stream. The conditions, at which there is no discontinuity connected to the point of impact, determine the value of critical angle, at which the flow comes onto the axis, and, ultimately, the minimum admissible value of the initial opening angle of the wedge-shaped hollow. Assume that according to the specified conditions angle β is higher than some critical value. Consider motions in the system associated to the point of contact of the hollow and axis of symmetry. It follows from Bernoulli law, that the flow velocity in the region between the rarefaction wave and the shock wave gone away from axis of symmetry equals to the velocity of the jet's head part at infinity:

319

$$\tilde{V}_{jet} = |W_y| / \sin(\alpha + 2\delta\varphi). \tag{16}$$

The velocity of the point of flow contact with axis of symmetry is

$$V_c = W_x + W_y / \operatorname{tg}(\alpha + 2\delta\varphi). \tag{17}$$

Turning to immobile system of coordinates we obtain the final expression for velocity of the shaped jet occurring on the wedge-shaped hollow axis:

$$V_{jet} = V_c + \tilde{V}_{jet} \approx U_{CII}(1 + \cos\alpha)/(1 + 2(\sigma - 1) \cdot \cos^2\alpha). \tag{18}$$

The resultant expression applicability can be extended to any form of perturbations periodical with respect to y using the following relation:

$$\operatorname{tg}\alpha = (\lambda/2)/(2a) = \lambda/(4a). \tag{19}$$

2. APPLICABILITY OF SOLUTION IN THE FORM OF SHAPED JETS

It is rather difficult, in general, to identify unambiguously the range of applicability for the solution found.

The condensed matter behavior behind the wave front, under shock waves of high rates, is determined by thermal properties of material. In the limit of $P \to \infty$ the condensed mater acts as an ideal gas. Equate the elastic and thermal components of pressure, $P_x(\sigma) = P_m(\sigma)$, in order to obtain a rough estimate of the upper limit for the rate of shock waves, at which instability in the form of a shaped jet is still possible. For iron this has been made with $P_{Fe,\kappa pum} \approx 800$ HPa.

Geometric parameters of boundary perturbations should be, first, so as to implement regular mode of SW/surface interactions (i.e. absence of Mach reflection of wave). Second, the opening angle of the hollow generated after the compression wave has passed across discontinuity point must be larger than the critical angle of jet shaping determined in classic theory of shaped jets.

Besides, it is required to take into account the rheological (flow) properties of materials. It is quite clear that the molecular viscosity and strength of material should have a stabilizing effect on the growing shaped jet.

3. NUMERICAL SIMULATION RESULTS

3.1. Computation Setting

Numerical simulation of cumulative effects was performed using Lagrangian-Eulerian code LEGAK [5, 6]. FS was specified either as sinusoidal, or as saw-toothed with the given amplitude a and wavelength λ. The left plane surface of a sample of length $L=0.5$ mm and width $H=0.3$ mm was loaded with pressure $P(t)$. The time profile of pressure $P(t)$ was assumed to be either a constant, or a dependence of the form:

$$P(t) = P_0 / (1 + t/\tau)^3. \tag{20}$$

3.2. Results of Numerical Experiments

3.2.1. Specification of the Expression for the Shaped Jet Velocity

The expression describing the velocity of jet was specified by introduction of additional factors to the equation (18), which were selected basing on numerical experiments:

$$V_{jet} = U_{CII}(1 + 2.7 \ \cos\alpha)/(1 + 2 \ (\sigma - 1) \ \cos^2\alpha \ f(a/\lambda)) \qquad (21)$$

where σ is compression at SW front, angle α correlates with initial amplitude a and wavelength λ according to (19), and function f is determined as

$$f(x) = 8.61x^2 - 13.92x + 6.31. \qquad (22)$$

3.2.2. Comparison with Classic Ricthmyer-Meshkov Instability

Fig.2 gives a typical pattern of jets being shaped, if the material properties are simulated using the equation of Mie-Grueneisen type. Shaping of jets is independent of the kind of initial perturbation boundary. The material mass, m, which is taken away by jets per unit surface, is close to the experimental value of this quantity [3], $m = 2\alpha \cdot \rho_0 \cdot a$, where a is the initial amplitude of perturbation, α is some factor varying within the range of values $\alpha = 3 \div 5$.

Jets that are shaped have a topology distinct from that of jets in classic R-M instability during simulation of material using EOS for gas, $\gamma = 1.4$ (Fig.2).

3.2.3. The Effect of Elastic-Plastic Properties of Materials

The material strength has a stabilizing effect on the development of jets. Let's take into account the strength properties for the problem of interest in the way similar to that described in [7] for classic cumulative (shaped) shells. Thus, the condition, under which jets are shaped, can be written, as follows:

$$|W_y| > W_{y,\kappa pum} = \sqrt{2\sigma_T/\rho_0} . \qquad (23)$$

Basing on the condition (23) one can obtain the expression for the minimum possible rate of shock wave, $P_{\kappa pum}$, with which a jet stream becomes possible on a rough free surface:

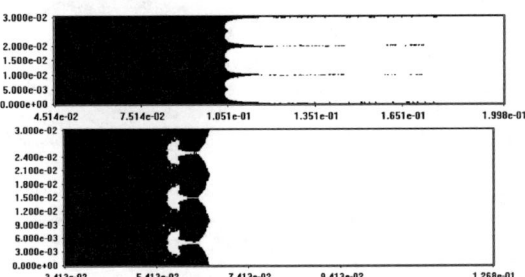

FIGURE 2. The distribution of materials in computations with Mie-Grueneisen EOS (on the top) and EOS for gas (on the bottom), t=0.3 μs.

$$P_{\text{крит}} \approx \sqrt{2\,\sigma_T\,\rho_0 c_0^2}\,/\sin 2\alpha = \sqrt{0.5 \cdot \sigma_T \rho_0 c_0^2}\,\left(\lambda/4a + 4a/\lambda\right). \tag{24}$$

The simplest way of taking into account the effect of strength properties is to make a substitution, $\left|W_y'\right| = \sqrt{W_y^2 - W_{y,\text{крит}}^2}$. With regard to this one can obtain an approximate dependence describing the jet velocity decrease owing to strength properties:

$$\frac{V_{jet}^{UP}}{V_{jet}} \approx 1 - \frac{\sigma_T}{\rho_0 U_{CII}^2} \cdot \frac{1 + \left[1 - 2(\sigma - 1)\ \sin^2\alpha\right]\ \cos\alpha}{\sin^2\alpha\ \cos\alpha\ (1 + 2.7\cos\alpha)}. \tag{25}$$

The method above for evaluation of the influence of material strength on jet formation is conditional, because it doesn't take in to account real properties of materials. Nevertheless, the solution modification in the form of (25) allows, to some extent, taking into account dissipation of energy to overcome the material strength. The simplest obtained estimates of the influence of strength on the processes of shaping jets are confirmed by results of computations.

CONCLUSION

It has been shown that the interface instability developed with arrival of a shock wave at the curved free surface of condensed matter may differ from classic Richtmyer-Meshkov instability and have the jet-shaping behavior. Basing on analytical estimates and results of numerical simulation using LEGAK code we obtained the expressions determining the parameters of resultant jets depending on the initial conditions and some rheological properties of materials. It has been shown also that material strength properties have a stabilizing effect on jet formation. The areas of the obtained solution applicability are identified. The obtained theoretical-and-computational results are in qualitative agreement with the experimental data on the condensed matter release from a rough surface available for the paper authors [3], [4].

The work was sponsored by RFRF (Project 05-01-00083).

REFERENCES

1. Inogamov N.A., Dem'yanov A.Yu., Son E.Ye. The Hydrodynamics of Mixing.- Moscow, Moscow Physics and Thechnology Institute Publishers, 1999, P.464.
2. Richtmyer R.D. // Commun. Pure and Appl. Math. 1960, v.13, N2, pp.297-319
3. Ogorodnikov V.A., Ivanov A.G., Mikhailov A.L., et al. On Particle Release from a Metal's Free Surface Upon a Shock Wave Arrival and the Methods of Diagnosing These Particles // FGV, 1998, V.34, N6, pp.103-107.
4. Lebedev A.I., Igonin V.V., Nizovtsev P.N., et al. Studies of Instabilities of Solid Free Surfaces Under Shock-Wave Loading // VNIIEF Scientific Materials, V.1, Sarov, 2001, pp. 590-597
5. Bakhrakh S.M., Spiridonov V.F., Shanin A.A. Method for Hydrodynamic Inhomogeneous Flow Computations in Lagrangian-Eulerian Variables // Doklady Academii Nauk SSSR. 1984. V.278, N. 4.pp. 829-833.
6. Avdeyev P.A., Artamonov M.V., Bakhrakh S.M., et al. LEGAK code complex for computation of time-dependent multicomponent continuum flows and the ideas of its implementation on distributed-memory multiprocessors // Voprosy Atomnoi Nauki I Techniki. Ser. Math. Model. Phys. Process. 2001. Iss. 3. pp. 14-18.
7. Kinelovsky S.A. Trishin Yu.A. Physical Aspects of Cumulation // FGV, 1980, N.5, pp.26-40.

Study of viscosity effect on turbulent mixing development at the gas-liquid interface

M.V. Bliznetsov, N.V. Nevmerzhitsky, E.A. Sotskov, L.V. Tochilina, V.I. Kozlov, A.K. Lychagin, V.A. Ustinenko

Russian Federal Nuclear Center-All Russia Scientific Research Institute of Experimental Physics, Sarov, Russia

Abstract: We present the results of the experimental study of viscosity effect on turbulent mixing development (TM) occurring at Raylegh-Taylor instability at the boundary of a liquid layer accelerated by compressed gas. In the experiments dynamic viscosity of liquid has varied from $\mu=1$ cP to $\mu=1,480$ cP. As liquid we used: water, glycerol, aqueous solution of glycerol having known viscosity. The value of acceleration of a liquid layer has amounted to: $g \cong 10^3 g_0$ and $g \cong 10^5 g_0$. As gas we used helium compressed previously up to pressure 4.5-500 atm. It has been demonstrated that when changing liquid viscosity a mixing zone structure changes. This influences on a mixing character of substances.

INTRODUCTION

So far as the authors know, in the course of interpretation of all experimental studies of turbulent mixing development at the gas/liquid interface and the liquid/liquid interface, stemming from R-T instability [1÷2] the influence of molecular viscosity is believed to be negligible [3÷4] thanks to a rather considerable value of turbulent Reynolds number ($Re=10^3-10^4$). However, when comparing dynamics of turbulent mixing zone (TMZ) at low ($g \cong 10^2 g_0$, Re of the order of 10^3) and increased ($g \cong 10^5 g_0$, Re of the order of 10^5) accelerations of water layer [5] it has been showed that with increasing g TMZ becomes small-scale, a penetration rate of a gas front into liquid decreases more than twice. To study these matters, we have performed a series of experiments, their set-up and results are presented in this paper.

EXPERIMENTAL TECHNIQUE

The experiments with the acceleration $g \approx 10^3 g_0$ have been conducted in conformity with the scheme of figure 1.a.

The instrument consists of a compressed gas chamber, a clear accelerating channel, transparent container made of Plexiglas and a failed diaphragm closing a lower channel end hermetically.

A layer mass together with a container has amounted to ≈ 70 g. A chamber and a channel beneath a container have been filled with compressed helium with equal

CP849, *Zababakhin Scientific Talks - 2005*,
edited by E. N. Avrorin and V. A. Simonenko
© 2006 American Institute of Physics 0-7354-0345-7/06/$23.00

pressure (P=5.5 atm gauge (gauge atmosphere)). Then a diaphragm is failed - gas from under the container goes to atmosphere, the latter accelerates down under pressure above the container.

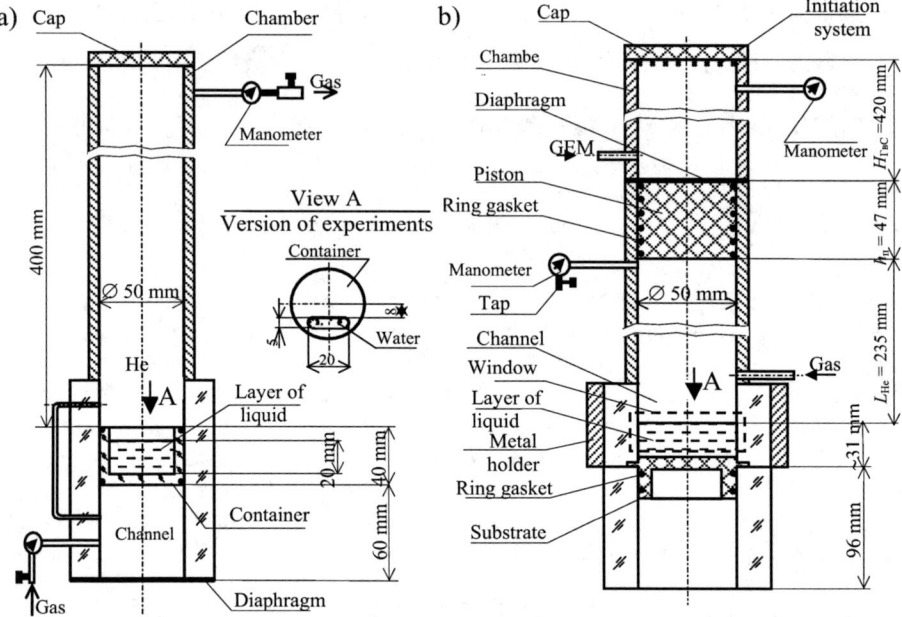

FIGURE 1. Schemes of experimental instruments a) on creation of quasi-constant acceleration of liquid layer, b) on creation of pulse acceleration

The experiments with the acceleration $g \cong 10^5 g_0$ have been conducted in conformity with the scheme of figure 1.b.

The instrument consists of a compound accelerating channel, a chamber of gas explosive mixture (GEM), a cap, a rigid piston (polyethylene), a failed diaphragm and a substrate (textolite). The lower end of the channel is open: it is connected with atmosphere.

A layer mass together with a substrate has amounted to ≈120 g.

A channel between a layer and a rigid piston has been filled with helium at the initial pressure P=3.5 atm gauge (gauge atmosphere)), a chamber with gas explosive mixture has been filled with a mixture of acetylene and oxygen of stoichiometric composition at the pressure P=4.1 atm gauge (gauge atmosphere)).

The acceleration of a layer in this instrument occurs as follows. After blasting GEM the diaphragm is failed, a piston is accelerated and compresses gas under the piston. When we have a gas pressure higher than a critical pressure (P≈100 atm), a shoulder/pad of a substrate is cut off, and a layer together with a substrate is accelerated vertically downwards.

The process of layer acceleration and instability growth is being recorded by the help of rapid motion-picture shooting in transmitted light.

To improve optical resolution of TM zone structure, in both experimental set-ups we have also performed the experiments with a narrow water layer (width is 5 mm) in a version of the view A in figure 1. A narrow groove of a transparent container has been filled with a layer of water in these experiments.

THE RESULTS OF THE EXPERIMENTS

Figures 2-4 present some separate shots of moving image frames of several performed experiments, figure 5 – inherent g(2S) diagrams, figures 6, 7 show the dependencies of gas front penetration into liquid h_{lh} and show the dependencies of growth of a total width of mixing zone H on layer displacement S as: $\sqrt{h_{lh}}\,(\sqrt{A\cdot 2S})$, $\sqrt{H}\,(\sqrt{A\cdot 2S})$, respectively, here A – Atwood number.

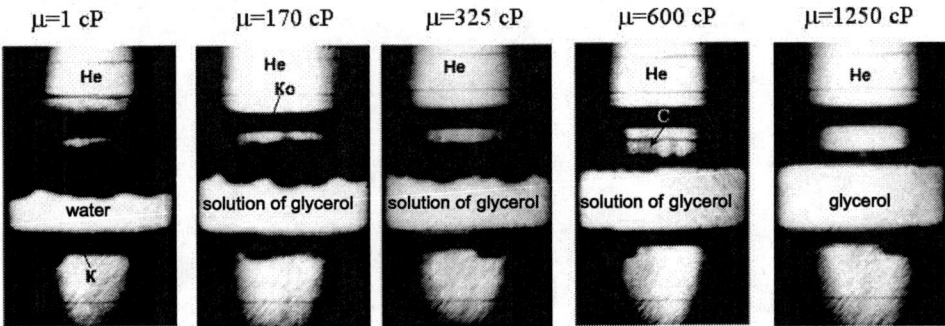

FIGURE 2. Some shots of moving image frames of experiments with g≈10³g₀ (displacement of the layer S≈24 mm) Designations: He – compressed helium; K – container; Ko – container ring; TMZ – turbulent mixing zone; C - jet.

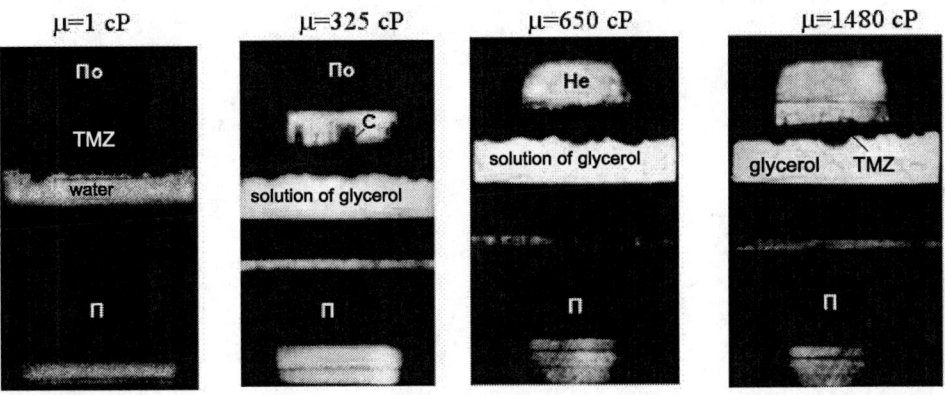

FIGURE 3. Some shots of moving image frames of experiments with g≈10⁵g₀ (displacement of the layer S≈17 mm) Designations: He – compressed helium; П – substrate; По – piston; TMZ – turbulent mixing zone; C – squirt; S – layer displacement

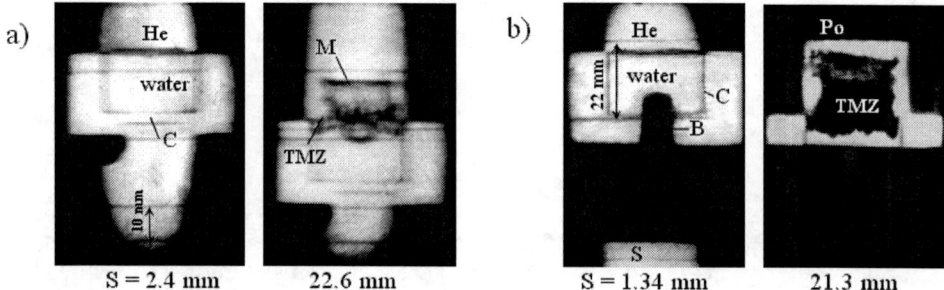

a) He | water | C | 10 mm | M | TMZ
b) He | 22 mm | water | C | B | Po | TMZ | S

S = 2.4 mm   22.6 mm   S = 1.34 mm   21.3 mm

FIGURE 4. Shots of moving image frames of experiments on mixing development in a narrow (width is 5 mm) layer of water (experimental set-up, figure 1, version view A): a) layer acceleration $g{\approx}10^3g_0$; b) layer acceleration $g{\approx}10^5g_0$. Designations: TMZ – turbulent mixing zone; Po – piston; C – container; S – substrate; B – screw (out of flow); He-compressed helium; M – separation of water meniscus

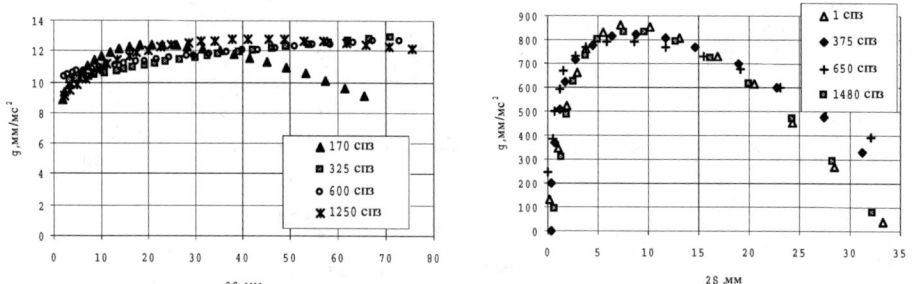

FIGURE 5. Dependencies of liquid layer acceleration g on its displacement 2S

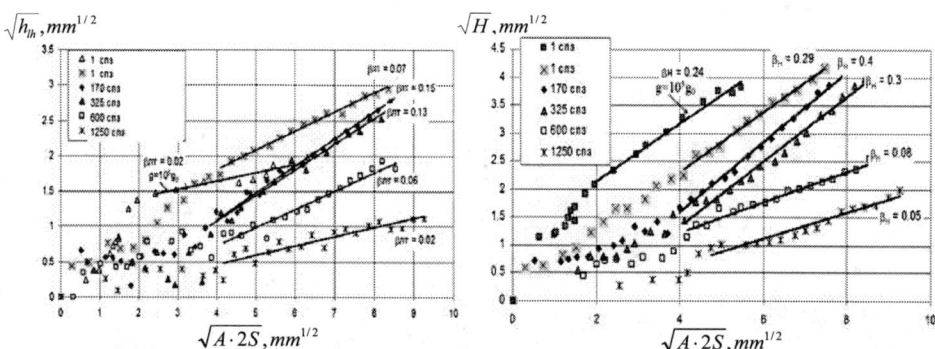

FIGURE 6. Gas front penetration into liquid   **FIGURE 7.** Growth of total width of mixing zone

On the dependencies $\sqrt{h_{lh}}(\sqrt{A \cdot 2S})$, $\sqrt{H}(\sqrt{A \cdot 2S})$ we observe the areas close to linear areas. They can be regarded as a steady-state mode of R-T instability growth (of perturbations or TM). A square of a slope ratio of these areas towards the abscissa axis characterizes a rate constant (pace) of gas front penetration into liquid – β_{lh} or growth rate of total width of TM zone (perturbations) – β_H: under conditions $h_{lh} = \beta_{lh} \cdot A \cdot gt^2$, $H = \beta_H \cdot A \cdot gt^2$.

Inaccuracies of measurements: $\Delta S = \pm 0.3$ mm; $\Delta h_{lh} = \pm 0.5$ mm; $\Delta H = \pm 1$ mm.

EXPERIMENTS WITH $G \approx 10^3 G_0$.

Referring to figure 2 it will be observed that perturbations growth occurs principally at $\mu > 600$ cP in an observed range of layer displacements, and at $\mu < 600$ cP – TM development.

Big bubbles are formed at the stage of TM from long-wave perturbations. The big bubbles grow into liquid faster than small ones. In the experiment with a narrow layer of water big bubbles and thick jets of a mixing zone are easily observable (s. fig. 4a).

Under small initial perturbations the viscosity inhibits the instability growth in time (s. fig. 2).

With decreasing dynamic viscosity of liquid:

The optically observable total width of the mixing zone increases (s. Fig. 2);

Penetration rate of a gas front into liquid β_{lh}(light into heavy) increases from 0.02 to 0.15 at first, and then decreases up to 0.07 (s. fig. 6); a growth rate of a total width of a mixing zone β_H behaves qualitatively similar to β_{lh}(light into heavy) (s. fig. 7).

EXPERIMENTS WITH C $G \approx 10^5 G_0$.

According to the results of these experiments one can say that:

- A mixing zone in glycerol and its aqueous solutions develops faster than in the experiments with $g \approx 10^3 g_0$ (s. fig. 2,3);
- In contrast with water, in the experiments with glycerol and its aqueous solutions, as in the first experiments bigger bubbles and jets are formed (s. fig. 3): in the experiment with a narrow water layer the turbulent mixing zone is small-scale (s. fig. 4b);
- value β lh(light into heavy) is about 0.02, β_H – 0.24.

ASSESSMENT OF CHARACTERISTIC PERTURBATION SCALES

In the case of R-T instability of viscous liquids the increment of growth of perturbations amplitude with the wave number k is equal in linear approximation to [1, 9]:

$$\gamma = \sqrt{Akg + v^2 k^4} - vk^2.$$

It would appear reasonable that at the instant of transition to TM, a maximum size of heterogeneities on the order of magnitude is equal to a wave length of the most rapidly growing harmonic, which is determined by the condition $\partial \gamma / \partial k = 0$, whence it follows that

$$d_{max} = 2\pi \sqrt[3]{v^2 / Ag}.$$

Hence it follows that viscosity growth of liquid at fixed values of acceleration (in accordance with the data of fig. 2) brings about a growth of a characteristic size of heterogeneities in TM zone. In this case, a penetration rate of gas bubbles into liquid may also increase, because a velocity of coming to surface (floating up) of gas bubbles in liquid is $V = \sqrt{gd} = \sqrt[3]{g\nu}$.

The transition time to the turbulent conditions of perturbation evolution on the order of magnitude is in inverse proportion to a maximum value of the increment of growth of perturbations amplitude:

$$\tau = \sqrt[3]{\nu/(Ag)^2}.$$

Therefore, the growth of liquid viscosity at fixed values of acceleration leads to the increase in the transition time towards TM. The increase in acceleration brings about the reduction of the transition time.

CONCLUSIONS

With increasing liquid viscosity (with decreasing Reynolds number) the transition time towards the turbulent conditions of mixing grows and large-scale structures arise in TM zone.

In relatively viscous fluids a penetration rate of a gas front into liquid and a growth rate of a total width of a mixing zone may increase because of the formation of big bubbles and jets, whose growth velocity is higher than the growth velocity of small ones.

ACKNOWLEDGMENTS

The authors are indebted to E.E. Meshkov and V.A. Raevsky for helpful comments on this work.

REFERENCES

1   Lord Rayleigh. *Proc. London Math. Soc.*, 1883, V. 14, P.170.
2   G.I. Taylor, "The instability of liquid surfaces when accelerated in a direction perpendicular to their planes". // *Proc. Roy. Soc.*, 1950, V. A201, P. 192.
3   K.I. Read. "Experimental investigation of turbulent mixing by Rayleigh-Taylor instability". // *Physica D12*, 1984, P. 45.
4   Yu.A. Kucherenko, G.G. Tomashev, L.I. Shibarshov. Experimental study of gravity turbulent mixing in automodel conditions. // *VANT*, iss.1, 1988, p.13.
5   N.V. Nevmerzhitsky, E.A. Sotskov, A.O. Drennov. Studies of turbulent mixing development at the gas/liquid interface at accelerations from $10^2 g_0$ to $10^5 g_0$. // *Abstracts of papers for International Conference V Khariton's Topical Scientific Readings, Substances, materials and constructions at intensive dynamic effects.* Sarov, March 17-21, 2003, p.190
6   D.H. Sharp. "An overview of Rayleigh-Taylor instability. Fronts, Interfaces and Patterns". // *Proc. of the Third Ann. Int. Conf of the Center for Nonlinear Studies.* Los Alamos, New Mexico, May 1983, North-Holland Physics publishing 1984. P.3-18.

Numerical Simulation Of The Self-Similar Problem On Shear Turbulent Mixing Using k-ε Model

M.I. Avramenko and A.N. Shushlebin

Russian Federal Nuclear Center – All-Russian Research Institute of Technical Physics named after acad. E.I. Zababakhin, P.O. Box 245, 456770, Snezhinsk, Chelyabinsk region, Russia

Abstract. The results are presented of numerical simulations of the self-similar problem on shear turbulent mixing of plain layers. The simulations used 2D k-ε model. The dependence was obtained of mixing zone growth rate on the layers density ratio in the incompressible limit. Influence of compressibility is also discussed.

Keywords: k-ε model, shear turbulent mixing
PACS: 47.27.em, 47.27.wj

INTRODUCTION

Shear turbulence is often observed in nature and technical devices. It arises as a result of the instability development when the flow contains jumps in tangential velocity component. This instability is also called as Kelvin-Helmholtz one. The self-similar problem on shear turbulent mixing is a basic idealization for that situation. The considered system consists of two homogeneous semi-infinite plain layers which slide as a whole relative each other along their interface. The problem presents simplest 1D example of shear turbulent zone. Experimental and calculation results on development of the sear layer mixing are important for validation of turbulence models.

Turbulent mixing of shear layers was studied in several experiments (see, for example, [1-4]). As the configuration of the considered self-similar problem on shear mixing is difficult for arrangement, the experiments had somewhat different design. Two plane streams with different velocities were initially separated by a thin partition. The streams came into contact at the downstream edge of the partition, so the mixing zone had a wedge shape.

Also 2D direct numerical simulations were done of development of the wedge-like mixing zone for incompressible layers [5] that corresponded to experiments [1]. The conclusions, which were made from these simulations for self-similar problem on shear mixing, have some uncertainties. They are connected to a) incomplete turbulence description in the fulfilled 2D simulations, b) approximate boundary conditions used on the outcome side of calculation domain, c) approximations done for transition to the considered self-similar problem. Remind that in all experiments and simulations [5], parameters of mixing zone are functions of downstream distance, and the mixing zone has a wedge-like shape, whereas for the considered self-similar problem mixing zone is a spreading with time plain layer.

In the present paper, the simulations are presented directly for the time-dependant self-similar problem on shear mixing. 2D k-ε model was used. These simulations are free of the abovementioned uncertainties, and their accuracy should be defined by the accuracy of k-ε model itself.

PROBLEM FORMULATION

The simulated problem is as following. At the initial moment $t = 0$, the lower layer ($y < 0$) is at rest and has a constant density $\rho_1 = 1$; the upper (lighter) layer of density $\rho_2 = 1/\delta$ moves in x-direction with velocity $U = 1$. Both mediums are an ideal gases with the same adiabatic exponent $\gamma = 5/3$. The system is under constant temperature and pressure, so the density difference is due to the difference in specific heat. In simulations for incompressible case, media's temperature remains (was set) constant. It also was high enough to ensure the speed of sound to be mach higher than velocity of relative movement that leads to incompressibility of the flow. For the case of compressible layers, additionally the balance equation for energy was solved. Density of internal energy and pressure in the mixed region were calculated by averaging by mass concentrations.

At the stage of the developed turbulence, the problem solution for the incompressible case has the form (c – volume concentration of the lighter gas)

$$
\begin{cases}
c(y,t) = \tilde{c}(\xi) \\
u_x(y,t) = U\,\tilde{u}_x(\xi) \\
u_y(y,t) = U\,\tilde{u}_y(\xi) \\
k(y,t) = U^2\,\tilde{k}(\xi) \\
\varepsilon(y,t) = \dfrac{U^2}{t}\,\tilde{\varepsilon}(\xi)
\end{cases}
\tag{1}
$$

where the self-similar variable is

$$
\xi = \frac{y}{Ut}.
\tag{2}
$$

The width of the mixing zone can be presented in the form similar to the case of the self-similar problem on gravitational mixing

$$
L^* = (\alpha_{sh,b} + \alpha_{sh,s})\,Ut.
\tag{3}
$$

Here $\alpha_{sh,b}(\delta)$, $\alpha_{sh,s}(\delta)$ are dimensionless parameters that define the depth of light gas penetration into heavy one, and respectively inverse. The width of the mixing zone also may be written without subdividing into the component contribution

$$L^* = 2\alpha_{sh} \, f(\delta) \, Ut \ . \tag{4}$$

In this form, factor f is normalized by $f(1) = 1$, and describes the dependence of width growth rate on the layer's density ratio.

In the frameworks of k-ε model, the boundaries of mixing zone refer to the common front of mutual nonlinear diffusion of turbulent energy density k, dissipation rate ε, tangential velocity v_x, and concentration c. Besides the dimensionless front positions $\alpha_{sh,b}$, $\alpha_{sh,s}$, the width of the velocity profile b is usually used as a mixing zone characteristic. It is defined by the coordinates where the excess velocity takes the value $0.1U$ and $0.9U$, and can be presented in the same form, as (3)

$$b = (\alpha_{U,b} + \alpha_{U,s}) \, Ut \ . \tag{5}$$

2D k-ε model simulations were performed with TIGR-72T code [6], in which the k-ε equations version [7] has been implemented. The computation mesh was comprised of $25\times(1000\div2000)$ intervals. Boundary conditions on the lateral sides of calculation domain were set approximately – they were taken from the previous time step. Nevertheless, it had no influence on the results, because each time step was supplemented with additional refinement procedure: at the end, all calculated y-profiles from central x-section were entered into whole calculation domain. Time step value was small enough to prevent perturbation from the lateral sides to reach the central section.

All calculations have been carried out using mass-coordinated uniform mesh. By the end of calculations, the mixing zone covered more than 300 mesh intervals. As was shown by calculations with mesh variation, it was enough for getting rather accurate results.

RESULTS FOR INCOMPRESSIBLE LIMIT

In all simulations, self-similar behavior of calculated quantities was observer. On figures 1 and 2 are presented the calculated self-similar profiles (1) for density ratio $\delta = 2$. In the case $\delta = 1$, profiles of k, ε and D, naturally, are symmetric with respect to the layer's interface. As δ increases, normal component of velocity v_y appears, and this symmetry breaks. Tangential velocity profile turned to be compressed and shifted relative to the concentration profile in direction to lighter medium. It's width b is about 1.3 times less than that of the mass concentration profile b_m. For volume concentration the similar relation takes place only if δ is not too large. This behavior of profile's width for velocity and concentration is due to rather high level of the quantity excess that was used for the width definition ($10\%\div90\%$). Remind that, at least in frameworks of k-ε model, width of any profile is the same, as the level for its definition goes to zero.

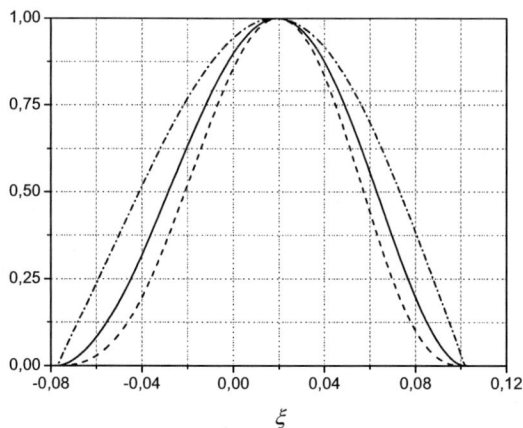

FIGURE 1. Profiles of turbulent energy density (solid line), dissipation rate (dashed line), and diffusion coefficient (dotted) for $\delta = 2.0$.

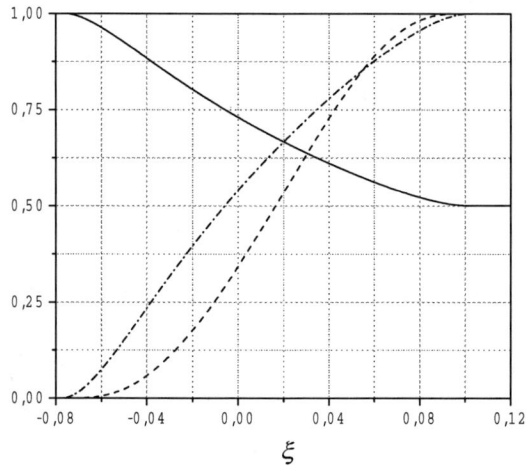

FIGURE 2. Profiles of density (solid line), velocity (dashed line) and volume concentration (dotted) for $\delta = 2.0$.

In table 1 are presented calculated peak values of self-similar quantities for a set of δ (or Atwood number $A = (\delta-1)/(\delta+1)$). The last column shows the displacement of the layer's interface. As it is seen, peak of dimensionless turbulent energy remains almost constant, as δ increases, while peak of dimensionless dissipation rate grows more obviously.

Table 2 presents the calculated mixing zone parameters. Relative to the shifted (Δy from table 1) position of the layer's interface, mixing zone spreads almost symmetric. In the case of equal densities, the obtain width of mixing zone is very near to the approximate analytical solution [7] according to which $\alpha_{sh} = 0.0906$.

332

TABLE 1. Peak values of self-similar quantities.

δ	A	k	ε	D	u_y	Δy
1.0	0	0.03297	0.0854	0.003055	0	0
1.5	0.666	0.03299	0.0867	0.003012	0.00793	0.00793
2.0	0.333	0.03303	0.0879	0.002981	0.01338	0.01338
3.0	0.5	0.03310	0.0905	0.002906	0.02060	0.02060
7.0	0.75	0.03330	0.0980	0.002719	0.03325	0.03325
20.0	0.9048	0.03357	0.1098	0.002464	0.04399	0.04399
40.0	0.9512	0.03372	0.1169	0.002335	0.04858	0.04858

TABLE 2. Mixing zone parameters.

δ	$\alpha_{U,b}$	$\alpha_{U,s}$	$\alpha_{U,b}+\alpha_{U,s}$	$\alpha_{sh,,b}$	$\alpha_{sh,,s}$	$\alpha_{sh,b}+\alpha_{sh,s}$
1.0	0.0473	0.0473	0.0946	0.0905	0.0905	0.181
1.5	0.03793	0.05619	0.09412	0.08220	0.09743	0.17963
2.0	0.03117	0.06200	0.09317	0.07622	0.10204	0.17826
3.0	0.02178	0.06933	0.09111	0.06787	0.10735	0.17522
7.0	0.004018	0.08120	0.08521	0.05089	0.11557	0.16646
20.0	-0.01259	0.09022	0.07763	0.03424	0.12152	0.15576
40.0	-0.02021	0.09368	0.07347	0.02570	0.12469	0.15039

The obtained growth rate of mixing zone is a slowly decreasing function of A. This is in a reasonably good agreement with the results of direct numerical simulations of the resembling problem on development of the wedge-like mixing zone [5], where the statement was made that $f(A) = 1$ for $A \le 0.75$.

RESULTS FOR COMPRESSIBLE LAYERS

In addition to the above presented simulations for incompressible shear layers, the simulations for $\delta = 1$ and 2.0 were also done in which the layer's temperature was decreased, and incompressibility approximation was no longer valid. As a major compressibility parameter, usually the relative Mach number is considered (see, for example [3, 4]): $M_r = U / a$, $a = (a_L + a_H) / 2$ – mean speed of sound. In the case of "incompressible layers" that was discussed in the previous section, $M_r \sim 2.\cdot10^{-2}$, while for the compressible layers simulations $M_r \sim 1$.

The fulfilled simulations exhibit a very slow reduction of the mixing zone growth rate. So the reduction of mixing zone was about 1% for $M_r = 0.77 \div 1.28$. It is sufficiently less, than that observed in the shear layer experiments with wedge-like mixing zone. For example, for $\delta = 1.75$ the observed velocity profile width was ≈ 0.56 of the incompressible value when $M_r = 1.37$ [3]. This discrepancy could be interpreted as that the used k-ε model cannot properly describe the effect of compressibility on shear mixing. Another possible reason could be the remaining bugs in the used code or in the simulation setup.

CONCLUSIONS

The fulfilled simulations of self-similar problem on shear turbulent mixing of two semi-infinite plain layers with 2D k-ε model have shown a very slow reduction of the mixing zone growth as the layers density ratio increases. So, for $\delta = 40$ (A ≈ 0.95) the growth rate is $\approx 17\%$ less, than for $\delta = 1$.

The simulations with k-ε model also exhibit sufficiently less influence of compressibility effects than what was experimentally observed for a very near shear layer configuration.

REFERENCES

1. G. L. Brown and A. Roshko, *Journal of Fluid Mechanics* **64**, 775-816 (1974).
2. L. P. Bernal and A. Roshko, *Journal of Fluid Mechanics* **170**, 499-525 (1986).
3. S. G. Goebel and J. C. Dutton, *AIAA Journal* **29**, 538-546 (1991).
4. M. R. Gruber, N. L. Messersmith, and J. C. Dutton, *AIAA Journal* **31**, 2061-2067 (1993).
5. N. S. Es'kov, A. S. Kozlovskikh, D. V Neuvazhayev, *PMTF* **412**, 77-83 (2000, in Russian).
6. A. N. Shushlebin, A. I. Gryaznykh, V. A. Lykov et al., "Testing of k-ε turbulence model with TIGR-3T and VH-1 codes on the convective instability problems", in *Proceedings of The VI Zababakhin scientific talks*, Snezhinsk, Russia, 2001.
7. M. I. Avramenko, *On k-ε Model of Turbulence*, Snezhinsk, RFNC-VNIITF Publishers, 2005 (in Russian).

On Instability of Hydrodynamical Flows

Yuriy G. Gubarev

*Lavrentyev Institute for Hydrodynamics, Siberian Division of Russian Academy for Sciences,
Lavrentyev Avenue 15, Novosibirsk, 630090, Russian Federation*

Abstract. A new analytical approach to studying a wide spectrum of linear problems in theory of hydrodynamical stability is described. The essence of this approach is constructing the regular Lyapunov functionals which grow in time on solutions to corresponding initial–boundary value problems for small perturbations. This approach allows either to prove absolute instability, or to derive sufficient conditions of instability with respect to small perturbations for the hydrodynamical flows under study. It is important that the concrete form of solutions to boundary value and/or initial–boundary value problems which describe the flows and perturbations is not needed. In the work, this approach is applied to a linear problem on stability with respect to small spatial perturbations for steady 3–D flows in a homogeneous ideal incompressible fluid. It is stated that the equilibrium states of inviscid incompressible fluid are absolutely stable with respect to small 3–D perturbations, whereas steady spatial flows are absolutely unstable. Lower a priori estimates are constructed; exponential growth in time is proven for small perturbations under study.

Keywords: ideal incompressible fluid, steady flows, Lyapunov functional, instability.
PACS: 47.20.–k, 47.20.Cq.

FORMULATION OF EXACT PROBLEM UNDER STUDY

We consider 3–D flows of a homogeneous inviscid incompressible fluid in a vessel τ with fixed solid impermeable walls $\partial\tau$. These flows are described by evolutionary solutions to the initial–boundary value problem [1, 2]:

$$\frac{\partial \vec{u}}{\partial t} + (\vec{u}\cdot\nabla)\vec{u} = -\nabla p, \quad div\,\vec{u} = 0 \text{ in } \tau\,;$$

$$\vec{u}\cdot\vec{n} = 0 \text{ on } \partial\tau\,; \tag{1}$$

$$\vec{u}(\vec{x},0) = \vec{u}_0(\vec{x}),$$

where $\vec{u}(\vec{x},t) = (u_1, u_2, u_3)$ is the velocity field; $p(\vec{x},t)$, the pressure field; $\vec{x} = (x_1, x_2, x_3)$, the Cartesian coordinates; t, the time; $\vec{n} = (n_1, n_2, n_3)$, the normal to the surface $\partial\tau$; $\vec{u}_0 = (u_{01}, u_{02}, u_{03})$, the initial velocity field.

Next, we assume that the initial–boundary value problem (1) has exact stationary solutions

$$\vec{u} = \vec{U}(\vec{x}) = (U_1, U_2, U_3), \quad p = P(\vec{x}) \tag{2}$$

which satisfy the relations

CP849, *Zababakhin Scientific Talks - 2005*,
edited by E. N. Avrorin and V. A. Simonenko
© 2006 American Institute of Physics 0-7354-0345-7/06/$23.00

$$(\vec{U}\cdot\nabla)\vec{U}=-\nabla P,\quad div\vec{U}=0 \tag{3}$$

inside the vessel τ and the condition

$$\vec{U}\cdot\vec{n}=0 \tag{4}$$

on its walls $\partial\tau$.

The aim of the subsequent study is to prove the absolute instability of the exact stationary solutions (2)–(4) with respect to isovortical small spatial perturbations.

FORMULATION OF THE LINEARIZED PROBLEM

With this aim in mind, we linearize the problem (1) in a neighborhood of exact stationary solutions (2)–(4), take into consideration the isovorticity condition [1, 2], and obtain the initial–boundary value problem of the form

$$\frac{\partial\vec{\xi}}{\partial t}=\vec{u}'+rot(\vec{U}\times\vec{\xi}),$$

$$\frac{\partial^2\xi_i}{\partial t^2}+2U_k\frac{\partial^2\xi_i}{\partial x_k\partial t}+U_k\frac{\partial}{\partial x_k}\left(U_m\frac{\partial\xi_i}{\partial x_m}\right)=-\frac{\partial p'}{\partial x_i}-\xi_k\frac{\partial^2 P}{\partial x_k\partial x_i}, \tag{5}$$

$$\frac{\partial\xi_i}{\partial x_i}=0,\quad \omega'_i=\Omega_k\frac{\partial\xi_i}{\partial x_k}-\xi_k\frac{\partial\Omega_i}{\partial x_k}\ \text{ in }\tau\,;$$

$$\xi_i n_i=0\ \text{ on }\partial\tau\,;\ \vec{\xi}(\vec{x},0)=\vec{\xi}_0(\vec{x}),\ \frac{\partial\vec{\xi}}{\partial t}(\vec{x},0)=\left(\frac{\partial\vec{\xi}}{\partial t}\right)_0(\vec{x}).$$

Here $\vec{\xi}(\vec{x},t)=(\xi_1,\xi_2,\xi_3)$ is the field of Lagrangian displacements [3]; $\vec{u}'(\vec{x},t)=(u'_1,u'_2,u'_3)$, $p'(\vec{x},t)$ and $\vec{\omega}'(\vec{x},t)=(\omega'_1,\omega'_2,\omega'_3)$ are small perturbations of the velocity, pressure and vorticity fields, correspondingly; $\vec{\Omega}(\vec{x})=(\Omega_1,\Omega_2,\Omega_3)\equiv rot\vec{U}$ is the stationary field of vorticity; $\vec{\xi}_0=(\xi_{01},\xi_{02},\xi_{03})$, the initial field of Lagrangian displacements; $(\partial\vec{\xi}/\partial t)_0$, the initial first partial time derivative of Lagrangian displacements field; the repeated vector indices mean the summation from 1 to 3.

The initial–boundary value problem (5) has an analog of the energy integral in the form

$$E\equiv\frac{1}{2}\int_\tau\left(u'_i u'_i+\omega'_i e_{ikm}U_k\xi_m\right)d\tau=const, \tag{6}$$

where e_{ikm} is the third order co–variant pseudotensor of the weight -1 [4]; $d\tau\equiv dx_1 dx_2 dx_3$, a volume element.

In a view of (6), it isn't difficult to conclude that the exact stationary solutions (2)–(4) to the problem (1) are absolutely stable with respect to isovortical small spatial perturbations (5) only if they describe equilibrium states in the fluid. For the other cases, the integral E on small perturbations (5) is neither of fixed nor of constant sign; by this, the solutions (2)–(4) can be absolutely unstable with respect to these perturbations.

Further we will use the following form of E (6):

$$E \equiv T + T_1 + \Pi = const.$$ (7)

Here

$$T \equiv \frac{1}{2} \int_\tau \left[\frac{\partial \vec{\xi}}{\partial t} + \left(\vec{U} \cdot \nabla \right) \vec{\xi} \right]^2 d\tau \geq 0,$$

$$T_1 \equiv -\int_\tau \frac{\partial \xi_i}{\partial t} \xi_k \frac{\partial U_i}{\partial x_k} d\tau, \quad \Pi \equiv -\frac{1}{2} \int_\tau \xi_i \xi_k \frac{\partial^2 P}{\partial x_i \partial x_k} d\tau.$$

THE LYAPUNOV FUNCTIONAL

For the sake of convenience, we introduce an additional integral

$$M \equiv \int_\tau \xi_i \xi_i d\tau \geq 0.$$ (8)

Differentiating the functional M twice with respect to the independent variable t and transforming the obtained integral with the use of (5) and (7), we easily derive the virial equality [3, 5, 6]

$$\frac{d^2 M}{dt^2} = 4(T + \Pi).$$ (9)

Let

$$\xi_i \xi_k \frac{\partial^2 P}{\partial x_i \partial x_k} \leq 0.$$

Then, with the help of the relation (7) and the equation (9), we obtain the key differential inequality [6] of the form

$$\frac{d^2 M}{dt^2} - 2\lambda \frac{dM}{dt} + 2\lambda^2 M \geq 0,$$ (10)

where λ is an arbitrary positive constant.

Integrating the relation (10) over the half–open intervals $\pi n/(2\lambda) \leq t < \pi(n+1)/(2\lambda)$ $(n = 0, 1, 2, ...)$ [7], provided that

$$M\left(\frac{\pi n}{2\lambda}\right) \equiv M(0)\exp\left(\frac{\pi n}{2}\right), \quad \frac{dM}{dt}\left(\frac{\pi n}{2\lambda}\right) \equiv \frac{dM}{dt}(0)\exp\left(\frac{\pi n}{2}\right),$$ (11)

$$M(0) > 0, \quad \frac{dM}{dt}(0) \geq 2\lambda M(0),$$

we arrive at the lower a priori exponential estimate

337

$$M(t) \geq C\exp(\lambda t) \tag{12}$$

(here C is known positive constant).

Now let

$$\xi_i \xi_k \frac{\partial^2 P}{\partial x_i \partial x_k} > 0.$$

In this case, in a view of boundedness of the domain τ and continuity of the function $P(\bar{x})$ and its derivatives up to the needed order, the left part of the last inequality can be estimated as follows:

$$\xi_i \xi_k \frac{\partial^2 P}{\partial x_i \partial x_k} \leq \alpha \xi_m^2,$$

where α is known positive constant.

As easily seen, a relation similar to the key differential inequality (10) can be derived from the expression (7) and virial equality (9), i.e.

$$\frac{d^2 M}{dt^2} - 2\lambda \frac{dM}{dt} + 2(\lambda^2 + \alpha)M \geq 0 \tag{13}$$

(here λ is an arbitrary positive constant as earlier). Presence of the constant α in the left–hand side of the relation (13) clearly requires some corrections while integrating this inequality as compared with integration of the inequality (10); it is important that the needed corrections are insignificant.

Based on this, we understand that if

$$M\left(\frac{\pi n}{2\sqrt{\lambda^2 + 2\alpha}}\right) \equiv M(0)\exp\left(\frac{\pi n\lambda}{2\sqrt{\lambda^2 + 2\alpha}}\right),$$

$$\frac{dM}{dt}\left(\frac{\pi n}{2\sqrt{\lambda^2 + 2\alpha}}\right) \equiv \frac{dM}{dt}(0)\exp\left(\frac{\pi n\lambda}{2\sqrt{\lambda^2 + 2\alpha}}\right) \quad (n = 0, 1, 2, ...), \tag{14}$$

$$M(0) > 0, \quad \frac{dM}{dt}(0) \geq 2\left(\lambda + \frac{\alpha}{\lambda}\right)M(0)$$

the lower a priori exponential estimate

$$M(t) \geq C_1\exp(\lambda t) \tag{15}$$

(C_1 is known positive constant) immediately follows from the inequality (13).

Finally, let the value

$$\xi_i \xi_k \frac{\partial^2 P}{\partial x_i \partial x_k}$$

have no concrete sign. Since the domain τ is bounded and the function $P(\bar{x})$ and its derivatives up to the needed order are continuous we easily obtain the upper and lower estimates:

$$-\beta\xi_m^2 \leq \xi_i\xi_k\frac{\partial^2 P}{\partial x_i\partial x_k} \leq \beta\xi_m^2,$$

where β is known positive constant.

After simple calculations, we derive another analog of the key differential inequality (10) using the right–hand side of the last inequality and the relations (7) and (9):

$$\frac{d^2 M}{dt^2} - 2\lambda\frac{dM}{dt} + 2(\lambda^2 + \beta)M \geq 0 \qquad (16)$$

(here λ is an arbitrary positive constant as before).

Integrating (16) by analogy with the inequalities (10) and (13), provided that

$$M\left(\frac{\pi n}{2\sqrt{\lambda^2 + 2\beta}}\right) \equiv M(0)\exp\left(\frac{\pi n\lambda}{2\sqrt{\lambda^2 + 2\beta}}\right),$$

$$\frac{dM}{dt}\left(\frac{\pi n}{2\sqrt{\lambda^2 + 2\beta}}\right) \equiv \frac{dM}{dt}(0)\exp\left(\frac{\pi n\lambda}{2\sqrt{\lambda^2 + 2\beta}}\right) \quad (n = 0, 1, 2, ...), \qquad (17)$$

$$M(0) > 0, \quad \frac{dM}{dt}(0) \geq 2\left(\lambda + \frac{\beta}{\lambda}\right)M(0),$$

we arrive at the lower a priori exponential estimate

$$M(t) \geq C_2\exp(\lambda t), \qquad (18)$$

where C_2 is known positive constant.

So, the relations (12), (15), and (18) clearly prove that the isovortical small spatial perturbations (5) with the initial data (11), (14), or (17) of the steady 3–D flows (2)–(4) in an ideal homogeneous incompressible fluid can exponentially grow in time.

Since the inequalities (12), (15), (18) have been constructed without additional restrictions on the exact stationary solutions (2)–(4), the fact from above justifies the absolute instability of such solutions with respect to the small spatial perturbations (5), (11), (14), (17).

Of special attention is the fact that exactly the functional M (8) is the desired Lyapunov functional [8–10] which grows in time on solutions to the initial–boundary value problem (5). Arbitrariness in the choice of the positive constant exponent λ in the right–hand sides of (12), (15), and (18) is a distinguishing characteristic of its behavior. In particular, this arbitrariness allows to treat every solution to the initial–boundary value problem (5), (11), (14), (17), satisfying either of the three estimates (12), (15), (18), as an ill–posedness Hadamard example [11].

CONCLUSION

A new analytical approach to studying a wide spectrum of linear problems on hydrodynamical stability is presented in the work; the approach is based on a specific algorithm for constructing Lyapunov functionals which grow in time on solutions to corresponding initial–boundary problems for small perturbations.

This approach has successfully been used in the problem on the linear stability with respect to isovortical small 3–D perturbations for steady spatial flows in a homogeneous inviscid incompressible fluid in a vessel τ with immovable solid impenetrable walls $\partial\tau$. It has been shown that the equilibrium states of the fluid are absolutely stable with respect to such perturbations, whereas the steady 3–D flows are absolutely unstable. Lower a priori estimates have been constructed, and the exponential in time growth of the isovortical small spatial perturbations has been proven with the use of these estimates.

The above described algorithm for constructing the growing in time Lyapunov functionals is valid for models of liquid, gas, and plasma of any type in the framework of the continuum mechanics because the single requirement while constructing the key differential inequality (10) and its analogs (13) and (16) is to formulate the initial–boundary value problem under study as a system of conservation laws. In particular, as far as the present work is concerned, the results formulated for the problem under study can be naturally extended to unbounded domains, to liquids with real physical properties (such as compressibility, viscosity, etc.), to unsteady main flows, to models with force fields, and so on.

At last, it should to note that, from the mathematical point of view, the results of the present work is of a priori character because theorems on existence of solutions to boundary value and initial–boundary value problems for systems of partial derivatives equations are not proven as yet.

ACKNOWLEDGMENTS

The work is financially supported by Russian Fund of Fundamental Researches (grant № 04–01–00900) and Presidium of Russian Academy for Sciences (integration program № 18, grant № 18.6).

REFERENCES

1. V. I. Arnol'd, *Prikl Matem. Mekh.* **29** (5), 846–851 (1965) (in Russian).
2. V. A. Vladimirov, *Prikl. Mekh. Tekhn. Fiz.* **28** (3), 36–45 (1987) (in Russian).
3. S. Chandrasekhar, *Ellipsoidal Figures of Equilibrium,* New Haven: Yale Univ. Press, 1969, 252 pp.
4. A. J. Mc Connell, *Introduction into tensor analysis and applications to geometry, mechanics, physics,* Moscow: Fizmatgiz, 1963, 412 pp. (in Russian).
5. V. A. Vladimirov and K. I. Ilin, *Moscow Mathematical Journal* **3** (2), 691–709 (2003).
6. Yu. G. Gubarev, *Prikl. Mekh. Tekhn. Fiz.* **45** (2), 111–123 (2004) (in Russian).
7. S. A. Chaplygin, *New approach to approximate integration of differential equations,* Moscow, Leningrad: GITTL, 1950, 104 pp. (in Russian).
8. A. M. Lyapunov, *The General Problem of the Stability of Motion,* London: Taylor & Francis, 1992, 270 pp.
9. N. G. Chetaev, *Stability of Motion,* Moscow: Nauka, 1990, 176 pp. (in Russian).
10. Yu. G. Gubarev, "On instability of steady 3–D flows in homogeneous ideal incompressible fluid" in International conference *"Mathematical hydrodynamics: models and methods", dedicated to 70–th anniversary of professor V.I.Yudovich, October 4–8 2004 ., Rostov upon Don, Russia, Book of abstracts, Rostov State University,* 2004, pp. 13–14 (in Russian).
11. S. K. Godunov, Equations of mathematical physics, Moscow: Nauka, 1979, 392 pp. (in Russian).

Study of Turbulent Mixing Development at the Gas-Gas Interface of Shock Wave at Mach Numbers from 2–9

M.V. Bliznetsov, N.V. Nevmerzhitsky, A.N. Rasin, E.A. Sotskov, E.D. Senkovsky, L.V. Tochilina, V.A. Ustinenko

Russian Federal Nuclear Center-All Russia Scientific Research Institute of Experimental Physics, Sarov, Russia

Abstract: The experimental technique and the results of the study into the turbulent mixing are presented. The turbulent mixing (TM) arises at Richtmyer-Meshkov instability at the gas-gas interface accelerated by a shock wave with Mach numbers up to 9. Helium (He) or air (Air) has been used as light gas, as heavy gas - six-fluorine sulfur (SF_6). In all the experiments a shock wave has propagated from light gas to heavy gas. Mach number of a shock wave in SF_6 has changed from 2 to 9. As a consequence the mixing zone width to be observed optically increases with increase in Mach number.

Keywords: Richtmyer-Meshkov instability, turbulent mixing, shock wave, Mach number, contact gaseous boundaries.
PACS: 47.20.Ma

INTRODUCTION

Hydrodynamic instabilities such as Rayleigh-Taylor [1], Richtmyer-Meshkov [2, 3] play a leading role in many fields of research. To calculate these instabilities and turbulent mixing (TM) connected with them both numerical methods and semi-empirical models are used. Both of them require the testing on experimental data.

A number of experimental works is known, in which development of turbulent mixing was studied at the interface of gases with different density at Mach numbers of a shock wave (SW) $M \leq 5$ (for instance, [3÷5]). To understand better the influence of compressibility of a medium on TM development, it is desirable to obtain the experimental data within a wider range of the change in Mach numbers of SW.

In this paper SW has been formed in a shock tube [6] as a result of detonation of combustible gas mixture (CGM) of acetylene and oxygen. Turbulent mixing was studied at the interface of heavy gas (SF_6) and light gas (He or air). Mach number of SW in SF_6 was changed in a range of 2 - 9.

EXPERIMENTAL TECHNIQUE

Shock tube scheme is presented at Figure 1. The shock tube consists of chambers with a high pressure (driver) and a low pressure. The chambers are separated from

CP849, *Zababakhin Scientific Talks - 2005*,
edited by E. N. Avrorin and V. A. Simonenko
© 2006 American Institute of Physics 0-7354-0345-7/06/$23.00

each other by the help of a diaphragm made of lavsan thick up to 100-150 μm. A measuring section with a silencer is connected with the low pressure chamber. The measuring section has two windows made of optically clear organic glass. It breaks away from the low pressure chamber by means of polymer film ≈ 0.3 μm in thickness, and from a silencer – membrane made of lavsan 50 μm thick.

In the experiments the low pressure chamber was filled with light gas: helium or air; the measuring section– SF_6 ($\rho_0 \approx 6.5$ g/l; $C_0 \approx 129.5$ m/s; $\gamma \approx 1,094$) under atmospheric conditions. There is air in the silencer at atmospheric pressure.

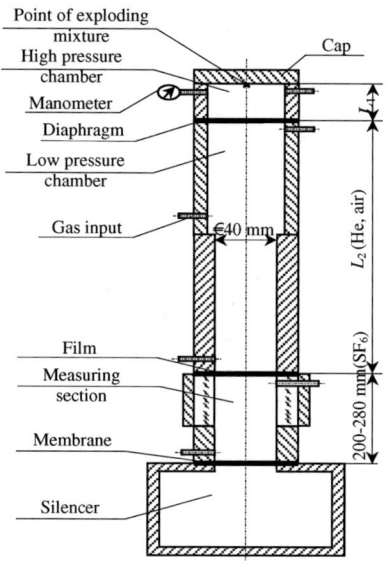

FIGURE 1. Shock tube scheme.

OPERATION OF THE INSTRUMENT

After initiating CGM a detonation wave propagates on the high pressure chamber. When this detonation wave enters a diaphragm the latter disrupts and a shock wave propagates in light gas. When a shock wave goes toward the light gas/SF_6 contact interface (CI), two waves are formed. A shock wave propagates on SF_6 and a reflected wave outgoes in light gas. After destructing a separating film CI of gases accelerates, with the result that Richtmyer-Meshkov instability arises in it, which causes the development of turbulent mixing zone (TMZ) of light and heavy gases.

The variation of CGM composition and CGM pressure has made it possible to obtain shock waves having a varied intensity.

The flow has been recorded through the schlieren method by the use of a rapid movie camera in the frame-by-frame mode and in the slot image scanning mode.

THE RESULTS OF THE EXPERIMENTS AND THEIR ANALYSIS

Figure 2 presents slot moving image frames and frame-by-frame moving image frames of the experiments on air, Figure 3 – experiments with helium, Figure 4 – calculation (without TM) and experimental $X(t)$ diagrams of the flow. The Atwood number is designated as A in figures with regard to gases compression. Gas dynamical calculations of the flow were conducted by using one-dimensional techniques «VIKHR» [7].

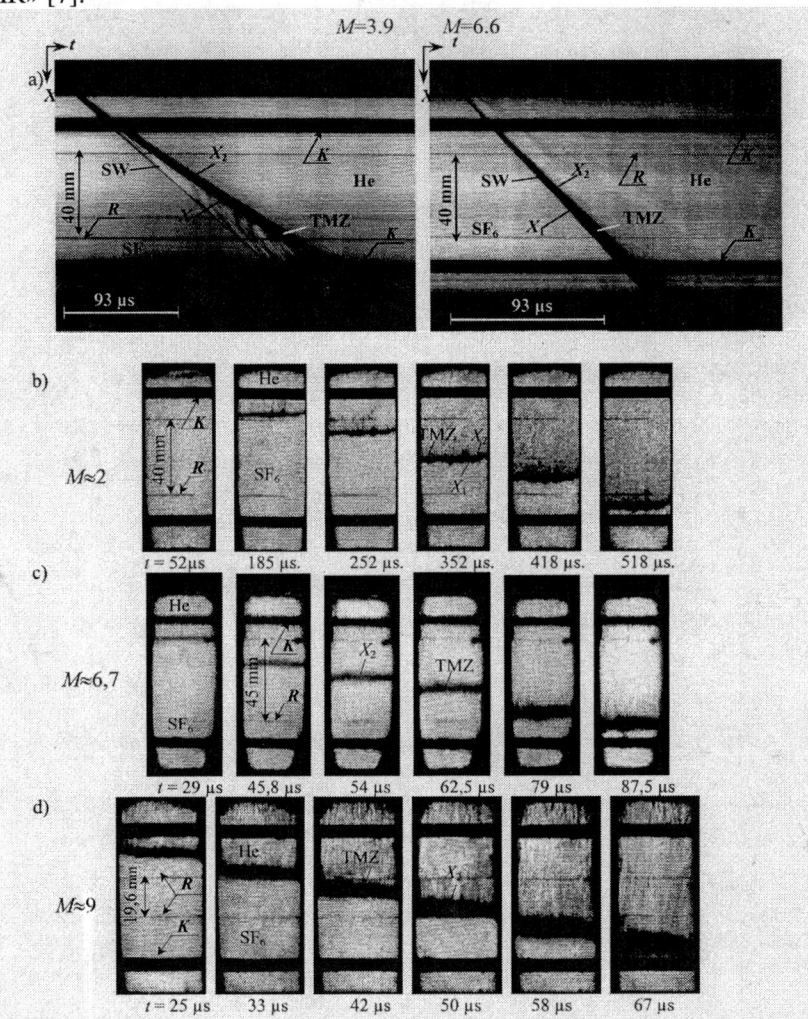

Designations: SW – shock wave; X_1 – front of penetration of «light» gas into «heavy» gas; X_2 – front of penetration of «heavy» gas in «light» gas; t – time is reckoned from output of SW to CI; K – structural element (out of flow); R – reference lines.

FIGURE 2. Moving image frames of the experiments on TMZ development at the *helium-SF$_6$* interface: a) experiment №1 (A≈0.95); b) experiment №2 (A≈0.99) – slot moving image frames; c) experiment №3 (A≈0.95); d) experiment №4 (A≈0.99); e) experiment №5 (A≈0.99)

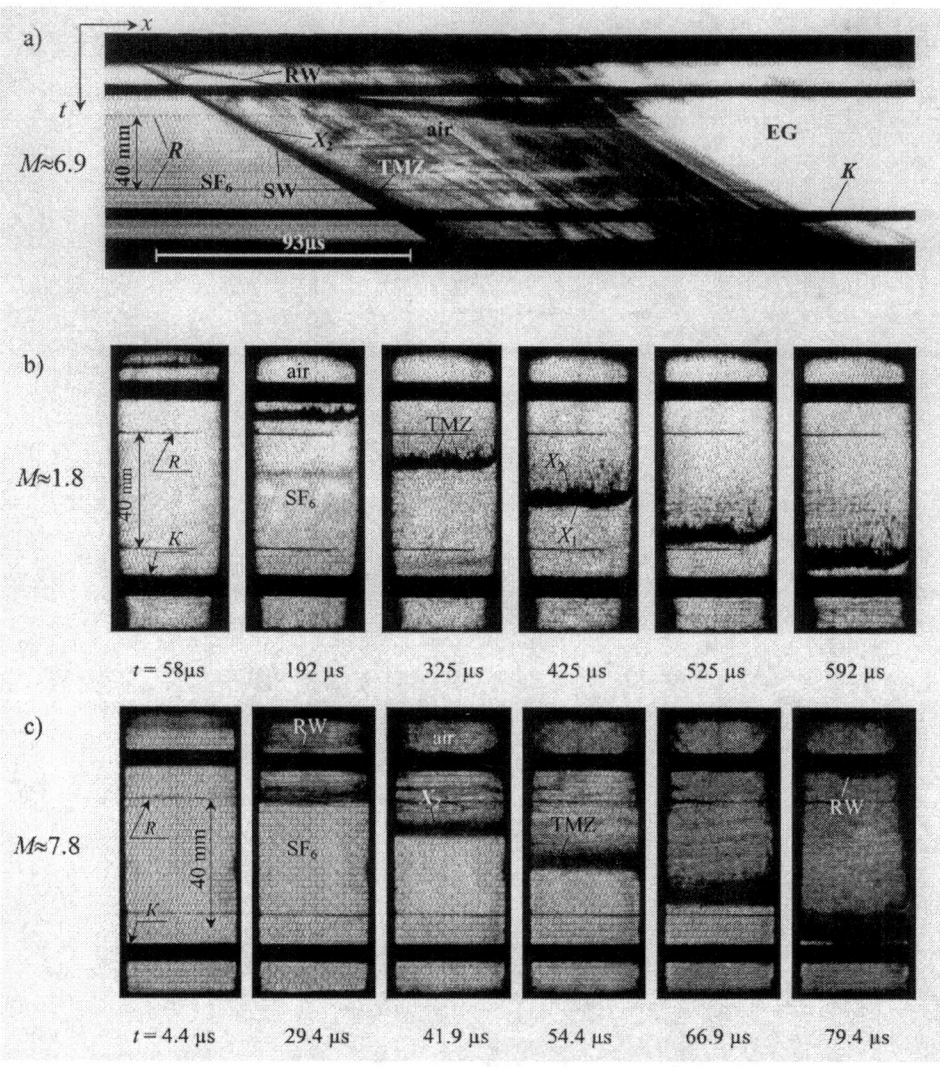

Designations: SW – shock wave; RW – reflected wave; EG – CGM detonation products; R - reference lines, K – structural element (out of flow); t – time is reckoned from the moment of coming SW to the contact interface.

FIGURE 3. Moving image frames of the experiments with TMZ development at the air/SF6 interface a) experiment №6 (A≈0.89) – slot moving image frame; b) experiment №8 (A≈0.83); c) experiment №7 (A≈0.89)

344

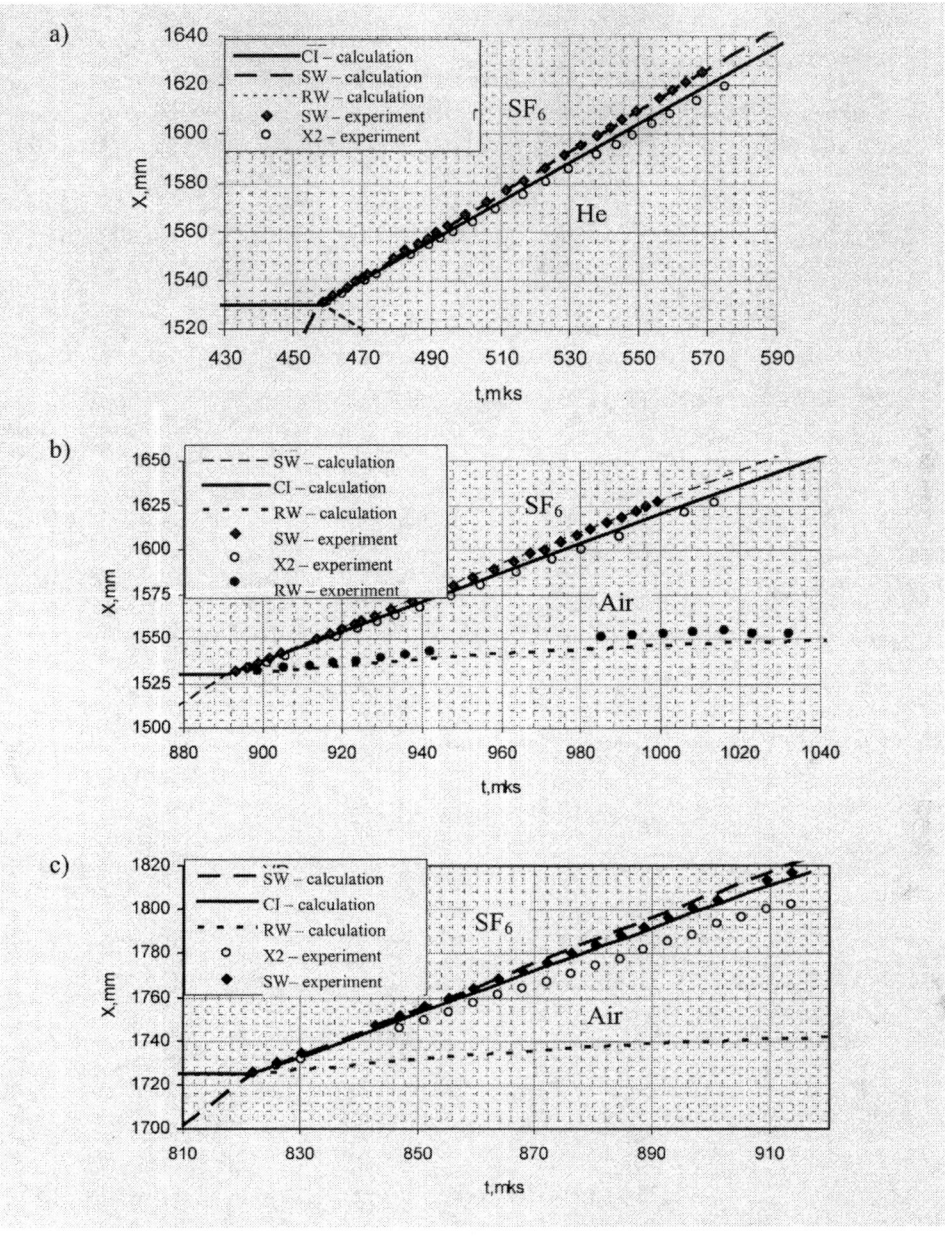

Designations: RW – reflected wave; t – time is reckoned from the beginning of CGM detonation; ordinate X is reckoned from the cap of the high pressure chamber.

FIGURE 4. X(t) diagrams of the flow in the shock tube. a) experiment №2 (He-SF6); b) experiment №6 (air-SF6); c) experiment №7 (air-SF6 L1=450 mm; L2=1,275 mm; CGM – C2H2+2.5O2 with P0=6,8±0,05 atm.)

From moving image frames and diagrams we notice that with increase in number M of SW in SF_6:

- contact interface of gases and, respectively, a forward front of TMZ approach a shock wave front (due to high compressibility of SF_6); at high M an optical gap is not observed between SW and a forward shock wave front of TM (X_1) in the experiments (s. Fig. 2b);
- optically observed thickness of TMZ increases

The proximity of the front of TM and SW may bring about their mutual influence yet to be studied in subsequent investigations.

CONCLUSIONS

Experimental technique was developed to study the turbulent mixing development at the gas-gas interface, which is accelerated by the shock wave with Mach number up to 9. As a consequence, the mixing zone thickness, which is observed optically, increases with increase in Mach number at the interfaces He-SF_6, Air-SF_6. It is seen that a forward front of TM approaches a shock wave. This is connected with SF_6 high compressibility. This circumstance can bring about mutual influence between SW and TM.

ACKNOWLEDGMENTS

The authors are indebted to E.E. Meshkov, V.A. Rayevsky, V.I. Kozlov, S.I. Gerasimov for helpful comments and proposals on this work. The authors are grateful to A.E. Egorushkina, O.L. Krivonos, V.I. Dudin, A.A. Nikulin for their assistance in the preparation and the performance of the experiments.

REFERENCES

1. G.I. Taylor. "The instability of liquid surfaces when accelerated in a direction perpendicular to their planes". *I. Proc.Roy.Soc.*, 1950, v.A201, p.192.
2. R.D. Richtmyer. Taylor instability in shock acceleration of compressible fluids. *Commun. Pure Appl.Math.*, 1960, v.13, p.297.
3. E.E. Meshkov Instability of the interface of both gases accelerated by shock wave. *NEWS AH Soviet Union, MZHG*, 1969, № 5, pp.151-158.
4. S.G. Zaytsev, E.V. Lazareva, V.V. Chernukha, V.M. Belyaev. Intensification of turbulent mixing at the boundary of media with different density as shock wave passes through the interface. *DAN Soviet Union*, 1985 , v. 283, №1, p.94-98.
5. E.A. Lasareva, A.N. Aleshin, S.V. Sergeev, S.G. Zaytsev and J.F. Haas. Shock-Induced Intensification of Turbulent Mixing. *The Proc of the 6th IWPCTM*, Marseille, France, 1997, p.295-300.
6. Kh.A. Rakhmatulina, S.S. Semionova. Shock tubes. M.: Foreign literature, 1962, p.699.
7. V.A. Andronov, V.I. Kozlov, V.V. Nikiforov, A.N. Razin, Yu.A. Yudin. Techniques for calculation of turbulent mixing in one-dimensional flows (VIKHR technique) // *Questions of Atomic Science and Engineering. Series: Mathematical simulation (modeling) of physical processes.* 1994, № 2, p.59.

SECTION 5

PROPERTIES OF MATTER
AT HIGH-INTENSIVE PROCESSES

COMPRESSION OF HIGH POROSITY ALUMINUM BY STRONG SHOCK WAVES

V.G. Vildanov, M.M. Gorshkov, V.M. Slobodenjukov,
A.O. Borshchevsky, A.V. Petrovtsev

Federal State Unitary Enterprise "Russian Federal Nuclear Center – Zababakhin All-Russian Research Institute of Technical Physics", P.O. Box 245, Snezhinsk, Chelyabinsk region, 456770, Russia

Abstract. Measuring results on shock compression of porous aluminum with initial density of $\rho_{00} = 0.6$ g/cm^3 up to pressures of 170 GPa are presented under shock wave velocity measurement scale of 40 mm. High underground explosion was used as a shock wave source. Obtained results were described in shock wave velocity (D) – particle velocity (u) coordinates by linear dependence of $D = 0.647 + 1.26\,u$ at $4.6 \le u \le 14.8$ km/s.

Keywords: shock compression, porous aluminum 0.6 g/cm^3, shock wave velocity, particle velocity.
PACS: 62.50.+p.

At laboratory conditions, increasing in shock compression pressure of porous specimens leads to measurement base decreasing. Moreover, necessary requirement to porous material specimens is that particles sizes have to be significantly lower than the measurement base. In this work, shock compression parameters of porous aluminum with initial density of $\rho_{00} = 0.6$ g/cm^3 have been measured in the pressure range of $\approx 18 \div 170$ GPa.

Porous aluminum specimens have been fabricated by FSUE VNIINM using technology which provided for powder *Al* content no less than 99 % and oxygen one no more than *0.9* % in mass. Particles size was less than 50 μ. For protection against oxidation, fabricated specimens were capsulated into *Al* container in argon medium. The specimens were of cylindrical form with sizes of $\varnothing \times H = 110 \times 40$ mm. They were mounted onto standard material screen, which was made of АД-1 aluminum alloy ($\rho_0 = 2.71$ g/cm^3) with sizes of $\varnothing \times H = 570 \times 125$ mm. Experimental units design is given in Figure 1.

Impedance match method [1] was used to obtain shock compression parameters of porous aluminum. For that, by electrocontact pins, shock wave velocities D in screen and in porous specimen were measured. When calculating of the velocities D in specimen and in screen, time dependence of electrocontact pins closing on pressure was taken into account, and small (<1%) computed correction of shock wave sphericity was introduced. When data processing, aluminum АД-1 shock adiabat has been taken in a table form from [2]. Mirror adiabat of P- u_{refl} (pressure – particle velocity) coordinates, with introducing of computed according to a equation of state (EOS) [3] correction Δu_{rel} was used as АД-1 release isentrope ($u = u_{\text{refl}} + \Delta u_{\text{rel}}$).

CP849, *Zababakhin Scientific Talks - 2005*,
edited by E. N. Avrorin and V. A. Simonenko
© 2006 American Institute of Physics 0-7354-0345-7/06/$23.00

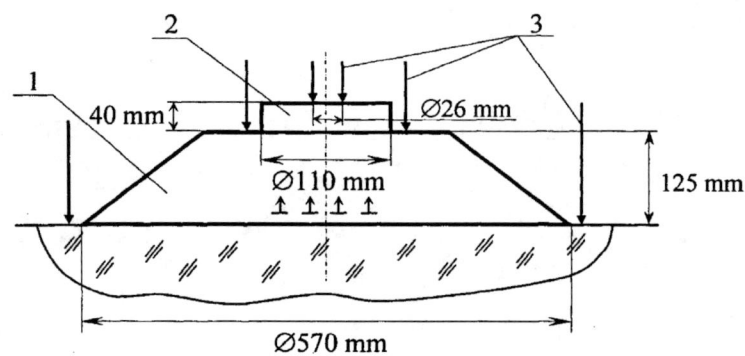

FIGURE 1. Design of experimental units. 1 – Standard material –АД-1 (aluminum alloy ρ_0 = 2,71 g/cm³), 2 – porous aluminum specimen ρ_{00}=0,6 g/cm³, 3 – electrical pins (3-4 pins at every level).

Experimental data on shock compression are presented in Table 1. Errors of shock compression parameters are given in the Table, as well. Measurement results of shock velocity in *Al* ρ_{00} = 0.6 g/cm³ were approximated by linear dependence on particle velocity.

$$D=0.647 +1.26u \quad \text{при } 4.6 \le u \le 14.8 \text{ km/s.} \tag{1}$$

Experimental data and dependence (1) approximating them are presented in Figure 2, at shock wave velocity (*D*) – particle velocity (*u*) coordinates. Data on shock adiabat of porous aluminum ρ_{00} = 0.35, 0.9 and 1.35 g/cm³ [4] and ρ_{00} = 1.66 g/cm³ [5] are given ibid. From Figure 2 it seems that shock adiabat of porous aluminum ρ_{00} = 0.60 g/cm³ was obtained in the region of more higher pressures as compared with laboratory experiments. However, its slope practically coincides with the slope of aluminum shock adiabat of ρ_{00} = 0.35 g/cm³, but the adiabat, ρ_{00} = 0.6 g/cm³, lies more gently with relation to other shock adiabats of the nearest densities of ρ_{00} = 0.9 and 1.66 g/cm³. In Figure2 there is also presented shock adiabat of aluminum ρ_{00} = 0.6 g/cm³ computed according to EOS [6]. It seems that computation and experiment practically coincide.

TABLE 1. Experimental data on shock compression of *Al* ρ_{00} = 0.6 g/cm³.

Shock wave velocity in screen D, km/s	Shock wave parameters in porous aluminum						
	D, km/s	Particle velocity *u*, km/s			Com-pression κσ	Density ρ, g/cm³	Pressure P, GPa
		u_{refl}	Δu_{rel}	$u = u_{\text{refl}} + \Delta u_{\text{rel}}$			
17.95± 0.17	19.242± 0.098	14.612	0.162	14.774± 0.24	4.30± 0.22	2.584± 0.13	170.6± 2.9
13.02± 0.08	12.260± 0.026	8.962	0.175	9.137± 0.18	3.93± 0.24	2.355± 0.14	67.2± 1.3
9.06± 0.11	6.415± 0.044	4.550	0.062	4.612± 0.18	3.558± 0.36	2.135± 0.22	17.75± 0.7

In Figure 3, experimental data are plotted in pressure (P) – density (ρ) coordinates. Shock adiabat of aluminum $\rho_{00} = 0.6$ g/cm^3 computed according to EOS [6] is given in the Figure 3, as well. It seems that computed shock adiabat and experimental data are practically coincide, their difference is within experimental error limits.

Thus, three experimental points of shock adiabat of porous aluminum with initial density of $\rho_{00} = 0.6$ g/cm^3 has been obtained. These new points expand the earlier data for close initial density by the order of magnitude (up to shock compression pressures of 170 GPa). The measurements were performed for aluminum specimens of high purity ($Al \geq 99$ %, $O \leq 0.9$ %), in conditions of loading by shock waves of high spatio-temporal homogeneity, that allowed to use large measurement bases (40 mm).

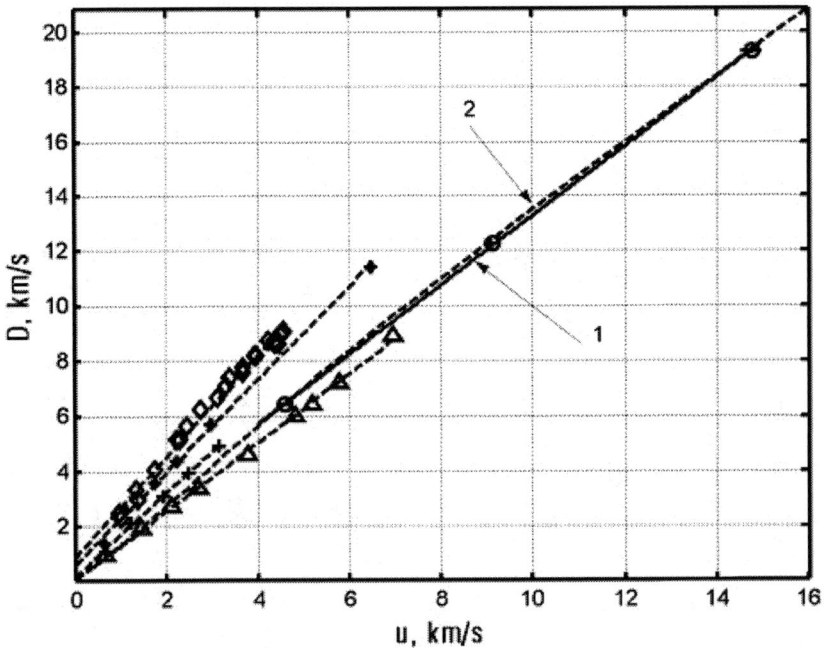

FIGURE 2. Shock adiabats of porous aluminum in $D - u$ (shock wave velocity– particle velocity) coordinates. Shock adiabat of porous aluminum, $\rho_{00} = 0.6$ g/cm^3: o – experimental data this work, 1– $D=0.647+1.26u$ linear approximation, 2– computed shock adiabat [6]. Experimental data on shock adiabat of porous aluminum: $\triangle$ –$\rho_{00} = 0.35$, + –$\rho_{00}=0.9$, * –$\rho_{00}=1.35$ g/cm^3 [4] and $\Diamond$ –$\rho_{00} = 1.66$ g/cm^3 [5]. Dashed lines – linear approximation of experimental data.

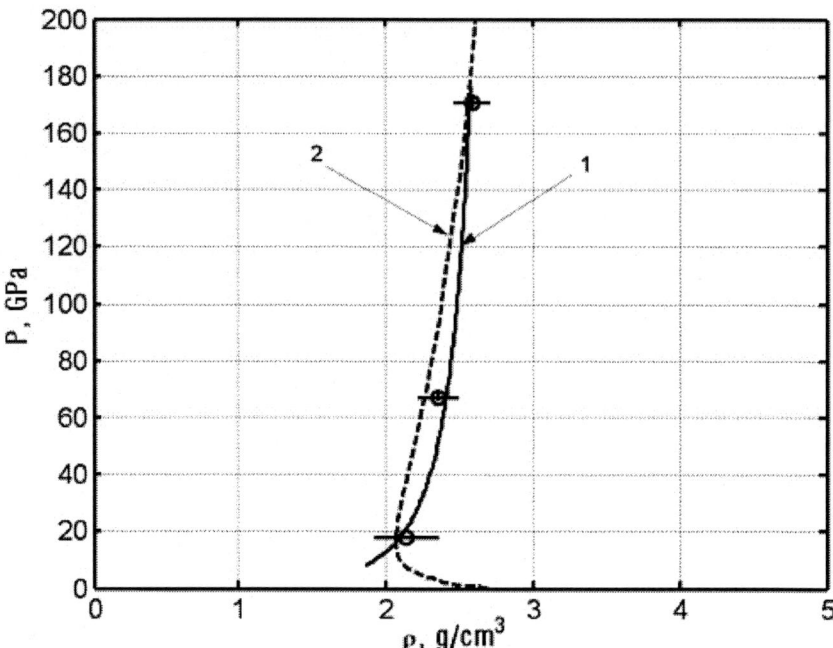

FIGURE 3. The shock adiabat of porous aluminum, ρ_{00} = 0.6 g/cm3, in $P - \rho$ (pressure– density) coordinates. o – Experimental data on shock adiabat of porous aluminum, ρ_{00} = 0.6 g/cm^3, this work. 1 – linear approximation D=0.647 +1.26u, 2 – computed shock adiabat [6]. Short lines – error bars.

REFERENCES

1. Ya. B. Zeldovich, Yu. P. Raizer, *Physics of Shock Waves and High-Temperature Hydrodynamic Phenomeno,*. Nauka, Moscow, 1963 (in Russian).
2. S. B. Kormer, M.V. Sinitzin, A.I. Funtikov, V.D. Urlin, A.V. Blinov, *JETP* **47**, issue 4 (10), 1202-1213 (1964) (in Russian).
3. V. F Kuropatenko, I.S. Minaeva, *Numerical Methods of Condensed Matter Mechanics* **18**, No. 6, (1983) (in Russian).
4. Experimental Data on Shock-Wave Compression and Adiabatic Expansion of Condensed Matter. edited by R.F.Trunin, RFNC-VNIIEF, Sarov, 2001 (in Russian).
5. LASL Shock Hugoniot Data, edited by Staney P, Marsh.University of California Press, Berkeley- Los Angeles- London, 1980.
6. A. T. Sapozhnikov, E. E. Mironova, "Equation of State for Aluminum Accounting Melting, Vaporization and Ionization" in *VIII Zababakhin Scientific Talks*, Snezhinsk, 2005.

Analysis of Typical Shock-Wave Experiments and Calculations of Thermodynamic Properties of Substances via Internet

P.R. Levashov, K.V. Khishchenko, I.V. Lomonosov

Institute for High Energy Densities RAS, Izhorskaya 13/19, Moscow 125412, Russia

Abstract. In this work we present the database on thermophysical properties of substances at high pressures and temperatures. The database contains experimental data, Hugoniot approximations and a number of caloric equations of state. A graphical Web-interface allows one to search data using different criteria and represent experimental points and calculation results in various forms including graphical. There is also an opportunity to simulate typical shock-wave experiments via the Internet using equations of state and approximations available in the database.

Keywords: shock-wave experiments, database, thermophysical properties, equation of state
PACS: 52.35.Tc, 64.30.+t

INTRODUCTION

Investigation of different processes at high pressures and temperatures usually requires an equation of state, which generalizes a lot of dissimilar experimental and theoretical data [1]. Recently created wide-range equations of state take into account melting, evaporation, sublimation, and ionization effects as well as the processes in metastable states of matter under both positive and negative pressures [1, 2]. Interpolation methods on a mesh adapted to phase and metastable region boundaries have been applied to simplify the incorporation of these equations of state into numerical codes [3]. We have used the tabular multiphase equations of state obtained in such a way in simulation of high velocity impact [4] and electrical explosion of wires [5].

Dynamic experiments on investigation of thermophysical properties of matter at high energy density almost always require an interpretation, as a direct measurement of most parameters experimentally is difficult. In particular, a temperature registration in shock-wave experiments is a complicated problem. Most often thermodynamic values such as adiabatic sound velocity or heat capacity required for the solution of some problems can be found only from thermodynamically complete equation-of-state calculation. Therefore it is reasonable to apply such equations of state to the interpretation of shock-wave experiments.

In this work we consider current possibilities of the database on thermophysical properties of substances at high energy density [6, 7] and its perspectives to simulate typical shock-wave experiments using multiphase equations of state. The database has been developing during the last decade and currently contains more than 20000 ex-

CP849, *Zababakhin Scientific Talks - 2005*,
edited by E. N. Avrorin and V. A. Simonenko
© 2006 American Institute of Physics 0-7354-0345-7/06/$23.00

perimental registrations on shock compression, isentropic expansion, double shock compression, adiabatic sound velocity measurements behind the shock front, isobaric expansion, and free surface velocity profiles of samples under shock loading. The database can be accessed via the Internet; all operations with data (selection, search, review, editing and treatment) are carried out remotely via the Internet with the help of common browsers by http://teos.ficp.ac.ru/rusbank/, http://www.ihed.ras.ru/rusbank/.

SIMULATION OF SHOCK-WAVE EXPERIMENTS

Recently we focus our efforts to the creation of interface for the simulation of shock-wave experiments by the collision and impedance matching techniques [7]. In the collision technique one accelerates a striker to some velocity W and hit it with a target [8]. As a collision result the shock wave having the velocity D is formed in the target. In the Web-interface we have provided for three variants of problem statement for the collision technique.

A1. Given are both values W и D as well as a shock Hugoniot or equation of state of striker. Determine the pressure P and the particle velocity U in shock-compressed target and striker, the density ρ and the specific internal energy E of target.

A2. Given are the striker velocity W and shock Hugoniots or equations of state of striker and target. Determine the shock wave velocity in the target D and the parameters P, U, ρ, and E in the target.

A3. Given are the shock wave velocity in the target D and shock Hugoniots or equations of state of striker and target. Determine the striker velocity W and the parameters P, U, ρ, and E behind the shock wave front in the target.

A Web-interface, worked out for the collision method, allows one to choose a substance and initial density of the striker and target as well as other parameters for calculations using A1–A3 variants. We have developed FORTRAN programming modules, which can be used to solve the discontinuity problem at any combination of initial conditions to a desired accuracy. The interface allows one to represent the results of simulation in a textual form or as a pressure–particle velocity (P–U) diagram. There are 4 types of shock-wave data approximations in the database and 4 models of caloric equations of state. Approximations and equations of state for the striker and target can be used in any combination; this possibility allows one to estimate the influence of the method of shock Hugoniot approximation to the experimental data interpretation.

In Fig. 1 the simulation result of the collision technique experiment is shown [9]. The aluminum striker (initial density $\rho_0 = 2.71$ g/cm^3) collides with the copper target ($\rho_0 = 8.93$ g/cm^3) and generates a shock wave with the velocity $D = 6.64$ km/s in the target. To simulate this experiment, for example, by variant A3 (the same experiment has been modeled by variant A1 in [10]) when the shock wave velocity in the target is only available we choose the equation of state [11] for both metals from 36 combinations of shock Hugoniot approximations for aluminum and copper. As a result of calculation the following parameters behind the shock wave has been obtained: $U = 1.771$ km/s, $P = 105$ GPa, $\rho/\rho_0 = 1.364$, and the striker velocity turned out to be $W = 5.49$ km/s. For comparison, in the work [9] the following values are given: $U = 1.82$ km/s, $P = 107.9$ GPa, $\rho/\rho_0 = 1.377$, $W = 5.6$ km/s. No wonder the disagreement is small, as the copper shock compressibility has been studied very well. Even more close parameters to the published ones in [9] can be obtained using shock Hugoniot approximations for aluminum and copper from the shock-wave compendium [12]. For instance, in this case the striker velocity 5.605 km/s practically coincides with the experimental value 5.6 km/s. This and other examples show that the results of treatment of shock-wave experiments are significantly determined by dependences $D(U)$, which are used for the interpretation; the cases of significant disagreement can be easily revealed for substances which properties are poorly investigated in shock-wave experiments.

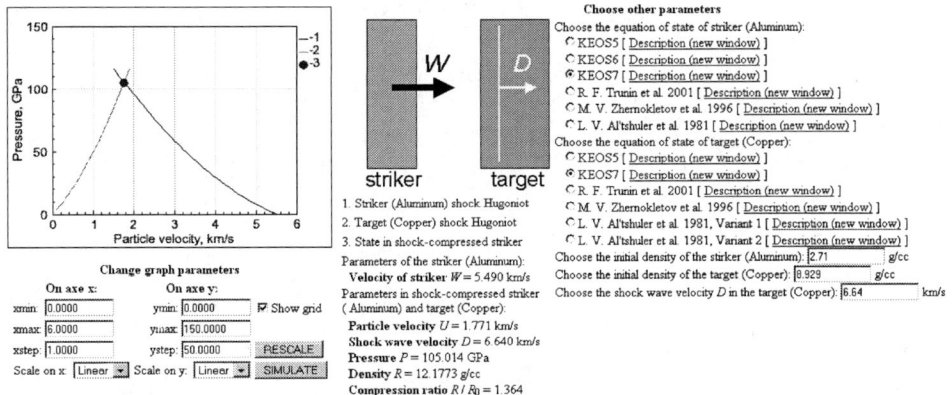

FIGURE 1. Simulation results (variant A3) of shock-wave experiment carried out by collision technique [9]. Designations in the diagram: 1 — striker shock Hugoniot (right curve), 2 — target shock Hugoniot (left curve), 3 — state in shock-compressed striker and target.

The impedance matching technique is applied far more often than the collision one for studying the shock-wave compressibility of matter. In this technique by some way one creates the shock wave in a screen behind which a sample is placed. In the database we have three variants of problem statement for this technique.

B1. Given are the shock wave velocity in the screen D_1 and in the sample D_2, and a shock Hugoniot approximation or equation of state of the screen. Determine the parameters P, U, ρ, and E in the sample behind the shock wave front.

B2. Given are the shock wave velocity in the screen D_1, and shock Hugoniots or equations of state of screen and sample. Determine the shock wave velocity in the sample D_2 and the parameters of sample P, U, ρ, and E behind the shock front.

B3. Given is the shock wave velocity in the sample D_2. Determine the shock wave velocity in the screen D_1 and the parameters P, U, ρ and E of the sample behind the shock front.

In contrast to the collision method, the treatment of experimental data by impedance matching technique in the general case requires not only the shock Hugoniot of the screen. It is also necessary to calculate by some way release isentropes and double shock Hugoniots originating from the state with pressure P_1 and particle velocity U_1; these curves, generally speaking, can be found only with the help of adequate equation of state models. However as an approximation of the curves mentioned above they often use the "mirror reflection" of original shock Hugoniot relative to the vertical line $U = U_1$ in the P–U diagram [12]. As there are both shock Hugoniot approximations and equation of state models in the database one can estimate the accuracy of such approach. The examples of simulation by the impedance matching technique can be found elsewhere [10, 13].

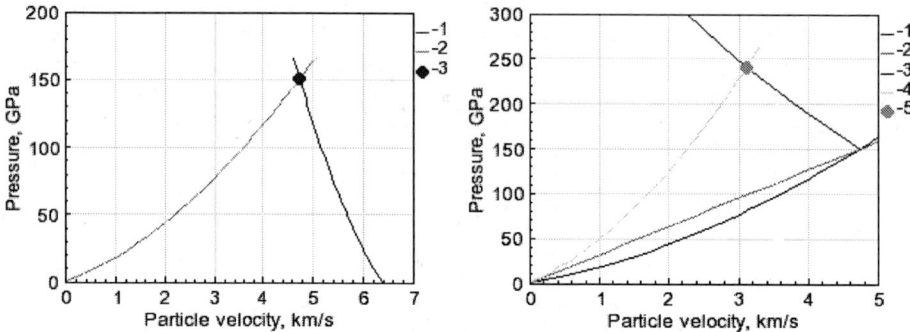

FIGURE 2. Pressure-particle velocity diagrams of experiment AlCu1 [14]. Left figure: 1 — striker shock Hugoniot (tantalum, right curve), 2 — sample shock Hugoniot (aluminum, left curve), 3 — state in shock-compressed striker and sample. Right figure: 1 — sample shock Hugoniot (aluminum, lower curve), 2 — sample wave ray (straight line), 3 — sample reshock, 4 — anvil shock Hugoniot (copper), 5 — state on the boundary aluminum-copper under sample double shock compression.

The interface for the simulation of shock-wave experiments allows the treatment of more complex experiments, e.g. on double shock compression. As an example we consider experiment AlCu1 from the work [14] (simulation of experiment CuAl1 is considered in [13]). In AlCu1 experiment a tantalum striker ($\rho_0 = 8.933$ g/cm^3) collides with an aluminum sample ($\rho_0 = 2.702$ g/cm^3) behind which a copper anvil ($\rho_0 = 8.93$ g/cm^3) is placed. The striker velocity is $W = 6.402$ km/s. The simulation of this experiment can be divided into two independent stages one of which is carried out by the collision technique and the other — by the impedance matching one. Therefore, to determine the parameters in the sample one should solve the discontinuity problem using A2 case (the tantalum shock Hugoniot has been calculated by the caloric equation of state [1], the aluminum one — by the equation of state [11]). The resultant shock wave velocity in the aluminum sample turned out to be 11.737 km/s, the particle velocity behind the shock front — 4.751 km/s (in [14] the value 4.747±0.012 km/s is reported). Further, knowing the shock wave velocity in the sample we can calculate its parameters under double compression by the shock wave reflected from the copper anvil. To do this we use variant B2 in which the aluminum sample is a screen and the copper anvil is a sample. Using the same equation of state model [11] for aluminum and copper we obtain the following parameters: the particle velocity of copper behind the shock wave or aluminum behind the reshock — 3.125 km/s (in [14] the value 3.130 ± 0.080 km/s is given), the pressure on the boundary aluminum–copper — 241.0 GPa (241.2 ± 8.4 GPa in [14]). It can be easily seen that the agreement with experimental values is excellent and discrepancies do not exceed the accuracy of measurements. The $P-U$ diagrams for the tantalum–aluminum and aluminum–copper discontinuities obtained with the help of the database interface via the Internet are shown in Fig. 2.

In the nearest future we plan to include multiphase thermodynamically complete equations of state in tabular form into the database. This step will allow users to obtain more information about the behavior of matter in different processes under high pressures and temperatures.

ACKNOWLEDGMENTS

This work is supported by Russian Foundation for Basic Research, Grants 03-02-16687 and 04-07-90310. The authors are also thankful to the Russian Science Support Foundation.

REFERENCES

1. A. V. Bushman, V. E. Fortov, and I. V. Lomonosov, "Wide-Range Equation-of-State Models for Matter" in *High Pressure Equations of State: Theory and Applications*, edited by S. Eliezer, R. A. Ricci, Amsterdam, North Holland, 1991, pp. 249–262.
2. K. V. Khishchenko and V. E. Fortov, "On Equation of State of Aluminum in Negative Pressure Region" in *Physics of Extreme States of Matter — 2002*, edited by V. E. Fortov et al., IPCP RAS, Chernogolovka, 2002, pp. 68–70.
3. P. R. Levashov and K. V. Khishchenko, "Tabular Equations of State Including Phase Transitions and Metastable Regions" in *Physics of Extreme States of Matter — 2004*, edited by V. E. Fortov et al., IPCP RAS, Chernogolovka, 2004, pp. 53–56.
4. M. E. Povarnitsyn, P. R. Levashov, and K. V. Khishchenko, "Simulation of High-Velocity Impact with Different Equations of State" in *Extreme States of Matter. Detonation. Shock Waves*, edited by A. L. Mikhailov, RFNC-VNIIEF, Sarov, 2005, pp. 577–582.
5. V. I. Oreshkin, R. B. Baksht, A. Yu. Labetsky, A. G. Rousskikh, A. V. Shishlov, P. R. Levashov, K. V. Khishchenko, and I. V. Glazyrin, *Tech. Phys.* **49**, 843–848 (2004).
6. P. R. Levashov, K. V. Khishchenko, I. V. Lomonosov, and V. E. Fortov, "Database on Shock-Wave Experiments and Equations of State Available via Internet" in *Shock Compression of Condensed Matter — 2003*, edited by M. D. Furnish et al., AIP Conference Proceedings 706, American Institute of Physics, Melville, NY, 2004, pp. 87–90.
7. P. R. Levashov, K. V. Khishchenko, I. V. Lomonosov, and V. E. Fortov, "Simulation of Typical Shock-Wave Experiments via Internet" in *Physics of Extreme States of Matter — 2004*, edited by V. E. Fortov et al., IPCP RAS, Chernogolovka, 2004. pp. 55–57.
8. Ya. B. Zel'dovich and Yu. P. Raizer, *Physics of Shock Waves and High-Temperature Hydrodynamic Phenomena*, New York: Dover Publications, Inc., 2002.
9. L. V. Al'tshuler, S. B. Kormer, A. A. Bakanova, and R. F. Trunin, *Sov. Phys.–JETP* **11**, 573–579 (1960).
10. P. R. Levashov and K. V. Khishchenko, "Tabular Multiphase Equations of State for Interpretation of Shock-Wave Experiments and Calculations of Thermodynamic Properties via Internet" in *Physics of Extreme State of Matter — 2005*, edited by V. E. Fortov et al., IPCP RAS, Chernogolovka, 2005, pp. 173–175.
11. K. V. Khishchenko, "Cold Curve and Caloric Equation of State for Copper" in *Physics of Extreme State of Matter — 2004*, edited by V. E. Fortov et al., IPCP RAS, Chernogolovka, 2004, pp. 45–48.
12. R. F. Trunin, L. F. Gudarenko, M. V. Zhernokletov, and G. V. Simakov, *Experimental Data on Shock-Wave Compression and Adiabatic Expansion of Condensed Substances*, Sarov: RFNC-VNIIEF, 2001.
13. P. R. Levashov and K. V. Khishchenko, "Application of Wide-Range Equations of State for Simulation of Typical Shock-Wave Experiments via Internet" in *Extreme States of Matter. Detonation. Shock Waves*, edited by A. L. Mikhailov, RFNC-VNIIEF, Sarov, 2005, pp. 234–239.
14. W. J. Nellis, A. C. Mitchell, and D. A. Young, *J. Appl. Phys.* **93**, 304–310 (2003).

TUR Software Package
for Constructing and
Investigating Equations of State

A.T. Sapozhnikov, E.E. Mironova

Russian Federal Nuclear Center – Zababakhin Institute of Applied Physics (RFNC-VNIITF),
Snezhinsk, Russia

The solution of problems in mechanics and high energy density physics requires equations of state (EOS) which differ in the applicability range, accuracy, mathematical form and computational efficiency.

RFNC-VNIITF developed a software package TUR which is used for constructing and investigating equations of state. Many RFNC-VNIITF workers contributed to its development, namely A.T. Sapozhnikov (principal investigator), A.B. Pershina, V.D. Dedova, G.V. Kovalenko, E.E. Mironova, E.L. Malyshkina, P.D. Gershchuk, L.N. Shakhova, N.K. Golubeva, E.P. Vakhrameyeva, L.V. Dyakina, M.E. Kotegova, T.E. ESkova, T.P. Rotko, I.N. Balandina, Y.V. Kaygorodtseva, M.S. Smirnova, V.V. Dryomov, F.A. Sapozhnikov, E.V. Pronina, and A.N. Krasnov.

TUR consists of four libraries, namely a library of special codes (SCL), a library of theoretical models (TML) describing thermodynamic material properties, a library of equations of state (EOSL), and a library of sets of constants (CSL) for equations of state for specific materials.

Special codes contained in the first library fall into three groups with respect to their purpose. Codes of group 1 are used for thermodynamic calculations with EOSs. These are

- KRAB-2 code which calculates thermodynamic functions (pressure, specific internal energy, heat capacity, volume expansion coefficient, Gruneisen coefficient, sound speed, entropy, free energy) along isolines (isotherm, isochore, isobar and constant energy line). It is necessary for the evaluation of accuracy with which experimental and theoretical data are reproduced.
- RKR-4 code which calculates critical parameters and vapor and liquid parameters in phase equilibrium. It is necessary for the evaluation of accuracy with which experimental data on vaporization are reproduced.
- RFD-2 code which calculates thermodynamic parameters for melting and polymorphic phase transitions. It is necessary for the evaluation of accuracy with which experimental data on melting and polymorphic phase transitions are reproduced.

CP849, *Zababakhin Scientific Talks - 2005,*
edited by E. N. Avrorin and V. A. Simonenko
© 2006 American Institute of Physics 0-7354-0345-7/06/$23.00

RKR-4 and RFD-2 calculate phase diagrams and provide information required for constructing EOSs for mixed phases in equilibrium. All the above codes are capable of automated mesh definition and generation. Uniform and log-uniform meshes are used.

Codes of group 2 are used for simple hydrodynamic calculations. These are

- UDAR code which calculates parameters of shock waves in continuous and porous materials and that of centered rarefaction waves. It is necessary for the evaluation of accuracy with which data from shock experiments are reproduced.
- DISK and DISK-2 codes treats experimental data on relative shock compression measured with the direct and back reflection method. They simulate shock propagation in two adjacent materials one of which has a known equation of state and is used as a reference material. The other material is studied. For back reflection, it is also necessary to have EOS for the study material, but its effect upon final results is small. The codes output parameters of the shock compressed material under study and their errors. They can be used to construct equations of state and to evaluate their accuracy.
- JUGE code which calculates hydrodynamic parameters in Jouguet point. It is necessary for the evaluation of accuracy of EOSs for explosion products.

Codes of group 3 are used to construct tabulated and analytical EOSs. These are:

- PODCON code which helps find optimal values for constants defined in analytical EOSs. These values are found through minimization of the sum of squared differences between the numerical values of thermodynamic functions from EOS calculations and the values obtained in experiment or evaluated from theoretical models. Input data (experimental or evaluated) include points of Hugoniots for continuous and porous materials, critical and vaporization parameters, values of thermodynamic functions on isolines, for example, experimental data on thermal expansion at ambient pressure.
- PTSOST code which calculates potential pressure and energy from experimental Hugoniots. The Hugoniot is presented as a linear relation between *material velocity* and *shock velocity*, $D = C_o + b \cdot U$, and Gruneisen coefficient as a function of density is defined from Landau-Slater, Dugdail-McDonald or Zubarev-Vashchenko models.
- SSHVK code which is used to construct wide-range EOSs by jointing local EOSs. Jointing is implemented through interpolation since it provides an optimal interpolation surface presented by a one-parameter family of curves, namely isochores and isotherms. The curves for different EOSs are jointed with cubic Hermitian polynomials. Jointing points are found from the condition of maximum proximity of jointed curves to the second derivatives of Hermitian polynomial, i.e., Hermitian polynomial must fit the curves to be jointed as well as possible.
- TDS-SSHVK code which is used to joint EOSs in a thermodynamically consistent way. Free-energy isotherms or isochores for different equations are jointed with the fifth-order Hermitian.
- TBLTR code tabulates thermodynamic functions of one or two variables using a user-specified mesh or a computer-generated optimal mesh with a least number of nodes providing the required accuracy of interpolation. Computer-generated and user-specified meshes may be uniform and log-uniform. In different tabulation

regions, thermodynamic functions may be described by different EOSs. The code has renewable libraries with types of arguments and functions, interpolations, optimization criteria, and types of test meshes which are used to calculate interpolation errors.

- FLOBER code smoothes functions of one or two variables by Fourier spectrum filtration. It is used to smooth numerical oscillations in initial functions, for example, in calculated data obtained with theoretical models. The code is equipped with visualization tools which help control smoothing quality by varying filter parameters.
- VSTDS-1 and VSTDS-2 codes check and recover thermodynamic consistency (TDC) of pressure and energy or their potential components by introducing relevant corrections, the residual in the TDC equation being distributed between pressure and energy in a specified proportion.

The library of theoretical models contains

- TFPK code which calculates thermodynamic parameters from Tomas-Fermi model with quantum and exchange corrections (TFP model) [1], with a model by Kopyshev [2] used to describe the contribution of thermal nuclei motion (TFPK model); the code was provided by its authors from RFNC-VNIIEF, namely V.P. Kopyshev and V.V. Khrustalyov;
- PLAZMA-4 code which calculates ionization equilibrium conditions in gases with Saha model [3]; the code was provided by N.N. Kalitkin;
- GESMGO code which calculates parameters of homogeneous and heterogeneous mixtures in thermodynamic equilibrium and parameters of heterogeneous mixtures of homogeneous components;
- A code which calculates properties of dense molecular fluids and their mixtures on the basis of Ross variational perturbation theory;
- PLLINTR code which calculates parameters of solids and liquids in melting with Lindemann law and Truton rule; it is used in cases where experimental data on melting parameters are lacking.

EOS Library contains equations of state of two types:

- Equations of purely scientific interest which cannot be used in codes because of some features, for example, too heavy on computer resources and time, but they can be used as a source of information on thermodynamic material properties for applied equations of state; and
- Applied EOSs, i.e., the equations that are continuous, satisfy Bethe-Weyl conditions at arbitrary densities and temperatures, and allow efficient implementation; they are used or can be used in continuous mechanics.

Figure 1 shows phase diagram and adiabatic curves for aluminum as an example of TUR calculations.

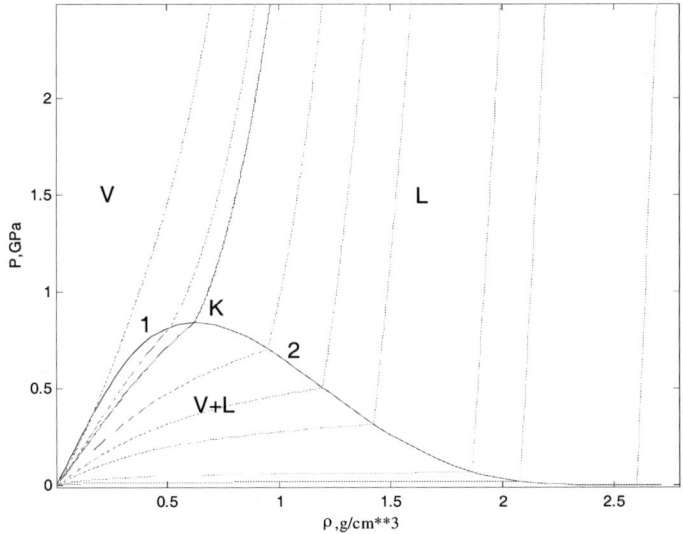

Figure 1 Phase diagram and adiabatic curves for aluminum; L – liquid, V – vapor, V+L – equilibrium
vapor-liquid mixture; 1 – condensation, 2 – vaporization; K – critical point

Figure 2 shows heat capacity of water in a wide range of densities and temperatures
obtained with a tabulated equation of state which describes vaporization, dissociation
and ionization [4].

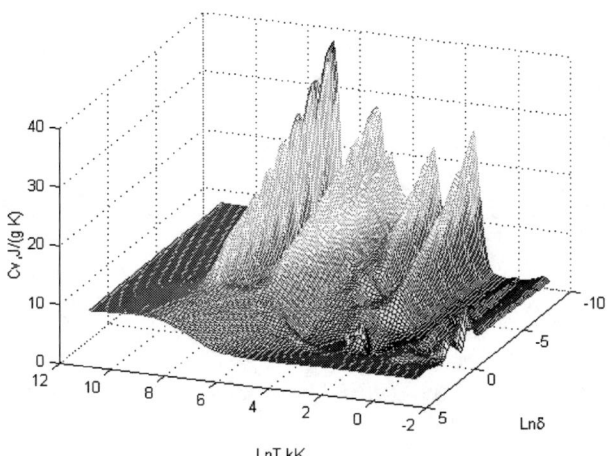

Figure 2 Heat capacity of water in a wide range of densities and temperatures

"Peaks, ridges and valleys" seen in the figure are the effects of hydrogen bond
breaks, excitation of molecular oscillations, dissociation and ionization.

361

Figure 3 shows isotherms of the function $\Pi = P_T / \rho \cdot T$ before (numerical oscillations are seen in some isotherms) and after smoothing (these data can be used, for example, for jointing). P_T is thermal pressure.

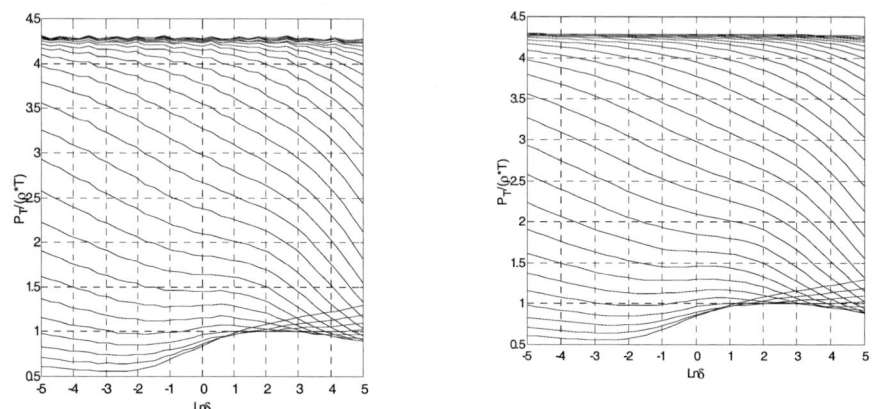

Figure 3   Non-smoothed and smoothed isotherms of the function

$$\Pi = P_T / \rho T \text{ obtained with the TFPK model.}$$

For ease of use, an interactive service shell based on state-of-the-art technologies was developed for the code; it implements the appropriate set of code functions through a window interface. The shell includes a database containing information on equations of state and sets of constants, a system to define input data and start calculational modules, and auxiliary tools to store results and construct graphs. The shell was developed by O.V. Verbitskaya, O.V. Kuznetsova and V.P. Sokolov.

TUR software package is a suitable and effective tool for constructing and investigating equations of state.

REFERENCES

1. Kalitkin, N.N., Kuzmina, L.V., "*Tables of thermodynamic functions for high energy densities*" // Preprint No. 35, Moscow, Institute of Applied Mathematics of USSR Academy of Sciences, 1975.
2. Kopyshev, V.P., "*Thermodynamics of monatomic nuclei*" // Numerical Methods of Continuum Mechanics, 1977, V.8, No. 6, p.54–67.
3. Kalitkin, N.N., Ritus, I.V., Mironov, A.M., "*Ionization equilibrium with electron degeneracy (Plazma-4)*", Preprint No. 46, Moscow, Institute of Applied Mathematics of USSR Academy of Sciences, 1983.
4. V.V.Dremov, A.T.Sapozhnikov, M.A.Smirnova, "Wide Range Equation Of State Of Water Taking Into Account Evaporation, Dissociation And Ionization", in *Shock Compression of Condensed Matter –2003* edited by M. D. Furnish, Y.M. Gupta, and J. W. Forbes, p.49.

Novel Technologies in Microfusion

Yu.A.Merkuliev[1], A.A.Akunets[1], N.G.Borisenko[1], N.A.Chirin[2],
V.M.Dorogotovtsev[1], V.V.Gorlevsky[2], A.I.Gromov[1], Yu.E.Markushkin[2],
V.G.Pimenov[3], P.A.Storozhenko[4], R.A.Svitsin[4], A.S.Vorontsov[2],
A.V.Zabrodin[2]

[1]Lebedev Physical Institute, 53 Leninsky pr., 119991, Moscow, Russia.
[2]Bochvar Institute of Inorganic Materials, 5a Rogova st., 123060, Moscow, Russia
[3]Zelinsky Institute of Organic Chemistry, 49 Leninsky pr., 119991, Moscow, Russia.
[4]GNIICHTEOS, 38 Shosse Entusiastov, 111123, Moscow, Russia

Abstract. The developed method of foaming for spherical shells made from metastable hydrides such as BeD_2 or ND_3BD_3 is discussed. The importance of low-density state of meta-stable hydrides due to the higher energy transfer rates is discussed. The properties of fuel - materials with high content of hydrogen isotopes, which have substantial importance for large fusion target fabrication technology and neutron generation in strong e.m. fields are analyzed. New laser targets and their characteristics are demonstrated.

Keywords: laser targets, new materials, neutron yield
PACS: 52.38.-r; 52.50.Jm; 52.57.Bc

1. INTRODUCTION

Solid materials with high content of hydrogen isotopes (for example BeD_2, $LiBeD_3$ and ND_3BD_3 or the same these substances with tritium) can be used as a wall of large (reactor-scale) fusion target instead of beryllium or polyimide. The burning reactor-size targets are shown to be profitable [1,2] as regards energy yield. An other variant of burning target is called "All DT" [3] have shell wall of polymer foams filled with solid DT-mixture (type: $CH(DT)_4$ or $CH(DT)_{64}$). Shells from BeDT or NT_3BD_3 can be used as surrogate cryogenic targets in experiments with large-scale lasers, Z-pinches or heavy ion drivers, when expensive DT cryogenic systems are not installed in interaction chamber. The targets from these materials are also used in the neutron generation research in super high intensity laser fields [4]. Low-density BeD_2 or $LiBeD_3$ foams layers can be used as absorber of laser radiation and for fast heat transfer on shell-target [5].

2. SOLID COMPOSITIONS OF LIGHT ELEMENTS WITH HYDROGEN AND SHELL FABRICATION

Basic properties of light elements' hydrides, needed for computer simulation of plasma heating, compressing and burn, are presented in Table 1.

CP849, *Zababakhin Scientific Talks - 2005*,
edited by E. N. Avrorin and V. A. Simonenko
© 2006 American Institute of Physics 0-7354-0345-7/06/$23.00

TABLE 1. Physical characteristics of the alternative solid fuels which are needed for heating and burn plasma simulations.

Properties	DT	LiBeD$_3$	LiBD$_4$	BeD$_2$	(C$_8$D$_8$)$_n$	ND$_3$BD$_3$	ND$_2$BD$_2$	(CD$_2$)$_n$
Density, g/cm^3	0.25	0.82	0.77	0.765	1.15	0.88	1.06	1.10
N$_i$ - ions concentration, 10^{23}cm^{-3}	0.5	1.12	1.065	1.06	0.97	1.14	1.1	1.23
I-average (Z+1) of fuel ions	2	5	4.5	4.5	9	4.33	5.5	5.5
I$_f$/I$_{DD}$	1	2.5	2.25	2.25	4.5	2.17	2.75	2.75
N$_i$ /N$_{DD}$	1	1.67	1.5	1.5	2	1.33	1.5	1.5
Y$_{DD}$/Y$_i$	1	41.7	25.6	25.6	390	20	51.8	51.8
Surface roughness, nm	300			10	3	50	20	
Compressibility, GPa^{-1}	5			0.05	0.15			

The parameter Y$_{DD}$/Yi in Table 1 indicates neutron yield decrease in situation where deuterium is substituted by the composition of light elements with deuterium (for example, BeD2, LiBeD3 or ND$_3$BD$_3$). The relation takes into consideration plasma temperature drop (neutron yield is proportional to temperature in power 7/2), and neutron yield decrease because of different ions adding. This approximate correlation is valid for DD and DT reactions at temperature up to 20 keV:

$$Y_{DD}/Y_i = (I_f/I_{DD})^{7/2}(N_i/N_{DD}) \qquad (1)$$

Of course, this correlation is only qualitative. We do not take into account the increasing of target mass and consequently the decreasing of shell thickness. The later will led to the decreasing of neutron yield but we do not consider it here. The data in Table 1 shows, that shifting from pure deuterium to the light elements deuterides will reduce the neutron yield by a factor of 20-40. But adding the 15%-20% of tritium to such composition will result in increasing of neutron yield by a factor of 20-30. Additionally, the energy of resulting neutrons and α-particles will be greater and this is important for plasma diagnostic. Of course, for adequate numerical simulations of heating and implosion processes and comparison with laser experiments one need to know the sound velocity and compression parameters of ammonia-borane and other meta-stable hydrides (see Table 1). But nevertheless it is clear, that such compositions are good alternative for surrogate cryogenic targets and the technology of shell fabrication is worth to develop. For such technology one of the most important issues is the physical parameters influencing the processes of the shell formation in drop-tower furnaces.

Beryllium (deuteride) hydride is meta-stable composition, it can not be produced directly as Be+H$_2$→BH$_2$ at room temperature or at higher temperatures. BH$_2$ starts to decompose slowly at 50°C and excrete intensively at 120°C. Dr. Yu. E. Markushkin with colleagues used original method of direct low-temperature syntheses [6] for beryllium hydride preparation. The method allows producing of beryllium deuteride as thin film – coating on special substrates. BeD$_2$ and BeDT coating on the inner surface of copper shells (diameter of 4 mm) with holes for laser beams were applied as laser targets in inverse corona experiments [7]. Flat films with thicknesses of 5-12 microns and plate with thickness of 0.6-0.9 mm from BeD$_2$ were used as targets in experiments on ps-lasers "Progress-P" and "Neodim" [4].

Usually beryllium hydride and beryllium deuteride does not oxidize in air at room temperature and does not interact with water and neutral solution. Pure BeD_2 is optical transparent material, but Be admixture adds color to the material.

Ammonia-borane is a white crystalline substance melting with dissociation at a temperature of 125oC. In GNIICHTEOS we have synthesized ammonia-borane samples of high purity (basic matter content 99.8% minimum that is proved by chemical analysis and 11B NMR spectra). Now GNIICHTEOS starts to produce the batches of ND3BD3. Density of ND3BD3 is 0.88 g/cm3 while deuterium content is 0.285 g/cm3. The investigated samples were obtained by diborane direct interaction with ammonia in ether environment: $B_2H_6+2NH_3 \rightarrow 2NH_3BH_3$ and purified by side-product hydrolyses and the following multiple recrystallization from solvents.

Thermal stability (in terms of gas emission) is determined by product purity. Thus ammonia-borane, containing 99.8% of basic matter, at a temperature of 20-25oC does not demonstrate gas emission for four months. At the same time a product sample with basic matter content ≤98.5% at the same temperature evolves hydrogen in amount of 0.6% of theoretically expected, but then gas emission has been terminated and was not observed during further 50 days of exposure. We have also found the conditions for fast isotope exchange (H↔D and D↔T) in ammonia-borane (about 10% during 6 hours). Plate from BH3NH3 turned into amorphous or amorphous-crystalline foams of BH3-xNH3-x at fast heating. The most important physical properties of commonly used for shell fabrication glass and polystyrene compared to ones for hydrides are shown in Table 2.

TABLE 2. The physical properties of substances used for target shell fabrication.

Parameter	Glass	LiBeD3	LiBD4	BeD2	(C8D8)n	ND3BD3	ND2BD2
Density, g/cc	2.5	0.82	0.77	0.765	1.15	0.89	1
Elastic modulus, GPa	60-70			27.3	2.5		
Melting (vitrifying) temperature, °C	(400)			(140)	(85)	104	150 (170)
Boiling (fast decomposition) temperature, °C				130 (350)	(570)	107 (320)	(470)
Furnance hot zone temperature, °C	1600			700	800-850	400-600	600-750
Viscosity at formation temperature, Pa·s	0.1-10			>1000	10÷1000	>100	>500
Surface tension, J/m²	0.2				0.02		
Permeability for H2, cm³(normal)·cm/cm²/atm/s, 10^{-10}	10^{-4}			<0.01	650	<50	<20
Roughness of surface, nm	3			10	3	50	30
State (amorphous or crystal)	amor	crys am-cr	crys	amor	amor	crys am-cr	amor

New technology of hollow microspheres fabrication from meta-stable hydrides is basing on LPI original high-temperature method of polymer shell formation from solid granules in vacuum drop-tower furnace [8] (see Figure 1a). Temperature of hot zone in drop-tower furnace is $300^\circ C$ higher than temperature of instantaneous decomposition of polymers. Granules of appropriate material are dropped through the hot zone in 0.5-2 seconds and then cooled during 0.3-1 second. As only small amount of particles matter dissociates through thermal abundance, the particles' temperature does not increase higher than temperature of instantaneous decomposition [9].

Before starting the work with beryllium we had tried to fabricate shells of ammonia-borane (see Figure 1b). The experiments show that crystalline H_3NBH_3 transformed to amorphous or amorphous-crystalline $H_{3-x}NBH_{3-y}$.

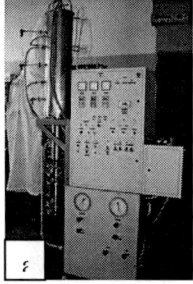 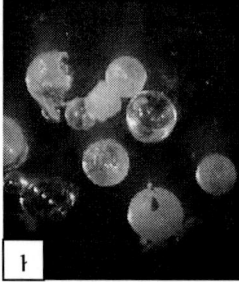

FIGURE 1. a) Automatic vacuum installation with 3 hot zones for formation of BeD_2 and $LiBeD_3$ shells; b) first shells from NH_3BH_3 fabricated on 12.04.03.

Basic distinctions between the meta-stable hydrides and polymer are connected with different dissociation temperatures (450-600°C for polymers and 100°C for the hydrides) and viscosity at temperatures higher then melting point (for the polymer viscosity is very high, for the meta-stable hydride it drops quickly). During our work, we realized the possibility to produce shells from meta-stable hydride in regime "fast heating – fast cooling".

3. LOW-DENSITY DEUTERIDE MATTER AND TARGETS FOR PS-LASERS

In our experiments [5] on picosecond lasers we primarily used plates or films from deuterated polymers. Soon we noticed, that in almost similar irradiation conditions utilizing similar targets the neutron yield varies. We start to analyze the shape and dimensions of craters and found, that with the nearly constant energy (measured with calorimeters) the shape and dimensions of craters vary (see Figure 2a). Preliminary explanation for this is unstable contrast and shape of prepulse of laser used.

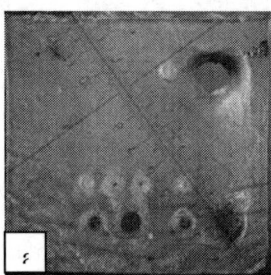 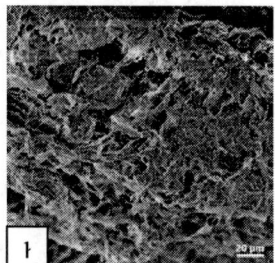 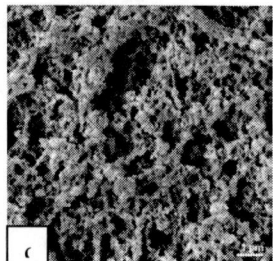

FIGURE 2. a) Target – CD_2 plate with thickness 350 μm after 10 shots; b) $(CD_2)_n$ foams, density 20 mg/cm³ scale 20 μm; c) deuterated epoxy resin foam, density 5 mg/cm³, scale 1 μm.

It was proved in simulation and was shown in experiment with intensive laser fields [5], that the neutron yield from a multi-layer deuterated polymer target is higher than that from a flat plate with thickness equal to thickness of multi-layer target. Of course, foams from deuterated material are similar to multi-layer targets and can be use in experiments with ps-(fs-)-lasers.

The samples of $(CD_2)_n$ foams with densities 10, 20, 30 and 40 mg/cc were used as targets for "Neodim" laser [5]. The foam has a close pore structure (see Fig. 2b). In published papers we failed to find the data concerning the fabrication of foam layers from $(CD_2)_n$ with densities lower than 50 mg/cc. Besides, the deuterated epoxy resin in the form of low-density (up to 4 mg/cm^3) 3D-network was synthesized (see Fig. 2c) by V.G. Pimenov (IOC RAS). Also the films from BeD_2 with thicknesses of 5-7 μm were used as single-layer of multi-layer targets. The multi-layer targets consisted of plates separated with 30-50 μm free space. The thick plates (0.7-0.9 mm) of BeD2 also were irradiated. The major achievement in development of fabrication methods of ultra-thin (0.1-0.2 μm) films from BeD_2 was made by N.A. Chirin (Bochvar Institute of Inorganic Materials). The development of such deuterated material allows us to propose the new target designs – models of ultra low-density foams, cavities filled with "BeD_2 gas" which could be irradiated in vacuum without utilizing protecting windows and so on.

4. CONCLUSION

New fabrication technologies of large fusion targets from deuteride-tritide light elements allow to organize the experiments with high gain by energy or the researches with surrogate cryogenic targets. Fabrication technologies of shell-targets (tritium system for DT filling shells, systems for collection and cleaning DT-mixture, D↔T isotopes exchanges in polymers and hydrides, cryogenic installations with DT-mixture etc.) give possibility to design unique pulsed neutron sources using powerful ps-lasers with optical pumping. Relatively cheap target system from deuteride-tritide light elements with apparatus of isotopes exchange can be used in experimental installations where there is needed a optimization targets-drivers (Z-pinches, heavy ions accelerators etc.).

REFERENCES

1. S. Yu. Gus'kov, N. V. Zmitrenko, Yu. E. Markushkin, Yu. A. Merkul'ev, Moscow, Preprint Lebedev Physical Institute, 2001, No 20, 45 p.
2. S.A. Bel'kov, G.V. Dolgoleva, G.G. Kochemasov, E.I. Mitrofanov., Quantum electronics. (Russian), 2002, V. 32, No 1, pp. 27-32.
3. S.E. Bodner, D.E. Colombant, A.J. Schmitt, M. Klapisch. High-gain direct-drive target design for laser fusion. // Physics of plasmas, (2000), Vol. 7, No 6, 2298-2301.
4. V.S. Belyaev, A.P. Matafonov, V.I. Vinogradov, et al. // IFSA 2005,
5. A.M. Khalenkov, N.G. Borisenko, V.N. Kondrashov, et al., Laser and Particles Beams (in press).
6. Yu. E. Markushkin, N. A. Chirin., J. Moscow Phys. Soc. Vol. 9, №1, (1999), pp. 76 - 81.
7. Yu.A.Abramov, A.V. Bessarab, A.V. Veselov, et al., Quantum electronics. (Russian), 1994, V. 21, No 2, pp. 155-157.
8. A.A. Akunets, V.M. Dorogotovtsev, Yu.A. Merkuliev, et al., Fusion Technology, 1995, Vol. 23, (11), pp. 1872-1877.
9. Yu.A. Merkuliev, S.A. Startsev., Fusion Technology, 1997, Vol. 31, (6), pp. 418-423.

Molecular Dynamics Modeling of Thermal Properties of Aluminum Near Melting Line

A.V. Karavaev, V.V. Dremov, F.A. Sapozhnikov

Russian Federal Nuclear Center – Zababahin Institute of Technical Physics, Snezhinsk, Russia

Abstract. In this work we present results of calculations of thermal properties of solid and liquid phases of aluminum at different densities and temperatures using classical molecular dynamics with EAM potential function. Dependencies of heat capacity C_V on temperature and density have been analyzed. It was shown that when temperature increases, heat capacity C_V behavior deviates from that by Dulong-Petit law. It may be explained by influence of anharmonicity of crystal lattice vibrations. Comparison of heat capacity C_V of liquid phase with Grover's model has been performed. Dependency of aluminum melting temperature on pressure has been acquired.

Keywords: Aluminum, heat capacity, anharmonicity, molecular dynamics.
PACS: 02.07.Ns,62.50.+p

INTRODUCTION

According to Debye theory of solid state specific heat capacity C_V is a function of θ_D / T (where θ_D – Debye temperature)

$$C_V = 3N v k_B \left[D(\theta_D / T) - (\theta_D / T) \cdot D'(\theta_D / T) \right], \tag{1}$$

where N is a number of unit cells, v is a quantity of atoms in one unit cell and D is Debye function

$$D(x) = \frac{3}{x^3} \int_0^x \frac{z^3}{e^z - 1} dz . \tag{2}$$

In the limit cases of high ($T \gg \theta_D$) and low ($T \ll \theta_D$) temperatures using equation (1) Dulong-Petit law and Debye law of heat capacity C_V can be obtained. According to Dulong-Petit law the molar heat capacity C_V is a constant $3R$ when the temperature $T \gg \theta_D$. When the temperature $T \ll \theta_D$ heat capacity is proportional to T^3.

Dulong-Petit law of heat capacity can be obtained using law of equidistribution of oscillation energy on degrees of freedom. According to that law the energy which is accounted for each degree of freedom is $k_B T$. However in practice deviations of experimental data from Dulong-Petit law are observed. These deviations are linked

CP849, *Zababakhin Scientific Talks - 2005*,
edited by E. N. Avrorin and V. A. Simonenko
© 2006 American Institute of Physics 0-7354-0345-7/06/$23.00

with anharmonicity of lattice vibrations and dispersion of acoustic phonons due to discrete structure of crystal.

In this work we present results of calculations of thermal properties of solid and liquid phases of aluminum at different densities and temperatures using classical molecular dynamics with EAM (Embedded Atom Model) potential function [1]. Dependencies of heat capacity C_V on temperature and density of model have been analyzed. It was shown that when temperature increases, heat capacity C_V behavior deviates from that by Dulong-Petit law. It may be explained by influence of anharmonicity of crystal lattice vibrations. Comparison of the dependency of the heat capacity C_V of the liquid phase on temperature with Grover's model [2] has been performed.

METHOD OF INVESTIGATION

While Dulong-Petit law formulation the law of equidistribution of oscillation energy on degree of freedom was used. In particular the fact that energy which accounted for each degree of freedom is $k_B T$ ($\frac{1}{2} k_B T$ is an average value of kinetic energy and $\frac{1}{2} k_B T$ is an average value of potential energy). This statement is correct only if the return force F which is acting on an atom is proportional to displacement of atom from the equilibrium position δx ($F = -C\delta x$, where $C = const$). However even in the cases when the temperature is not so high (it means that the displacements are also not so big) deviations of this law from experimental data are observed.

Let consider the model linear chain of atoms in which atoms interact according to the law by Lennard-Johns

$$\varphi_{ij}\left(r_{ij}\right) = 4\varepsilon\left(\left(\frac{\sigma}{r_{ij}}\right)^{12} - \left(\frac{\sigma}{r_{ij}}\right)^{6}\right). \tag{3}$$

In Figure 1 dependence of potential energy of atom number 2, which is located in point of origin, at the position x. One can see in Figure 1 (a) that when $r = r_0$ dependence of total potential energy of atom number 2 U on displacement from equilibrium position is quite well approximated by square function. It means that in this case return force $F_x = -\partial U / \partial x$ in a wide range can be described by linear function ($F_x = -Cx$, where C is a constant value in a wide range of displacement x). In this case indeed the average values of potential and kinetic energies are equal to $\frac{1}{2} k_B T$. And molar heat capacity C_V comply with Dulong-Petit law. Taking into account one dimension of the chain molar heat capacity is $C_V = Nk$.

Now let reduce linear density of our crystal by the value of 10%. One can see in Figure 1 (b) that in this case some neighborhood near equilibrium position $x = 0$ exists where insignificant increase of total potential energy of interaction takes place while displacement is considerable. It means that when the temperature is small the average value of potential energy is lower than $\frac{1}{2}k_BT$. In the limit case of rectangular potential well $(U(x)=0$, if $x\in[-x_0,x_0]$ and $U(x)=\infty$ if $x\notin[-x_0,x_0])$ the average value of potential energy of atom placed in this well is equal to 0 although the average value of kinetic energy as before $\frac{1}{2}k_BT$. According to this argumentation one can expect that in the states with reduced densities at low temperatures value of heat capacity C_V is lower than that predicted by Dulong-Petit law but in the cases of increased densities heat capacity increases. Certainly pair potential function of Lennard-Johns too simple for description of thermal properties of real materials however it can be used for receiving of qualitative assessment of dependencies of heat capacity C_V on density and temperatures when the temperatures are higher than Debye temperature and all oscillations are excited.

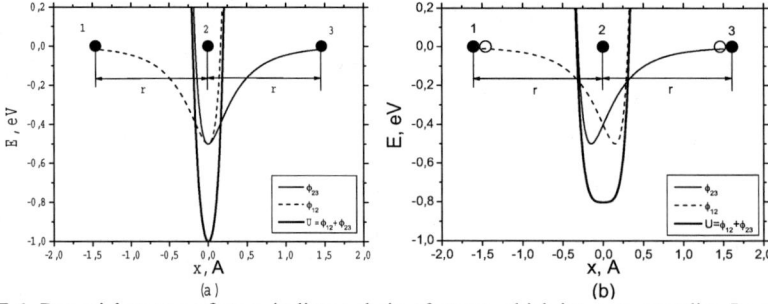

(a)   (b)

FIGURE 1. Potential energy of atom in linear chain of atoms which interact according Lennard-Johns law (a) $r=r_0$, (b) $r=r+\Delta r$.

So the goal of this work was investigation of dependencies of value of nonlinear contribution (anharmonicity of lattice vibrations) to specific heat capacity C_V of metal on temperature and density. As object of this research was taken aluminum because it has relatively low in comparison with melting temperature Debye temperature. And also it has simple phase diagram. Calculations were performed using classical molecular dynamics method with EAM (Embedded Atom Model) potential [1]. Calculation cells consist of 32,000 (20x20x20 unit cells) or 108,000 (30x30x30 unit cells) atoms. For investigation of behavior of heat capacity calculation of isochors at various densities ($\rho=0.9\rho_0$, $\rho=\rho_0$, $\rho=1.5\rho_0$, $\rho=2.0\rho_0$) were performed. For obtaining computation points NVT-ensembles at various temperatures and densities were modeled. In the initial moment the model was an ideal crystal structure with appropriate density. Then velocities of atoms were drawn with Maxwell distribution. After that molecular dynamics modeling was performed until total energy and pressure in system became stable. Duration of modeling in time scale of model was several tens of picoseconds. Sufficient duration of numerical experiment near the melting point is extremely important. We can talk about melting if we observe jerky increase of pressure in the system which correspond to phase transition from solid state to liquid. From the beginning pinpointing the melting line was not main goal of this work. Special methods for determination of melting points exist in the context of classical molecular dynamics modeling.

In Figure 2 (a) dependencies of pressure of system on temperature at various densities are presented. These dependencies were obtained by using classical molecular dynamics (without taking thermal excitation of electrons into consideration). Melting temperature at the given density is an abscissa of the curve break. In Figure 2 (b) dependencies of specific total energy on temperature at various densities are shown. These results were approximated and then acquired dependencies were differentiated for obtaining the dependencies of specific heat capacity C_V on density and temperature. They are shown on Figure 3.

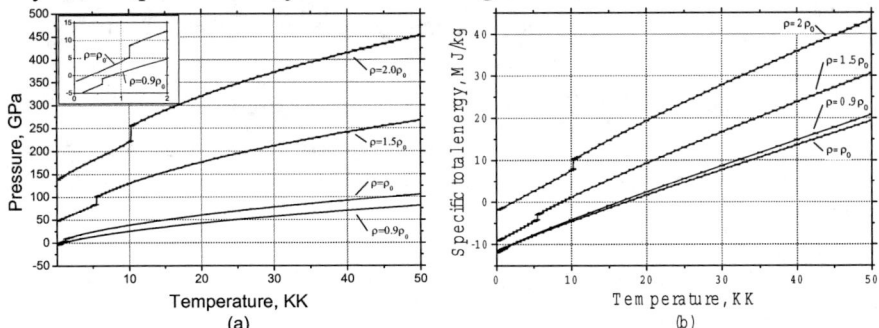

FIGURE 2. Dependencies of pressure (a) and specific total energy (b) on temperature at various densities.

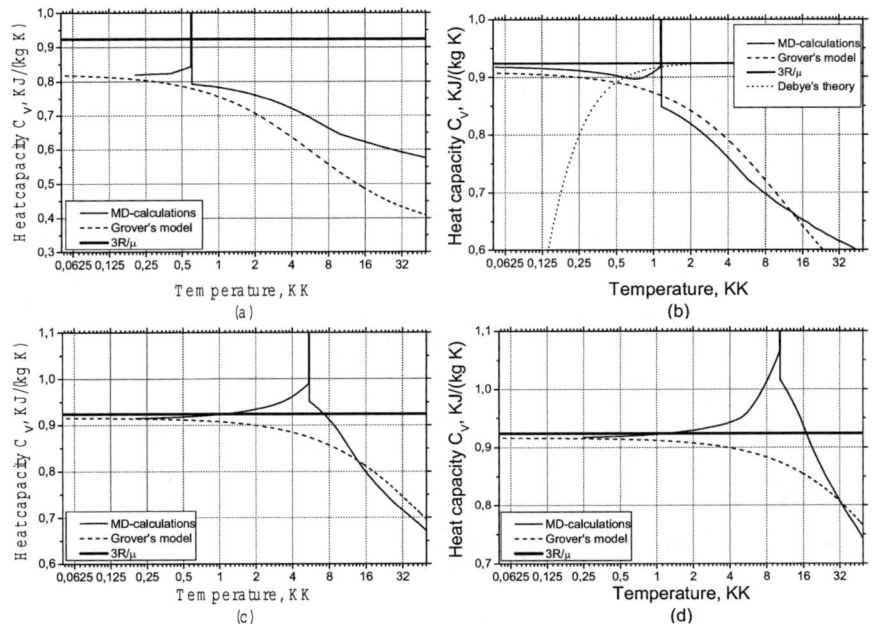

FIGURE 3. Dependencies of specific total energy on temperature at various densities (a) $\rho = 0.9\rho_0$, (b) $\rho = \rho_0$, (c) $\rho = 1.5\rho_0$, (d) $\rho = 2.0\rho_0$.

As it was shown before by using potential of Lennard-Johns specific heat capacity C_V in the case with reduced density is noticeably lower than $3R/\mu$ at all temperatures.

In the samples with normal and increased densities curves close to value of Dulong-Petit law. As a matter of fact classical molecular dynamics does not know about existence of Debye temperature and all oscillation degrees of freedom are excited even at very low temperatures. However aluminum has relatively low Debye temperature θ_D in comparison with the melting point. That is why we can ignore this fact. In Figure 3 (b) dependency of heat capacity on temperature which calculated using equation (1) also shown (Debye temperature was taken $\theta_D = 394K$ as for normal density). We can see that at increased densities noticeable growth of specific heat capacity C_V is observed when the temperature of a sample increases. It may be explained by influence of anharmonicity of crystal lattice vibrations. So in linear approximation when $F_x = -\partial U/\partial x = -C\delta x + o(\delta x^2)$ we ignore the terms the order of which are higher than first. In this case potential energy is a square function of the displacement δx and heat capacity is a constant. At high temperature (it means at high displacement δx) influence of the terms order of which is higher than first increase what leads to growth of heat capacity.

Also in Figure 3 dependencies of heat capacity on temperature of liquid phase are presented which were calculated using Grover's model [2]. In the context of this simple empirical model relation for heat capacity of liquid phase of material has the following form

$$C_{V,l} = C_{V,s} - 1.5R\frac{\alpha\tau}{1+\alpha\tau}, \tag{4}$$

where $C_{V,l}$ and $C_{V,s}$ are heat capacities of liquid and solid phases at the given specific volume respectively, R is gas constant, α is an empirical constant (for metal it usually is taken 0.1) and $\tau = T/T_m$ (where T_m is melting temperature at the given specific volume). As initial data for equation (4) we have taken values of melting temperature and heat capacity of solid phase at $T = 300K$ which were obtained in numerical molecular dynamic experiment at given density. Breaks of heat capacity C_V in melting points have typical for solid-liquid phase transition λ-form.

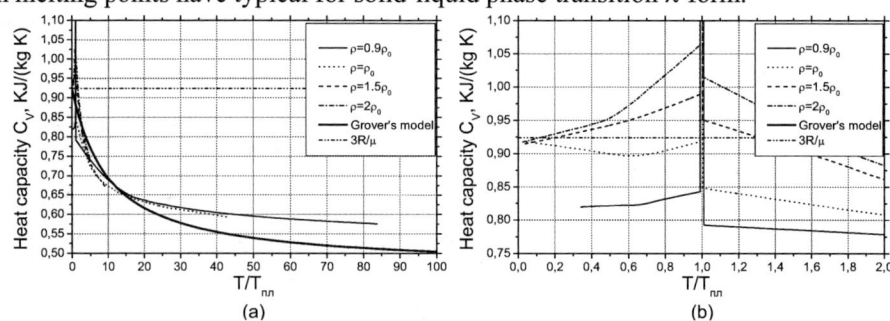

FIGURE 4. Dependencies of specific heat capacities C_V on temperature at various densities.

In the frames of this work the calculations of Grüneisen coefficient (5) dependencies on temperature and density of solid state (Figure 5) were performed.

$$\gamma(T,\rho) = \frac{1}{\rho} \cdot \frac{\left(\dfrac{\partial P}{\partial T}\right)_V}{\left(\dfrac{\partial E}{\partial T}\right)_V} \qquad (5)$$

As it was expected in solid state Grüneisen coefficient is practically independent on temperature and decreases as the density increases. However acquired values of Grüneisen coefficient are greatly higher than those calculated using the model in paper [3].

In Figure 7 melting line of aluminum which was calculated by molecular dynamics is presented in comparison with known works [4]-[9]. As one can see our calculated curve is in a good agreement with published data.

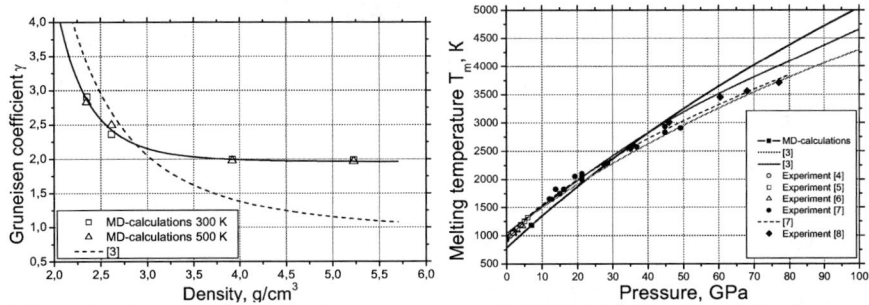

FIGURE 5. Dependencies of Grüneisen coefficient of solid phase on density.

FIGURE 6. Melting line of aluminum.

CONCLUSION

In the frames of this work it was shown that heat capacity C_V of solid phase of aluminum differs from Dulong-Petit law due to the anharmonicity of lattice vibrations. At reduced density it leads to noticeable ($\sim 11\%$) reduction of heat capacity. At normal density molar heat capacity is about 3R. At increased densities noticeable growth of specific heat capacity C_V is observed when the temperature of a sample increases. Such growth is steeper when the density of a sample is higher (Figure 4 (b)).

In the liquid phase heat capacities which were taken from molecular dynamics and Grover's model have similar behaviors. However one can see (Figure 4 (a)) that reduction of heat capacity with the growth of temperature occurs slower than in Grover's model. But changing parameter α one can obtain exact coincidence of MD curves with Grover's model. Melting line calculated by using molecular dynamics method is in a good agreement with published experimental data. Calculations of Grüneisen coefficient at high densities give values much higher than that published in [3]. Model in [3] for Grüneisen coefficient was constructed with relation to the fact that with growth of density Grüneisen coefficient must amount to the values typical for one component plasma. But molecular dynamics calculations show that this reduction takes place slower than it is predicted in [3].

REFERENCES

1. *Voter A.F., Chen S.P.,* Accurate interatomic potentials for Ni, Al and Ni$_3$Al. In Characterization of Defect in Materials, MRS Symposia Proceedings #82, edited by R.W. Siegal, J.R. Weertman, R. Sinclair (Materials Research Society, Pittsburgh, 1987), p.175.
2. *Grover R.,* 7th Symp. On Thermodynamic Properties, Cezairian, Ed. (Amer. Soc. Mech. Eng., New York, 1977). p. 67.
3. *Burakovsky L., Greeff C. W., Preston D. L.*, Phys. Rev. B 67, 094107 (2003).
4. *Moriarty J. A., Yang D. A., Ross M.*, Phys. Rev. B 30, 578 (1984).
5. *Gonikberg M. G., Shahovsky G. P., Butuzov V. P.,* J.Ch.Ph., 1957, v. 31, p. 1839.
6. *Lees J., Williamson B. H. J.*, Nature, 1965, v. 206, p. 278.
7. *Jayaraman A., Klement W., Newton R. C., Kenedy G. C.,* J. Phys. Chem. Solids, 1963, v. 24, p. 7.
8. *Hängström A., Lazor P.*, J. Alloys Compd. 305, 209 (2000).
9. *Boehler R., Ross M.*, Earth Planet Sci. Lett. 153 (1997) 233.

Ab initio Calculation of Cold Curves for FCC, BCC and HCP Nickel to Ultrahigh Pressures

G.V. Sin'ko, N.A. Smirnov

*Federal Nuclear Center - Institute of Technical Physics, PO BOX 245, Snezhinsk 456770
Russia*

Abstract. The paper presents cold curves for a magnetic *fcc* and three nonmagnetic (*fcc*, *bcc* and *hcp*) structures of Ni up to ~ 800 Mbar, obtained from ab initio calculations done with the full-potential scalar-relativistic method of electronic structure calculation FPLMTO with gradient corrections to the exchange-correlation functional. Our calculations confirm the results that were earlier obtained and suggest that nickel dielectrizes at very high pressures. A gap in the energy spectrum of electrons is formed for all the above Ni structures that is also in agreement with the results presented earlier. Our research suggests that *fcc* Ni dielectrizes in the pressure interval 300–720 Mbar, and other Ni structures dielectrize in roughly the same interval.

Keywords: Electronic structure calculation, nickel, ultrahigh pressures.
PACS: 71.15.Nc, 64.30.+t, 62.50.+p, 71.30.+h

Recently, a possible loss of electrical conductivity by some alkaline metals (Li and Na) under pressure has been intensively discussed. As theoretically shown in [1,2], compressed Li and Na crystals undergo a series polymorphic transitions which radically reconstruct their energy bands up to the formation of a gap between occupied and unoccupied states, and hence to the loss of electrical conductivity. However, the hypothesis that simple metals become insulators under pressure was proposed in theoretical papers much earlier. As early as 1963, the authors of [3], following from very simplified calculations on the electronic structure of *fcc* nickel, proposed that nickel dielectrized under ultrahigh pressures (~250 Mbar). In that case, the band reconstruction with the formation of the gap occurred at no changes in the crystal structure. As shown in [3,4], under increasing pressure, the 4*s* band of *fcc* nickel which is below the 3*d* band at ambient conditions, is quickly lifted above the 3*d* band. Therefore, electrons from the 4*s* band occupy the unoccupied states of the 3*d* band and fill it completely. As a result, nickel becomes insulator. The authors of [3,4] come to the conclusion that the energy gap in nickel exists in the pressure range 250-1500 Mbar. Then the reverse transformation into a metallic state occurs due to band overlap recovery. Almost two decades later the issue on possible metal-insulator transition in *fcc* nickel under pressure was investigated with the more accurate ab initio methods APW and LMTO-ASA in [5]. Those calculations also showed the insulating gap resulted from the fast lift of the 4*s* band above Fermi level. By data from [5], the gap exists in the pressure range 340-510 Mbar. Moreover, the authors of [5] note that the same effect was observed for other (*bcc* and *hcp*) structures they studied.

CP849, *Zababakhin Scientific Talks - 2005*,
edited by E. N. Avrorin and V. A. Simonenko
© 2006 American Institute of Physics 0-7354-0345-7/06/$23.00

This paper presents cold curves for some nickel crystals calculated with the contemporary ab initio method FP-LMTO [6] and investigates the issue on the loss of conductivity by these crystals at high pressures. We studied four nickel structures, namely ferromagnetic fcc (fcc FM) and non-magnetic fcc, bcc and hcp. Our scalar-relativistic calculations were done with a local density functional taken in the form [7] with gradient corrections [8]. Electrons which occupied 3s, 3p, 3d and 4s levels in an isolated atom of nickel were taken to be valence electrons. The basis set included linear muffin-tin orbitals of s-, p-, d- and f-type. We used ten linearization centers to derive basis functions in muffin-tin spheres and three tail energies in the interstitial region [6]. The values of these parameters for different specific volumes were selected with a special automatic algorithm which allowed for a pressure dependence of the crystal energy spectrum. In the prism-shaped Brillouin zone a mesh for integration over $\bar{k}$ space with the linear tetrahedron method was constructed by dividing each edge into the same number of parts. For all structures, calculations were done with a 50×50×50 mesh. For the hcp structure, the equilibrium axial ratio c/a for each specific volume was calculated from the condition of specific energy minimum as a function of c/a at constant volume. Nuclear zero-point vibrations were not taken into account. The pressure versus volume dependence at T=0 K was calculated by differentiating specific energy as a function of volume.

Calculated results are sensitive to the approximation chosen for the exchange-correlation energy, especially for densities lower than or close to ambient. Before having opted for the approximation [7], we considered approximations proposed in [7, 9-12] and used them to calculate specific volume V_{00}, magnetic moment μ_{00}, bulk modulus B_{00} and its pressure derivative B'_{00} for ferromagnetic fcc nickel at P=0 and T=0 K. Calculated results are presented in Table 1 along with experimental values for μ_{00}, B_{00} and B'_{00} measured at room temperature, and a value of V_{00} obtained through extrapolation of the experimental relationship $V(T, P = 0)$ to T=0 K. The table also contains the calculated minimum pressure in the crystal, characterizing its tensile strength. It is seen from Table 1 that compared to the others the approximations [7] and [11] give roughly identical and better fits to the experimental values of nickel characteristics at ambient pressure and T=0 K. We preferred the approximation [7] because our calculations are made with no account for zero-point vibrations.

To evaluate accuracy of our calculations, we compared the specific volume of nickel at P=0 obtained in this paper and reported earlier in paper [5] with experiment (see Table 2). It is seen that the values we obtained for fcc FM agree well with experiment. It also follows from table that the significant difference of the value reported in [5] from experiment cannot be explained by neglected magnetic effects.

In Figure 1a we compare our calculated dependence of pressure on compression for fcc FM nickel with the same dependence from the diamond anvil experiments at room temperature [15], and the isotherm T=0 K obtained from shock experiments [16]. Hereafter $\rho_0 = 1/V_0$, V_0=73.65 (a.u.)^3/atom is the experimental specific volume of fcc FM nickel at ambient conditions. As seen from Fig. 1a, our curve is in good agreement with experiment data.

TABLE 1. Calculated V_{00}, μ_{00}, B_{00} and B'_{00} for ferromagnetic *fcc* nickel obtained with different exchange-correlation energy approximations, corresponding experimental data and calculated minimum pressures in bulk tension.

	V_{00} (a.u.)^3/at	μ_{00} (μ_B /at)	B_{00} (Mbar)	B'_{00}	P_{min} (Mbar)
Experiment	73.29	0.61 [13]	1.86 [13]	6.20 [14]	—
			1.61 [15]	7.55 [15]	
Exchange-correlation energy [7]	72.28	0.627	2.09	4.90	-0.315
Exchange-correlation energy [9]	73.42	0.628	1.99	4.90	-0.300
Exchange-correlation energy [10]	73.37	0.628	2.00	4.87	-0.301
Exchange-correlation energy [11]	73.00	0.621	2.04	4.90	-0.307
Exchange-correlation energy [12]	73.66	0.654	1.98	4.91	-0.300

TABLE 2. Specific volume V_{00} at P=0 and T=0 K calculated for *fcc* FM and *fcc* nickel, corresponding experimental data obtained at ambient conditions, and the result calculation [5].

	Experiment fcc FM, T=293 K	Result from this paper		Result from [5]
		fcc FM	fcc	fcc
V_{00} (a.u.)^3/at	73.65 [13]	72.28	71.83	69.93

Figure 1b shows the pressure dependence of specific magnetic moment for *fcc* FM nickel. The structure is seen to gradually lose magnetic properties under pressure and above 12 Mbar the magnetic moment vanishes. Therefore, we studied only three nickel structures, namely *fcc*, *bcc* and *hcp* at ultrahigh pressures.

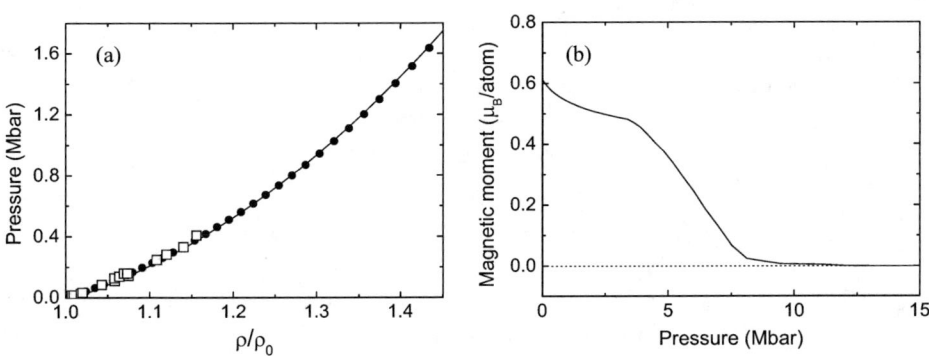

FIGURE 1. a) Pressure vs. compression at T=0 K for *fcc* FM nickel: — calculation, • - evaluation of data from the shock experiments [16], □ – data from static experiments at room temperature [15]. b) Specific magnetic moment vs. pressure in *fcc* FM nickel.

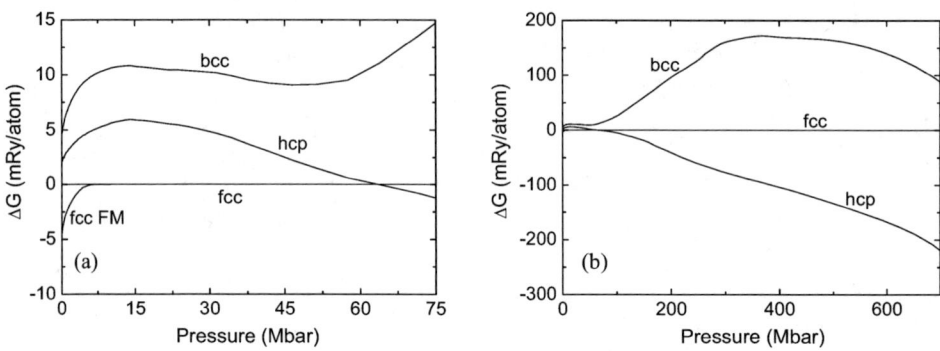

FIGURE 2. Gibbs potential differences vs. pressure for considered structures of nickel. a) high pressures, b) ultrahigh pressures.

In Figures 2, Gibbs potentials $G = E + PV$ at T=0 K for all studied structures are plotted relative to Gibbs potential of the *fcc* structure. It follows from Fig. 2a that the *fcc* FM structure is most stable at ambient condition, and this is in agreement with experiment. When pressure reaches about 12 Mbar, the structure loses magnetic properties and transforms into the non-magnetic *fcc* structure which remains most stable to 63.5 Mbar where the *fcc* → *hcp* transition occurs. As seen from Fig. 2b, of the three considered structures, the *hcp* one remains most stable up to pressure 700 Mbar.

Figure 3 shows pressure dependence of densities of states on Fermi surface for *fcc*, *bcc* and *hcp* Ni structures. It is seen for all considered structures that the density of states on Fermi surface vanishes at certain pressure, i.e., the energy gap appears in the energy spectrum which separates fully occupied and unoccupied bands. Our calculations show that the pressure range where the energy gap exists are 302–720 Mbar for *fcc* nickel, 365–560 Mbar for *bcc* nickel, and 300–445 Mbar for *hcp* nickel. The data we obtained for *fcc* nickel qualitatively agree with previous calculations [3-5]. However, the pressure ranges where nickel dielectrizes are rather different. That may be a result of our more accurate calculations.

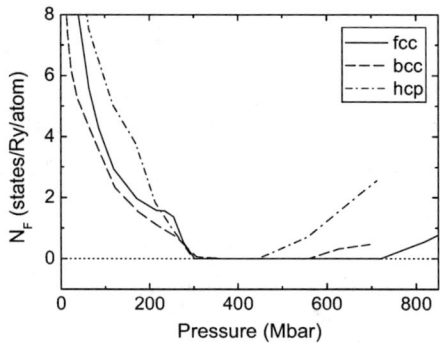

FIGURE 3. Density of states on Fermi surface vs. pressure for considered structures of nickel.

TABLE 3. Calculated pressure and compression ranges where considered structures of Ni show dielectric properties, and the maximum width of the gap in the dielectrization region.

		Results of this paper	Results from [3,4]	Results from [5]
Pressure range, Mbar	*fcc*	(302 - 720)	(250 - 1500)	(340 - 510)
	bcc	(365 - 560)	—	—
	hcp	(300 - 445)	—	—
ρ / ρ_0 range	*fcc*	(7.14 - 10.00)	(6.5 - 15)	(6.9 - 8.8)
	bcc	(7.69 - 9.09)	—	—
	hcp	(7.14 - 8.33)	—	—
Maximum width of the gap, eV	*fcc*	6.1	—	(2.3 - 3.7)
	bcc	8.8	—	—
	hcp	5.7	—	—

By our data, the maximum magnitude of the energy gap in nickel has following values: 6.1 eV for *fcc* Ni, 8.8 eV for *bcc* Ni, and 5.7 eV for *hcp* Ni. For the *fcc* structure, the gap is about twice as large as the value reported in [5]. Our calculations show that among the considered structures the *hcp* one, which is most stable under ultrahigh pressures, exhibits the narrowest range where the gap exists and the lowest maximum magnitude of the gap.

Our results and the literature data on the position and magnitude of the dielectric gap in nickel at high pressures are summarized in Table 3.

ACKNOWLEDGMENTS

The authors express gratitude to the academician V.E. Fortov, who initiated this research and to the Russian Foundation for Basic Research for support (Grant # 04-02-17292).

REFERENCES

1. J. B. Neaton, N. W. Ashcroft, *Nature* **400**, 141 (1999).
2. M. Hanfland, K. Syassen, N. E. Christensen, D. L. Novikov, *Nature* **408**, 174 (2000).
3. G. M. Gandelman, V. M. Ermachenko, Y. B. Zeldovich, *ZhETF* **44**, 386 (1963).
4. A. I. Voropinov, G. M. Gandelman, V. G. Podvalny, *Usp. Fiz. Nauk* **100**, 193 (1970).
5. A. K. McMahan and R. C. Albers, *Phys. Rev. Letters* **49**, 1198 (1982).
6. S. Yu. Savrasov and D. Yu. Savrasov, *Phys. Rev. B* **46**, 12181 (1992).
7. O. Gunnarsson, B. I. Lundqvist, *Phys. Rev. B* **13**, 4274 (1976).
8. J. P. Perdew, J. A. Chevary, S. H. Vosko et al., *Phys. Rev. B* **46**, 6671 (1992).
9. J. P. Perdew and Y. Wang, *Phys. Rev. B* **45**, 13244 (1992).
10. S. H. Vosko, L. Wilk, and M. Nusair, *Can. J. Phys.* **58**, 1200 (1980).
11. U. von Barth, L. Hedin, *J. Phys. C: Solid State Phys.* **5**, 1629 (1972).
12. J. F. Janak, V. L. Moruzzi, and A. R. Williams, *Phys. Rev. B* **12**, 1257 (1975).
13. C. Kittel, *Introduction to Solid State Physics*, Wilye, New York, 1996.
14. M. W. Guinan, D. J. Steinberg, *J. Phys. Chem. Solids* **35**, 1501 (1974).
15. P. Lazor, S. K. Saxena, *Terra Abstracts* **5**, 363 (1993).
16. S. B. Kormer, A. I. Funtikov, V. D. Urlin, A. N. Kolesnikova, *ZhETF* **42**, 686 (1962).

Equation of state and phase diagram of quartz

A.V. Petrovtsev, V.V. Dremov, V.G. Vildanov, M.M. Gorshkov,
V.T. Zahikin, Yu.N. Zhugin

Russian Federal Nuclear Center - Institute of Technical Physics
named after academician E.I.Zababakhin

Abstract. On the basis of the latest experimental data and ab-initio calculations the multi-phase equation of state of quartz has been constructed. Quartz is a basic rock forming mineral. This fact initiated a comprehensive investigation into its thermodynamic, mechanical and shock-wave properties. Quartz has complicated phase diagram and one of the very interesting problems having fundamental character is the study of the polymorphous and phase transformations in quartz at high pressures and temperatures realizing when dynamic loading. Numerical modeling of these processes requires multi-phase equation of state. Here we present such an equation of state including α-quartz, stishovite and liquid quartz. Calculations carried out with the equation of state are in good agreement with the experimental data on static and shock compression of solid and porous quartz including the temperature measurements along Hugoniots. On the basis of an analysis of the calculated phase diagram and porous quartz Hugoniots the conclusions about what phase the legs of the experimental Hugoniots belong to have been done.

INTRODUCTION

Quartz is a basic rock forming mineral. This fact initiated a detail and comprehensive investigation into the varied properties of this mineral. Most part of these studies refers to the dynamic investigations, that is due to the different applications associated with the study of the shock and blast events.

One of the most interesting problems having fundamental character is the study of the phase transformations in quartz and quartzite being practically a pure polycrystalline quartz. These materials have a complicated phase diagram (Fig.1), at which there are two solid phases of high pressure of coesite (C) and of stishovite (S) besides the phase of low pressure of quartz (Q).

All mentioned phases have complicated crystalline structure and differ greatly in density [1]. The polymorphous transformations in quartz in the shock-wave experiments are realized in a significantly nonequilibrium way [2-11]: the states corresponding to coesite are not realized practically and the transformation into stishovite-like phase is characterized by very strong metastability with large extent along Hugoniot curve in the phase-mixture region and by relatively long time of transformation. This situation is close to the situation of graphite-diamond transformation in carbon, for which a large delay of transformation is also typical, contrary to the α–ε transformation in iron, for which metastability appears essentially weaker than for the mentioned above cases [12,13].

CP849, *Zababakhin Scientific Talks - 2005,*
edited by E. N. Avrorin and V. A. Simonenko
© 2006 American Institute of Physics 0-7354-0345-7/06/$23.00

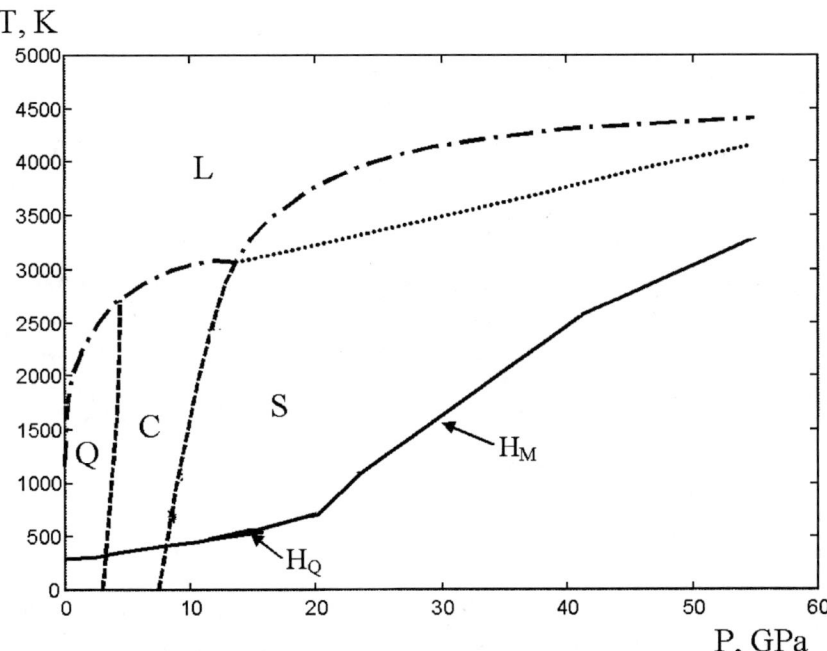

FIGURE 1. Phase diagram of quartz [1] (dashed lines - Q-C and C-S equilibrium curves; dashed and dotted line – melting curve, static measurements; dotted line – melting curve, shock wave measurements) and location of calculated Hugoniot (solid line: H_Q – quartz Hugoniot, H_M – mixture region)

It should be noted that at the present time there is no one point of view on the character and mechanism for the phase transformation progress in quartz in the shock-wave processes. In particular, the mechanisms associated with the formation of high-dense amorphous quartz phases (see, [14]) are examined.

Numerical modeling of the above mentioned processes requires the construction of the multi-phase equation of state adequately reflecting thermodynamic properties for the different phases of quartz. In this work we present such multi-phase equation of state including α–quartz, stishovite and liquid quartz. We compare calculative and experimental data on static and shock compression of solid and porous quartz.

EOS FORMULATING AND ITS PARAMETERS

When developing the equation of state for different quartzite phases, an approach was used that was implemented in multiphase equation of state for iron [15]. Free energy for α-quartz and stishovite is written down as a sum of two terms

$$F = F_p + F_T - S_0 T, \tag{1}$$

corresponding to potential and thermal contributions determined from lattice oscillations. To present the potential energy, a modified Berch formula was used [16]

$$F_p = -\int P_p dV + F_0, \tag{2}$$

$$P_p = 1.5 B_{0K} \left\{ -B_K \eta^3 + (1 + 2B_K)\eta^{7/3} - (1 + B_K)\eta^{5/3} \right\} \Theta(\eta),$$

where B_{0K} is bulk modulus at T=0K, P=0, $B_K = 0.75(4 - B'_0)$, B'_0 is derivative of bulk modulus with respect to pressure, and $\eta = V/V_0$ is dimensionless specific volume,

$$\Theta(\eta) = \begin{cases} 1, & \text{at } V/V_{0K} \leq 1, \\ 0.5(\eta^2 + \eta^{-2}), & \text{at } V/V_{0K} \geq 1. \end{cases} \tag{3}$$

Free phonon energy is described with the Debye model. The Debye temperature θ_D is related to volume through the Gruneisen coefficient in the following form:

$$\Gamma = -\frac{d \ln \theta_D}{d \ln V} \tag{4}$$

Volume dependence of the Gruneisen coefficient is given by the formula from [17]

$$\Gamma(V) = \Gamma_0 \frac{\tilde{\Gamma}(V)}{\tilde{\Gamma}(V_0)}, \tag{5}$$

$$\tilde{\Gamma}(V) = -\frac{V}{2} \frac{\partial^2 (P_x V^{2t/3})/\partial^2 V}{\partial (P_x V^{2t/3})/\partial V} + \frac{1}{3}(t - 2),$$

which defines volume dependence of the Gruneisen coefficient by Landau-Slater at t=0, by Dugdale-McDonald at t = 1 and by free volume theory at t = 2.

As in [15], equation of state for liquid quartz was based on the Grover model of liquid state [18] according to which thermodynamic functions for liquid are obtained by integrating the following presentation of heat capacity for liquid at constant volume over temperature in the area of melting

$$C_{V,L} = C_{V,S} - 1.5R \frac{\alpha \tau}{1 + \alpha \tau}, \tag{6}$$

where $C_{V,L}, C_{V,S}$ is heat capacity of liquid and solid phases, respectively, R is gas constant, $\tau = T/T_m(V)$, $T_m(V)$ is melting point at the specified volume and α is empirical constant (~0.1). Integrating (6) over the area of melting, we obtain the following relationships for energy E and entropy S of liquid:

$$E_L(V,T) = E_S(V,T) + RT_m \left\{ \frac{\Delta S_m}{R} - \frac{3\alpha}{4} + \frac{3}{2}\tau \left[\frac{\ln(1 + \alpha\tau)}{\alpha\tau} - 1 \right] \right\}, \tag{7}$$

$$S_l(V,T) = S_S(V,T) + \Delta S_m - 1.5R\ln(1 + \alpha\tau) ,$$

where functions related to the solid phase considered basic for EOS of liquid have indices s. In the EOS of liquid quartz the stishovite phase was considered basic.

Volume dependence of the melting point is calculated from the Lindemann law. Free energy of liquid is obtained as

$$F_L(V,T) = E_L(V,T) - TS_L(V,T). \tag{8}$$

Parameters of EOS for crystalline and liquid phases of quartz are given in Table 1. In the vicinity of the triple point α-quartz-stishovite-liquid, liquid state of quartz has a mixed nature in terms that it shows properties of both stishovite and α-quartz. Therefore, to improve description of the liquid properties in the vicinity of the triple point, some corrections were introduced in the appropriate parameters of stishovite. As a result, liquid became less dense and more compressible, i.e. got properties typical of α-quartz, that led to steepening of the melting curve of stishovite in the vicinity of the triple point and to the quick formation of a plateau with pressure growth that agreed with the experimental data.

Table 1. EOS parameters for α-quartz, stishovite and liquid quartz phases

Parameter	α-quartz	γ- quartz	Liquid
V_0, cm^3/g	0.3774	0.225	0.24
Θ_D, K	999.0 [44]	1037.0 [44]	1037.0 [44]
Γ_0	0.7	2.0	1.5
t	0	0	0
B^*, GPa	37.7	300.0	175.0
B_1^*	5.5	3.8	3.9
F_0, kJ/g	0.0	0.501	1.31
S_0, kJ/g·K	-0.0004	-0.00065	0.0004
α			0.1
ΔS_m, kJ/g·K			0.00017

COMPARISON WITH EXPERIMENTAL DATA

Static measurements

Static compressibility of α-quartz is rather well studied experimentally [19-24]. Figure 2 shows experimental data compared to calculations done with the proposed equation of state. Good agreement is observed between the experimental data and calculations with the developed equation of state.

As for stishovite, data on static compressibility obtained by different authors in X-ray diffraction measurements differ significantly. This, as mentioned in [1], means that at high pressures structural properties of stishovite are extremely sensitive to the emerging stresses. Stishovite shows higher compressibility along the *a* axis than along the *c* axis.

When selecting EOS parameters we relied on the experimental data [25,26,28,29] since they better correspond to the data of shock experiments. The calculated curve of static compression for stishovite is given in Figure 3 as well as the experimental data.

Phase diagram

Phase diagram of quartzite calculated with the developed EOS and presented in Fig.1 includes two high-pressure phases - coesite and stishovite.

As noted above, presently there are no reliable evidences for coesite production in the shock waves, therefore, we have not so far included this phase in the developed multi-phase EOS.

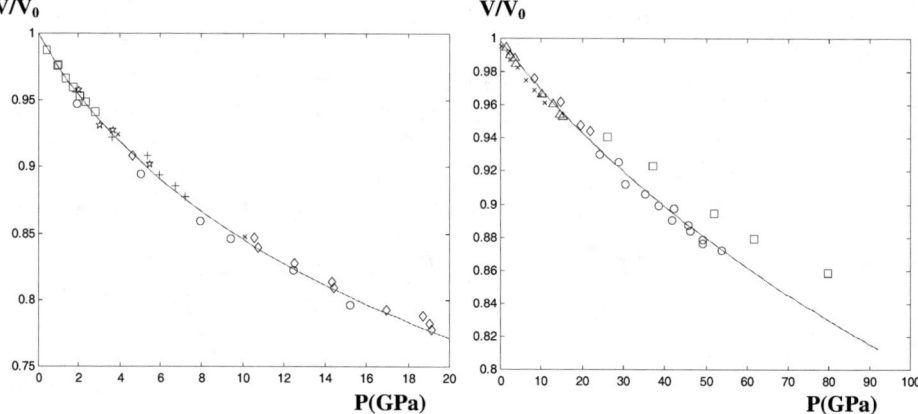

FIGURE 2. Static compressibility of α-quartz. Solid curve corresponds to the calculation done with the developed equation of state. Experimental data are shown with squares – [19], pluses – [20], stars – [21], circles – [22], × – [23], rhombs – [24]

FIGURE 3. Static compressibility of stishovite. Solid curve is calculation done with the developed EOS. The experimental data are shown with squares – [27], circles – [29], rhombs – [25], triangles – [28] and crosses – [26]

Figures 4 and 5 show phase diagrams of quartzite under high and low pressures, respectively. It is obvious from Fig. 4, that melting curves of stishovite obtained in static [32] and shock [33-34] experiments differ significantly in the pressure range of 20-60 GPa. According to the shock measurements, stishovite melts at lower temperatures. The experimental data on the temperatures of shocked α-quartz and melted quartz [30, 31] allow the melting curve of stishovite to be constructed up to the pressures of ~150 GPa.

In the pressure range of 20-60 GPa the melting curve of stishovite calculated with the developed EOS shows better agreement with the shock-wave experiments [33,34]. At higher pressures of 100-150 GPa the calculated melting curve is in good agreement with the data on temperatures of shock-compressed quartz [30,31] (see Fig. 4)

Because of a large volume jump at the α-quartz-stishovite transition it was impossible to obtain the melting curve of α-quartz fitting the experimental data. The reason for this is that liquid was based on stishovite and inherited its properties, though some properties of α-quartz were also accounted for in the development of the EOS for liquid. Analysis of the experimental data shows that at the pressures >10 GPa liquid is indeed stishovite-like.

Issue on the correct description of the quartzite properties at relaxation to low pressures can be solved by accounting for kinetics of phase transformations at relaxation. To make simulation of relaxation down to pressures <10 GPa correct, the following can be done. Figure 5 shows an isentrope calculated from the triple point "α-quartz – stishovite – liquid". By prohibiting "liquid – α-quartz" transition above the isentrope, we obtain rather accurate description of material behavior and properties in the low-pressure area.

Figure 4 shows calculated Hugoniots of solid and porous quartzite, α-quartz at the initial state and Hugoniots of melted quartz. There are experimental data on temperatures of shock-compressed quartz for melted quartz and solid quartz with initial state of α-quartz [30,31]. From the comparison of the calculated and experimental data a conclusion can be drawn that the proposed equation of state reproduces well the liquid and stishovite properties at the pressures and temperatures corresponding to the experiment. For example, according to the equation of state points of the start and end of stishovite melting are located on the Hugoniot at pressures of 80 GPa and 120 GPa, respectively, that is in good agreement with the data from [30, 31].

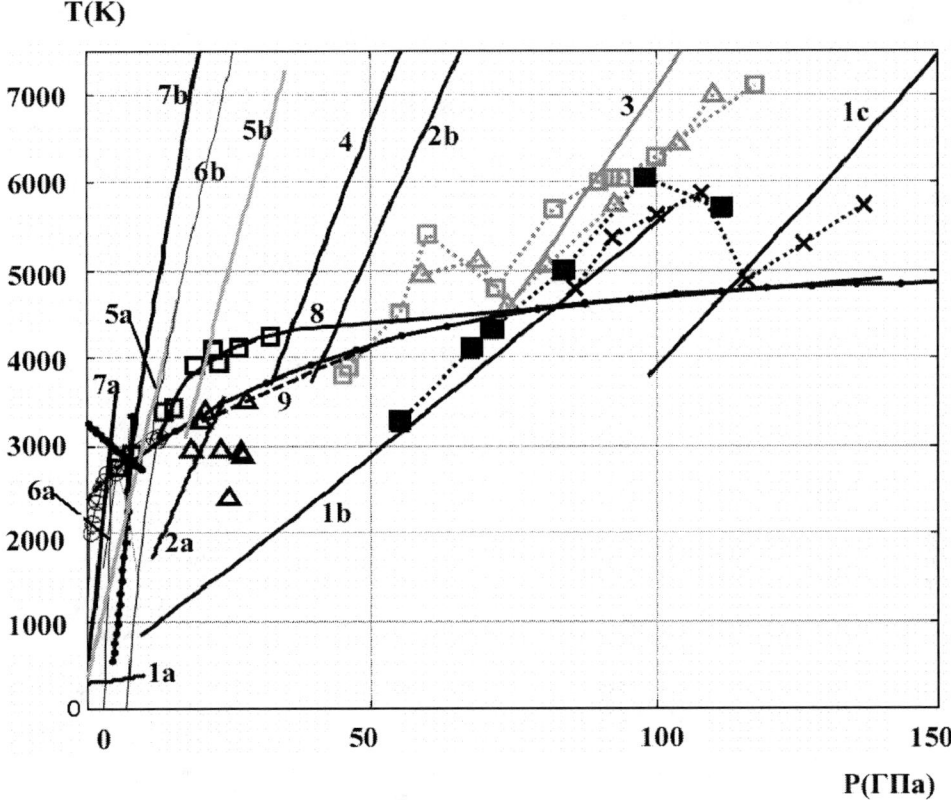

FIGURE 4. Phase diagram and Hugoniots of quartz. Curves 1a, 1b, and 1c give Hugoniots of solid α-quartz, stishovite and liquid quartz calculated with the developed EOS; 2a and 2b are Hugoniots of stishovite and liquid with initial state of α-quartz with the density of 1.65 g/cm^3; 3 is Hugoniot of liquid at the initial density of 2.2 g/cm^3 (fused quartz); 4 is Hugoniot of liquid at the initial density of 1.4 g/cm^3; (5a, 5b), (6a, 6b) and (7a, 7b) are Hugoniot of α-quartz and liquid at the initial densities of 1.0, 0.7 and 0.52 g/cm^3, respectively; solid curves with dots correspond to the equilibrium lines between the phases calculated with the developed EOS; 8 is approximation of the static experiments on the melting curve of stishovite; 9 is approximation of the shock experiments on the melting curve of stishovite [1]; gray squares and triangles show the experimental data on temperatures of shock-compressed fused quartz [30,31]; ■ and crosses are the experimental data on temperatures of shock-compressed solid quartz (ρ_0=2.65 g/cm^3) [30,31], □ and triangles are data of static and shock experiments (see [1]) on the melting curve of stishovite.

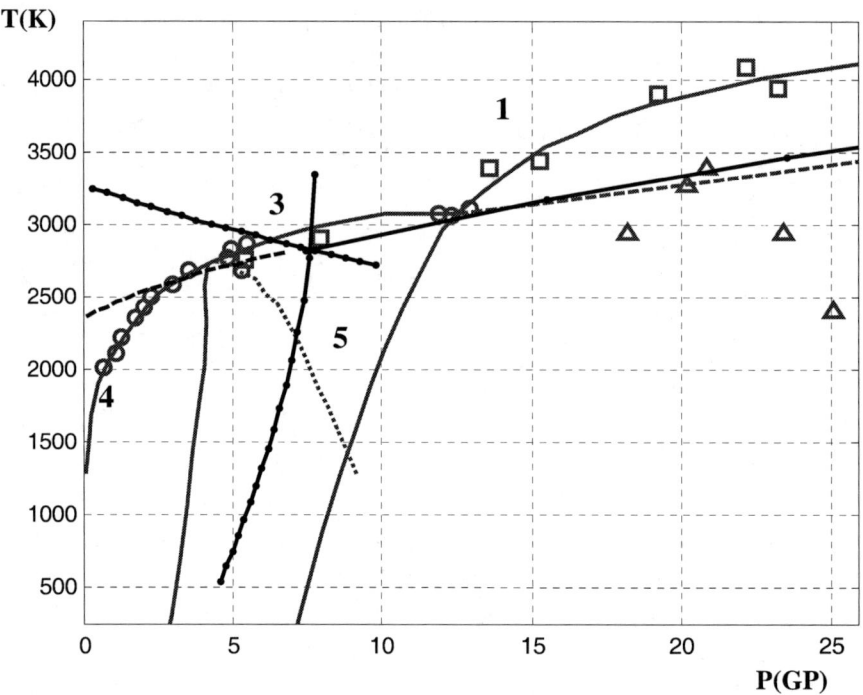

FIGURE 5. Phase diagram of quartzite at low pressures. Solid curves with dots show phase equilibrium borders calculated with the proposed EOS; 1 is approximation of the static experiments with respect to the melting curve of stishovite; 2 is approximation of the shock-wave experiments with respect to the melting curve of stishovite [36]; 3 is approximation of the experimental data with respect to the melting curve of α-quartz and coesite; 4 and 5 are curves of α-quartz – coesite and coesite - stishovite equilibrium; dash line is extrapolation of the melting curve of α-quartz [1], 6 is isentrope calculated with EOS for liquid from the triple point α-quartz-stishovite-liquid; blue circles stand for experimental data on the melting curve of α-quartz and coesite [35, 37]; and blue squares and triangles are data of static [32] and shock [33, 34] experiments studying the melting curve of stishovite

Shock wave experiments data

Figure 6 presents the experimental data on shock compression of solid α-quartz. An area corresponding to the polymorphous transition α-quartz-stishovite (from ~6 to ~55 GPa) is well seen on the experimental Hugoniot. Figure 6 shows the experimental data of two types: data obtained in experiments with solid quartz and data obtained in the so-called mixture (quartz with paraffin and fluoroplastic) experiments realizing hydrodynamic compression of quartz. Comparison of data from two types of the experiments leads to the conclusion on a strong effect of shear stresses on the processes running in quartz under shock compression.

The proposed in Ref. [38] model of elastic-plastic properties and kinetics of non-equilibrium polymorphous α-quartz-stishovite transition was used in this work to calculate shock compression of quartzite with the proposed multiphase equation of state. Good agreement of calculated and experimental data was reached (see Fig.6).

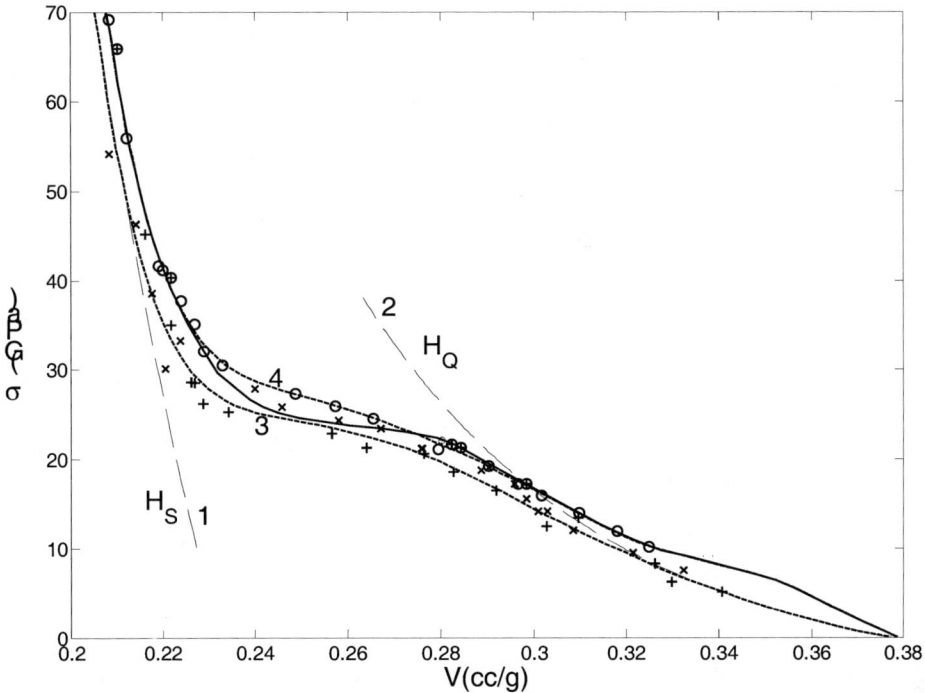

FIGURE 6. Shock compressibility of quartzite. Dashed (4) and solid curves show approximation of the experimental data O and ⊕ [4, 5, 8, 11] for solid quartzite done with model [38] and the proposed EOS. Dashed line (3) shows approximation of data obtained in the mixture experiments +, and × [6, 7]. Dashed lines (1,2) correspond to the calculated hydrodynamic Hugoniots of pure quartz and stishovite

In addition, Fig. 6 shows Hugoniots of pure α-quartz and stishovite (the initial point for stishovite was selected in compliance with the usual state of α-quartz), calculated with the proposed equations of state. The Figure proves that calculations with the EOS are in good agreement with the data of hydrodynamic mixture experiments.

Figure 7 compares the experimental data with the calculations done with the proposed multiphase EOS and model [38] of pressure dependence of longitudinal speed of sound on Hugoniot of solid quartzite. Dash curve in the plot marks the stishovite-liquid transition area. For solid phase area the data correspond to the longitudinal speed of sound. Good agreement with the whole set of experimental data can be noticed.

It is of great interest to analyze shock experiments carried out with porous quartz [39] with the help of multiphase equation of state. From the data presented in the phase diagram (Fig. 4) a conclusion can be drawn on phase-related nature of different sections of the experimentally obtained Hugoniots.

C_L, C_B (km/s)

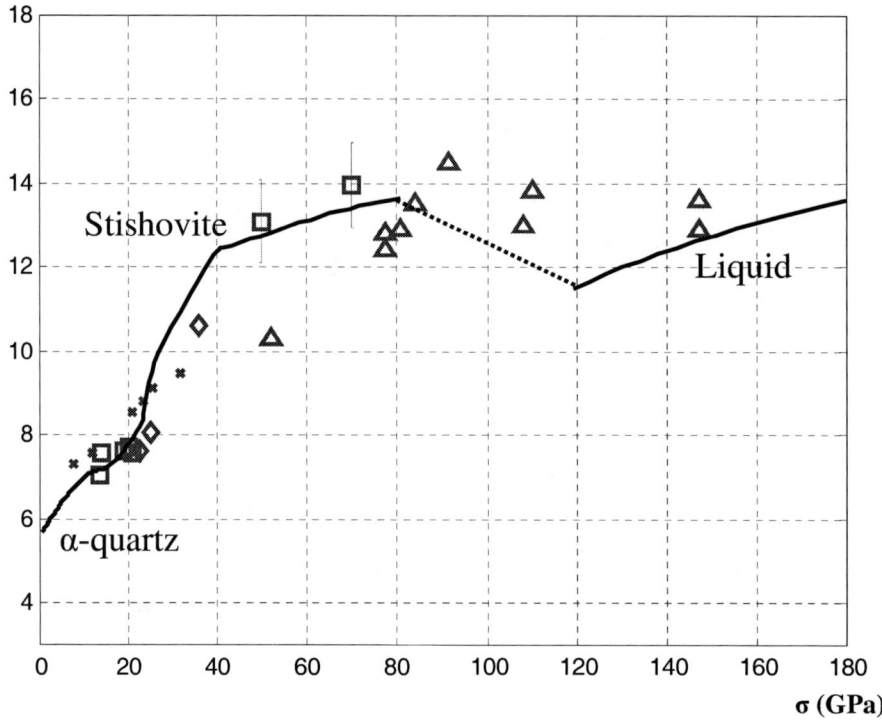

FIGURE 7. Longitudinal (for solid states) and bulk (for liquid states) sound speed of shocked quartz versus wave strength. ◊ - [42], × - [43], - [10], Δ - [41]. Solid curve stands for calculations carried out with the considered quartzite model (EOS+[38]), dash line shows area of phase mixture (stishovite-liquid)

Transition α-quartz–stishovite on the Hugoniot of solid quartzite was described above. Deviation of Hugoniots for α-quartz and stishovite (at initial density ρ_0=2.65 g/cm^3) from the experimental data in Fig.8 is explained by the fact that calculation of the Hugoniots ignored the effect of shear stresses. This effect is much smaller on the Hugoniots of porous quartz. As mentioned above, solid quartz, according to the Hugoniot, starts melting at ~80 GPa and finishes at ~120 GPa. This agrees with the data on measuring temperature on the Hugoniot [30,31].

The Hugoniot at initial density ρ_{00}=1.65 g/cm^3 intersects the melting curve at P=22 GPa and stays in the mixture area up to ~43 GPa. Thus, the vertical section of the Hugoniot (see Fig. 8) is in the area of liquid. The same is true for the data for initial density ρ_{00}=1.45 g/cm^3. The Hugoniots for initial densities ρ_{00}=1.65 g/cm^3 and ρ_{00}=1.45 g/cm^3 calculated under condition of equilibrium polymorphous transformation have the sections in the area of stishovite stability. However, due to significantly metastable nature of phase transition on the experimental Hugoniots up to the intersection with the melting curve the stishovite concentration seems to be far from 1.

P(GPa)

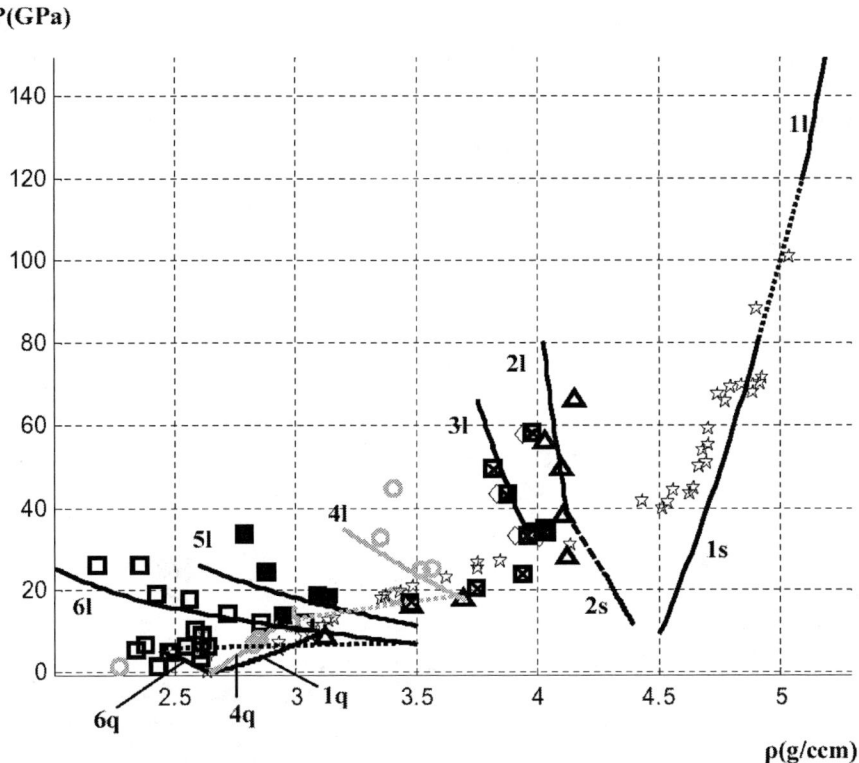

FIGURE 8. Hugoniots of porous quartz. Labels 'q','s' and 'l' are for calculated Hugoniots of α-quartz, stishovite and liquid. Dash curves show the regions of mixed phases when melting. The curves numbered 1-6 correspond to the calculated Hugoniots with initial densities 2.65, 1.65, 1.4, 1.0, 0.7, 0.52 g/cm³ respectively. Experimental data [39] are shown for different initial densities: ☆ - (2.65 g/cm³), △ - (1.65 g/cm³), ▢ - (1.4 g/cm³), circles (1.0 g/cm³), ■ - (0.7 g/cm³), ▢ - (0.52 g/cm³).

At lower initial densities ρ_{00}=1.0 g/cm³, ρ_{00}=0.7 g/cm³ and ρ_{00}=0.52 g/cm³ transition to stishovite seems not to happen. It is α-quartz which experiences melting and all experimental data at pressures >10 GPa correspond to the area of liquid according to the proposed multiphase EOS.

It should be noted that the issue of phase-related nature of the upper sections of Hugoniots of porous material was discussed in Ref. [39] and [40]. According to [40], at initial densities <1.35 g/cm³ the upper sections of Hugoniots are in the coesite phase, and at higher densities they are in the stishovite phase. According to [39], there is no division of Hugoniots into two groups, and states corresponding to the upper sections of all Hugoniots of porous materials are in the stishovite phase. Calculations with the proposed multiphase equation of state prove that these states can only belong to the area of liquid.

Figure 8 shows that the proposed equation of state describes rather accurately the set of the experimental data on shock compression of porous quartz.

CONCLUSION

Quartz and quartzite are of great interest to develop methods of modeling of elastic-plastic behavior and kinetics of phase transitions of materials under dynamic loading. Multiphase equation of state is an important element of such modeling.

In this paper multiphase equation of state of quarts has been evaluated. This equation of state together with model, proposed in [38], allow to acquire accurate description of available experimental data on shock wave loading of solid and porous quartz and quartzite.

REFERENCES

1. Hemley R.J., High-Pressure Behavior of Silica. In Silica: Physical Behavior, Geochemistry and Materials Applications. Reviews in Mineralogy, Vol. 29, pp.41-82, Mineralogical Society of America, Washington, D.C. (1994).
2. Wackerle, J. J.Appl.Phys. 33 (2), 922-937 (1962)
3. Fowles, R. J. Geoph. Res, 72 (22) 5729-5742 (1967)
4. Trunin R.F., Simakov G.V. et al. News of the AS of the USSR, Physics of the Earth, 1, 13 (1971)
5. LASL Shock Hugoniot Data. Berkly, Los Angeles, London: University California Press, 1980, p.499
6. Zhugin, Yu. N. The behaviour of α-Quartz under High Dynamic and Static Pressures: New Results and Views. In Shock Compression of Condensed Matter-1995, Woodbury, New-York: AIP Press, 1996. Part I, pp.97-100.
7. Yu. N. Zhugin, K.K. Krupnikov, et al., On Some Specifics of Dynamic Compressibility of Quartz. News of the AS of the USSR, Physics of Earth, #10, pp. 16-22 (1994)
8. Yu. N. Zhugin, K.K. Krupnikov, et al., On Peculiar Features of Shock Wave Front in Quartz in the Region of Phase Mixture. Proceeding of IUPAP-IUTAM Symposium on Non-linear Acoustics "Problems of Non-linear Acoustics" edited by E.G. Kedrinsky (Novosibirsk, USSR AS SB, 1987), part 2, pp. 196...200
9. Grady, D.E., Zhugin, Yu. N. Critical Transition Stress in Shock Compression of SiO_2. Bull. Amer. Phys. Soc. Ser. 2, V.39, No1, pp.410-411.
10. Chhabildas, L.C. Shock Loading and Release Behaviour of X-Cut Quartz. In Shock Waves in Condensed Matter -1985, edited by Y.M. Gupta (Plenum Press, New-York, 1986), pp.601-605.
11. Vildamov, V.G., Gorshkov, M.M., Slobodenjukov, V.M., Rushkovan, E.H. Shock compression of low initial density quartzat pressures up to 100 GPa. In Shock Compression of Condensed Matter-1995, Woodbury, New-York: AIP Press, 1996. Part I, pp.121-124.
12. Gust, W.H., Phys. Rev. B 22 (10), 4744-4756 (1980).
13. Zhugin, Yu.N., Krupnikov, K.K., Tarzhanov, V.I., Investigation into Kinetics of Natural Ceylon Graphite Transformations in Shock Waves, in Shock Wave in Condensed Matter – 1998, edited by I.Yu. Klimenko et al., High Pressure SIC Press, St.-Petersburg, 1998, pp.163-166

14. Kuznetsov N.M., Kinetics of Shock-Induced Phase Transition in Quartz. In: Shock Waves and Extremes States of Matter, edited by V.E.Fortov, L.V. Al'tshuler, R.F. Trunin, A.I. Funtikov, M.: Nauka, 2000, 425p.

15. V.V.Dremov, A.L.Koutepov, A.V. Petrovtsev, A.T. Sapozhnikov Equation of State and Phase Diagram of Iron In Proceedings of the APS Conference on Shock Compression of Condensed Matter, Atlanta, 2001 AIP Conference Proceedings 620. Ed. by M. Furnish, pp.87-90

16. M. van Thiel, F.H. Ree, Thermodynamic properties and phase diagram of the graphite diamond-liquid carbon system. High Pressure Research, vol.10, pp.607-628

17. V.N. Zharkov, V.A. Kalinin, Equations of state for solids under high pressures and temperatures. Moscow: Nauka, 1968

18. Grover R., High-temperature equation of state for simple metals", in Proceedings of Seventh Symposium on Thermodynamic Properties, p.67, (1977)

19. J.D. Jorgensen, Compression mechanisms in α-quartz-SiO_2 and GeO_2. J. Appl. Phys., V.49, pp.5473-5478, (1978)

20. H. d'Amur, W. Denner, H. Schulz, Structure determination of α-quartz up to 6.8 GPa, Acta Crystallogr B, V.35, pp.550-555, (1979)

21. L. Levien, C.T. Prewitt, D.J. Weidner, Structure and elastic properties of quartz at pressure, Am. Min., V.65, pp.920-930, (1980)

22. R.M. Hazen, L.W. Finger, R.J. Hemley, H.K. Mao, High-pressure crystal chemistry and amorphization of α-quartz, Solid State Comm., V.72, pp.507-511, (1989)

23. J. Glinnemann Jr., H.E. King, H. Schulz, T, Hahn, S.J.L. Placa, F. Dacol, Crystal structures of the low-temperature quartz-type phases of SiO_2 and GeO_2 at elevated pressure, Z. Kristallogr, V.198, pp.177-212, (1992)

24. K.J. Kingma, Pressure-induced transformations in SiO_2, PhD Dissertation, Johns Hopkins University, Baltimore, MD

25. L-G. Liu, Silicate perovskite from phase transformations of pyrope-garnet at high pressure and temperature, Geophys. Res. Lett., V.1, pp.277-280, (1974)

26. Y. Sato, Pressure-volume relationship of stishovite under hydrostatic compression, Earth Planet Sci. Lett., V.34, pp.307-312, (1977)

27. Y. Tsuchida, T. Yagi, A new post-stishovite high-pressure polymorph of silica, Nature, V.340, pp.217-220, (1989)

28. N.L. Ross, J-F. Shu, R.M. Hazen, High-pressure crystal chemistry of stishovite, Am. Min., V.75, pp.739-747, (1990)

29. J-F. Shu et. al. (data were taken from [1], where they were given with reference to this paper, which had been in print, when [1] had been published)

30. R.G. McQueen, J.N. Fritz, J.W. Hopson, in New ways for looking for phase transitions at multi-megabar dynamic pressures. Physics of solids under high pressures, 1981, 99-108

31. G.A. Lyzenga, T.J. Ahrens, A.C. Mitchell, Shock temperatures of SiO_2 and their geophysical implication, J. Geoph. Res., 1983, v.88, N33, pp.2431-2444

32. G. Shen, P. Lazor, Melting of minerals under lower mantle conditions: I. Experimental results, J. Geophys. Res. (data were taken from [1], where they were given with reference to this paper, which had been in print, when [1] had been published)

33. K-I. Kondo, T.J. Ahrens, Shock-induced spectra of fused quartz, J. Appl. Phys., V.54, pp.4383-4385, (1983)

34. D.R. Schmitt, T.J. Ahrens, Shock temperatures in silica glass: implications for models of shock-induced deformations, phase transformations and melting with pressure, J. Geophys. Res., V.94, pp.5851-5871, (1989)

35. I.Jackson, Melting of silica isotypes SiO_2, BeF_2, GeO_2 at elevated pressures, Phys. Earth Planet Int., V.13, pp.218-231, (1976)

36. J. Zhang, R.C. Liebermann, T. Gasparik, C.T. Gerzberg, Y Fei, Melting and subsolidus relations of SiO_2 at 9-14 GPa, J. Geophys. Res., V.98, pp.19785-19793, (1993)
37. M. Kanzaki, Melting of silica up to 7 GPa, J. Am. Ceram. Soc., V.73, pp.3706-3707, (1990)
38. G.V. Kovalenko, Yu.N. Zhugin, A.V. Petrovtsev, Modeling Polymorphic Transformations of Quartzite in Dynamic Processes. In Proceedings of the APS Conference on Shock Compression of Condensed Matter, Portland, 2003, AIP Conference Proceedings 706. Ed. by M. Furnish, Y.M. Gupta ans J.W. Forbes, pp.255-259
39. V.G. Vildanov, M.M. Gorshkov, K.K. Krupnikov, Shock compression of porous quartz, RFNC-VNIITF report, VINITI G83219, 1987
40. R.F. Trunin, G.V. Simakov, M.A. Poduretz, Compression of porous quartz with strong shock waves, News of the AS of the USSR, Physics of the Earth, #2, pp.33-39, (1972)
41. McQueen, R.G. The Velocity of Sound behind String Shocks in SiO_2. In Shock Waves in Condensed Matter -1991, edited by S.C.Schmidt, R.D.Dick, J.W.Forbes, D.G.Tasker (Elsevier Science Publishers B.V., 1992), pp.75-78
42. Grady, D.E., Murri, W.J., DeCarli, P.S. J. Geophys. Res., **80** (35), 4857-4861 (1975)
43. Pavlovsky M.N. Russ. PMTF (J. of Appl. Mech. & Tech. Phys.), 1981, No. 5, pp.136-139
44. Swegle, J.W., J. Appl. Phys., Vol. 68, No. 4, pp. 1563-1579 (1990).

Registration of Free Surface Velocity Dispersion of Iron and Steel Samples by Optical Lever Method

E.A. Kozlov, V.I. Tarzhanov, I.V. Telichko, D.G. Pankratov,
A.V. Vorobyov, D.M. Gorbachev, V.I. Stavrietsky

Federal State Unitary Enterprise "Russian Federal Nuclear Center – Zababakhin All-Russian Research Institute of Technical Physics", P.O. Box 245, Snezhinsk, Chelyabinsk region, 456770, Russia

Abstract. Results of explosive experiments are presented with time-resolved and space-resolved diagnostics of free surface velocity profiles of wedge samples by optical lever method. The objective of this work is to record free-surface velocity dispersion of samples made of high purity iron with fine and coarse grains, as well as 30KhGSA steel, in delivery state and hardening up to HR_c 50. It is demonstrated, that the value of velocity dispersion of free surface of the samples under identical loading modes is varied according to initial microstructure and grain size of investigated metals.

Keywords: stress-waves, strong plastic waves, shock-induced plasticity, shock-induced solid-solid transformation, microstructure, meso-scale, time-resolved and space-resolved diagnostics, free surface velocity dispersion.
PACS: 46.40.Cd, 46.80.+j, 07.10.-h, 47.11.St

INTRODUCTION

With advances in laser interferometric methods of diagnostics of free surfaces of construction material samples at their shock-wave loading a possibility has evolved of recording the features of material's elastic-plastic behavior, related to their structure heterogeneity [1-7]. The iron with fine and coarse grains, as well as ferrite-pearlitic steels in delivery state and hardened, are the materials being interesting for studying such features in connection with the presence of α-ε – polymorphic transformation in them. Multi-wave configurations are registered in these materials under their shock-wave loading in the wide range of longitudinal stresses σ_{xx} [8-11]:

- single-wave, elastic wave at $\sigma_{xx} \leq \sigma_{xx}^{HEL} \sim$1-3 GPa;
- two-wave, elastic-plastic configuration in α-phase at $\sigma_{xx}^{HEL} \leq \sigma_{xx} \leq \sigma_{xx}^{\alpha-\varepsilon} \sim$12.5-13.5 GPa;
- three-wave, elastic-plastic with phase precursor observation at $\sigma_{xx}^{\alpha-\varepsilon} \leq \sigma_{xx} \leq \sigma_{xx}^{*}$=35 GPa;
- two-wave, elastic-plastic at $\sigma_{xx}^{*} < \sigma_{xx} \leq \sigma_{xx}^{**}$=70 GPa;

CP849, *Zababakhin Scientific Talks - 2005*,
edited by E. N. Avrorin and V. A. Simonenko
© 2006 American Institute of Physics 0-7354-0345-7/06/$23.00

‒ single-wave, plastic wave in ε-phase at $\sigma_{xx} \geq 70\,\text{GPa}$.

The purpose of this work was a verification of phenomena discovered in [4-7] and consisted in relationship of *in-situ* registered free surface velocity dispersion with micro-and mesostructures of materials investigated. Verification must be carried out with using independent diagnostic method and in more wider range of stress waves, where not only strong plastic wave in initial α-phase is realized, but also shock-induced α-ε-phase transition is observed.

MATERIALS, SAMPLES AND REGIMES OF IT'S EXPLOSIVE LOADING

The steel 30KhGSA hardened up to HRc 50 and in delivery state, as well as high purity unalloyed iron with fine and coarse grains, i.e., 80 and 275 μm, were investigated. The samples were made in the form of wedges [9-12].

Experimental set-up is shown in Figure 1.

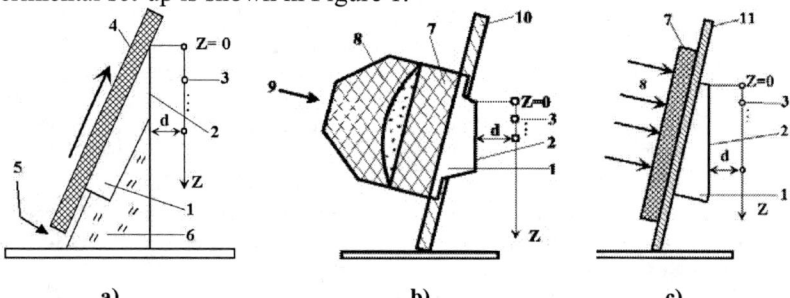

a) b) c)

FIGURE 1. Experimental assemblies (a, b, c) for loading of wedge samples by sliding and normal detonation of HE layers.
1 – tested wedge samples; 2 – polished free surface of the sample; 3 – slit diaphragm system – raster; 4 – sheet HE charge; 5 – linear detonation-wave generator; 6 –caprolon rest; 7 – main charge; 8 – plane wave generator; 9 – electric detonator; 10 – protective plate; 11 – base plate.

The states $\sigma_{xx} \leq 15\,\text{GPa}$ were realized by explosive loadings of wedge samples by sliding detonation of PETN-based composition with 1.5 mm thickness and states $\sigma_{xx} \leq 25\,\text{GPa}$ – by sliding detonation of HMX-based composition with 20 mm thickness. Loading up to 25-35 GPa was realized by normal detonation of HMX-based composition, which was in direct contact with the sample (b), or through the base plate made of 12Kh18N10T steel of 5 mm thick (c). Experiments set-up data are given in the table with the obtained results.

The optical lever method [9-12] was used for time-resolved and space-resolved diagnostics of wave process. In the optical lever method, the free surface of the tested sample is analyzed by narrow (~1°) light beams, which come through raster slits of width from 0.15 up to 0.6 mm. The turn of structural elements of the material and the ensuing dispersion of free-surface velocity result in multiple local displacements of reflected beams of light being registered, in addition to expected average displacement. A scattered component, by which parameters the velocity dispersion is

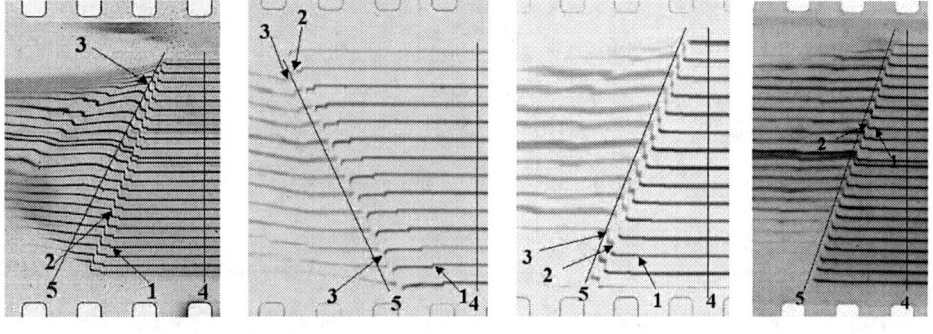

a)   **b)**   **c)**   **d)**

FIGURE 2. Typical streak camera records of explosive experiments with samples made of steel and armco-iron.

a) – experiment No176, 30KhGSA steel HR$_c$ 50, d=44.0mm; b) – experiment No16, 30KhGSA steel in delivery state, d=100.0mm; c) – experiment No21, armco-iron 80 µm, d=80.0mm; d) – experiment No199, armco-iron 275 µm, d=29.9mm.

1 – elastic precursor, 2 – phase precursor, 3 – strong plastic wave, 4,5 – directions of photometric scanning.

determined, appears in the reflected light. Streak camera records obtained for some tested materials are given in Figure 2. It is obvious that after the output of plastic front the blurring of raster lines takes place due to light scattering in the line.

RESULTS

To obtain quantitative information on the scattered component photometric scannings were withdrawn from the digital pattern of streak camera records in the directions being parallel to the slope of the breakdown of raster lines at the output of the waves fronts. The scattering function L, being the function of ideal registration channel response to the light of infinitely narrow slit was found from photometric profiles (Figure 3):

$$L = \frac{1}{y_L \sqrt{2p}} e^{\frac{-(y-y_i)^2}{2y_L^2}} .$$

Here y is a coordinate on the photographic film along the vertical, y_i is a position of the i- raster line, σ_L is a mean-square width of the function L.

The quantity σ_L was calculated from the values of the width of displaced and initial raster lines $\Delta l_0'$ and Δl_0 at zero width of raster slits δ:

$$2\sigma_L = \sqrt{\Delta l_0'^2 - \Delta l_0^2} .$$

The quantities $\Delta l_0'$ and Δl_0 were determined by extrapolation to δ=0 of the dependencies of the width of the initial and displaced lines on streak camera records versus the width of raster slits (Figure 4).

395

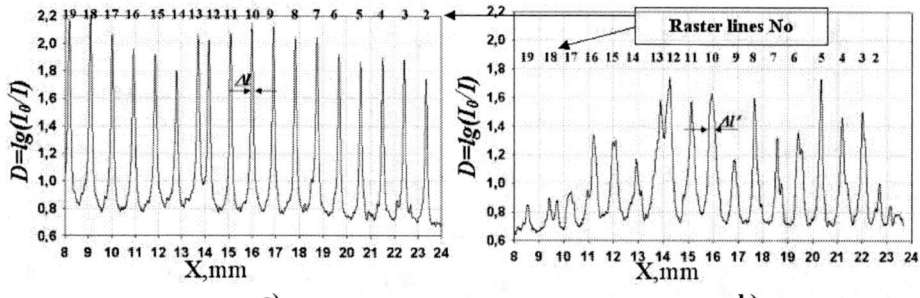

FIGURE 3. Photometric scannings with profiles of raster lines in experiment No199 (armco-iron 275 μm). a) – profiles of initial raster lines; b) – profiles of displaced raster lines. Δl is measured on the level $0.607(D_{max}-D_{min})$.

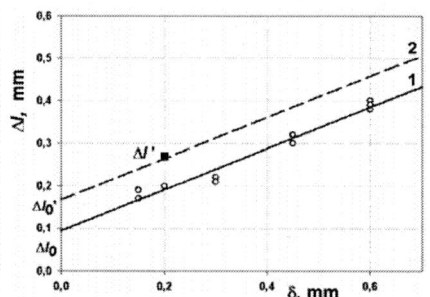

FIGURE 4. Dependence of width of raster lines Δl on streak camera records vs width of raster slits for initial and displaced raster lines δ.
1 – initial lines (separate set of experiments); 2 – displaced lines.

TABLE 1. Experimental set-up and results.

Material	30KhGSA HR_C 50	30KhGSA in delivery state		Armco-iron 80 μm		Armco-iron 275 μm
ρ_0, g/cm^3	7.768	7.768		7.85		7.88
Experiment No	176	16	17	13	21	199
H_{HE}, mm	5.0	20	40	1.5	60	5.0
Scheme on Fig.1	c)	a)	b)	a)	b)	c)
Wedge angle	12° 00′	24° 01′	9° 10′	24° 00′	9° 09′	12° 07′
Optical lever, mm	44.0	100.0	80.0	300.0	80.0	29.9
a, mm	1.33	1.73	1.98	2.03	1.90	0.672
W, km/s	0.85	0.74	1.08	0.22	1.33	0.58
σ_{xx}, GPa	16.7	14.0	20.8	4.40	25.8	12.0
$\pm S_W$, km/s	0.010	0.016	0.041	0.003	0.057	0.054

Mean-square fluctuation of free-surface velocity S_W (and consequently the velocity dispersion S_W^2) was determined from σ_L as a portion of free-surface velocity itself W, realized after the output of plastic front:

$$\pm S_W = \frac{\sigma_L}{a} W .$$

Here a is a jump of raster line, which defines the quantity W.

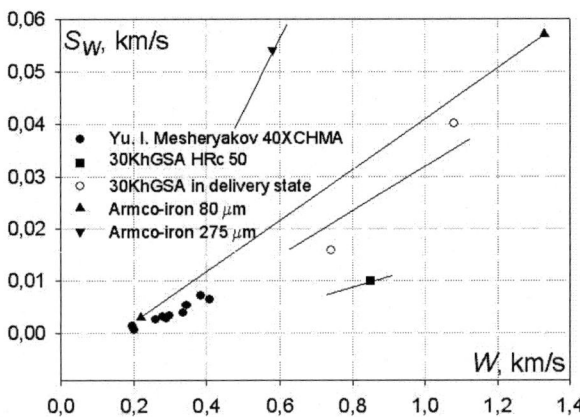

FIGURE 5. Mean-square fluctuations of free-surface velocity for samples made of armco-iron and steels depending on free-surface velocity.

It is evident from the Table 1 and Figure5 that the value of mean-square fluctuation of free-surface velocity S_W is increased in the row: 30KhGSA hardened steel, the same steel in delivery state, armco-iron with grains size 80 μm, armco-iron with grains size 275 μm.

DISCUSSION

Yu. I. Mesheryakov proposed an explanation for a decrease in mirror reflection component of laser emission from initially polished surface of the polycrystalline sample at the cost of disordered orientation of grain blocks and arising of velocity dispersion of analyzed area of sample's free surface at the output of intense plastic waves [4-7]. For registration of two-wave configurations in high strength steels authors of [4-7] applied the local (single-point [2], in contrast to multi-channel one [1]) VISAR–diagnostics using the low-powered laser. For high-rate loading of flat plate targets the laboratory gas gun was used allowing the acceleration of plane steel impactors ∅37×2 mm up to 0.2-0.4 km/s. A threshold of structural stability loss for steels was estimated in [4-7] by plastic wave amplitude being only several times higher than Hugoniot elastic limit of the material $\sigma_{xx}^* \geq \sigma_{xx}^{HEL}$. The so conservative conclusion was made by Yu.I. Mesheryakov evidently owing to impossibility of studying under laboratory conditions the behavior of structural materials in wider loading range, with the use of more powerful laser in VISAR-diagnostics.

In Figure 5 the results obtained with optical lever method in this work as well as the data [7] are presented in S_W,W-coordinates. For the purpose of comparison with mean-square fluctuations, the maximum velocity fluctuations published in [7], were divided by 3 under the assumption that maximum fluctuations correspond to $3\sigma_L$. It is apparent that the first values of fluctuations of free-surface velocity in iron and steels, which were obtained by the optical lever method, are not contradictory to the data obtained using the laser-interferometric diagnostics.

CONCLUSIONS

Application of optical lever method and samples of unalloyed high pure iron with fine and coarse grain sizes (80 and 275 μm), as well as 30KhGSA steel, in delivery state and hardened up to HR_c 50, under its intensive explosive loading in the region of realization of three wave configuration, was permitted ones to validate an observations of dispersion of free surface velocities of metal samples under reflection of strong plastic wave from initially polished surface.

It's demonstrated, that under identical loading modes the minimal dispersion of free surface velocity is observed for samples of 30KhGSA hardened up to HR_c 50, some more – for samples of 30KhGSA steel in delivery state. Larger dispersions are observed for samples of unalloyed high pure iron with fine grain size and the largest one – for high pure iron with coarse grain size 275μm.

Our results obtained by using optical lever method conformed with Yu.I.Mescheryakov's measurements by using VISAR-diagnostics. Differences in dynamic behavior of investigated materials and values of dispersion of free surface velocities of samples under identical loading are connected with peculiarities of its initial micro- and mesostructure, as well as with α-ε–phase transformation in stress waves under compression and release.

ACKNOWLEDGMENTS

We would like to thank Yu.I.Mescheryakov (IIEP-S.Petersburg) for reprints and fruitful discussion. Some results for high pure iron with fine grains were obtained in the course of Contract # I005300009-35 between RFNC-VNIITF and LANL, 1997-2000.

REFERENCES

1. L.M.Barker, "Multy-Beam VISAR for Simultaneous v/s Time Measurements" in *Shock Compression of Condensed Matter-1999*, edited by M.D.Furnish et al., APS, Melville, New York, 2000, pp.999-1002.
2. L.M.Barker and R.E.Hollenbach, *J.Appl. Phys.***45**(11), 4872-4887 (1974).
3. J.R.Asay and L.M.Barker, *J. Appl. Phys.* **45**(6), 2540-2546 (1974).
4. Yu.I.Mescheryakov and A.K.Divakov, *Dymat. J.* **1**(4), 271-287 (1994).
5. Yu.I. Mescheryakov, "Mesoscopic Effects and Particle Velocity Distribution in Shock Compressed Solids" in *SCCM-1999*, edited by M.D.Furnish et al., AIP, Proc. 505, NY, 1999, pp 1065-1070.
6. Yu.I.Mescheryakov, "Meso-Macro-Energy Exchange in Shock Deformed and Fractured Solids" in *High Pressure Shock Compression of Solids VI, Old Paradigms and New Challenge,* edited by Y.Y.Horie et al., Springer-Verlag, Berlin, 2002, pp.169-213.
7. Yu.I.Mescheryakov, A.K.Divakov and N.I.Zhigacheva, *Inter. J. Solids & Structure* **41**, 2349-2362 (2004).
8. D.Bancroft, E.L.Peterson, and S.F.Minshall, *J. Appl. Phys.***27**, 291 (1956).
9. E.A.Kozlov, *High Pressure Research* **10**, 541-582 (1992).
10. E.A.Kozlov, E.S.Buslova, V.A.Boboedova and D.M.Gorbachev, "Macrokinetics of α-ε Phase Transition for some Steels under Different Conditions of Explosive Loading" in *Shock Compression of Condensed Matter-1991*, edited by S.C. Schmidt et al., Elsevier Science Publishers B.V., 1992, pp.173-176.
11. E.A.Kozlov, V.I.Tarzhanov, I.V.Telichko, D.M.Gorbachev, D.G.Pankratov, A.V.Petrovtsev, G.V.Kovalenko, et al., "Kinetics of Elasto-plastic Deformation, Phase Transfer, and also Spallation, Structure and Properties of Highly Pure Iron and 30KhGSA Steel. The Results of Experimental and Theoretical Investigations into Behavior and Properties of Iron and 30KhGSA Steel at Dynamic Loading", Report of RFNC-VNIITF on Contract #I00530009-35 with LANL, Task Order 001, 2000, 00.7739 of September 26, 2000, 336 p.
12. G.R.Fowles, *J. Appl. Phys.* **32**(8), 1475-1482 (1961).

Optical Analyzer Technique for Spall Investigations in Metal Plates

G.V. Kovalenko, E.A. Kozlov, V.N. Nogin,
D.G. Pankratov, A.K. Yakunin

Federal State Unitary Enterprise "Russian Federal Nuclear Center — Zababakhin All–Russia Research Institute of Technical Physics", 456770, Snezhinsk, Chelyabinsk region, Russia

Abstract. Possible experimental set-up for investigation of the stage of spalls closing in plates before impact on base plate are considered. Multi-wave configurations in indicator-matter situated on the base plate are observed by optical analyzer technique. Oscillograms and results of their processing are presented. Thickness and average density of the spall layer in the plate immediately before its impact on the base plate are estimated.

Keywords: stress-waves, spall and shear fractures, explosive loading, kinetics of nucleation, grow and recompaction of damages, time-resolved diagnostics, numerical simulation.
PACS: 62.50.+p, 46.15.-x, 47.40.-x, 83.60Uv.

INTRODUCTION

Modernization and development, in addition to [1-5], of the new strength models and software having the enhanced predictive capabilities, in particular, allowing description not only the kinetics of nucleation, grow, but also the recompaction and healing of the spall and shear micro-, meso- and macrodamages, is an urgent problem at the present stage of the dynamic properties investigation of structural materials. Simple approximations are usually used for description of the process of recompaction spall and shear damages in 1D-numerical simulation. For example, the assumption of the strength properties recovery in the damaged material when the compression stress occurs in it and when it reaches some level. Thus one-dimensionality of the spall formation and its closing is implied. At the same time, a line-imaging laser-Doppler velocimeter results, as well as the recovery experiments showed, that spall surface, as a rule (except for the cases of smooth spalls formation with participation of the shock rarefaction waves), essentially differs from the flat one and contains the traces of both tensile and local shear fractures connecting these surfaces. That is why we should not expect that the closing of the surfaces of the tensile damages will occur in the same manner as in 1-D- simulations when spalls occur and the spalled layer flies at least through its several thicknesses up to the beginning of the closing.

Experiments with the steel plates acceleration by charges of high explosives (⌀200mm, weight of 15-25 kg) were implemented for more detail investigation of the peculiarities of nucleation, grow and recompaction of the spall and shear damages. In addition to [6], the optical analyzer technique was applied for recording of the wave processes in the base plate of different initial thickness during spall plate deceleration

on it. The optical analyzer technique has a number of advantages, versus to the manganine sensors technique [7-9] frequently used in the similar measurements (at the stage of the well-developed spalls without their closing), such as:
–   higher time and amplitude resolution;
–   absence of the necessity to protect electric outputs and to ensure the absence of conductivity in insulator material of manganine sensors, especially in intensive compression-rarefaction waves;
–   absence of mechanical disturbance into the being analyzed process of damage recompaction.

We should note the advantages of optical analyzer technique over the techniques of pulse X-ray radiography [10], multi-frame pulse proton radiography [11] for studying of the stage of spall closing. Being much more simple and economic, the optical analyzer technique, from the point of view of reliable recording of spall fractures in plates and shells, is close to laser interferometry by its opportunities. However, optical analyzer technique permits simple development of the multi-channel analogue measuring complex. Implementation of the multi-channel laser-interferometric measurements is possible (for example [12-15], etc.), but requires appreciably heavy spending.

The purpose of this work is demonstration of capabilities of optical analyzer technique for acquisition of new experimental data for investigation of the closing of the spall damages in plates and using the results for verification of recent strength models, as well as 1-D and 2-D–numerical codes.

EXPERIMENTAL SET-UP AND RESULTS

The design of experimental assemblage is given in Figure 1, the recorded oscillograms are presented in Figure 2.

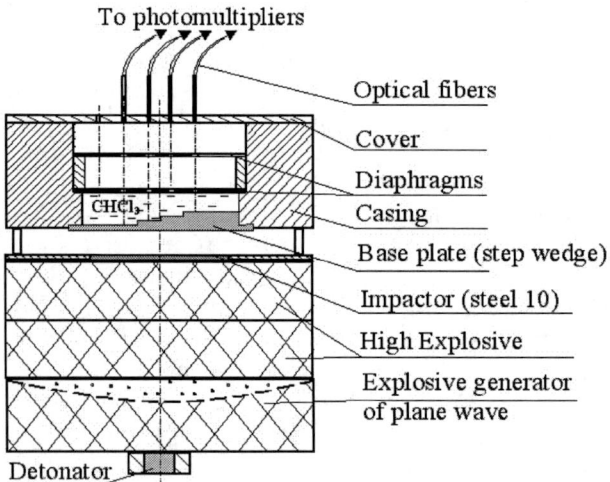

FIGURE 1. Design of measuring assemblage (not in a scale).

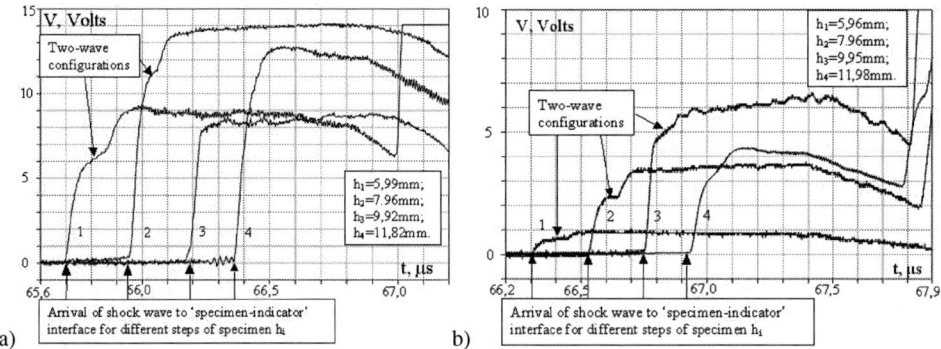

FIGURE 2. Oscillograms of explosive experiments.

Spall damage in a steel plate (initial thickness 3mm, $\rho_0 = 7.83 \, \text{g}/\text{cm}^3$) was studied in process of it acceleration by high explosive products. Experiments differed only by the magnitude of the projectile flight distance. In the first experiment the projectile flight distance was 64.97 mm, in the second experiment it was increased by 3 mm and was 67.89 mm.

Two-wave configurations, evolving into one-wave configurations during propagation over the base step wedge, are well discernable in the profiles given in Figure 2a, recorded on the first two thinnest steps of the wedge. The increase of the flight distance results in regular change of the two-wave configuration parameters. Two-wave configurations in the second experiment, Figure 2b, were recorded behind the steps of all four-thickness values. Apparently, they occur because of consequent deceleration on the barrier of the porous spall layer and the basic part of a plate practically catching up the spall. By estimates, change of the two-wave compression mode by one-wave mode occurs at a distance $l \approx 9$ mm from the left boundary of the aluminum step wedge in the first experiment and at some larger distance in the second experiment.

CALCULATIONS SET-UP AND RESULTS

"VOLNA" 1-D software, allowing to trace motion of shock waves and weak ruptures in hydrodynamic approximation [16], was used for calculations. The single-phase equation of state for steel 12Kh18N10T, ignoring realisation in steel 10 at compression and unloading of reversible α-ε phase transformation, was used for the description of the steel impactor properties. The equation of state for aluminum, Hugoniot adiabat of which is defined by a linear relation $D = 5.231 + 1.467u$, was used for description of properties of the aluminum stepped wedge. The equation of state for chloroform is taken from ref. [17]. Preliminary computations of the dynamic of impactor acceleration illustrated, that formation of multiple spall damages is possible in it, Figure 3.

We can assume, that incomplete spall closing in the process of impactor acceleration is the reason of generation of the observed two-wave configurations. The elementary model was used to estimate the parameters of the spall layer in impactor.

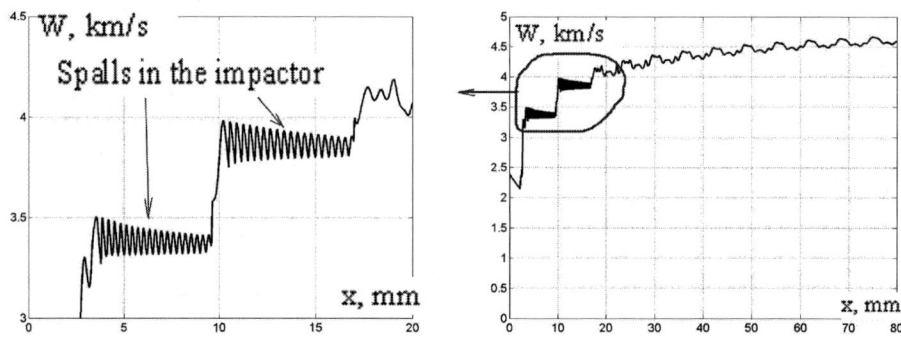

FIGURE 3. Free surface velocity of flying plate versus distance.

We assume, that the plate after spall fracture and subsequent spall closing consists of two parts: damaged part with an effective density ρ_{00} and thickness Δ , and undamaged with an initial density ρ_0 and thickness $\Delta_1 = \Delta_0 - \dfrac{\rho_{00}}{\rho_0}\Delta$, where Δ_0 is the initial thickness of the projectile. The ratio of pressures in these waves can be evaluated from the ratios of signal amplitudes of the first and the second shock waves, taking into account the data of [18]. The corresponding value is $\dfrac{P_{00}}{P_1} \approx 0.9$. By analyzing the wave processes in (P-u)-coordinates one can take such value of porosity that it meets the obtained above pressure amplitudes of the shock waves fronts, Figure 4. Note, that at small porosity the amplitude of second wave P_1 is close to the amplitude P_0 of the shock wave occurring in chloroform at the impact of undamaged projectile. This follows both from the results of numerical calculations, and from the analysis in (P-u)-coordinates. As a result of the analysis the effective value of density in the damaged part of the plate is evaluated as $\rho_{00} \approx 7.1$ g/sm^3. Thickness of the damaged part of the plate $\Delta \approx 1.1\, mm$ is determined from the density and velocity of the projectile, and also from coordinate of the point of transformation of two-wave configuration into one-wave ones. Calculations with "VOLNA" code, which confirmed the above estimates, were performed with the obtained Δ value, Figure 5.

Note, that close Δ values were also obtained for another simple model of the plate failure: the impactor consists of two plates having the initial density and equal velocity, but by the moment of impact with the base plate moving with a gap Δ_z between them, Figure 6. This model also has two parameters: thickness of the first plate Δ and gap width Δ_z. In this case the shock wave caused by impact of the first spall plate on a base plate, will begin to decay. This complicates analytical estimates. Nevertheless, the model parameters are estimated here by the results of numerical calculations. The ratio Δ_z / Δ was varied so that it allowed description of a pressure jump at transformation of the two-wave configuration into one-wave configuration.

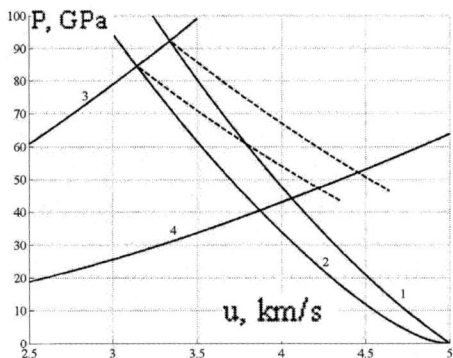

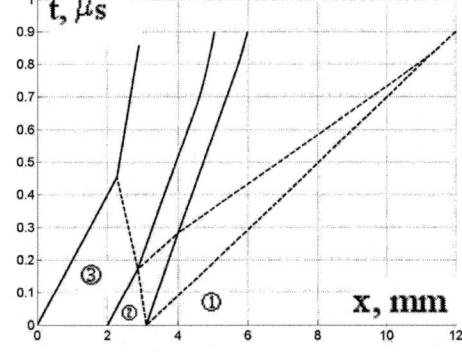

FIGURE 4. (P, u)-diagram. Solid curves – Hugoniot adiabats, dotted curves – isentropes. 1 – full density iron, 2 – porous iron, 3 – aluminum, 4 – CHCl₃

FIGURE 5. (x, t)-diagram of wave process in aluminum target (1) under its loading by impact of layer of porous iron (2) and layer of full density iron (3).

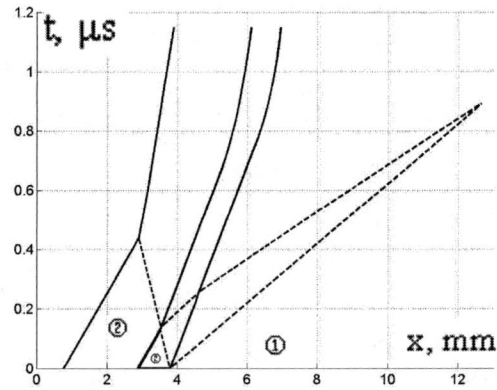

FIGURE 6. (x, t)-diagram of wave process in aluminum target (1) under its loading by impact of two layers of full density iron (2) with small distance between it.

Then all sizes were changed proportionally to combine points of transformation of wave configurations in calculation and in the experiment. Thus spall thickness was equal to $\Delta \approx 0.9mm$ when gap width was $\Delta_z = 0.3mm$.

CONCLUSIONS

1. Capabilities of optical analyzer technique for recording of the parameters of the multi-wave configurations occurring in the base stepped wedge at metal plate deceleration on them in the mode of spall closing were demonstrated.
2. Parameters of a flying plate were determined at its arrival to the barrier based on the obtained experimental results and the data of the numerical simulation of the wave processes for the conditions of the described explosive experiments.

3. It has been illustrated that the decay of the disturbances in the stepped wedge barrier, caused by small spall damages of the flying plate. This circumstance is to be taken into account at the analysis of the results of experiments, especially with application of contact sensors.
4. The observed high sensitivity of optical analyzer technique in diagnostics of the structure and parameters of multi-wave configurations of shock waves in the base plate suggests that it should be applied in series of additional validation experiments, including experiments with application of laser interferometry diagnostics for determination of spall parameters in a steel projectile plate at initial stage of its movement.

REFERENCES

1. D.R.Curran, L.Seaman and D.A.Shockey, "Dynamic failure of solid", *Phys.Rep.* **147**, 253 (1987).
2. L.Seaman, D.R.Curran and D.A.Shockey, *J.Appl.Phys.* **47**, 4814-4824 (1976).
3. J.N.Johnson, *J.Appl.Phys.* **52**, 2812-2825 (1981).
4. J.N.Johnson and F.L.Addesio, *J.Appl.Phys.* **64**, 6699-6712 (1988).
5. Fundamental Issues and Applications of Shock-Wave and High-Strain-Rate Phenomena (EXPLOMET-2000), edited by K.P.Staudhammer, L.E.Murr, M.A.Meyers, Elsevier Science Ltd.., Oxford, 2001, pp.V-XX..
6. E.A.Kozlov, S.A.Brichikov, D.M.Gorbachev, I.G.Brodova and T.I.Yablonskih, "Peculiarities of nucleation, grow and recompaction of spall and shear fractures in the shells of unalloyed iron and some steels at their spherical explosive loading" in *VIII-ZST,* RFNC-VNIITF, Snezhinsk, 2005, pp.172-173.
7. A.N.Dremin, G.I.Kanel and S.A.Koldunov, *III All-Union Symposium on Combustion and Explosion,* Moscow: Nauka, 1972, pp.567-574.
8. A.N.Dremin, A.M.Molodets, A.I.Melkumov and A.V.Kolesnikov, "On Anomalous Increase of Steel Spall Strength and Relationship to Martensitic Transformation" in *Shock-Wave and High-Strain-Rate Phenomena in Materials (EXPLOMET-90),* edited by M.A.Meyers et al., Marcel Dekker Inc., New York, 1992, pp.751-757.
9. V.A.Bychenkov, Y.N.Zhugin, G.V.Kovalenko, E.A.Kozlov, S.V.Lobachyov and A.V.Petrovtsev, "Spallation of armco-iron and 30CGSA steel under explosive loading in the region of α-ε transition: experiment and calculation" in *V HDP Symposium*, Saint-Malo, France, 2003, v. I, pp.431-434.
10. B.R.Bread, C.L.Mader and C.J.Venable, *J.Appl.Phys.* **38**, 3271 (1967).
11. D.B.Holtkamp, D.A.Clark, E.N.Ferm, R.A.Gallegos, D.Hammon, W.F.Hemsing, G.E.Hogan, V.H.Holmes, N.S.King, R.Lijestrand, R.P.Lopes, F.E.Merrill, C.L.Morris, K.B.Morley, M.M.Murray, P.D.Pazuchanics, K.P.Prestridge, J.P.Quintana, A.Saunders, T.Schafer, M.A.Shinas, H.L.Stacy., "A survey of high explosive-induced damage and spall in selected metals using proton radiography" in *Shock Compression of Condensed Matter (SCCM-2003)*, edited by M.D.Furnish, Y.M.Gupta, J.W.Forbes, AIP Conference Proceeding 706, Melville, New York, 2004, pp.477-482.
12. D.B.Holtkamp, D.A.Clark, M.D.Crain, M.D.Furnish, C.H.Gallegos, I.A.Garcia, D.L.Hammon, W.F.Hemsing, M.A.Shinas, and K.A.Thomas, "Development of a non-radiographic spall and damage diagnostic" in *Shock Compression of Condensed Matter (SCCM-2003)*, edited by M.D.Furnish, Y.M.Gupta, J.W.Forbes, AIP Conference Proceeding 706, Melville, New York, 2004, pp.473-476.
13. P. Mercier, J. Veaux, J. Benier, M. Vincent and S. Basseuil, "Doppler Laser Interferometry improvements in detonics" in *V HDP Symposium*, Saint-Malo, France, 2003, v.II, pp.99-106.
14. M.D. Knudson, D.L. Hanson, J.E. Bailey, C.A. Hall, C. Deeney, and J.R. Asay, "Equation of State Measurements in Liquid Deuterium to 100 GPa" in *V HDP Symposium*, Saint-Malo, France, 2003, v.II, pp.353-364; *SCCM-2003*, pp.81-86.
15. W.M. Trott, M.D. Knudson, L.C. Chhabildas and J.R.Asay. "Measurements of spatially resolved velocity variations in shock compressed heterogeneous materials using a line-imaging velocity interferometer" in Shock Compression of Condensed Matter-1999, edited by M.D.Furnish et al., AIP 2000, pp.993-998.

16. V.F. Kuropatenko, G.V. Kovalenko, V.I. Kuznetsova, etc., "The Software Complex VOLNA and non-uniform difference method for calculation of movements of compressed matter" in *Problems of Atomic Science and Technics. Techniques and Programs for Numerical Solving of the Problems of Mathematical Physics,* 1989, Issue 2. pp. 9-25.
17. V.V.Dremov and D.G.Modestov, *Russ. J. Chem. Phys.* **17**, #4, 109-115 (1998).
18. M.F.Gogulya, Temperaltures of shock compression of the condensed matte, Moscow: MEPhI, 1988.

Application of optical analyzer technique for measurements of sound velocities in shock-compressed Al-Mn alloy for calibration of recent elastic-viscous-plastic models

E.A. Kozlov, V.I. Tarzhanov, D.G. Pankratov, A.K. Yakunin,
V.M. Yelkin, V.N. Mikhailov

Federal State Unitary Enterprise "Russian Federal Nuclear Center – Zababakhin All–Russia Research Institute of Technical Physics", P.O. Box 245, Snezhinsk, Chelyabinsk region, 456770, Russia

Abstract. Registration results of longitudinal $C_L(\sigma_{XX})$ and volume $C_B(\sigma_{XX})$ sound velocities in shock-compressed aluminum alloy are presented. Experimental data were obtained in wide range of longitudinal stress, including the stress, corresponding to solid-liquid shock-induced transformation. By using experimentally measured values of sound velocities, the changes of Poisson ratio and shear modulus were calculated along the shock adiabat. These data are needed for calibration of resent elastic-viscous-plastic models.

Keywords: Explosive loading, optical analyzer technique, longitudinal sound velocity, volume sound velocity, elastic constants, Al-1.25% Mn alloy.
PACS: 62.50.+p, 62.20.Dc, 64.70.Dv

INTRODUCTION

Experimental data on longitudinal C_L and volume C_B sound velocities in the shock-compressed matter are important for adequate numerical description of the features of elastic-viscous-plastic behavior of metals and alloys in stress waves, running of polymorphous phase transitions, correct prediction of shock wave attenuation, needed for adequate calculation of stress gradients.

The following methods can be used to measure C_L and C_B: laser interferometry [1-3], magnetoelectric method [4], manganin sensors method [5-7], optical analyzer technique [8, 9]. An optical analyzer technique is the one being operable in the widest range of σ_{XX}.

The method is based on registration of a change in intensity of the shock wave front in the indicator-matter, which is placed after the sample being tested, as it is overtaken by rarefaction waves. Here the strong dependence of intensity of thermal radiation caused by compression in the shock wave front is applied. According to [9], under optimal conditions of experiment set-up and in the case of obtaining qualitative oscillograms such method allows one to determine the velocity of rarefaction waves

CP849, *Zababakhin Scientific Talks - 2005,*
edited by E. N. Avrorin and V. A. Simonenko

with an accuracy up to tenth fraction of percent, even in the case of unknown shock-wave and spectral characteristics of the indicator.

The method offers the following advantages:

– high sensitivity of registration of weak unloading waves against the background of strong shock wave;

– high time resolution of registration channels;

– simplicity of multi-channel measurements;

– operability in considerably wider range of longitudinal stresses, which far exceeds the operability ranges of other techniques.

The purposes of this work are:

– to obtain new experimental data on sound velocities for Al-1.25% Mn alloy using optical analyzer technique and to compare of the obtained results with experimental estimations of C_L, available in literature and measured by another methods;

– to determine changes along the Hugoniot adiabat for Al-1.25% Mn alloy of Poisson ratio and shear modulus for calibration of elastic-viscous-plastic model of this alloy.

SET-UP OF EXPLOSIVE EXPERIMENTS

To determine the sound velocity in the shock-compressed material the optical analyzer technique was applied in the variant of overtaking unloading method. The method is described in some papers, for example [9]. In our paper the Al-based alloy, Al-1.25% Mn alloy, was investigated. The choice of Al-1.25% Mn alloy as an object of investigation was made on the grounds that aluminum and alloys on its basis have been studied in details using other methods [1-15]. The available data on longitudinal sound velocities in aluminum and its alloys have been collected and analyzed for consistency [10].

The tested samples represented the disks with two or four steps. The Al-1.25% Mn alloys, 12Kh18N10T steel, Steel 10 were used as an impactor material. Chloroform $CHCl_3$ was employed as an indicator-matter. The initial data of setting-up explosive experiments are presented in Table 1.

TABLE 1. Initial data on explosive experiments.

Experiment No		1	2	3	4	5	6	7
Impactor material		12Kh18N10T	Al-Mn	Steel 10	Al-Mn	Steel 10	12Kh18N10T	Steel 10
ρ_0, g/cm^3		7.89	2.73	7.83	2.73	7.83	7.89	7.83
$h_{impactor}$, mm		1.5	2	3	2	3	2	2
$\varnothing_{impactor}$, mm		50	70	116	70	116	50	108
$W_{impactor}$, km/s		1.9	5.55	4.85	5.55	4.85	2.37	5.77
Target material		Al-1.25 ± 0.25 wt % Mn alloy						
ρ_0, g/cm^3		2.73						
h_{target}, mm	h_1	4	3	6	3	2	6	4
	h_2	7	5.5	12	4.5	3.5	8	6
	h_3	-	-	-	6	5	10	8
	h_4	-	-	-	7.5	6.5	12	10
Indicator		$CHCl_3$						
ρ_0, g/cm^3		1.483						

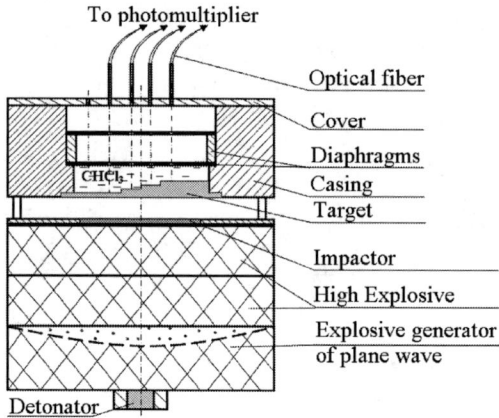

To photomultiplier

Optical fiber
Cover
Diaphragms
Casing
Target

Impactor
High Explosive
Explosive generator
of plane wave

CHCl₃

Detonator

FIGURE 1. Typical design of measuring assembly (not in a scale).

Figure 1 gives the typical design of the experimental assembly used in explosive experiments.

Plane impactors are launched by a detonating high explosive charges, ⌀60 mm in diameter and with mass up to 300 g of TNT, or by HE charges of ⌀120 mm and ⌀200 mm in diameters and with HE mass from 5 to 25 kg of TNT. The optical fibers are applied as channels to transmit light signals from the experimental assembly to photodetectors. They have the following parameters and characteristics:
- type of optical fiber   quartz-polymeric, multi-mode, stepwise;
- diameter of optical fiber core   0.35 or 0.7 mm;
- specific light energy losses   < 30 dB/km.

EXPERIMENTAL RESULTS AND TEORETICAL ESTIMATIONS

Typical oscillograms of explosive experiment on measurement of longitudinal sound velocity in shock-compressed Al-1.25% Mn alloy are presented in Figure 2a. The obtained values of longitudinal sound velocity as compared to previously published data [1-17] are given in Figure 2b. The results of estimation of volume sound velocity C_B, given by other authors, are also shown in Figure 2b, as well as our results from experiments in which indicator thickness was sufficient for registration of intensity change of shock wave front when it is overtaken by head characteristic of plastic rarefaction wave.

Solid lines in Figure 2b present the calculational dependencies $C_L(\sigma_{XX})$ and $C_B(\sigma_{XX})$ ($C_L = \sqrt{(B + 4/3 \cdot G)/\rho}$, $C_B = \sqrt{B/\rho}$) for Al-1.25% Mn alloy along the Hugoniot adiabat. The bulk compression modulus B was calculated using the equation of state, which included the potential part in the form of the universal equation of state VRF [18] and contribution of atomic thermal vibrations in Debye approximation.

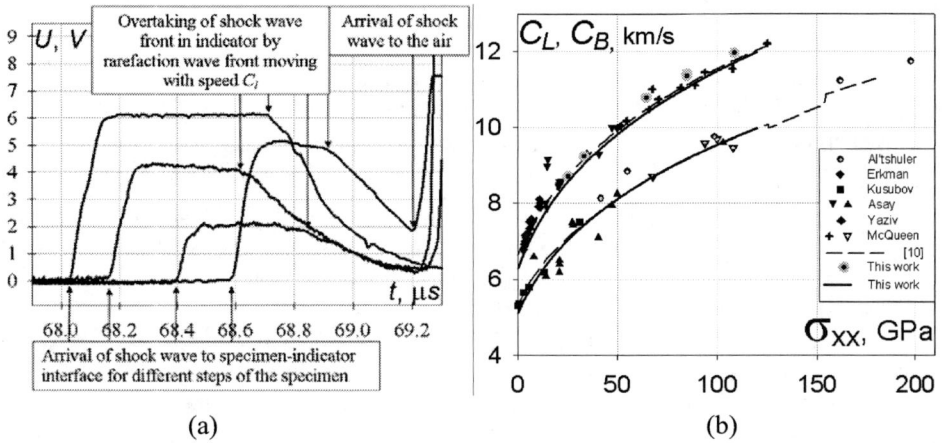

(a)   (b)

FIGURE 2. Typical oscillograms of explosive experiment (Experiment No 5) (a) and longitudinal and volume sound velocities in shock-compressed aluminum and alloys on its basis, which were measured using different methods, as compared to calculations by two-phase equation of state for Al, formulated in [10] and [17], as well as in this work (b).

For calculations of the shear modulus G the functional dependence on volume and temperature, being the same as for the bulk compression modulus, was used, but with another set of parameters matched to experimental temperature and pressure dependencies of the shear modulus [19]. The dependencies obtained for Poisson ratio $v = \dfrac{3B - 2G}{2(3B + G)}$ and shear modulus on shock compression pressure in this work are given in Figure 3 in comparison with results of work [10].

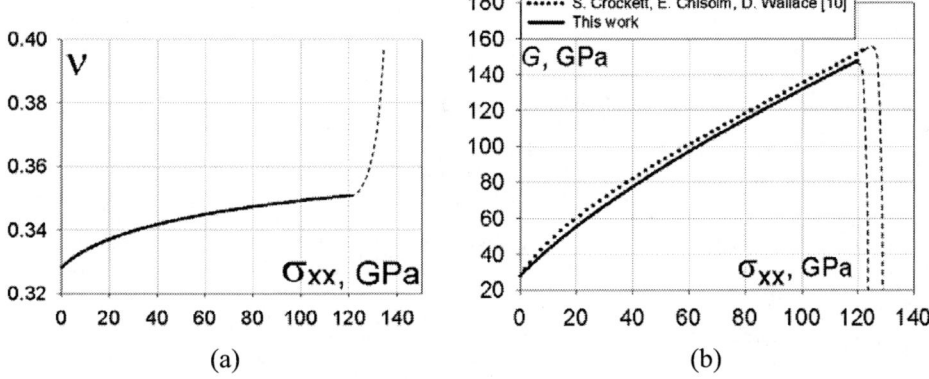

(a)   (b)

FIGURE 3. Changes along shock adiabat of Poisson ratio v (a) and shear modulus G (b) for Al and alloys on its basis calculated from the results [1, 12-16] and data obtained in the current work.

These results are needed for verification and calibration of recent elastic-viscous-plastic models. Based on all obtained experimental [1-9, 11-15, 19] and calculation - theoretical data [10, 16-18] the start of melting of aluminum and its alloys is estimated as 125-150 GPa.

CONCLUSIONS

1. Seven explosive experiments were performed on registration of longitudinal and volume sound velocities in the shock-compressed domestic Al-1.25% Mn alloy in the range of longitudinal stresses σ_{xx}=25-110 GPa, i.e., almost up to the stress of the beginning of alloy melting in shock wave front at σ_{xx}=125-150 GPa.

2. In the conducted experiments with using the optical analyzer technique, the measurement error was estimated for longitudinal sound velocity in shock-compressed materials. If the target and the impactor are of the same material, the error is equal to $\approx$ ±1.8 %, and in another ones – ±2.8%.

3. The experimental data obtained in this work were compared to previously published ones. It is shown that these data are in close agreement.

4. As compared to the low-resistance manganin sensors method the optical analyzer technique possesses higher sensitivity in registration of weak (elastic) disturbances against the background of strong plastic waves and it is operable in considerably wider and higher range of longitudinal stresses and pressures.

5. In future the optical analyzer technique will be applied for refining the shock wave amplitude, which corresponds to the beginning of Al-1.25% Mn alloy melting directly in the shock wave front for verification of recent two-phase equation of state for Al-1.25% Mn alloy with regard to exhibition of elastic-viscous-plastic properties by the alloy.

REFERENCES

1. J.R. Asay and L.C. Chhabildas, "Determination of the Shear Strength of Shock-Compressed 6061-T6 Aluminum" in *Shock-Waves and High-Strain-Rate Phenomena in Metals,* edited by M.A.Meyers and L.E.Murr, Plenum, New York, 1981, pp.417-431.
2. R.A. Graham and J.R. Asay, *High Temp. High Press*. **10,** #2, 355-390 (1978).
3. L.C. Chhabildas and R.A. Graham, "Development in Measurement Technique for Shock Loaded Solids" in *Techniques and Theory of Stress Measurements for Shock Wave Applications*, edited by R.R. Stout et al., American Society of Mechanical Engineers, New York, 1987, pp.1-18.
4. A.A. Vorob'ev, A.N. Dremin and G.I. Kanel, *Zh. Prikl. Mekh. Tekh. Fiz*. **15,** #5, 94-100 (1974).
5. A.N. Dremin and G.I. Kanel, *Zh. Prikl. Mekh. Tekh. Fiz*. **17,** #2, 146-153 (1976).
6. A.N. Dremin, G.I. Kanel and O.B. Chernikova, *Zh .Prikl .Mekh. Tekh. Fiz*. **22,** #4, 132-138 (1981).
7. G.I. Kanel, S.V. Razorenov and V.E. Fortov, *Zh. Prikl. Mekh. Tekh. Fiz*. **29,** #6, 67-70 (1988).
8. I.M. Voskoboynikov and M.F. *Gogulya, Khim. Fiz*. **3,** # 7, 1036-1041 (1984).
9. R.G. McQueen, J.W. Hopson and L.N. Fritz, *Rev. Sci. Instr.* **53,** #2, 245-250 (1982).
10. S. Crokett, E. Chisolm and D. Wallace, "A Comparison of Theory and Experiment of the Bulk Sound Velocity in Aluminum Using a Two-Phase EOS" in *Shock Compression of Condensed Matter-2003*, edited by M.D. Furnish et al., AIP Conference Proceedings 706, American Institute of Physics, Melville, NY, 2004, pp. 45-49.
11. L.V. Al'tshuler, S.B. Kormer, M.I. Brazhnik, L.A. Vladimirov, M.P. Speranskaya and A.I. Funtikov, *Zh. Exp. Teor. Fiz.* **38,** #4, 1061-1072 (1960)
12. J.O. Erkman and A.B. Christensen, *J. Appl. Phys.* **38,** 5395 (1967).
13. A.S. Kusubov, and M. van Thiel, *J. Appl. Phys.* **40,** #2, 839-899 (1969).
14. R.G. McQueen, J.N. Fritz and C.E. Morris, "The Velocity of Sound Behind Strong Shock Waves in 2024 Al" in *Shock Waves in Condensed Matter,* edited by J.R.Asay et al., Elsevier, Science Publishers, 1984, pp. 95-98.
15. D. Yaziv, Z. Rozenberg and Y. Partom, *J. Appl. Phys.* **53,** 353 (1982).
16. D.J. Steinberg and R.W. Sharp, *J. Appl. Phys.* **52,** #8, 5072-5083 (1981).
17. A.T. Sapozhnikov, E.E. Mironova and L.N. Shakhova, VIII-ZST, September, 5-10, 2005, RFNC-VNIITF, Snezhinsk.
18. P. Vinet, J.H. Rose, J. Ferrante, J.R. Smith, *J. Phys. Condens. Matter* **1,** 1941-1963 (1989)
19. I.N. Frantsevich, F.F Voronov, S.A. Bakuta, Elastic constants and elastic modulus of metals and non-metals, Kiev, Naukova Dumka, 1982.

Using The Plasticity Theory With Taking Into Account Rate Sensitivity To Monoaxial Tension Analysis Of Porous Bodies

L.A. Ryabicheva

Department of Applied Material Science,East-Ukrainian Volodymir Dal National University, Molodiozhny block, 20A, Lugansk, 91034, Ukraine

Abstract. Within the framework of modified model of plastic yielding the behavior of porous bodies observed at a tension. The kinematic characteristics of process are defined as dependences of porosity from axial deformation and average properties of solid phase and dependence of axial force from deformation too. According to experimental data, such dependence has an extremum for a fixed value of initial porosity.

Keywords: Plastic Deformation, Tension, Strain Degree, Rate, Monoaxial, Porosity, Yielding Stress, Hard Phase
PACS: 81.20.Ev

INTRODUCTION

The calculation of limiting loading is carried out on the basis of the volume conservation law at plastic deformation. In contrast to compact materials, the tension test of porous bodies is accompanied by forming of defects and their development. The reasons of the indicated phenomenon are well known and it is possible brought them to two basic reasons: presence of generically stipulated defects (residual pores in raw materials) and pores arising during deformation (for example, as a result of alienation of hardening phase particles from the matrix material). Therefore, during calculation of limiting loading it is necessary to take into account change of volume in accordance with a tension [1, 2].

For simplification of performances calculation at tension of porous material, it is necessary to use a variant of the plasticity theory of porous bodies offered in paper [3], where more appropriate material's functions implemented. Two factors, causing hardening of porous material are taking into consideration: hardening of a porous body matrix and change of its density [4, 5]. Both factors affect in opposite directions, because hardening of a hard phase promotes of resistance to deformation increase (as well as at compression), and loosening stipulates softening.

For simplification of performances calculation at tension of porous material, it is necessary to use a variant of the plasticity theory of porous bodies offered in paper [3], where more appropriate material's functions implemented. In accordance to this

CP849, *Zababakhin Scientific Talks - 2005,*
edited by E. N. Avrorin and V. A. Simonenko
© 2006 American Institute of Physics 0-7354-0345-7/06/$23.00

approach, the preference given to the approach closed to physical representations concerning deformation of porous bodies at a tension. Two factors, causing hardening of porous material are taking into consideration: hardening of a porous body matrix and change of its density [4, 5].

THE POROSITY CHANGE AND RELATION OF MACROSCOPIC STRESS WITH YIELDING STRESS OF HARD PHASE AT TENSION

According to a model [3], the dependence of porosity on accumulated strain of hard phase defined. Then, at the basis of the «mean square» concept [6], the accumulated strain of matrix phase is defined. The obtained results will be use directly for the further analysis of stress-strain dependence at a tension.

The monoaxial tension of porous cylindrical sample of current height h (initial h_H) is considered. It is supposed that deformation of the sample is always macroscopically homogeneous and the measure of deformation is the magnitude:

$$\varepsilon_z = \ln \frac{h}{h_H} \tag{1}$$

On the basis of results [3] a porosity change can be described by an exponential type dependence:

$$\theta = \theta_H \exp(k_1 \varepsilon_z) \tag{2}$$

where θ_H - is an initial porosity; k_1- is a parameter of experimental data approximation sensitive to a strain rate.

Then we introduce a concept concerning a yielding stress of hard phase, which is a local performance that is various in different points of matrix material. Further keeping in mind its average value, because it will be use for macroscopic characterisation. In correspondence with data [3], the ratio σ_z / σ_s, where σ_s - is yielding stress of hard phase, looks like:

$$\sigma_z = \sigma_s (1 - k_2 \theta), \tag{3}$$

THE ESTIMATION OF STRAIN AND DEFORMATION HARDENING OFTHE MATERIAL OF MATRIX

To estimate the contribution of deformation hardening of the material of matrix to tension resistance of porous material, it is necessary to connect yielding stress in it with a macroscopic strain. According to conventional concepts, the indicated yielding stress is function of accumulated strain in the material of matrix. The aspect of this function is defined by the mechanism yielding, and the dependence has a local character. Therefore at macroscopic characterisation an average value used, that is necessary to connect with a macroscopic strain for solving of the problem formulated.

With this purpose, the postulate of V.V. Scorokhod about the uniqueness of dissipation function [6], is implemented. Its substance is reduced to the following statement: the specific velocity of energy dissipation at deformation of porous body is equal to specific velocity of dissipation of its hard phase, averaged to the representative element. In case of monoaxial deformation in the conjecture, that the sample is always macroscopically homogeneous, the specific velocity of dissipation can be calculated as follows.

At the beginning, we shall estimate power of sample deformation, which is equal by a definition:

$$N = Fv, \tag{4}$$

where F - is a deformation force; v - is a velocity of deformation, $F = \sigma_z S$, $v = \dot{h}$, S - is a current section area; $\dot{h}$ - is a velocity of upper surface displacement.
The specific power on definition is equal N/m, where m - is mass of a sample, $m = \rho S h$. The expression for a specific power becomes:

$$\frac{N}{m} = \frac{1}{\rho} \sigma_z \frac{\dot{h}}{h}. \tag{5}$$

Now we may estimate the power of a hard phase deformation. This estimate based on the following reasons. In the volume of representative element strains distribution is non-homogeneous. Besides the local tensor of strain rates can`t be reduced to one component. Therefore, it is necessary to operate with average performances. As the strain of a hard phase is reduced to shears, scalar performance of its deformed state change is the intensity of shear strain rates γ. The relation between average strain rate of hard phase of porous body W and γ is possible to represent as:

$$W = \langle \gamma \rangle = \frac{1}{V} \int_V \gamma dV, \tag{6}$$

where V - is volume of hard phase of representative element.
The power of hard phase deformation:

$$N_0 = \frac{1}{\rho_k} \sigma_s \langle \gamma \rangle. \tag{7}$$

413

Having estimated power of sample deformation (5) and power of its hard phase (6), it is possible to come to the solution of the main problem - definition of average strain rate of a hard phase.

According to postulate of V.V. Scorokhod [6], a specific power of hard phase is equal to a specific power of all porous body strain. Therefore, taking into account (2) and (3), finally, we have following expression:

$$\omega = \varepsilon_z + \frac{1-k_2}{k_1} \ln \frac{1-\theta}{1-\theta_H} . \tag{8}$$

It is possible, that at the beginning of process the strain of a hard phase is close to macroscopic, thus for lower values of a porosity the relevant values are closer to each other.

THE LIMITING STRAIN DEGREE OF POROUS SAMPLES AT MONOAXIAL TENSION

We shall start from the following definition: the limiting degree of strain at monoaxial tension of porous samples is such value of axial strain, with what the value of deformation reaches a maximal value. Therefore, required strain degree will be defined from the analysis of force - strain curve. It is impossible to present solutions of the following equation in elementary functions:

$$\frac{\partial F}{\partial \varepsilon_z} = 0 . \tag{9}$$

Therefore, consequent analysis is based on the analysis of dependence $F = F(\varepsilon)$, Constructed on sequential use of expressions obtained with the account of the definition $F = \sigma_z S$. The expression for S may be obtained from the mass conservation law and by using the equation (2) it is possible to write:

$$S = S_H (\frac{\theta_H}{\theta})^{\frac{1}{k_1}} \frac{1-\theta_H}{1-\theta} . \tag{10}$$

During calculations the law of hardening in the Ludwig's formulation used. The typical graph for F presented in a Figure 1 ($\theta_H = 0.3$).

The indicated calculations done for values of initial porosity from 0.001 up to 0.45. During calculations the dependence $F = F(\varepsilon)$ were tabulated, and the value of ε_z is fixed, when the maximal value of F reached. Further obtained limiting values ε_{lim} compared to the initial value of porosity. The relevant graph presented in Figure 2.

414

The obtained result allows define dependence of limiting strain degree on initial porosity. The data presented in fig. 2, are corresponding to known experimental results and may be used during processing of limiting strain and limiting stress before fracture values for porous samples.

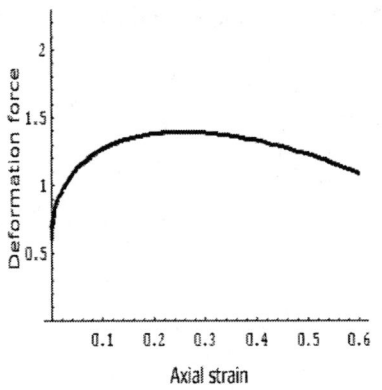

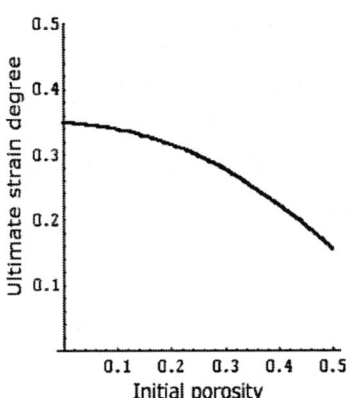

FIGURE 1. The calculated dependence of deformation force on axial strain at monoaxial tension

FIGURE 2. Dependence of limiting strain degree of porous samples ε_{lim} from initial porosity

CONCLUSIONS

The existing techniques for definition of plasticity resource at monoaxial tension are not taking into account possible irreversible change of volume. Such account is possible only on the basis of the plasticity theory of porous bodies. Among other models it is necessary to use the model where the formulation of material's functions is reasonable and convenient for the analysis of monoaxial loading. The definition of limiting performances of deformation is reduced to evaluation of dependences of porosity and accumulated strain in the matrix of material from axial strain. The accumulated strain in a matrix defined on the basis of postulate about uniqueness of dissipation function. The limiting strain degree in such conditions should be sufficient to ensure an extremum of dependence conditional stress - true strain. The limiting strain changes under the almost parabolic law with growth of initial porosity.

REFERENCES

1. M. B. Shtern, Powder Metallurgy **5**, 5-14, (1989).
2. M. B. Shtern, Powder Metallurgy **6**, 13-18, (1989).
3. L. A. Ryabicheva, Yu.V. Kravtsova, «The Most New Technologies in Powder Metallurgy and Ceramics», Conference Proceedings, Kyiv, Academperiodica, 2003.
4. I. F. Martinova, V. V. Scorokhod, Powder Metallurgy, **5**, 14-17, (1976).
5. M. B. Shtern, V. D. Dudunov, Powder Metallurgy, **11/12**, 31-40, (1999).
6. V. V. Scorokhod, "Reological basics of sintering theory", Kyiv, Naukova Dumka, 1972, 142 p.

About Influence of Ionic Beams of Metals of IVB-VIB Groups on Structure and Properties of a Target at Various Modes of Irradiation

E.I. Kurbatova, V.A. Klimanov, A.I. Ksenofontov, I.N. Fridlyander

Moscow Engineering Physics Institute, 31 Kashirskoe shosse, Moscow, 115409, Russia

Abstract. A complex of special properties of materials and a threshold of their working temperatures which is frequently limited 500-550⁰C define the level of modern technological development and, first of all, for nuclear-industrial plants. The alloys on the basis of iron are the most widespread kind of constructional materials, and therefore an increase of their special properties (high-temperature strength, corrosion stability, durability and other characteristics) is of great value. The ionic beams with a low energy (300-1,500 eV) which can change the structure and the properties of a target, in particular iron and its alloys were used for the solution of these problems in the present work. In this work theoretical and experimental results of the research of the process of impact interaction of the ionic beams with iron are also given. The properties of a iron surface with the help of modern methods of physical and chemical analyse are investigated. The opportunities to produce new surface properties of iron using the ionic beams are shown. On the basis of the results obtained the supplementary work directions are analyzed.

INTRODUCTION

Interaction of ionic beams with a target, leading to a change of its structure, is one of the finest processes of handling surface properties [1-3]. Taking into account the complexity of physical and chemical phenomena proceeding on the surface, the solution of each particular problem demands carrying out of some preliminary research for which both the theoretical and experimental methods of investigation were used.

As iron possesses an extremely limited set of properties the irradiation was carried out by the ionic beams of metals of IVB-VIB groups. These groups cover practically a full range of high-strength, high-melting and heat-proof metals and consequently can greatly influence the initial properties of iron.

METHODS AND RESULTS OF RESEARCH

The research of the process of interaction of these ionic beams started from the calculation of a penetration depth of ions into iron, normal to the surface. The energy range of ions amounted to 300-1,500 eV.

At the first stage the calculation was carried out for the simplest case of interaction of unary plasma with iron, arising from "a direct penetration" and stopping of the ions

CP849, *Zababakhin Scientific Talks - 2005,*
edited by E. N. Avrorin and V. A. Simonenko

[4]. The calculation results for various energy values of ions and degrees of ionization of plasma are demonstrated in Table 1.

TABLE 1. The penetration depth of ions of titan in iron, cm $*10^{-8}$.

Charge	Energy of ions, eV		
	500	1000	2000
+1	0.09	0.32	0.92·
+2	0.45	0.98	2.6
+3	0.81	1.72	4.3

Thus, for the investigated energy range the calculated value of penetration depth of ions of titanium into iron reaches the amount of about $0.09\text{-}4.3 \cdot 10^{-8}$ cm. For heavy metals (molybdenum and tungsten) the penetration depth becomes even less.

A complex of methods of physical and chemical analysis was used to compare theoretical and experimental data the research of interaction process of the ionic beams of metals of IVB-VIB groups with iron.

The ionic beams were produced when evaporating metal cathodes in plasma of an arc discharge in vacuum [5].

As a result of the studies involved it was found that practically in all cases of an irradiation the interaction between the metal of an ionic beam and iron at initial stages mainly proceeds on an iron surface. Then the interaction extended deep into the target. The character of proceeding reactions in the zone of interaction is significantly individual for each kind of ions and is accompanied by the formation of various phases.

The α-solid solutions on the basis of iron and the intermetallic compounds (Laves-phases - λ and σ-phases) were identified by the X-ray spectral analysis in zones of interaction. Their structure is demonstrated in Table 2.

TABLE 2. The phase structure of interaction zones of ions of IVB-VIB groups metals with iron.

Metal of an ionic beam	Titanium	Vanadium	Molybdenum
Phase structure of zone of interaction	$\alpha_{Fe}+\lambda(TiFe_2)$	$\alpha_{Fe}+\sigma(TiV)$	$\alpha_{Fe}+\lambda(MoFe_2)$

Thus, the phase structure of zones formed during the interaction of ions with the target is various. The phase structure defines the physical and chemical properties of the zones. This behavior allows to form the preset characteristics of a surface with an optimum combination of ions.

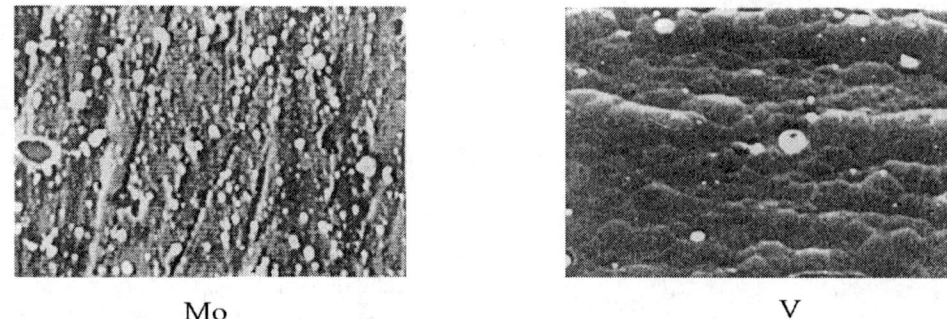

Mo   V

FIGURE 1. The microstructure of zones of interaction of the ionic beams of vanadium and molybdenum with iron, *1000.

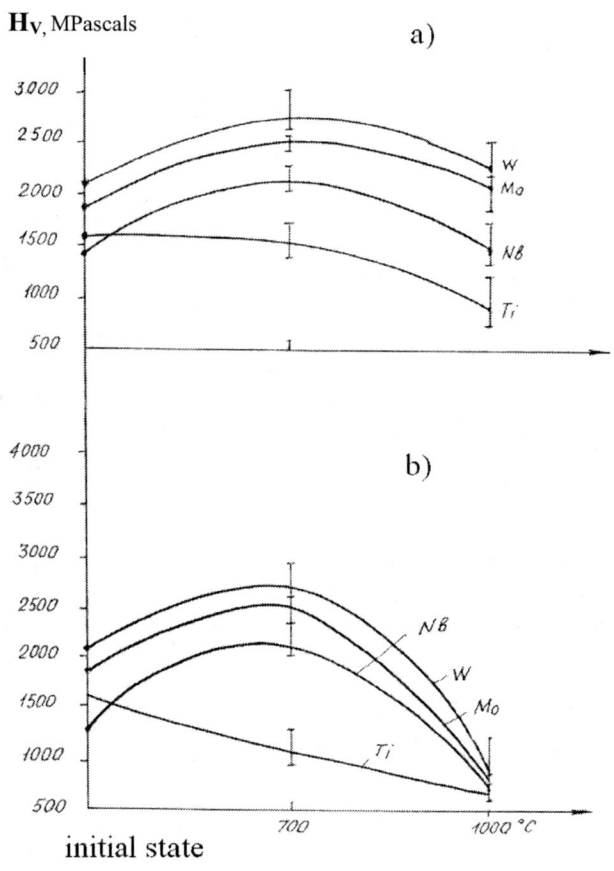

FIGURE 2. The influence of bakeout on hardness of iron after an irradiation by metals of IVB-VIB groups. The cure time is 15 min (a) and 60 min (b).

The structure of the synthesized zones of interaction of ions with iron by the electron-microscopic and microstructural analysis was investigated. It was established that the type of their structures is associated with the process of physical and chemical interaction of iron with metal ions. In most cases the microstructure has an irregular character, thus the average size of a grain reaches 800-900 microns. The most coarsely dispersed structure is observed in the zones formed during the interaction of iron with ions of vanadium (Figure 1), and the fine dispersed structure — by ions of high-melting metals (molybdenum on Figure 1).

The method of local X-ray spectral analysis established, that the length of zones of interaction of metal ions with iron in its depth is on the average 2-10 microns, which is considerably in excess of the expected value. It follows that the penetration depth of ions into iron is only to a small extent defined by the energy range of ions. Probably, this depth is defined by the character of proceeding physical and chemical and radiation-dynamic processes to a much greater extent.

The research of the properties of the irradiated iron was carried out in two directions: the increase of high temperature strength and corrosion stability. Corrosion tests of the samples of iron irradiated with the ionic beams of metals of IVB-VIB groups were carried out by the express technique. The comparative criterion of estimation of sample stability to oxidation was the time of reaching the potential of the basic material (iron) after immersion of the samples into the 20 % solution of H_3PO_4.

The time of the incubatory period necessary for the achievement of iron potential for several tests is given in Table 3.

TABLE 3. The research results of corrosion stability of iron samples after irradiation with ions of titanium, vanadium and molybdenum.

Kind of a sample	Time of the incubatory period, min
The iron irradiated with ions of titanium	3
The iron irradiated with ions of vanadium	4
The iron irradiated with ions of molybdenum	5

The results of the tests testify to the increase of corrosion stability of iron after an ionic irradiation, it especially concerns the ions of molybdenum. During the research of high temperature strength of the irradiated iron the mechanical tests were carried out and the changes of structure and properties during high-temperature bakeout were investigated. In particular, the changes of microhardness of interaction zones of iron with ions of metals of IVB-VIB groups after bakeout at temperatures 700-1000^{0}C were investigated. It was established, that the zones of interaction containing the Laves-phases, possess greater microhardness which is kept at temperatures 700^{0}C. The zones containing the Laves-phase of high-melting metals in their structure (Figure 2) for a short period of time possess increased microhardness at higher temperatures.

CONCLUSION

The increase of a threshold of working temperatures of iron and its alloys is an urgent problem from the point of view of their ability to resist shape changer and heat emission under conditions of various kinds of radiation doses.

The results obtained confirm large-scale feasibilities of the use of an ionic irradiation process to increase properties of iron and its alloys, and also to produce materials with preset properties.

This work was fairly supported by the International Science and Technology Center (ISTC), the ISTC project #2942.

REFERENCES

1. *Physics and Technology of the Ionic Sources,* edited by J. Brown, Moscow, Mir, 1998, 448p.
2. *Ionic Implantation,* edited by J. K. Hirvoneka, Moscow, Metallurgy, 1985, 390p.
3. N. V. Pleshivtsev and A. I. Bazhin, *Physics of Influence of Ionic Beams on Materials,* Moscow, Vuzovskaya kniga, 1998, 392p.
4. M. Tompson, *Defects and Radiation Damages in the Metals,* Moscow, Mir, 1985, 367p.
5. L.T.Bugaenko and L.S.Polak, *Chemistry of High Energies,* Moscow, Nauka, 1988, 135p.

Investigation of Destruction of Functional Gradient Barriers at Shockwave Loading

V.P. Glazyrin, M.Yu. Orlov, Yu.N. Orlov

Scientific Research Institute of Applied Mathematics and Mechanics,
Tomsk State University,
Tomsk, Russia

Abstract: Nowadays, functional gradient materials (FGM) are successfully used to manufacture parts and structures which are subject to strains and temperature loadings. Use of FGM as well as other advanced materials to develop and manufacture shock-resistant protective structures is conditioned by continuous upgrade of shock and explosive devices. In this paper, computational investigation of shockwave loading of barriers with strength gradient properties is presented. In particular, the effect of linear change in characteristics responsible for spallation and shear strains over the whole barrier thickness from the surface loaded through its back surface on strain and destruction processes was investigated. The effect on barriers was defined as plane shock wave loading, and as impact of compact steel oblong strikers against the barriers along the normal.

INTRODUCTION

In this paper, computational investigation of shockwave loading of barriers with a strength gradient properties is presented. In particular, the effect of linear change in characteristics responsible for spallation and shear strains over the whole barrier thickness from the surface loaded through its back surface on strain and destruction processes was investigated.

In our investigation, we used the lagrangian simulation technique of hyper-velocity strain of compressed elastic-plastic porous body taking into account partial destruction referred to two-dimensional axial symmetry [1, 2]. This technique assumes that spallation occurs when principal tensile stress σ_1 achieves spallation strength value σ_k, and shear destructions occur when specific work of shear plastic strain A_p reaches its critical value A_k.

The selected intervals of change in strength parameters correspond to real steel alloys. In our calculations, spallation strength value σ_k was varied from 2.1 to 3.5 GPa, and A_k value was varied from 35 to 55 kDj/kg. The change in the given parameters over barrier thickness was defined either by increasing or decreasing linear function. For homogeneous barrier, the following were the values: $\sigma_k =2.8$ GPa, $A_k = 45$ kDj/kg.

CP849, *Zababakhin Scientific Talks - 2005,*
edited by E. N. Avrorin and V. A. Simonenko

The appropriate physical and mechanical characteristics of the materials were borrowed from the works [2, 3]. The effect on barriers was defined as plane shock wave loading, and as impact of compact steel oblong strikers against the barriers along the normal. The barriers were 1.2 sm thick with diameter of 4 sm.

Total nine scenarios for combinations of changes in σ_k and A_k values over barrier thickness were considered. For convenience, the following symbols were introduced "−", "↑", and "↓" to designate constancy, increase, and decrease (correspondingly) of the above-mentioned characteristic over barrier thickness. For example, if a barrier is defined by $[\sigma_k^-, A_k^-]$, thus, σ_k and A_k values are constant over the whole barrier thickness. If a barrier is defined by $[\sigma_k^\uparrow, A_k^\downarrow]$, thus, σ_k value increases, and A_k value decreases from the barrier surface loaded through its back one. The remaining seven barriers are defined in a similar way.

PLANE SHOCK PULSE LOADING

At first, a series of computational experiments on barriers loading with plane shock pulse (pulse width of 1.2 µs, and amplitude of 175 m/s) was carried out. Stress-strain state, strain volumes and shapes, and back surface velocity at the axis of symmetry were calculated at each time point.

The calculation results for homogeneous steel barrier are given in fig. 1. It is found out that the first destruction areas induced by strain stress effect appear at the second micro-second. Then, at the forth micro-second, localization of the destruction areas in plane is observed as a line area covered with cracks which directs separation of the spall part from the sample. These destructions are mostly treated as spallations with average rate (in fig. 1a, t = 20 µs).

The calculated velocity profiles of the barrier back surface are provided in fig. 1b. It is evident that when the shock wave leaves the free surface, its amplitude decreases by 18.6% if compare with the initial amplitude under effect of upcoming unloading wave. In the plot, a minor drop and increase in velocity followed by its decrease is observed within time gap from 5 to 8 µs. At time point of 20 µs, back surface velocity is equal to 150 m/s.

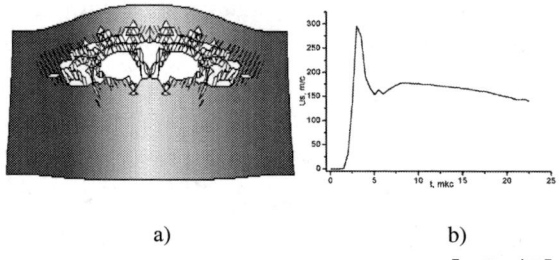

a)   b)

FIGURE 1. Calculation results for barrier $[\sigma_k^-, A_k^-]$.

The calculations of shock loading of barriers $[\sigma_k^-, A_k^\downarrow]$ and $[\sigma_k^-, A_k^\uparrow]$ show that numerical values of stress-strain state and back surface velocity differ from the values calculated for barrier $[\sigma_k^-, A_k^-]$ no more than by 0,5%. Patterns of destructions and back surface velocity profiles for these three scenarios are almost identical by eye.

The results of simulation of loading of barrier $[\sigma_k^\uparrow, A_k^-]$ with plane shock pulse show that, like in the previous scenario, destruction points localize in plane. The destructions are mostly treated as spallations. In this scenario, near-surface layer of the spall taking shape is less destroyed if compare with the scenario for barrier $[\sigma_k^-, A_k^-]$. In the middle part of the sample, light destructions can be observed that were absent in the previous scenario. When comparing shapes of back surface velocity profiles for barriers $[\sigma_k^-, A_k^-]$ and $[\sigma_k^\uparrow, A_k^-]$ it is found out that in the latter scenario, oscillations behind shock wave front are more distinct that is probably connected with less severe destruction of near-to-surface layer of the spall.

The calculations of shock loading of barriers $[\sigma_k^\uparrow, A_k^\uparrow]$ and $[\sigma_k^\uparrow, A_k^\downarrow]$ show that patterns of strains and back surface velocity profiles are almost identical if compare with the scenario for barrier $[\sigma_k^\uparrow, A_k^-]$.

Then, loading of barrier $[\sigma_k^\downarrow, A_k^-]$ was modeled. It is evident that spallation is followed by severe destruction of near-to-surface layer of the spall, and, thus, oscillations behind shock wave front are very weak. In this scenario, the internal destruction is the most severe if compare with other scenarios. The calculations results obtained from loading of barriers $[\sigma_k^\downarrow, A_k^\downarrow]$ and $[\sigma_k^\downarrow, A_k^\uparrow]$ are almost identical to the results obtained from barrier $[\sigma_k^\downarrow, A_k^-]$. The divergences do not exceed 0.5 %.

Thus, the calculation results prove that spallation mechanism is prevailing in destruction process at shockwave loading of the barriers with strength gradient properties that is why change in critical value of A_k affects neither strain pattern nor destruction volume and shape. At increase of σ_k value over barrier thickness, spallation is followed by less severe destruction of near-to-surface layer of the spall if compare with the scenario for homogeneous barrier.

EFFECT OF COMPACT STRIKER

Shock interaction of compact steel strikers against the above-mentioned barriers along the normal with velocity of 800 m/s was simulated. Axial section of cylindrical strikers was 5×5 mm. Sliding condition was defined at the contact surface.

When counting, the following values were recorded: V_c - velocity of striker center of mass of, L_k - striker penetration depth, strain pattern, and destruction areas. At time point of 20 µs, the counting was stopped since decrease in striker velocity was minor, and destruction volume and shape were almost the same. At this time point, areas of the material destroyed at the interface "striker – target" (S_k), and at the barrier back surface (S_b) were calculated for the considered section of the interacting bodies.

423

The calculated typical positions of the interacting bodies at time point t = 20 μs are described in fig. 2.

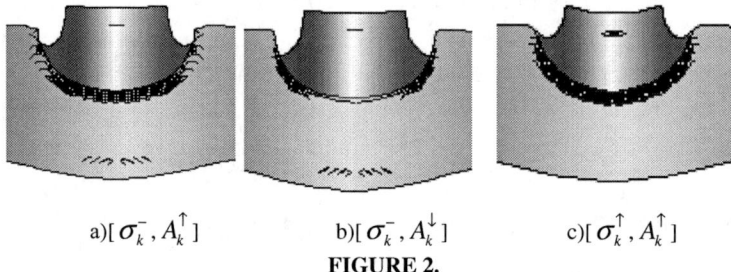

a)$[\sigma_k^-, A_k^\uparrow]$ b)$[\sigma_k^-, A_k^\downarrow]$ c)$[\sigma_k^\uparrow, A_k^\uparrow]$

FIGURE 2.

All the values of V_c, L_k S_k, S_b, and S_Σ obtained from the nine barriers at time point of 20 μs are given in Table 1. Total destruction volume i.e. $S_\Sigma = S_k + S_b$ is provided in the last column.

In the table, it is observed that decrease in velocity of the striker center of mass as far as it penetrates into the barrier is the fastest when shear strength value A_k decreases from the barrier front surface through its back one. If A_k decreases, and σ_k increases, significant increase in velocity of the striker center of mass is observed.

TABLE 1.

Compact striker effect.

Barrier type	V_c [m/s]	L_k [mm]	S_k [mm²]	S_b [mm²]	S_Σ [mm²]
Barrier #1 $[\sigma_k^-, A_k^-]$	25.95	279	1.9	0.45	2.35
Barrier #2 $[\sigma_k^\uparrow, A_k^-]$	23.23	3.00	2.64	0	2.64
Barrier #3 $[\sigma_k^\downarrow, A_k^-]$	19.74	2.79	2.11	3.16	5.27
Barrier #4 $[\sigma_k^-, A_k^\uparrow]$	23.39	2.95	2.84	0.43	3.27
Barrier #5 $[\sigma_k^-, A_k^\downarrow]$	9.55	2.71	0.64	0.48	1.12
Barrier #6 $[\sigma_k^\uparrow, A_k^\uparrow]$	23.56	2.75	2.6	0	2.6
Barrier #7 $[\sigma_k^\downarrow, A_k^\downarrow]$	9.77	2.77	1.08	3.6	4.69
Barrier #8 $[\sigma_k^\uparrow, A_k^\downarrow]$	14.71	2.91	0.3	0	0.3
Barrier #9 $[\sigma_k^\downarrow, A_k^\uparrow]$	23.75	2.74	2.84	2.60	5.44

Striker penetration depth for all the scenarios varies within range of 10%. The lowest value of L_k = 2,7 mm was obtained from barrier $[\sigma_k^-, A_k^\downarrow]$.

The destructions at the interface of the interaction bodies are mostly treated as shear strains that is why the lowest value of S_k was obtained from the barriers for which the value A_k at the front surface was maximum. The destructions at the barrier back surface are mostly treated as spallations that is why zero values of S_b were obtained from the scenarios with the increase in σ_k.

The highest total destruction volume was obtained from the barrier for which σ_k decreased, and A_k increased from the front surface through its back one. The lowest destruction volume was obtained from the barrier for which σ_k increased, and A_k decreased.

Thus, as a result of mathematical modeling, it was found out that the decrease in shear strength over barrier thickness results in faster decrease in striker velocity as far as it penetrates into the barrier. The less severe destructions were observed at the increase in spall strength of the barrier material, and the decrease in its shear strength.

EFFECT OF OBLONG STRIKER

Shock interaction of 6,1Smk-bullet core part (Germany) with gradient barriers was modeled to investigate the effect of the striker shape on penetration process. The striker was 1.6 sm long with diameter of 0.61 sm. The dimensionless criterion was introduced for quantitative description of barrier shock resistance. This criterion was expressed via V_c - initial velocity of the striker's center of mass, and its velocity at time point of 42 μs

$$R = \frac{(V_0 - V_c)}{V_0};$$

where V_0 - striker initial velocity (800 m/s), V_c - velocity of striker center of mass at time point $t = 42$ μs.

The calculated values for criterion R for each scenario are given in the table and analyzed below. All of them are within the range of 0.49 to 0.67. It is clear that the closer the R to unity, the higher the barrier shock resistance.

The visual comparison of barrier destruction patterns resulted in the following conclusions. At the 12th μs, striker partially penetrated into the barrier, and as result, volume of the barrier material round the ogival striker part got more compact. Light plastic strain of the striker along radial and axial directions was observed. After the calculations were performed, it was found out that velocity of the striker center of mass was decreasing much slowly within time gap from the 20th through the 44th μs than within time gap from 0 to the 20th μs.

The calculations results obtained from the series of computational experiments on shock loading of oblong striker with ogival head part against one homogeneous and eight gradient barriers are given in Table 2. The calculated values for R criterion are provided in the last column.

TABLE 2.

Oblong striker effect.

Barrier type	V_c [m/s]	R
Barrier #1 [σ_k^-, A_k^-]	286.3	0.64
Barrier #2 [$\sigma_k^\uparrow, A_k^-$]	270.0	0.66
Barrier #3 [$\sigma_k^\downarrow, A_k^-$]	272.0	0.66
Barrier #4 [$\sigma_k^-, A_k^\uparrow$]	263.0	0.67
Barrier #5 [$\sigma_k^-, A_k^\downarrow$]	409.1	0.49
Barrier #6 [$\sigma_k^\uparrow, A_k^\uparrow$]	318.7	0.60
Barrier #7 [$\sigma_k^\downarrow, A_k^\downarrow$]	397.0	0.50
Barrier #8 [$\sigma_k^\uparrow, A_k^\downarrow$]	363.5	0.55
Barrier #9 [$\sigma_k^\downarrow, A_k^\uparrow$]	263.1	0.67

In the table, one can see that those barriers for which value of A_k is constantly increasing following the linear law from barrier front surface through its back one are the most highly-resistant to the effect of oblong striker. In fact, in all the scenarios when the value of A_k reached its maximum at the back surface, the value of resistance criteria R was equal to more than 0.55.

Thus, in the scenario for oblong striker, shear strain mechanism is prevailing like in the scenario for compact striker. However, this is the increase in shear strength over barrier thickness that results in faster decrease in the striker velocity.

ACKNOWLEDGMENTS

The present work was performed under the auspices of the Russian Fundamental Research Fund under Contact RF FI (05-08-01196a).

REFERENCES

1. V.P. Glazyrin, Yu. N. Orlov, G.G. Ol'shanskaya. Modeling of penetration of combined strikers into barriers, *Computing Technologie* edited by Yu. I. Shokin, Special Issue, 2002, pp. 144-153.
2. V.P. Glazyrin, Yu. N. Orlov. Computational method of shock interaction of solid bodies, Up-to-date methods of design and development of missile and artillery armaments-2000, 2000, pp. 164-167.
3. Physics of explosion, edited by K.P. Stanukovich, Moscow, Science, 1975, pp. 704.
4. G.I. Canel, S.V. Razorenov, A.V. Utkin, V.E. Fortov. Shockwave phenomena in condensed media, Moscow, Yanus-Co, 1996, p. 407.

Comment to a Problem of Two-Wave Configurations Existing in Unalloyed Depleted Uranium and its Alloys with Molybdenum in a Region of 50.5-57.0 GPa

E.A. Kozlov

Federal State Unitary Enterprise "Russian Federal Nuclear Center – Zababakhin All-Russian Research Institute of Technical Physics", P.O. Box 245, Snezhinsk, Chelyabinsk region, 456770, Russia
E-mail: kozlov@gdd.ch70.chel.su

Abstract. An overview of experimental and theoretical investigations of features of shock and isothermal compressibility for uranium and thorium in the regions of their polymorphous, electronic and phase transitions are presented. Experimental set-up and results of time-resolved diagnostics and shock-wave recovery experiments are published. The absence in uranium of two-wave configuration within the range of 50.5-57.0 GPa is confirmed. No significant alterations of microstructures, phase composition and properties of recovered samples were observed in the loading region up to 60-70 GPa under the materials science investigations. Alterations in microstructure of recovered samples of unalloyed uranium shock compressed up to σ_{xx}=85 GPa and released to atmospheric pressure were observed and interpreted by the α–γ–phase transition proceeding in uranium under shock compression to σ_{xx}=85 GPa and $T \approx 2000$ K, as well as by recrystallization in the process of reverse transformation under unloading. Signs of the beginning uranium melting on isentrope – in the process of unloading to atmospheric pressure from a state of single shock compression up to 100 GPa – were discovered in investigations of recovered samples. It seems to be actual to refine the shock wave melting conditions for unalloyed uranium and its alloys based on registration of the longitudinal $C_\ell(\sigma_{xx})$ and volume $C_B(\sigma_{xx})$ sound velocities on the Hugoniots by optical analyzer technique, as well as by direct pyrometrical measurements in shock-wave experiments.

Keywords: Hugoniot adiabat; phase transitions; time-resolved diagnostics; shock-wave recovery experiments; uranium; *U-Mo*-alloys; high explosives.
PACS: 62.50.+p, 64.30.+t, 81.30.-t, 81.40.Vw

INTRODUCTION

Paper [1] is the first publication devoted to shock compressibility of the unalloyed uranium. J.Viard used the discrete method of time intervals registration to measure shock adiabat of this 5f–metal within 25-60 GPa. At 50.5 GPa, on the Hugoniot he observed a weak bending and related it to the polymorphous transformation in uranium under compression. Based on data from [1], within 50.5–57.0 GPa the two-wave configuration is expected to arise in unalloyed uranium.

Investigation results of isothermal compressibility of unalloyed uranium using diamond anvil cells (DAC) and synchrotron light source were published in 1985 in article [2]. Some reflections in XRD–pattern having no relation to α–U were observed under room temperature and pressure 50-70 GPa [2]. A more recent study by the same authors [3,4], as well as studies [5,6] with uranium, which used the same technique but a wider range of pressures and temperatures ($P \leq 100$ GPa, $T \leq 4500$ K) revealed no phases distinct from α–U under isothermal compression at room temperature. In contrast to uranium, experimental studies of metallic thorium [7, 8] did possible to observe a new theoretically predicted phase transition under static compression up to 300 GPa at room temperature.

Hypothesis about possible electronic transitions in uranium shock-compressed up to 80 GPa was forwarded in [9]. *Ab initio* calculations in [10] predicted polymorphous transformation from the rhombohedral lattice α–U into the double body-centered tetragonal (dbct)–lattice of a new high-pressure phase. But in [11] published practically simultaneously with [10], the same authors excluded feasibility of this transformation. Recently published paper [12] made it possible to settle up these contradictions in theory and provided both detailed consideration of possible structural transformations in uranium, and evaluated for the first time all its elastic constants in a wide range of α–U existence.

The purposes of this work realized 25 years ago were: (i) principal verification by using time-resolved diagnostics of possible existence of two-wave configuration in unalloyed depleted uranium and its alloys with molybdenum in a region of 50-70 GPa, as well as (ii) material science investigations of recovered samples after their shock-wave loading within 35-125 GPa, i.e. below and above the expected transformation.

MATERIAL, SAMPLES, AND CONDITIONS OF THEIR EXPLOSIVE LOADING

Consideration is given to the unalloyed depleted cast uranium and two U-based alloys with Mo.

Experiments with time-resolved diagnostics used two types of samples:
– wedges fabricated from parallelepipeds with the sizes of 30×10×70 mm³ and the vertex angle of 12°00′. They were cut out from the ⌀85mm cast rods. Thickness of the wedge sample changed within 0.1–10mm;
– wedges fabricated from ⌀100×25mm disks with the angle of 5°00′. They were made from the ⌀120mm cast rods. Thickness of this type of samples changed within 10-15 mm or 15-20 mm.

First-type and second-type samples were loaded by the normal detonation of HE charges having different type and height. As a basic charge one used pressed:
– TNT, $\rho_{HE} = 1.53$ g/cm³,
– RDX-based composition, $\rho_{HE} = 1.72$ g/cm³,
– HMX-based composition, $\rho_{HE} = 1.86$ g/cm³.

The basic HE charges of different size were used in our experiments. When the diameter of the cylindrical HE charge was 120 mm, its height varied from 20 to 120 mm. An explosive lens – the plane-wave generator – was used for initiation. With the three above-mentioned explosive compositions, the pressure of 45 GPa, 52.5 GPa, and 68 GPa, respectively, was created on the loading surface of unalloyed-uranium samples.

The required range of loading on wedge samples was attained through varying both the type of the explosive composition, and the ratio between the sample thickness and the basic HE charge height.

RESULTS AND THEIR DISCUSSION

The modified optical-lever method [13-16] was used to register wave processes directly under explosive loading of samples. Results were obtained as streak-camera records. Typical records for wedges of the first and the second types are presented in Figure 1. Measurements were taken on the freshly polished samples without any protective light-reflecting coatings used.

Given streak camera records show that one-wave configurations propagate within the range of our interest, i.e. 50.5–57.0 GPa, in uranium and its alloy with molybdenum. The variant of the shock adiabat proposed in [1] was not confirmed in our experiments, as well as hypotheses on polymorphous or electron transformations in uranium within the above–mentioned pressure range.

The method used in [1] failed to register onset in uranium of the elastic precursor reflecting the elastic–viscous-plastic character of the uranium high rate deformation. Structure and parameters of the elastic precursor in unalloyed uranium and its alloy with molybdenum in the range up to $\sigma_{xx} \leq 4$ GPa were registered using the capacitor method [17]. Later and in a significantly wider range of longitudinal stresses, the elastic-viscous-plastic configurations in uranium and its alloys with Mo were registered using the optical-lever method when wedge samples were loaded by the sliding and normal detonation of HE charges having different power and thickness.

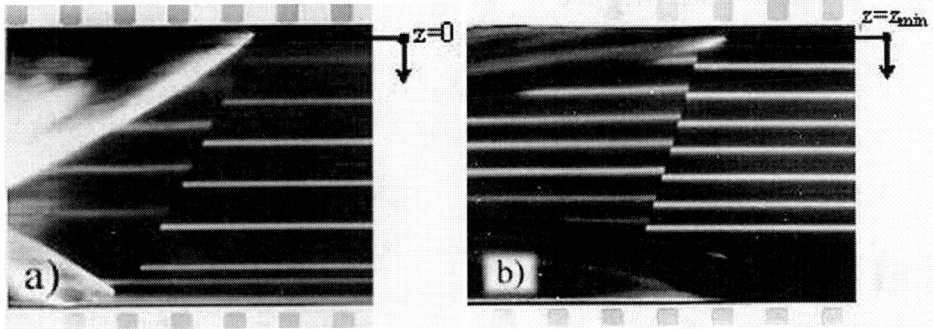

FIGURE 1. Typical streak camera records under loading of first-type (a) and second-type (b) samples by normal detonation of HE charges having different power and height. a) RDX-based composition, ρ_{HE}=1.72 g/cm^3, Ø120×40 mm, b) HMX-based composition, ρ_{HE}=1.86 g/cm^3, Ø120×120 mm.

With the best amplitude- and time-resolutions the elastic-viscous-plastic free surface velocity profiles of unalloyed uranium samples in the region of $\sigma_{xx} \leq 15$ GPa were registered by laser interferometer method in [18].

Weak feature on the Hugoniot adiabat for unalloyed uranium observed in [1] at $\sigma_{xx} \approx 50.5$ GPa is possibly connected to weak entropy jump at the overdrive stress – so called *strong shock threshold* – corresponding to the overdriving of the elastic precursor by the strong plastic wave front, but not with polymorphic $\alpha-\beta$ [1], $\alpha-U-$ dbct [10], or electronic [9, 19] transitions in uranium.

Peculiarities in the behavior of unalloyed uranium were investigated not only in a region of 50.5–57.0 GPa, but also in a wider region 35–125 GPa. I used not only time–resolved diagnostics, which permitted me to analyze the behavior of investigated material under its shock-wave loading and release, but also soft recovery experiments, followed by detail materials science investigations of the recovered matter.

Series of explosive and shock-wave recovery experiments with the samples of unalloyed uranium and its alloys with molybdenum was carried out 25 years ago. Materials science investigations of shocked, released and soft recovered samples of uranium and its alloys was performed using optical microscopy and XRD–analysis, as well as hardness measurements. Research was realized with participation of scientists from Bochvar All-Russian Research Institute of Nonorganic Materials (VNIINM): Z.P.Nikolaeva, V.M.Teplinskaya, A.A.Kruglov, N.T.Chebotarev, V.K.Orlov.

This series of explosive and shock-wave recovery experiments permits determining the loading conditions of unalloyed uranium, which correspond to
- significant microstructure alteration at $\sigma_{xx}=85$ GPa after deformation in intensive stress waves in solid state, as well as to,
- beginning of melting on the isentrope – in the process of releasing to atmospheric pressure from a state of single shock compression up to 100 GPa.

According to theoretical estimations based on data [20-24], shock wave amplitudes up to 175-225 GPa are required for melting of unalloyed uranium on the Hugoniot – immediately at the shock wave front.

Refinement of shock-wave melting conditions for unalloyed uranium and its alloys, in addition to the spherical-shock-waves-recovery experiments [25], requires the systematic measurements along the Hugoniot of longitudinal $C_\ell(\sigma_{xx})$ and volume $C_B(\sigma_{xx})$ sound velocities by the optical analyzer method, as well as direct pyrometric measurements.

It should be noted, that difference in estimations of shock adiabat temperatures, calculated according to models [23, 24], is as much as 1.5 times in the vicinity of intersection point of unalloyed uranium Hugoniot and its melting curve.

CONCLUSIONS

1. Data on the existing peculiarity on the Hugoniot for unalloyed uranium at σ_{xx}=50.5 GPa presented in [1] were not confirmed in our investigation. At least, the use of high–sensitive time-resolved optical lever method has not permitted us to register two-wave configuration in unalloyed uranium and its two alloys with molybdenum in a region 50.5-57.0 GPa, where the two-wave configuration would be observed if the type of the uranium Hugoniot presented in [1] is correct.

2. No significant alterations of microstructures, phase composition and properties of recovered samples were observed in the loading region up to 60-70 GPa under the materials science investigations.

3. Alterations in microstructure of recovered samples of unalloyed uranium shock compressed up to σ_{xx} = 85 GPa and released to atmospheric pressure were observed and interpreted by the α–γ–phase transition proceeding in uranium under shock compression to σ_{xx} = 85 GPa and $T \approx$ 2000 K, as well as by recrystallization in the process of reverse transformation under unloading.

4. Signs of the beginning uranium melting on isentrope – in the process of unloading to atmospheric pressure from a state of single shock compression up to 100 GPa – were discovered in investigations of recovered samples.

5. It seems to be actual to refine the shock wave melting conditions for unalloyed uranium and its alloys based on registration of the longitudinal $C_\ell(\sigma_{xx})$ and volume $C_B(\sigma_{xx})$ sound velocities on the Hugoniot by optical analyzer technique, as well as by direct pyrometric measurements in shock-wave experiments.

REFERENCES

1. J.Viard, "Adiabatique Dynamique de l'Uranium" in *Les Ondes de Détonation, Colloques Internationaux*, 28 août - septembre 1961, Gif-sur-Yvette, Paris, 1962, pp.383-390.
2. J.Akella, G.S.Smith and H.Weed, *J.Phys.Chem.Solids* **46**, #3, 399-400 (1985).
3. J.Akella, G.S.Smith, R.Grover, Y.Wu and S.Martin, *High Pressure Research* **2**, 295-302 (1990).
4. J.Akella, S.Weir, J.M.Wills, and P.Söderlind, *J.Phys.Condens.Matter* **9**, L549-L555 (1997).
5. C.S.Yoo, H.Cynn, and P.Söderlind, *Phys.Rev.B* **57**, 10359-10362 (1998).
6. T.Le Bihan, S.Heathman and M.Idiri, *Phys. Rev.B* **67**, 134102 (2003).
7. J.Akella, Q.Johnson, R.S.Smith and L.C.Ming, *High Pressure Research* **1**, 91-95 (1988).
8. Y.K.Vohra and J.Akella, *Phys.Rev.Lett.* **67**, #25, 3563-3566 (1991).
9. M.Pénicaud, "A Theoretical Equation of State for Uranium" in *Shock Waves in Condensed Matter-1983*, edited by J.R.Asay et al., Elsevier Science Publishers B.V., 1984, pp.61-64.
10. J.M.Wills and Olle Eriksson, "Theoretical Studies of the Crystal Structure of Rare Earths and Actinides at Zero Temperature" in *High-Pressure Science and Technolog-1993*, edited by S.C.Schmidt et al., American Institute of Physics, New York, 1994, pp.175-178.
11. J.M.Wills and Olle Eriksson, *Phys. Rev. B* **45**, #24, 13879-13890 (1992).
12. P.Söderlind, *Phys.Rev.B* **66**, 085113 (2002).
13. G.R.Fowles, *J.Appl.Phys.* **32**(8), 1475-1482 (1961).
14. E.A.Kozlov, *High Pressure Research* **10**, 541-582 (1992).
15. E.A.Kozlov, E.S.Buslova, V.A.Boboedova and D.M.Gorbachev, "Macrokinetics of α-ε Phase Transition for some Steels under Different Conditions of Explosive Loading" in *Shock Compression of Condensed Matter-1991*, edited by S.C. Schmidt et al., Elsevier Science Publishers B.V., 1992, pp.173-176.

16. E.A.Kozlov, V.I.Tarzhanov, I.V.Telichko, D.M.Gorbachev and D.G.Pankratov, "Stress Relaxation on the Elastic Precursor in the Unalloyed Uranium and Some U-based Alloys" in *Workshop on Fundamental Properties of Plutonium,* RFNC-VNIITF, Snezhinsk, September 12 – 17, 2005, pp.144-152

17. S.A.Novikov and V.A.Sinitsyn, "Shock Compressibility of Uranium in the Elastoplastic Area at 10-250°C" in *Workshop on Fundamental Properties of Plutonium,* RFNC-VNIIEF, Sarov, August 30 – September 2, 2004, pp.181-186.

18. D.E.Grady, "Steady-Wave Risetime and Spall Measurements on Uranium (3-15 GPa)" in *Metallurgical Applications of Shock-Wave and High-Strain-Rate Phenomena,* edited by L.E.Murr et al., Marcel Dekker INC, New York and Basel, 1986, p.763-781.

19. B.A.Nadykto, "Effect of High Static Pressure on the Crystal and Electron Structure of Transition Metals" in *Workshop on Fundamental Properties of Plutonium,* RFNC-VNIIEF, Sarov, August 30 – September 2, 2004, pp.117-124.

20. L.V.Altshuler, A.A.Bakanova, M.I.Brazhnik, V.I.Zhuchikhin, S.B.Kormer, K.K.Krupnikov and R.F.Trunin, *Russ. J. Chemical Physics* **14**, №2-3, 65-67 (1995).

21. Los Alamos Series on Dynamic Material Properties, LANL SHOCK HUGONIOT DATA, edited by S.P.Marsh, University of California Press, Berkeley • Los Angeles • London, 1980.

22. Los Alamos SHOCK WAVE PROFILE DATA, edited by Charles E. Morris, University of California Press, Berkeley • Los Angeles • London, 1982, 487p.

23. A.B.Medvedev, *Model of the Equation of State for Metals with the Regard for Evaporation, Thermal Ionization, and Melting,* VANT. Series: Theoretical and Applied Physics, 1992, issue 1, p.12.

24. I.V.Lomonosov, "Phase Diagrams and Thermodynamic Properties of Metals under High Pressures and Temperatures", Dissertation for doctorial degree, RAS Inst. Problems of Chemical Physics, Chernogolovka, 1999.

25. E.A.Kozlov, B.V.Litvinov, I.G.Kabin, E.V.Abakshin, V.K.Orlov, V.M.Teplinskaya, S.S.Kislyakov and A.A.Kruglov, " Microstructure, Phase Composition and Microhardness of the Balls of Unalloyed Uranium and Two its Alloys after Loading in Spherical Stress Waves" in *Int. Conf. Shock Waves in Condensed Matter-1994,* St.-Petersburg, RUSSIA, 1994, Abstract 18-6, p.31.

Stress Relaxation of Elastic Precursor for Unalloyed Uranium and Some Uranium-Based Alloys

E.A. Kozlov, V.I. Tarzhanov, I.V. Telichko,
D.M. Gorbachev, D.G. Pankratov

*Federal State Unitary Enterprise "Russian Federal Nuclear Center – Zababakhin All-Russian Research Institute of Technical Physics", P.O. Box 245, Snezhinsk, Chelyabinsk region, 456770, Russia,
E-mail: kozlov@gdd.ch70.chel.su*

Abstract. The new data are presented on relaxation of an elastic precursor in unalloyed depleted uranium and two its alloys. Results were obtained under low-intense explosive loading. Statistic thermofluctuational model was used for approximation of experimental data. The inversion of strength properties of the tested *U-Mo* and *U-Fe-Ge* alloys at their quasi-static and high-rate loading was revealed.

Keywords: stress waves; high elastic limit; elastic precursor relaxation; uranium and its alloys; static and dynamic loading; deformation mechanisms; high explosive.
PACS: 62.20.-x, 62.20.Fe, 62.50.+p, 81.70.Bt

INTRODUCTION

First results related to the structure and parameters of elastic precursors in the unalloyed depleted uranium and its alloy with molybdenum were obtained more than 40 years ago using the capacitor method [1-4]. The best time and amplitude resolution was realized for the elastic precursor and a weak plastic wave within the range of $\sigma_{xx} \leq 15$ GPa during laser interferometer measurements in paper [5]. Some data on the stress relaxation of the elastic precursor for the *U*-0.75% Ti alloy are published in [6-8]. *The purpose of this work is* to present data on stress relaxation of elastic precursor in the unalloyed depleted coarse-grained uranium and two alloys on its basis in the case of their low-intensive explosive loading, as well as to approximate published and new experimental data within the statistical thermofluctuation (STF) model [9].

MATERIAL, SAMPLES, AND CONDITIONS OF THEIR EXPLOSIVE LOADING

Consideration was given to the unalloyed depleted uranium and two alloys on its basis: an alloy with molybdenum and an alloy with iron and germanium.

CP849, *Zababakhin Scientific Talks - 2005*,
edited by E. N. Avrorin and V. A. Simonenko
© 2006 American Institute of Physics 0-7354-0345-7/06/$23.00

TABLE 1. Results of quasi-static tests of investigated blanks at 22°C

Material	Density	Mechanical tension[*]			Mechanical compression properties[***]	
	g/cm^3	σ_R, GPa	$\sigma_{0.2}$, GPa	δ, %	σ_R^C, GPa	ε_{comp}, %
Cast U	18.9	0.51	0.27	11		
U-Mo	18.6	0.97	0.63	20		
U-Fe-Ge	17.8	0.42	–[**]	0.5	1.78	2.1

[*]   Velocity of active capture shifting during tensile tests was 1mm/min. Samples geometry met requirements of GOST 1497-84 (type 9, Table 3).

[**] Destruction of all samples of the U-Fe-Ge alloy was brittle. No yield strength under static tensile was observed.

[***]Upsetting rate for cylindrical samples with ∅6×6 mm was 1 mm/min.

Samples in the form of wedges were milled out of the ∅80mm rod-like cast blanks. The same blanks were used to prepare *samples for quasi-static tensile and compression tests*. Measured mechanical characteristics (values averaged over tested 3-4), as well as data on the density of studied blanks are given in Table 1.

Explosive loading of wedge samples was direct or through the 2mm thick uranium based plate by the sliding detonation of PETN-based plastic HE layers with 1.5…3 mm thickness. Thin coatings were used to protect a polished free surface of samples against oxidation.

RESULTS

The optical lever method was used to register elastic-plastic wave configurations and to obtain data on stress relaxation of elastic precursors [9-11]. The results are obtained in the form of streak camera records. Typical streak camera records for alloy of depleted uranium with molybdenum and the alloy with iron and germanium are presented in Figure 1.

It is clear that under realized explosive loading conditions, the two-wave elastic-plastic configuration arises in studied alloys. States on each wave are unstable and decrease with increasing the distance of analyzed region of the wedge sample from the loading surface.

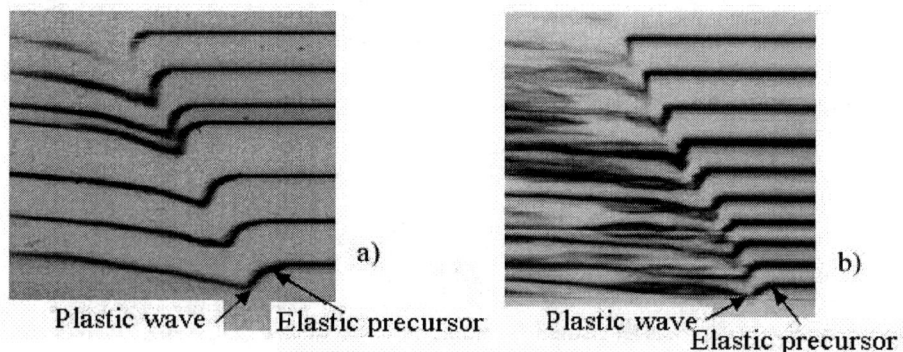

FIGURE 1. Streak camera records of experiments with samples of the *U– Mo* alloy (a) and the *U–Fe–Ge* alloy (b).

434

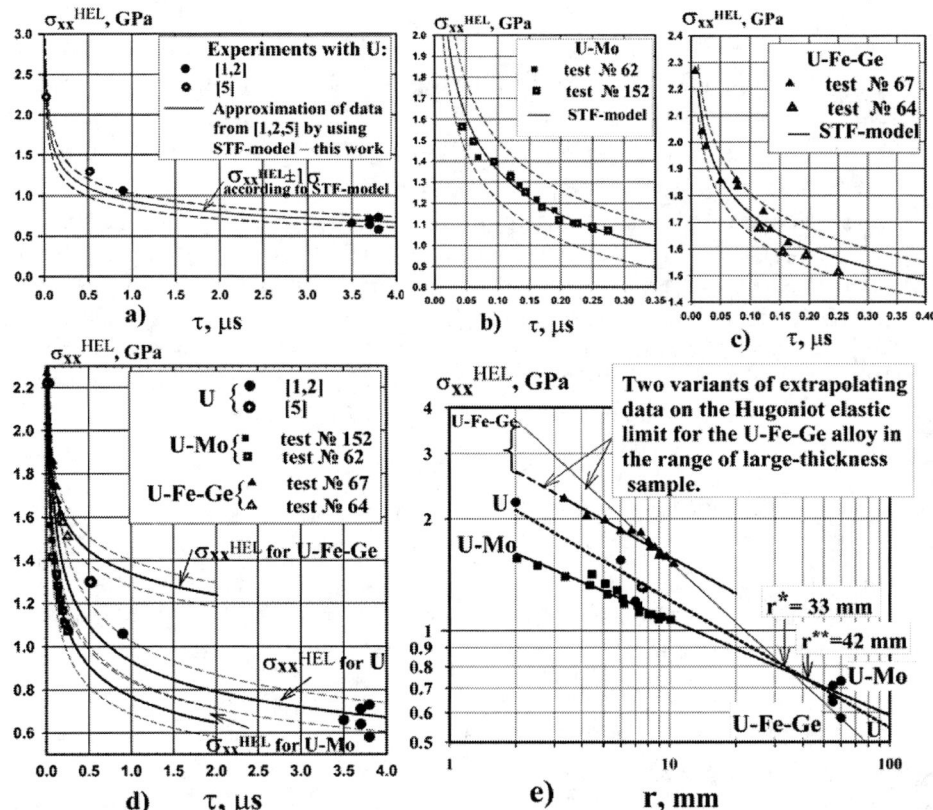

FIGURE 2. Data on stress relaxation kinetics for elastic precursor in studied alloys and results of these data approximation within the STF-model. Symbols – are experimental data. Solid curves – are their approximations within the STF-model. Dashed line shows 1σ-intervals of average values of Hugoniot elastic limit. a) – *U*; b) – alloy *U-Mo*; c) – alloy *U-Fe-Ge*; d) – summary data on elastic precursors relaxation vs time of separation of the elastic precursor head-part from the plastic-wave front, e) – the same but depending on the distance passed by the elastic precursor r, mm.

Figure 2 gives results of streak camera records processing with the purpose to get data on stress relaxation of the elastic precursor for each studied sample of unalloyed uranium and its alloys. Early published data [1,2,5] are also presented.

Data on Fig. 2 show that experimental results on the kinetics of stress relaxation for elastic precursors of three considered materials are well approximated within the one-parametric STF-model when the Weibull distribution function shape parameters are:

$$\alpha_U = 12.06 \pm 0.10(1\sigma); \quad \alpha_{U\text{-}Mo} = 11.58 \pm 0.04(1\sigma); \quad \alpha_{U\text{-}Fe\text{-}Ge} = 28.48 \pm 0.20(1\sigma).$$

Alloying of uranium by molybdenum plasticizes the material, and Fe and Ge introduced into uranium, vise versa, embrittle it. According to the STF-model, the plasticized material (alloy *U-Mo*) has a somewhat lower average value of the dynamic yield strength under high-strain rate deformation if compared with the unalloyed uranium, but it has a more narrow distribution of averaged values. This ensures better repeatability of mechanical testing results [2–4] including data on the kinetics of stress relaxation under explosive loading. Experimental points for different experiments with samples cut out of the same blank are found practically on the same curve, which corresponds to the average Hugoniot elastic limit (HEL) versus of τ (Fig.2b). In addition to the average y_{xx}^{HEL} versus τ (time between the elastic precursor head-part and the plastic wave front) in all plots the dashed curves show one-sigma intervals of this average values calculated by ratios of the STF-model, which are given in [9].

Small amounts of *Fe* and *Ge* (total content ≤ 3 %) introduced into uranium cause embrittlement of the material, growth of its dynamic strength as well as the Weibull distribution function shape parameter $\alpha_{U-Fe-Ge}=28.48\pm0.20(1\sigma)$. This gives (Fig.2c) more significant scatter of data obtained in different explosive experiments even with samples fabricated from the same blank. Experimentally registered scatter of averaged values y_{xx}^{HEL} also falls within the $\pm1\sigma$-band for the *U-Fe-Ge* alloy.

Comparisons of data from Table 1 and Figure 2d,e demonstrate the cardinal change in the mechanical characteristics of considered materials under quasi-static and explosive loadings. According to Table 1 as well as to [2–4], under quasi-static and low-rate loading down to $\dot{e} \leq 2\cdot10^3$ 1/s, alloying of uranium by molybdenum increased strength characteristics and if by iron and germanium – decreased them. Under high-rate explosive loading, alloying changed the sign. In identical explosive loading conditions, the *U-Fe-Ge* alloy has the largest Hugoniot elastic limit y_{xx}^{HEL} and dynamic yield strength $Y=y_{xx}^{HEL}(1-2\nu)/(1-\nu)$, and the *U-Mo* alloy – the least ones. Unalloyed uranium has the intermediate position between them. Data on Figure 2e show that the noted «inversion» y_{xx}^{HEL} exists only with the small-thickness samples, i.e. in the area of strong plastic waves and high rate deformations of materials, and vanishes at $r \geq r^* \approx 33$mm or $r \geq r^{**} \geq 43$ mm, which are marked with arrows and correspond to the points of intersection between extrapolated relationships $\lg(y_{xx}^{HEL}) - \lg(r)$ for alloys and unalloyed uranium. With large-thickness samples (smaller amplitudes of plastic waves and lower rate deformations), the seeming «inversion» of properties vanishes and in the area of relatively low rate deformations, the *U-Mo* alloy again has a significantly larger than unalloyed uranium Hugoniot elastic limit or the conventional yield strength under quasi-static tension.

DISCUSSION OF RESULTS

Physical mechanisms of the crystalline matter plastic deformation are based on nucleation, multiplication, and movement of both dislocations, and disclinations. Alloying (especially by embedded elements) reduces mobility of these elementary carriers of plastic strain. Motion of dislocations and disclinations is also impeded by such structural heterogeneities as accumulation of dislocations, grain boundaries, and nonmetallic inclusions. Not all those structural factors having strengthening effect under the low strain rate can be efficient under the high-rate, i.e. explosive, loading. Moreover, since nucleation of dislocations and disclinations takes place near stress concentrators, which are violations in the crystalline long range, these structural imperfections is a strengthening factor, on the one hand, and are sources of plastic strain carriers, on the other hand. This means that the same defects can increase resistance to deformation in quasi-static conditions and serve sources of plastic strain carriers under high rate deformation and high stresses, thus reducing resistance of materials to plastic deformation. Not only dislocations, but also dislocations-related microstructural specifics, such as stacking faults and twins, participate in plastic deformation of uranium-based alloys. Paper [12] shows that in contrast to the dislocations – induced plastic flow causing metals hardening with the increase of the equivalent plastic strain, the twins–induced deformation can both strengthening and softening the alloy. For metals and alloys with low-symmetric crystalline lattices (orthorhombic for U), twinning is representative even under relatively low rate deformation. Tendency of metals and alloys to deformation localization and formation of adiabatic shear microbands are also notable under high-intensity explosive or shock-wave loading.

Under identical explosive loading, the above-mentioned decrease in the Hugoniot elastic limit and the dynamic yield strength of the U–Mo alloy, if compared with the unalloyed uranium, is apparently associated with (i) the tendency of the high rate deformation to localization, (ii) formation of shear microbands, (iii) more significant high-temperature softening of this alloy despite its higher strain-and rate–hardening, compared with U, observed in [2-4] within $P \leq 5GPa$, $\varepsilon \leq 0.12$, $\dot{\varepsilon} \leq 2 \cdot 10^3$ 1/s, 293 K $\leq T \leq$ 873 K, i.e. in the range where materials are in the process of homogeneous deformation.

CONCLUSIONS

New data on the structure, parameters, and kinetics of stress relaxation of the elastic precursor in the unalloyed depleted cast uranium and two U-based alloys with molybdenum, as well as with iron and germanium are obtained.

Uranium alloying by molybdenum allows material plastification and ensures better repeatability of both its shear strength, and data on stress relaxation kinetics.

Fe and Ge admixtures (total content - 3%) introduced into uranium allow efficient embrittlement of the material and ensure the required mechanical properties under quasi-static and explosive loading.

Data on the kinetics of stress relaxation of elastic precursor in the unalloyed uranium, its alloy with molybdenum and its alloy with iron and germanium were realistically described within the statistical thermofluctuation model with the following shape parameter of the Weibull distribution function:

$$\alpha_U = 12.06 \pm 0.10(1\sigma); \quad \alpha_{U\text{-}Mo} = 11.58 \pm 0.04(1\sigma); \quad \alpha_{U\text{-}Fe\text{-}Ge} = 28.48 \pm 0.20(1\sigma).$$

ACKNOWLEDGMENTS

We'd like to express our sincere gratitude to A.L.Zapysov and A.V.Koshkin for their help in deposition of thin-film coatings onto wedge samples and to D.I.Shestakov for the provided results of quasi-static tests of reference samples for tension and compression (Table 1).

REFERENCES

1. S.A.Novikov and V.A.Sinitsyn, "Shock compressibility of uranium in the plastoelastic range at 10-250°C" in *International Workshop on Fundamental Properties of Plutonium*, RFNC-VNIIEF, Sarov, August 30 – September 2, 2004, pp.181-186.
2. A.P.Bolshakov, G.A.Kvaskov, S.A.Novikov, V.A.Pushkov and V.A.Sinitsyn, *Mechanical Properties of Uranium under Quasi-Static and Shock-Wave Loading*, Preprint 54-97, RFNC-VNIIEF, 1997, 44p.
3. A.P.Bolshakov, A.S.Girin, S.A.Novikov, V.A.Pushkov and V.A.Sinitsyn, *Russ.J.Applied Mechanics and Technical Physics* **40**, № 6, 197-203 (1999).
4. S.A.Novikov, V.A.Pushkov, O.N.Ignatova, V.A.Sinitsyn and B.L.Glushak, *Russ.J.Chemical Physics* **18**, №10, 22-25 (1999).
5. D.E.Grady, "Steady-Wave Risetime and Spall Measurements on Uranium (3-15 GPa)" in *Metallurgical Applications of Shock-Wave and High-Strain-Rate Phenomena*, edited by L.E.Murr et al., Marcel Dekker INC, New York and Basel, 1986, p.763-781.
6. D.P.Dandekar, A.G.Martin and J.V.Kelley, *J. Appl. Phys.* **51**(69), 4784-4789 (1980).
7. Los Alamos SHOCK WAVE PROFILE DATA, edited by Charles E. Morris, University of California Press, Berkeley • Los Angeles • London, 1982, 487p.
8. B.Hermann, V.Favorsky, A.Landau, D.Shvarts and E.B.Zaretsky, "1-D and 2-D Modeling of U-Ti Alloy Response in Impact Experiments" in *7th International Conference on Mechanical and Physical Behaviour of Materials under Dynamic Loading*, September 8-12, 2003, Porto, Portugal; *Journal de Physique IV* **110**, 269-274 (2003).
9. E.A.Kozlov, *High Pressure Research* **10**, 541-582 (1992).
10. G.R.Fowles, *J. Appl. Phys.* **32**(8), 1475-1482 (1961).
11. E.A.Kozlov, E.S.Buslova, V.A.Boboedova and D.M.Gorbachev, "Macrokinetics of α-ε Phase Transition for some Steels under Different Conditions of Explosive Loading" in *Shock Compression of Condensed Matter-1991*, edited by S.C. Schmidt et al., Elsevier Science Publishers B.V., 1992, pp.173-176.
12. M.H.Yoo, *Metall. Trans.* **12A**, 409-418 (1981).

Shock-Wave Synthesis
of Novel Superhard Materials

A. Yunoshev, V. Sil'vestrov

Lavrentyev Institute of Hydrodynamics, Lavrentyev Av., 15, Novosibirs, Russia

Abstract. We have used shock-wave method to synthesize the high-pressure phase of silicon nitride and diamond-like boron carbonitride. We have synthesized new silicon nitride phase as nano-powder and then consolidated the received matter into the bulk sample using the special high-quality HPHT apparatus. In the result we have obtained the bulk samples of cubic silicon nitride up to 6 mm in size. Vickers microhardness of this superhard material equals 30-50 GPa, which exceeds significantly the data of other investigations. We could not synthesize boron carbonitride. The discussion of the reasons is given.

Keywords: Shock synthesis, cubic silicon nitride, superhard material.
PACS: 62.25.+g, 62.50.+p, 64.70.Kb

INTRODUCTION

Two polymorphic phases of silicon nitride are known, α- and β-phases. In 1999 at compression in diamond anvil cell of α and β phase the new high pressure phase, cubic silicon nitride, was synthesized. In result sample 15 micron in size were received [1]. Later c-phase was obtained by a shock-wave technique [2]. In this case an initial sample represented a mix of a micron β-Si_3N_4 powder with a copper powder. Result of synthesis is a powder with the size of coherence area 10-50 nm. But about compaction or sintering of nano-scale powder, received by a shock-wave method, in a "massive" bulk sample it is not known.

Boron carbonitride BC_xN at $x=2$ and 4 for the first time was synthesized as a bulk sample by the size of 1.5 mm at static pressure of 20 GPa, temperature 2200-2250 K and endurance 5 minutes [3]. The initial sample was a powder mix of hexagonal boron nitride and graphite, amorphized by multihour mechano-chemical processing. The microhardness of a material has reached HV= 60-80 GPa, and on this parameter boron carbonitride is on the second place, after diamond, among known superhard materials.

The purpose of the given work is synthesis of cubic silicon nitride and boron carbonitride by a shock-wave technique in amount, sufficient for sintering or compaction in a bulk sample and study of physico-mechanical properties of a material.

CP849, *Zababakhin Scientific Talks - 2005,*
edited by E. N. Avrorin and V. A. Simonenko

EXPERIMENTAL PROCEDURE

The synthesis was performed in a container made from copper. Shock loading was induced by impact of duralumin flyers by thickness 5 or 8 mm accelerated by explosion up to 5.3 and 3.4 km/s respectively. Distinguishing feature of the assembly is the application of a lateral "supporting" charge from loose RDX or ammonite, which was initiated through pieces of a detonating cord from the main charge of explosive (Fig. 1). Use of an additional charge allows to reduce a radial deformation of the container and up to 5 times to increase useful internal volume of container in comparison with a metal "momentum trap" of the same size.

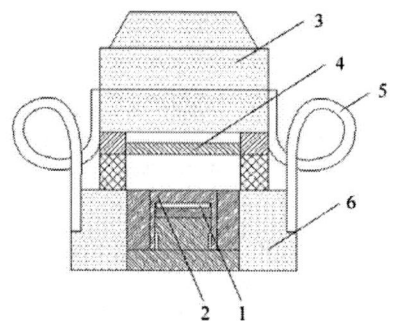

FIGURE 1. Experimental assembly: 1 – sample; 2 – container; 3 – main HE charge; 4 – flyer plate; 5 – detonating cord; 6 – lateral HE charge.

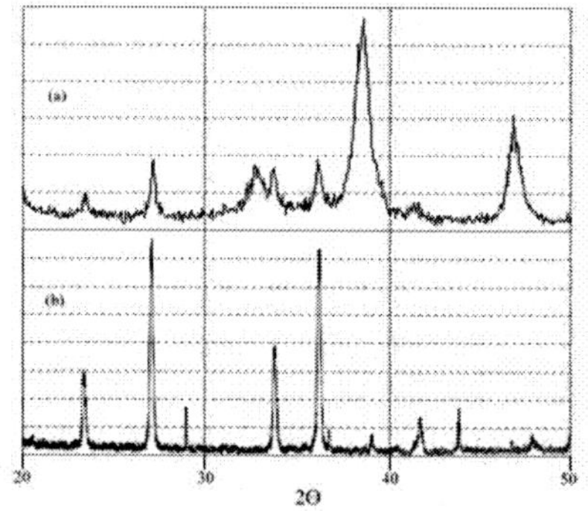

FIGURE 2. XRD spectra of the synthesized cubic silicon nitride (a) and an initial $\beta - Si_3N_4$ (b).

Synthesis of cubic silicon nitride

To synthesize the cubic silicon nitride the β-Si_3N_4 was mixed with copper and then was pressed into the container in vacuum. The sample was 38 mm in diameter and 3 mm thick. Bulk density of the sample was 60-85 % from theoretical value. The part of copper varied in a range 0 up to 92 %.

It was supposed, that the pressure in a sample as a result of repeated circulation of shock waves in it is close to pressure in a cover of the container. In some experiments the 0.1-ohm manganin gauges were used to measure a pressure in container cover and on 3 mm below the sample. The measurements of pressure show, that in a cover of the container the pressure reaches 52-53 GPa. The pressure under a sample does not exceed 36 GPa. This value corresponds to the conclusion [4] that the shock-induced phase transition of the β-Si_3N_4 to cubic phase occurs at pressure not less than 36 GPa. Probably, at longer duration of shock pulse the phase transition can be completely finished at pressure close to the beginning of phase transformation.

After the experiment, the container was opened, and copper was removed by washing of the product in nitric acid. Then the residue was washed out in water and dried. To analyze the synthesized matter we used XRD structural analysis with CuK_α emission line with wavelength of 1.5418 Å.

At dynamic loading of pure hexagonal silicon nitride in the pressure range of 40-75 GPa no phase transition was observed. Apparently, negative factor here is the high temperature resulting in annealing of the new phase at rarefaction.

For the residual temperature to be lower, the silicon nitride powder was mixed with copper powder. In a case if the part of copper exceeded 80 %, sample porosity of about 1.43, and shock pressure in container cover of 53 GPa, the yield of the new cubic phase was close to 90-95 %.

Figure 2 shows the X-ray spectra of the initial hexagonal and the synthesized cubic silicon nitrides. The new material displays three broad lines with maximum at $2\theta = 32.2$, 38.2, and 46.5°. These lines are most intense in the interval $2\theta = 20-50°$; and their positions close to the data for shock synthesized cubic silicon nitride with a spinel structure reported in [2].

We also have used a transmission electron microscopy (TEM) to study our product. TEM-pictures show the presence of both crystalline and amorphous phase. Local chemical analysis of amorphous phase gave an average atomic composition close to Si_3N_4. Thus, the product, synthesized by us, contains crystalline and amorphous phases of silicon nitride. More than 95% of crystalline phase is the c-Si_3N_4 phase and the rest is the β-Si_3N_4 phase. The amount of amorphous phase is hard to estimate but we suppose that it is not more then 20 %. We can not separate the phases. Presence of intermediate amorphous phase can resulting in formation of the more strong bulk sample. Superhard nanocrystallites embedded in an amorphous matrix are currently the most promising concept for the synthesis of novel superhard material [3].

Thus, the synthesized c-Si_3N_4 is an ultrafine powder with a size of the coherent-scattering region of 10-20 nm, which is about one tenth of the size of the corresponding region for the initial β-phase (0.1-0.15 µm). Subsequently, both explosive compaction and high-pressure high-temperature (HPHT) sintering were used to obtain the bulk samples.

HPHT Sintering and Microhardness

Before sintering the specimens were pressed to density of about 60 % of monolith density and were wrapped up platinum foil 10 µm in thick. The HPHT sintering of β-Si_3N_4 and c-Si_3N_4 powders was realized at static pressure of 5-6 GPa and temperature 1100°C during 5 hours [5]. In result the bulk specimens 3-6 mm in diameter and 2.5-3 mm in height are obtained (Fig. 3).

The Vickers microhardness of the bulk samples was measured at indentation force from 2 till 0.3 N. At each force loading was carried out 5-10 measurements. The results are given in Fig. 4.

The bulk sample of β-Si_3N_4 was tested first. Its density was 3.3±0.1 g/cc; its Vickers microhardness is 20±2 Gpa and almost independent of the indentation load. This result points to strong bonding between grains. So, our HPHT technique may be used to consolidate of superhard nano-powder into bulk sample.

Density of the sintered bulk sample of a c-Si_3N_4 is 3.6±0.1 g/cc that is significantly lower than calculated XRD-value of 4 g/cc [4]. XRD analysis of the sintered sample shows the appearance up to 20 % of β-Si_3N_4. According to [6], cubic silicon nitride synthesized by shock-waves is stable up to temperature of 1370°C. Therefore, the appearance of β-Si_3N_4 into the sintered sample is caused, probably, by transition of a part of an amorphous phase to hexagonal one under HPHT conditions.

Despite the content of β-Si_3N_4 in sintered cubic silicon nitride sample, its microhardness is high and exceeds 30 Gpa at the indentation load less than 2 N, and reaches 55±13 Gpa at indentation force of 0.3 N. Microhardness at 1 N is two times greater than values listed in [7] at the same load. Unfortunately, we can not measure the microhardness at indentation load less than 0.3 N.

We also have tried to compact the c-Si_3N_4 powders by explosion. The sample with a plenty of cracks was received. In this case the appearance of β-phase was not observed. Its microhardness reaches HV=27±2 GPa at the indentation load of 1 N (Fig. 4, point 3).

FIGURE 3. Sintering bulk sample of cubic silicon nitride (scale factor – 1 mm).

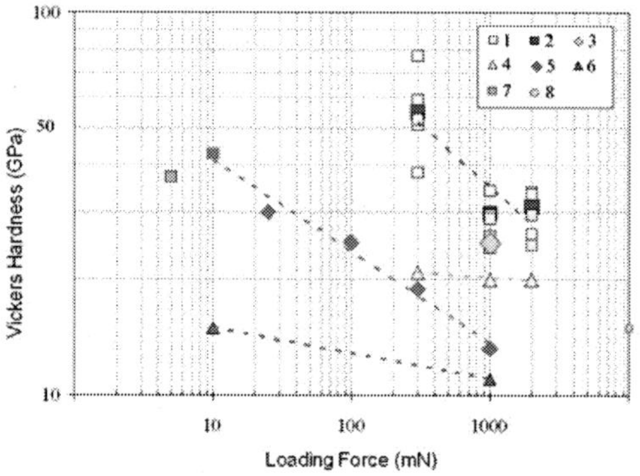

FIGURE 4. Microhardness of silicon nitrides. Our data: 1-3 – c-Si_3N_4, 4 – β-Si_3N_4; 1 – separate measurements, 2-4 – average values. Other authors: c-Si_3N_4 – 5 [7], 7 [7]; β-Si_3N_4 – 6 [7], 8 [7].

ATTEMPT OF SYNTHESIS OF BORON CARBONITRIDE

We have tried also to synthesize boron carbonitride by the shock-wave method. The complete enough reviews of works on numerous attempts of synthesis of these materials are given in [3]. At use of the method of shock-wave synthesis the decision of the given task much more difficult, than in a case of silicon nitride. It was necessary at fast shock loading to realize conditions for diffusion phase transformation, as the atoms of synthesized material are included into structure of various substances.

We tried to synthesize a composition BC$_2$N and BC$_4$N, by analogy with [3]. Our starting material was a mixture of graphite and hexagonal boron nitride in necessary atomic proportions. The mixture of powders was exposed to mechano-chemical activation in a high-energy planetary mill AGO-2 with hardened steel balls within 20 minutes. X-ray diffraction studies of the activated mixture showed that the starting material was completely amorphized. Thus, the close contact between components for acceleration of reaction in a short time of shock loading was supplied.

The activated material mixed with a copper powder and was pressed to the container in vacuum up to density 80 % from theoretical value. Then the sample was shock-loaded up to pressure 30-52 GPa, by the method similar for silicon nitride.

After experiment the cleared of copper sample, represented the graphite-like material. The XRD analysis has shown, that the contents of hexagonal modifications of carbon and boron nitride more than 90 %. Unfortunately, we have failed to extract the diamond-like phase from the sample.

The reasons of failure are not clear yet, it is possible: 1) the small duration of the shock compression period, and phase transformation has not time to occur; 2) temperature of shock compression is too high, and the material decompose at unloading; 3) we have not created good physical contact between components, and it is necessary to change a method of preparation of an initial mixture.

CONCLUSIONS

At shock pressure of 35 GPa in a mixture of copper and β-Si_3N_4 powders almost the whole of silicon nitride transforms to a cubic phase. The realized scheme of explosive synthesis allows receiving up to 1.5 grams of cubic phase of silicon nitride in one shot in laboratory conditions. HPHT sintering allows receiving specimens of nano-crystal silicon nitride up to 6 mm in size with microhardness much greater, than was known earlier.

The attempts of synthesis by a shock-wave method of new high pressure phases of boron carbonitride are unsuccessful, but the work is continued.

ACKNOWLEDGMENTS

The author thanks Dr. Yu.N. Pal'yanov and A.A. Kalinin for performing the HPHT experiments, T.S. Teslenko for XRD analysis and Pr. A.A. Deribas for usefull discussions.

REFERENCES

1. Zerr A., Miehe G., Serghiou G., Schwarz M., Kroke E., Riedel R., Fuess H., Kroll P. and Boehler R. *Nature*. **400,** 340-342 (1999).
2. Sekine T., Hongliang He, Kobayashi T., Zhang Ming, Fangfang Xu. *Appl. Phys. Lett* **76,** 3706-3708 (2000).
3. Zhao Y., He D.W., Daemen L.L., Shen T.D., Schwarz R.B., Zhu Y., Bish D.L., Huang J., Zhang J., Shen G., Quin J., and Zerda T.W. *J. Mater. Res*. **17,** 3139-3145 (2002).
4. Hongliang He, Sekine T., Kobayashi T., Hirosaki H. *Phys. Rev. B*. **62,** 11412-11417 (2000).
5. Yu.N. Pal'yanov, A.G. Sokol, Yu.M. Borzdov, A.F. Fhokhryakov, V.A. Gusev, N.V. Sobolev *Translations (Doklady) of the Russian Academy of Science / Earth Science* **355,** 856-858 (1997).
6. Sekine T., Mitsuhashi T. *Appl. Phys. Lett*. **79,** 2719-2721 (2001).
7. Tanaka I., Oba F., et al. *J. Mater. Res*. **17,** 731-733 (2002).

High-Pressure Polymorphic Modifications of Minerals in the Products of Impact Metamorphism of Polyminerals Rocks

V.I. Feldman[*], L.V. Sazonova[*], E.A. Kozlov[**]

[*]M.V.Lomonosov Moscow State University, Geological Department, Chair of Petrology,
119899 Moscow, Russia
[**]Russian Federal Nuclear Center –E.I.Zababakhin Research Institute of Technical Physics,
456770 Snezhinsk, Chelyabinsk region, Russia

Abstract. High-pressure phases of olivine and pyroxene composition were obtained in laboratory experiments with spherical converging shock waves at Zababakhin All-Russian Research Institute of Technical Physics. These experiments were conducted with three samples of different composition. The high-pressure phases developed after biotite and garnet, i.e. minerals that had compositions different from those of the high-pressure phases. The shock pressure under which these newly formed minerals crystallized were approximately three times higher than the static pressures needed to synthesize these minerals. In nature (in astroblemes these minerals can be preserved only in ejects from the craters, during the super fast quenching of the products of shock metamorphism.

INTRODUCTION

When studying natural impact rocks in astroblemes, geologists ubiquitously find high-density polymorphic modifications of minerals. The high-pressure minerals found as of yet in astroblemes include polymorphs of silica (SiO_2: monoclinic coesite an tetragonal stishovite) carbon (cubic diamond and hexagonal lonsdaleite), orthopyroxene $MgSiO_3$ (cubic majorite), and olivine $Mg_2Si^{iv}O_4$ (cubic ringwoodite). Coesite and stishovite are the two most widely spread of these minerals. All samples that were processed with the use of necessary enrichment operations were determined to contain impact diamond and lonsdaleite [1,2 e.a.]. Majorite was found only once: it was identified in impact cinder from the Zhamanshin astrobleme [3]. This also pertains to ringwoodite, which was found only in impact pumice from El Gasko in Spain [4].

It was determined that high-pressure polymorphs of minerals can be produced by three mechanism that act when minerals and rocks are affected by shock-wave loading in laboratory experiments: this is the crystallization of impact melt under pressure, martensite phase transition, and migrational phase transition in the solid state [5]. The first of these mechanisms is known in to occur in nature and was reproduced experimentally for all of the four minerals; martensite transitions were documented for stishovite and lonsdaleite (also in nature and experiment); and migrational phase transitions have still been identified only for coesite (both in nature and in

CP849, *Zababakhin Scientific Talks - 2005*,
edited by E. N. Avrorin and V. A. Simonenko
© 2006 American Institute of Physics 0-7354-0345-7/06/$23.00

experiment), diamond (togortite, in nature), ringwoodite, and a phase of pyroxene composition (in experiment).

High-pressure phases of olivine and pyroxene composition were obtained in laboratory experiments with spherical converging shock waves at Zababakhin All-Russia Research Institute of Technical Physics (VNIITF) of the Russian Federal Nuclear Center [6-8].

METHODS

A sphere 48-51 mm in diameter prepared from the rock was welded in vacuum into sealed casing of 6-mm-thick stainless steel, which was covered by a layer of plastic explosive (HMX/Plastic Binder–9/1, $\rho_{ex} = 1,86$ g/cm^3) 10 mm thick. In order to simultaneously initiate a shock wave at several points at the surface of the sample at a radius (R_{ex}) of 40 mm, we utilized a spherical system for explosion initiation [9,10]. The shock pressures (P_{sh}) reached approximately 20-25 GPa at the sample surface and were as high as 300 GPa at the front of the converging shock wave at a radius $R = 1$ mm. The duration of the compression did not exceed 2-2.5 µs. The initial cooling rate during the relaxation was 10^8-10^9 °C/s, which ensured, on the one hand, the preservation of all of the newly formed phases and, on the other, eliminated the possibility of annealing, low-temperature hydrothermal alterations, etc. After this the sample cooled under atmospheric pressure, and the cooling rate decreased to 10^3-10^4 °C/s. The pressures and temperatures, which increased toward the center of the sphere along its radii were calculated by the VOLNA computer program [11], which makes use of the Hugoniot adiabat, calculated from the density of the rock and its chemical composition [12].

After spherical-wave loading, relaxation, and cooling, the sphere was cut by diamond-impregnated circular saw into slices parallel to the meridional plane. The polished sections of the spheres were used to analyze the transformations in minerals in the schist along the sphere radium. The composition of minerals that were affected by a converging spherical shock wave was analyzed on a CamScan-4DV electron microscope equipped with a Link-AN 10000 EDS analytical system at an accelerating voltage of 15 kV and a beam current of $(1–3) \times 10^{-9}$ A.

The Raman spectra of ringwoodite were obtained on a LabRam spectrometer at the Bayerisches Geoinstitut in Bayreuth [7,13].

PRISTINE-ROCK SAMPLES AND THEIR TRANSFORMATIONS IN THE SPHERICAL CONVERGING SHOCK WAVE

Sample 1 was staurolite-garnet-plagioclase-quartz-biotite schist, which was sampled from the target rocks of the Janisjarvi astrobleme in Karelia, Russia. It acquired the following concentric zoning (from periphery to center) in the course of the experiment. The first (outermost) zone had a width of 12.5 mm and was characterized by the preservation of the original rock texture and clear boundaries between mineral grains. The shock pressure in this zone was close to 25-20 GPa. The second zone (5-7 mm thick) also showed easily discernible features of the pristine rock texture, but the boundaries between minerals became diffuse because of the migration of chemical elements due to the effect of the shock wave [14-18]. The shock pressures in the inner part of this zone amounted to 50 GPa. The zone was characterized by widespread solid-state transformations of its minerals: amorphization of quartz and plagioclase, the development of shock-thermal aggregates (STA) after biotite, garnet, and staurolite. The third zone was approximately 1.5-2 mm thick. The rock in it displayed partly obliterated boundaries between mineral grains, because the increase in the shock pressure in this zone to 70-80 GPa resulted in the selective melting of its minerals. The fourth zone (in the central part of the sphere) had a diameter of 7-9 mm and was marked by the complete melting of the material under shock pressures as high as 80 GPa. The boundaries between the third and fourth zones are gradational and uneven, through a 100-μm transitional zone.

Sample 2 was quartz-biotite-plagioclase-garnet schist form metapelites of the Archean Taratash metamorphic complex in the Southern Urals. The pristine rock contained large (up to 2-4 mm) porphyroblasts of garnet in a granoblastic matrix.

Upon its shock-loading, the sample acquired zoning consisting of four zones. The outer zone had a thickness of 9-11 mm and had a preserved texture of the unaltered rock, with sharp boundaries between mineral grains. The shock pressure in this zone did not exceed 27 GPa. Garnet grains in this zone were cut by thick networks of randomly oriented cracks and fractures, whose number increased toward the center of the sphere. The second zone had a thickness of 2-3 mm and was marked by the development of veinlets, lenses, and irregularly shaped patches of newly formed minerals in garnet. The amount of these aggregates and their sizes increased form the outer to inner parts of this zone (the sizes of the aggregates increased form a few fractions of a micrometer to a few hundred micrometers). The shock pressures in the inner parts of this zone reached 27-37 GPa. The third zone was characterized by the complete melting of garnet and was 2-3 mm thick. Grain boundaries in this zone were often diffuse and unclear. The shock pressure varied in this zone from 37 GPa in its outer part to 58 GPa in the inner part. Finally, the fourth zone (in the central part of the sphere) was 9-12 mm in diameter and developed under shock pressures of more than 60 GPa. The zone consisted of porous glass of mixed composition, although the melt was not completely homogenized.

Sample 3 was biotite-garnet-plagioclase gneiss from the Archean rocks hosting the Popigai astrobleme in the Anabar Massif in northern central Siberia. After its shock-loading, the sample displayed five zones (from margins to center), which were not spherical but elongated along the banding of the rock, with the "lengths" of the zones 15-30% greater than their "widths". The outer zone was 15 mm thick (here and below, the thicknesses of zones are given in the direction parallel to the rock banding) and was characterized by the development of mostly randomly oriented cracks in minerals. The shock pressure in this zone was no higher than 25 GPa. The second zone, ~1 mm thick, was affected by pressures as high as 27 GPa. Thin (1-10 μm) fractures in garnet contained newly formed phases (STA). The third zone (approximately 3.5 mm thick) suffered shock loading to 40 GPa (in the inner parts of this zone) and contained STA that replaced the whole volumes of garnet grains. Other minerals in this zone were affected by selective melting, but the melts typically did not mingle. The fourth zone was 2-2.5 mm thick, was affected by shock pressures of up to

446

66 GPa, and contained melts of garnet and quartz-feldspar composition, which practically did not mix. The rock contained "grains" having garnet shapes but consisting of glass. This zone practically did not extend across the rock banding. The fifth (central) zone had an amoeba-shaped morphology and did not exceed 4 mm across. It was affected by shock pressures of more than 66-70 GPa and underwent complete melting.

The comparison of the alterations induced in the three distinct rocks by shock waves in our laboratory experiments demonstrates that these alterations in all of the samples were principally analogous.

DEVELOPMENT OF RINGWOODITE AFTER BIOTITE

Ringwoodite developing after biotite was identified in sample 1, in the shock thermal aggregates (STA) that developed in thin fractures and cracks in biotite already at shock pressures of 20-21 GPa. STA likely developed under these relatively low pressures in zones of adiabatic shear. During the increase in the pressure to >25 GPa at the front of the shock wave, STA completely replaced biotite grains. Ringwoodite in these aggregates composes small (no more than 20 μm, long and up to 7-8 μm wide) crystals, together with equally small hercynite crystals submerged in an amorphous mass replacing biotite. The crystals are platy, sometimes with hollow sheath-shaped ends. As the shock pressures increased, the amount of ringwoodite in the shock-metamorphic aggregates replacing biotite increased. The mineral was identified by Raman spectroscopy [7,13] and was determined to contain such components as Al, Ti, K, and, sometimes, Ca, which are generally atypical of this phase. Their presence can be explained by the origin of ringwoodite via the solid-state transformations of the rock under the effect of a shock wave, a process coupled with the migration of chemical elements without melting of minerals. The compositional mapping of STA developing after biotite indicates that an increase in the shock pressure was associated with the activation of the removal of K and Fe and the introduction of Ca, Na, and Si. Although the ringwoodite contains foreign admixtures, its composition is well enough approximated by the formula $Si^{vi}(Fe,Mg)_2O_4$ of spinel of olivine composition. The high Al_2O_3 concentration (up to 14%) in the ringwoodite allowed us to identify it as aluminous ringwoodite. The Fe mole fraction of this mineral [its FeO/(FeO + MgO) molar proportion] ranges from 0.5 to 0.73.

Plotting the parameters under which ringwoodite was synthesized in our experiments (P_{sh} = 25-30 GPa, T = 1060-1500°C) in the diagram for olivine-ringwoodite polymorphic modifications [19], one can see that ringwoodite of this composition was obtained in our experiments under shock pressure much higher than those under static conditions (8-11 GPa).

DEVELOPMENT OF HIGH-PRESSURE MINERALS AFTER GARNET

High-pressure minerals replacing garnet were identified in the experimental products of samples 2 and 3. Both of these samples contained the same high-pressure phases of olivine and pyroxene composition, which were formed simultaneously. The phase of olivine composition was identified as aluminous ringwoodite, and the nature of the other phase (of pyroxene composition) remains uncertain because of the very small sizes of its grains.

In sample 2, these phases crystallized in the second zone, in thin cracks. These are tiny grains ranging from 1.0 to 8.0 μm of equant polygonal morphology. The shapes of the aggregates of these minerals suggest that they were spatially restricted to bands of adiabatic shear. In addition to the "pyroxene" phase, the cracks also contained anhedral and subhedral grains of aluminous ringwoodite. Both the ringwoodite and "pyroxene" have elevated alumina concentrations: up to 13 and 16 wt % (and sometimes more), respectively. As the shock pressure increased with the transition to the third zone, the amount of the newly formed phases increased. Near the boundary of the third zone, ringwoodite and "pyroxene" occur as acicular, often skeleton crystals, and garnet is mostly replaced by glass (solidified melt).

The parameters under which the aluminous ringwoodite (f = 0.72–0.80) and "pyroxene" were produced in the experiment were estimated at 30-35 GPa and 1750-2000°C. Comparing them with those in the diagram [19], one can readily note that these pressures are much higher than those required to form ringwoodite under static conditions (10-13 GPa).

In sample 3, aluminous ringwoodite and the phase of pyroxene composition were produced (as in the previous sample) already in the second zone, within thin cracks (zones of adiabatic shear). An increase in the shock pressure was associated with an increase in the amount of the cracks with high-pressure phases, and these cracks occured as a randomly oriented network at pressures of 26-27 GPa. At pressures of about 30-33 GPa, shock-thermal aggregates usually completely replaced garnet. In the third and, particularly, fourth zones, garnet is melted, with this process the more intense, the higher the shock pressure. The high rates of pressure relaxation were favorable for the preservation of the shapes of former garnet grains and precluded the spreading of their melt and its mingling with the nearby quartz-feldspar melt. As can be seen in large (1.5-2.0 mm) garnet grains, these grains were partly replaced by STA and partly melted and quenched into glass. The aluminous ringwoodite (f = 0.57–0.71) and the phase of pyroxene composition were formed in sample 3 at 30-40 GPa and 1750-2400°C. These parameters are, again, much higher than the static conditions in the diagram [19] and also than the parameters of ringwoodite origin in sample 2.

CONCLUSIONS

1.Our experiments were the first to produce aluminous ringwoodite and a high-pressure phase of pyroxene composition in a rock under the effect of spherical converging shock wave. 2. The high-pressure phases synthesized in our experiments developed after biotite and garnet, i.e., minerals that had compositions different from those of the high-pressure phases. 3. The shock pressures under which these newly formed minerals crystallized were approximately three times higher than the static pressures needed to synthesize these minerals. 4. In nature (in astroblemes), these minerals can be preserved only in ejecta from the craters, during the superfast quenching of the products of shock metamorphism.

ACKNOWLEDGMENTS

The petrological study of the results of the shock-wave experiments was financially supported of the Russian Foundation for Basic Research, project no. 05-05-64778.

REFERENCES

1. Val'ter, A.A., Er'omenko, G.K., Kvasnitsa, V.N., and Polkanov, Yu.A., , Shock-metamorphic minerals of carbon, Kiev, Naukova Dumka, 1992172 p. [in Russian]
2. Vishnevskii S.A., Afanasiev K.P., Argunov K.P., and Pal'chik N.A., Impact diamond: theirfeatures, origin, and significance, United Institute of Geology, Geophysics and Mineralogy, Siberian Branch, Russian Academy of Sciences, Novosibirsk, Is. 835, SB RAS Press. 1997, 53 p.
3. Badyukov D.D., High-pressure phases in impactites of the Zhamanshin crater (USSR), Abstracts XVI Lunar and Planetary Science, Houston, 1985, p. 21 – 22.
4. Diaz-Martinez E., Sanz-Rubio E., Fernandez C., Martinez-Frias J., Evidence for a small meteorite impact in Extremadura (W.Spain), Impact Markers in the Stratigraphic Record, 6 ESF – Impact Workshop, Granada (Spain), Abstracts, 2001, p.21-22.
5. Feldman, V.I., Petrology of Impactites, Moscow, Moscow State University, 1990, 299 p., [in Russian].
6. Kozlov, E.A., Sazonova, L.V., Fel′dman, V.I., , Crystallochemical Structure of Rock-Forming Minerals and Peculiarities, Sequence and Completeness of Physicochemical Transformations in Weak and Strong Shock Waves: Fifth International Symposium High Dynamic Pressures, June 23-27, Saint-Malo, France, Proceedings, 2003a, Tome 1, p.389-397.
7. Kozlov, E.A., Sazonova, L.V., Fel′dman, V.I., Dubrovinskya, N.A., and Dubrovinsky, L.S., Formation of ringwoodite during shock-wave loading of two-mica quartz schist: experimental data, Doklady RAS, v.390, no. 4, 2003b, p.571-573, [in Russian].
8. Kozlov, E.A., Fel′dman, V.I., Sazonova, L.V., Sizova E.V., Beliatinskaia I.V., High-pressure phases, that produced by shock-wave loading of garnet. III International Conference "Phase transitions by high pressurs", Abstracts, Chernogolovca, 2004, p. O – 41, [in Russian].
9. Litvinov B.V., Kozlov E.A., Zhugin Yu.N., Korepanov Yu.M., Abakshin E.V., Kabin I.G., Simonenko V.A., Petrovithev A.V., Kuropatenko V.F., Kovalenko G.V., Sapozhnikova G.N. On new experimental possibilities in studying polymorphic and phase transitions, solid-statechemical reactions in minerals and rocks, Doklady of the Academy of Sciences of the USSR, v.319, N. 6, 1991, p. 1428 – 1429, [in Russian].
10. Kozlov, E.A., Metals, Minerals and Meteorites Research in Spherical Shock-Isentropic Recovery Experiments: Polymorphous and Phase Transitions, Spall and

Shear Fractures, Physicochemical Transformations (Review): Proceedings International Conference, V Zababakhin Scientific Talks, September 21-25, 1998, Snezhinsk, Russia, Russian Federal Nuclear Center – Research Institute Of Technical Physics, Part II, 1999, p. 413 – 424.

11. Kuropatenko V.F., Model of a multicomponent medium, Doklady Akademii Nauk, vol. 403, N 6, 2005, p.761 – 763, [in Russian].

12. Telegin G.S., Trunin R.F. et al., Evaluation of the Hugoniots for the rock and minerals. Proceedings of the Academy of Sciences of the USSR, ser. Physics of the Earth, 1980. N. 5, p.22 – 30, [in Russian].

13. Kozlov, E.A., Sazonova, L.V., Fel′dman, V.I., Dubrovinskya, N.A., Dubrovinsky, L.S., Formation of ringwoodite in high-explosive experiments on muscovite-biotite-quartz states: Bayerisches Forschungsinstitut für Experimentelle Geochemie und Geophysik Universität Bayreuth, Annual Report. Jahresbericht, 2002, p.100-101.

14. Kozlov E.A., Zhugin Yu.N., Litvinov B.V., Feldman, V.I., Sazonova, L.V., Transformation of chemical composition of minerals by shock-wave metamorphism, Snezhinsk, Russia, Russian Federal Nuclear Center – Research Institute of Technical Physics. Preprint, N 152, 1998, 35 p. [in Russian].

15. Kozlov E.A., Zhugin Yu.N., Sazonova L.V., Fel′dman, V.I., Migration of Chemical Components of Minerals under Shock-wave Loading of Janisjarvy Astrobleme Target Rocks (Karelia, Russia), International Conference "VI Zababakhin Scientific Talks", September 24 – 28, 2001, Russian Federal Nuclear Center – Research Institute of Technical Physics. Abstracts, 2001. p. 173 – 174, [in Russian].

16. Fel′dman V.I., Sazonova L.V., Kozlov E.A., Zhugin Yu.N., Migration of some chemical elements in spherical stress waves. Abstracts XVI Lunar and Planetary Science, Houston, 1997, p. 351 – 352.

17. Fel′dman V.I., Sazonova L.V., Transformation of chemical composition of minerals by shock metamorphism. In "Geology, geochemistry and geophysics on the boundary XX and XXI century", Moscow, Svias′- print, 2002, P. 340 – 341, [in Russian].

18. Fel′dman V.I., Sazonova L.V., Kozlov E.A., Conformities of agility of rock-forming components by shock metamorphism (experimental data). Doklady RAS, v.393, N 6, 2003, p.813 - 815, [in Russian].

19. Hemley, R.J., editor, Ultrahigh-pressure mineralogy, Reviews in Mineralogy, V.37, 1998, 671 p.

SECTION 6

NUMERICAL METHODS, ALGORITHMS, CODES, AND ACCURATE SOLUTIONS

Numerical Technique LEGAK-3D for Computation of 3D Time-Dependent Multicomponent Continuum Flows on Multiprocessor Systems

Bakhrakh S.M., Velichko S.V., Spiridonov V.F., Avdeyev P.A., Artamonov M.V., Bakulina E.A., Bezrukova I.Yu., Borlyaev V.V., Volodina N.A., Naumov A.O., Ogneva N.E., Rezvova T.V., Rezyapov A.A., Starodubov S.V., Taradai I.Yu., Tikhonova A.P., Tsiberev K.V., Shanin A.A., Shirshova M.O., Shuvalova E.V.

RFNC-VNIIEF, 37 Mir Avenue, Sarov, 607190 Nizhny Novgorod region

Abstract. The paper presents the ideas of finite-difference Lagrangian-Eulerian technique LEGAK-3D for computation of time-dependent multicomponent continuum flows in 3D geometry. The specific feature of this technique is in the use of concentrations for multicomponent medium computations. The paper also describes main principles of this technique implementation on distributed-memory multiprocessor systems.

FUNDAMENTALS OF LEGAK-3D TECHNIQUE

LEGAK-3D technique is a finite-difference Lagrangian-Eulerian technique using a regular grid. In 3D geometry, this is a grid of hexahedrons. LEGAK-3D technique is a version of LEGAK technique [1] for 2D computations of multicomponent continuum flows with axial symmetry, which has been extended to a 3D case.

LEGAK-3D uses
- a Lagrangian-Eulerian computational grid, which is partially carried away by material; it is assumed therewith that surfaces of the computational grid can be either coincident, or not coincident with material interfaces; in the latter case there occur grid cells containing several materials and concentrations are taken into consideration [2];
- continuous and consistent representation [3] of flows of mass, energy, momentum, and some other quantities, when approximating the convection terms of the original system of equations;

CP849, *Zababakhin Scientific Talks - 2005*,
edited by E. N. Avrorin and V. A. Simonenko

- donor-acceptor algorithm [2,4] for computation of convective flows to prevent computational diffusion, which determines over the field of concentrations of materials near a donor cell, what materials and in what proportion do fly out of a cell containing several materials.

The difference scheme is constructed using the system of conservation laws written for an arbitrary component of space Ω bounded by surface S (notations are commonly used):

$$\frac{dF}{dt} + \int_S G(\overline{u} - \overline{u}^\times)\,d\overline{s} = -\int_\Omega H\,dv, \tag{1}$$

The system (1) is divided into two auxiliary systems.

We obtain the first of them – in Lagrangian phase –assuming that surface S moves with the velocity of material:

$$\frac{dF^{(1)}}{dt} = -\int_\omega H\,dV \tag{2}$$

In the second, Eulerian phase we assume that the material is at rest and surface S continues moving:

$$\frac{dF}{dt} + \int_S G^{(1)}(u^{(1)} - u^*)\,dS = 0. \tag{3}$$

Difference equations for the systems (2) and (3) are the generalization to a 3D case of the corresponding relations used for computation of flows with axial symmetry [1].

In LEGAK-3D technique, edges of computational cells (hexahedrons) may be not coincident with material interfaces.

If it is the case, there are cells containing several materials (mixed cells) and we introduce quantities $\alpha_i = M_i / M$ and $\beta_i = V_i / V$, which are the mass and volume concentrations of components (materials), respectively [4]. Each material has its own equation of state $P_i = P_i(\rho_i, e_i)$.

It is assumed that changes of the values of quantities for each material contained in a mixed cell are described by equations similar to that of the system (2) in Lagrangian phase and system (3) in Eulerian phase.

To determine changes of the component densities, we assume that

$$div\,\overline{u}_i = div\,\overline{u}. \tag{4}$$

In view of the assumption (4), the rule for calculation of pressure P in a mixed cell follows from the requirement of additivity of specific energies and the adopted method for approximation of equations describing energies of components:

$$\rho_i = \alpha_i \rho / \beta_i \qquad P = \sum_{i=1}^{N} \beta_i P_i(\rho_i, e_i), \tag{5}$$

where N is the number of materials in a mixed cell, ρ_i is the i-th material density.

The difference scheme (two-layered, explicit) is of the first order of accuracy and conditionally stable with the following limitation on the time integration step: $\tau(\tilde{u} + c) < kh$, where h is a typical linear size of a computational grid cell, c is sound speed, $\tilde{u}$ is velocity of grid moving relative to material, $k = 0.5$.

LEGAK-3D uses the "per sheet" data organization. One of the families of the computational grid surfaces forms planes that are either intersected along one line (the system axis), or parallel.

Implementation of LEGAK-3D technique in codes allows computations of time-dependent continuum flows in Lagrangian-Eulerian variables, including

- computation of time-dependent gas-dynamic flows;
- computation of elastic-plastic flows;
- computation of the propagation of detonation waves at a constant velocity and with regard to the kinetics of disintegration of an explosive;
- consideration of material destruction.

Codes for computation of the processes above are implemented as separate computational modules.

The complex of codes implementing LEGAK-3D technique consists of a computational and a service parts. The computational part is written in Fortran-90. It can be operated both on PCs and other computer systems that support Fortran-90 and MPI standard.

The service part intended for preparation of jobs and processing of results is implemented as Windows-application in C++ and operates on PC.

LEGAK-3D PARALLELIZATION

Here are the main ideas of LEGAK-3D parallelization:

- no program limitations on the number of processors in use (such limitation may concern only the size of a particular problem to be solved);
- a possibility to scale (to change the number of processors, when solving a problem);
- physical results of computations are independent of the number of processors in use;
- the main load related by computations in multiprocessor mode can be shifted to service routines;
- the technique implementing codes are portable to various distributed-memory computer systems that support MPI (message passing interface) standard;
- textual unity of the LEGAK-3D technique implementation in codes for computations on various single-processor and multiprocessor computer systems in various modes;
- parallelization of codes by placing minimum modifications to their texts and arrangement of computations in multiprocessor mode similar to computations in single-processor mode.

The matrix-type geometric decomposition of job over processes is used for LEGAK-3D parallelization, with a computational fragment for one processor having a form of several closely located computational grid cells in columns and rows with full data from all sheets.

Each processor calculates its own fragment, one of the processors have additional functions of control and timing. Arrays of values, which sizes depend on the number of rows and/or columns of the job, are subjected to decomposing and the main memory of each process stores only those sections of the arrays that are required for a given process.

For implementation of the ideas above, data structures and service tools have been developed to provide:

1. Decomposition of a job and data. Decomposition computations are performed automatically during the process of reading a section independently for each region by each process of the job. The distribution of rows and columns in processes is calculated independently.

2. Reading and recording a section. A section that stores data on the job consists of multiple files. A head file contains all the general data on the job and its current state, as well as the table of data location in additional files. The additional files contain blocks of arrays and concentrations.

3. Matching the grid data in multiprocessor mode. Computational modules of LEGAK-3D change their fragments of the grid arrays. To provide coincidence of data in cells with overlapped data upon execution of programs changing the grid data arrays, the values of quantities of these arrays are updated in the neighboring fragments, for this purpose the mechanism of renewable requests and derivative data types are used.

4. Calculation of a time step. The LEGAK-3D algorithms for selection of a time step of one or another level are similar to those used in the 2D technique. A time step size common for all processes is selected using group communications.

5. Calculation of global values. As in the 2D technique, each process calculates global values for its own job fragment. The global values of quantities over the job, as a whole, are calculated using group communications. A control process calculates the mean values of quantities.

6. Operation of the tools noting the execution time. Data on the execution time for various components (computational modules, interrupt directories, etc.) and data on parallelization overheads (operation of data communication programs, timing, etc.) are collected during computation process. Upon completion of computations the textual data obtained from all processes is recorded to a file of special format. In so doing, the efficiency of computation parallelization and the extent to which computations on various processors have been balanced are estimated automatically.

EXAMPLES OF COMPUTATIONS BY LEGAK-3D

Series of test and methodical computations that confirm the LEGAK-3D technique serviceability for computation of time-dependent continuum flows has been carried out.

Results of some test computations are given in [5]:

- Adiabatic expansion of a three-axis gaseous ellipsoid
- Development of 3D Richtmyer-Meshkov instability
- Acceleration and compression of a spherical shell by an HE layer

In this paper we give a more detailed formulation of the problem of acceleration and compression of a spherical shell by an HE layer with initiation at an "equatorial" point and results of computations for this problem.

The initial geometry of this problem with the field of concentrations is shown in Fig.1.

Region   0<r<8.14cm: vacuum. Region 8.14cm<r<8.8cm: a shell with Mie-Grueneisen equation of state and parameters

$$\rho_{00} = 7.82 g / cm^3, \quad \gamma = 3.5474, \quad c_0 = 4.9 km / s, \quad n = 3$$

HE is inside the region 8.8cm<r<15cm. The equation of state of HE and EP is

$$P = (\gamma - 1)\rho e, \quad \gamma = 3$$

Region 15cm<r<30cm: vacuum.

Initial conditions:

$$\rho_{shell} = 7.82 g / cm^3, \rho_{HE} = 1.67 g / cm^3.$$

$$E_{shell} = E_{HE} = U_{shell} = U_{HE} = 0$$

HE parameters: caloricity Q=3.61kJ/g, detonation rate D = 7.6 km/s.

Detonation is initiated at a point with coordinates x=0.0, y=15.0cm, z=0.0 (Fig.1, point B). The job was run up to the time of the maximum compressed shell.

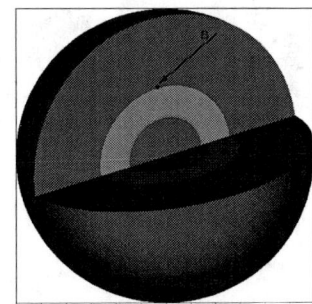

Fig. 1. The initial geometry

The computational domain was partitioned into 2 000 000 cells. The type of interpolation: uniform angular interpolation - along rows, HE separation from the shell by Lagrangian surface – along columns, iteration-less uniform partition – at the remaining points. Fig. 2 shows the of detonation wave development pattern.

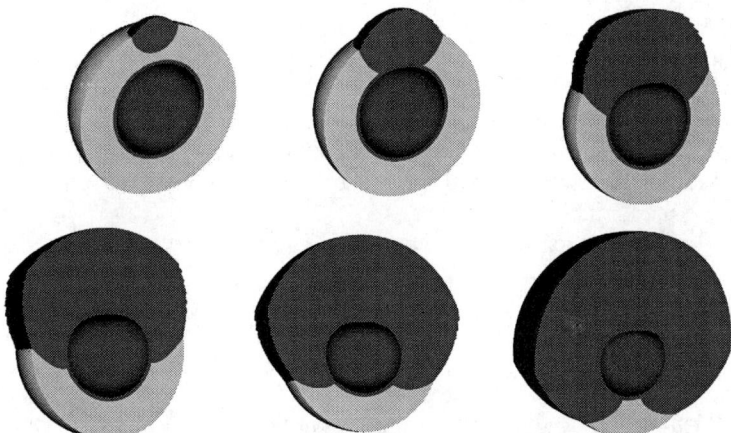

Fig. 2. The detonation wave development and acceleration of the shell.

This process results in the collapsed shell, at some time its maximum density is achieved. Computations fixed this time point and the corresponding maximum value of the shell density.

Fig. 3 shows the collapsing shell.

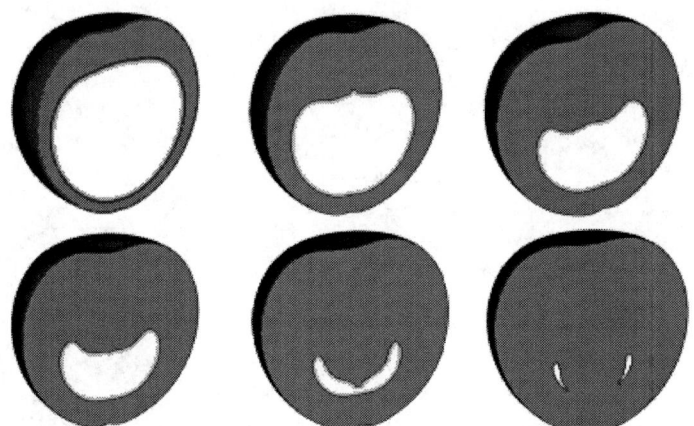

Fig. 3. The process of collapsing the shell

The problem was also solved using 2D LEGAK technique [6] with initiation on the system's axis of symmetry. The difference between the maximum values of the mean density of shell in 3D (initiation from equatorial point) and 2D (initiation on the system's axis of symmetry) computations doesn't exceed 0.1%.

CONCLUSION

Implementation of numerical technique LEGAK-3D in codes for computation of 3D time-dependent multicomponent continuum flows operates on distributed-memory multiprocessor systems.

Computations for various types of problems confirmed serviceability both of LEGAK-3D technique and its implementation in codes.

Computations carried out in multiprocessor mode demonstrated a high efficiency of the technique parallelization.

The work was sponsored by Russian Fundamental Research Foundation (Projects 02-01-0796 and 05-01-00083).

REFERENCES

1. Bakhrakh S.M., Spiridonov V.F., Shanin A.A. The method for computation of gas dynamic inhomogeneous flows in Lagrangian-Eulerian variables //Doklady Academii Nauk SSSR. 1984. V. 276, No.4. pp. 429-433.
2. Bakhrakh S.M., Spiridonov V.F. The method of concentrations for computation of time-dependent continuum flows // Voprosy Atomnoi Nauki I Tekhniki. Ser. Math. Model. Phys. Process. 1999. Issue 4. pp. 32-36.

3. Bakhrakh S.M., Spiridonov V.F. The scheme of consistent flow approximation used in LEGAK technique // Voprosy Atomnoi Nauki I Tekhniki. Ser. Math. Model. Phys. Process. 1988. Issue 4. pp. 38-43.
4. Bakhrakh S.M., Glagoleva Yu.G., Samigulin M.S., et al. Gas dynamic computations based on the method of concentrations //Doclady Academii Nauk SSSR. 1981. V. 257, No.3. pp. 566-569.
5. Bakhrakh S.M., Velichko S.V., Spiridonov V.F. Avdeyev P.A., Artamonov M.V., Bakulina E.A., Bezrukova I.Yu., Borlyaev V.V., Volodina N.A., Naumov A.O., Ogneva N.E., Rezvova T.V., Rezyapov A.A., Starodubov S.V., Taradai I.Yu., Tikhonova A.P., Tsyberev K.V., Shanin A.A., Shirshova M.O., Shuvalova E.V. LEGAK-3D technique for 3D time-dependent multicomponent continuum flows and principle of its implementation of distributed-memory multiprocessor systems // Voprosy Atomnoi Nauki I Tekhniki. Ser. Math. Model. Phys. Process. 2004. Issue 4. pp. 32-36.
6. Avdeyev P.A., Artamonov M.V., Bakhrakh S.M., et al. LEGAK complex for computation of time-dependent multicomponent continuum flows and principles of its implementation on distributed-memory multiprocessors // Voprosy Atomnoi Nauki I Tekhniki. Ser. Math. Model. Phys. Process. 2001. Issue 3. pp. 14-18.

Mathematical Simulation of Flows
In the Field of the Flow Separation

Marina A. Kartasheva, and Alexander L. Kartashev

The Chelyabinsk State University, Kashirinyh Brothers street, 129, Chelyabinsk, Russia

Abstract. The modern space vehicles intended for flight in the space and returns to the Earth, operate in conditions of the considerable gas dynamic and heat loads. At designing of a thermal protection of space vehicle it is necessary to determine gas dynamic properties of streamlining of the vehicle, that calculate thermal effects on devices of its construction. For solution of a considered task it is carried out mathematical simulation of gas–dynamic properties of separation base field of space vehicle.

One of the major problems of mathematical simulation of flows of gas is simulation of processes in separation zones behind streamlined of different configurations. On Figure 1 the scheme of separation base field behind a space vehicle, presenting blunted axisymmetrical cone, streamlined by the uniform supersonic flow under a zero angle of attack at the moderated Reynold's numbers is shown.

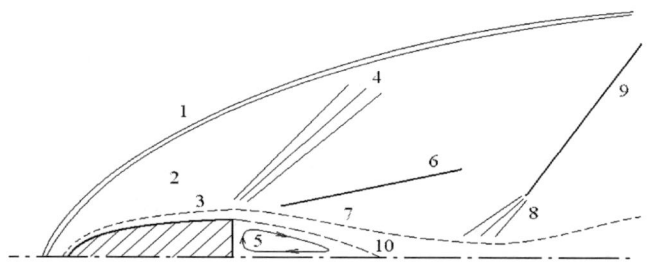

FIGURE 1. The scheme of flow about a space vehicle.

The disturbed area from an undisturbed flow is separated go away head shock wave (1). The boundary layer (3) having strongly pronounced viscous properties borders on to frontal and lateral surfaces of a cone. The boundary layer is broken from a trailing edge of a cone, generating a viscous layer of mixing or the free boundary layer (7) which separates the closed area of a return-circulation flow generated for a base edge (5) and area inviscid (low viscid) flow (2). For a trailing edge of a cone the flow swivels aside axis of flow in a depression wave (4). Below a rarefaction wave the hanging after shock wave (6) places. On some distance from a end surface of a space vehicle there is an attachment of the separated viscous layer. Thus the series of wave of compression (8), passing in the closing shock wave (9), derivates. The separating streamline (10) separates a flow which has an impact pressure, enough for passage through flow of compression, from a flow which impact pressure is not enough for passage of flow of compression and which is forced to turn in the opposite direction. This streamline leaves from a streamlined body in a separation point and comes in a point of attachment of a separated flow.

In the present investigation the model of separation base field using a method of separation streamline, based on model of body of the displacement consisting of two fields: separated base field, pressure in which is supposed to stationary values and equal base, and a remote track, pressure in which is equal to pressure of inviscid flow is offered. A basic parameter determining boundary lines of fields of such flow is value of a base pressure. For its definition flow in the self–similar viscous layer developing along boundary line of an inviscid flow in isobaric area is considered. The method builts on the basis of method of Korst [1] with using of theses developed in article [2], with the taking into account of axisymmetrical of flows.

The scheme of flow is shown on Fig. 2. The flow, incident on a base along a two-dimensional surface, is sonic or supersonic and also remains supersonic after separation from a corner.

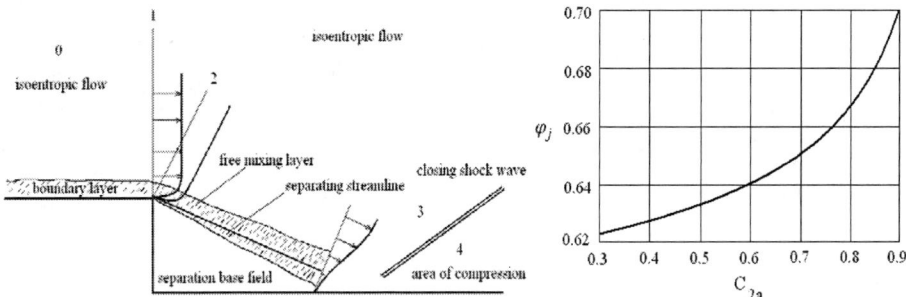

FIGURE 2. The scheme of flow for calculation of a values of a base pressure.

FIGURE 3. The dimensionless velocity along a separated streamline.

Basic theses of theory of Korst is following: in the field of a viscous flow static pressure is equal to pressure in an undisturbed flow: $p = p_H$; expansion of an external flow between sections 1 and 2 take place in correspondence to the solution of Prandtl and Mayer; mixing on a stream boundary between sections 2 and 3 take place at a constant pressure: $p_2 = p_3$; the dimensionless velocity profile (Fig. 3) in a layer of mixing is described by ratio:

$$\varphi = \frac{1}{2}(1 + erf\eta), \ \eta = \sigma(y/x), \ C_{2a} = \left\{1 + \frac{2}{(k-1)M_H{}^2}\right\} \tag{1}$$

Increase of pressure in the field of recompression in the end separation field of flow is determined by adjoining external flow, and compression in an external flow – flat slanting shock wave between areas 3 and 4; in separation field the mass of gas should be saved, the condition an attachment streamlines receives from a condition of conservation of mass in a separation zone and is applied to a streamline which comes in a stagnation point of area of attachment.

The described two–dimensional scheme of flow is modified for axisymmetrical case with the help of an equation of the consumption of gas and definition of axial and radial coordinates of a point of attachment of a separating streamline. The further modification of considered model is connected in view of effect thickness an initial boundary layer on value of a base pressure [2,3] which consists of the taking into account of mass of the gas contained in a boundary layer, with aid of a law of conservation of mass of gas in the field of separation, and also updating the dimensionless velocity profile of gas φ in a layer of mixing.

Basic theses of the mathematical model of the viscous – inviscid interaction which have been set up on consideration of a viscous flow in approaching of a boundary layer and inviscid flow at the same value of a base pressure p_d concluding in following: velocity profiles of a boundary layer before and behind a depression (compression) wave are described by power laws, value an impact pressure is constant along a line in a boundary layer in within the bounds of a wave; behind a ledge, up to a attachment point of flows, the base pressure p_d and a base enthalpy H_d are constant; for each pair values p_d and H_d parameters on external bounds of zones of mixing and maximum pressure in area of attachment of two flows are equal to parameters of corresponding inviscid flows.

The dimensionless velocity profile in a layer of mixing is described by ratio:

$$\varphi = \frac{1}{2}[1 + erf(\eta - \eta_x)] + \frac{1}{\sqrt{\pi}} \int_{\eta - \eta_x}^{\eta} \left(\frac{\eta - \beta}{\eta_x}\right)^{1/n} e^{-\beta^2} d\beta \qquad (2)$$

where $\eta = \eta_x \dfrac{y}{\delta_1}$; $\eta_x = \dfrac{\sigma\delta}{1.5\left(x^2 + 4.4\sigma^2 x \delta\varepsilon_0^{-}\right)^{0.5}}$; β– parameter of integer; δ_1– a boundary layer thickness in a separation point.

In the capacity of conditions of attachment of a flow, in a model the condition which is taking into account work of frictional force in a boundary layer and a viscous layer of mixing will be used:

$$\frac{p_o}{p_d} = \frac{p_c}{p_d} \frac{1}{\overline{p}_{cr}} \qquad (3)$$

where p_o – an impact pressure of gas in a layer of mixing (on a separating streamline), p_d – a base pressure on a end surface of a central body, p_c – static pressure behind a slanting shock wave in the field of attachment of a flow, $\overline{p}_{cr}$ – a critical pressure differential on a shock wave, defined as a measure of work of frictional force in a boundary layer and a zone of mixing.

The approximate method of calculation of a base pressure and enthalpy for flat or the axisymmetrical ledge streamlined by a supersonic flow, set up on methodologies of article [2] is introduced. The scheme of computational area is shown on Fig. 4.

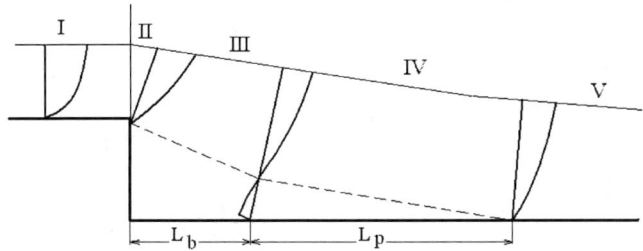

FIGURE 4. The scheme of a separated flow behind an axisymmetrical base ledge.

Flow behind a ledge contains following typical areas:
I – an incident supersonic flow with a boundary layer;
II – transition of a flow and a boundary layer through a depression waves or a shock wave for an edge of a ledge;
III – isobaric area in length L_b, with pressure, approximately equal to a base pressure p_d;
IV – recompression in length L_p; V – pressure equalization.

Recompression in area of attachment takes place smoothly, and the length of this area in some times is more, than length of isobaric area behind a separation point of flow. The length L_p of this area is determined with the help of an equation of impulses of boundary layer in the integral form for a axisymmetrical flow – as a balance equation of forces of pressure and the friction written for a point of attachment, where $u \approx v \approx 0$.

Thus, in the present article the model of flow in the separation base area, set up on modified model of Korst, with the taking into account of axisymmetrical of flows, thickness of an initial boundary layer in a separation point, work of frictional force in a layer of mixing and length of a zone of recompression is introduced in the field of attachment of a flow.

Numerical parametric researches of relation of gas–dynamic properties in axisymmetrical of separation base area of a high-speed space vehicle are carried out at different characteristics of a flow in a point from separation of flow. The received results are shown on Fig. 5.

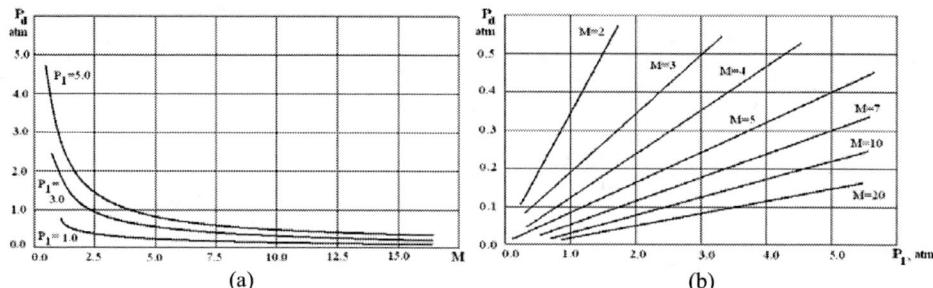

FIGURE 5. Relation of a base pressure: from a Mach number at different pressure in a separation point (a); from pressure in a separation point at different Mach numbers (b).

On Fig. 6 results of calculations are shown: a thick line – by introduced model, a light line – by scheme of S.K. Godunov–V.P. Kolgan [4], a dotted line – by a method of «large particles» [5].

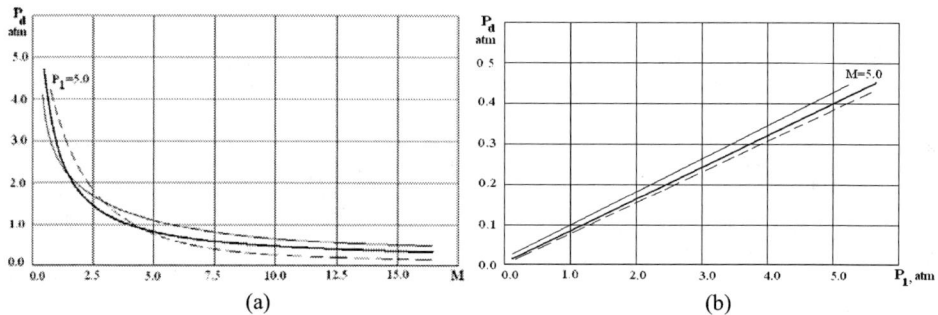

FIGURE 6. Relation of a base pressure from a Mach number (a) and pressure in a separation point (b).

REFERENCES

1. Chang P. Separation of flow. - Moscow.: The World, 1973. - V. 3. - 333 p.
2. Aukin M.K., Тагиров R.K. Calculation of a base pressure and enthalpy behind the flat or axisymmetrical ledge streamlined by a supersonic flow, taking into account of effect of an initial boundary layer // News of The Russian Academy of Science. Mechanic of fluids and gases. - 1999. - № 2. – pp. 110-119.
3. Masalov V.K., Tagirov R.K. Calculation of a base pressure and enthalpy behind a ledge streamlined two supersonic flows, taking into account of effect of boundary layers and heat flows // News of The Academy of Science of USSR. Mechanic of fluids and gases. - 1991. - № 5. – pp. 167-176.
4. Godunov S.K., Zabrodin A.V., Ivanov M.J., Krajko A.N., Prokopov G.P. Numerical solution of many–dimensional problems of gas–dynamics. - Moscow.: The Science, 1976. - 400 p.
5. Belotserkovskii O.M., Davydov Yu.M. Method of «large particles» in gas dynamics. - Moscow: The Science. Main edition of the physical and mathematical literature, 1982.-392 p.

A Numerical Method of Solving Hydrodynamic Equations with Non-Structured Meshes

E.M. Vaziev, A.D. Gadzhiev, S.Y. Kuzmin, A.V. Skovpen

Federal State Unitary Enterprise "Russian Federal Nuclear Center - Zababakhin All-Russia Research Institute of Technical Physics", 456770, Snezhinsk, Russia

Abstract. The paper offers a method to solve hydrodynamic equations on unstructured triangular and quadrilateral meshes which are generated with a Q-Morph-based algorithm. Provided are results of test problems including a test on Rayleigh-Taylor instability.

Keywords: Lagrangian hydrodynamics, unstructured mesh
PACS: 02.60.Cb, 02.70.Bf, 47.11.-j.

INTRODUCTION

The simulation of sophisticated layer structures which experience strong deformations is difficult with solely Lagrangian formalism. A Lagrangian mesh frozen into the flow strongly deforms, the finite-difference approximation fails and the calculation stops. The Arbitrary Lagrangian-Eulerian (ALE) approach offers a good way out of such situations. With this approach the calculation is run on a Lagrangian mesh until it gets strongly deformed; after that the flow is remapped onto a new, Eulerian mesh. However, even with this approach the number of emergency stops remains large. A promising way of reducing emergency stops is to combine structured and unstructured meshes. Unstructured meshes are easier to generate than structured ones; they are more adequate to the complicated configuration of a system to be simulated and can be made finer or coarser where necessary. This paper presents a traditional staggered scheme (the explicit D-scheme [1]) to solve 2D hydrodynamic equations on unstructured triangular and quadrilateral meshes generated with an algorithm described in [2].

DIFFERENCE SCHEME ON UNSTRUCTURED MESHES

Let D be a physical domain containing the flow of interest; D is bounded by a closed smooth surface $\sum$. The flow is characterized by the following hydrodynamic equations:

$$\frac{d\upsilon}{dt} - \upsilon \cdot div(\vec{u}) = 0, \quad \frac{d\vec{u}}{dt} + \upsilon.grad(p) = 0, \quad \frac{d\varepsilon}{dt} + p\frac{d\upsilon}{dt} = 0. \tag{1}$$

$$p = p(\upsilon, T), \quad \varepsilon = \varepsilon(\upsilon, T) \text{ are equations of state.} \tag{2}$$

CP849, *Zababakhin Scientific Talks - 2005*,
edited by E. N. Avrorin and V. A. Simonenko
© 2006 American Institute of Physics 0-7354-0345-7/06/$23.00

Here υ is specific volume $(\upsilon = 1/\rho,\ \rho$ - density$)$, $\vec{u} = (u_r, u_z)$ is velocity vector, p is pressure, ε is internal energy, T is temperature, $\vec{r} = (r, z)$ is radius vector in Lagrangian coordinates.

Cover D by a mesh of triangular and quadrilateral cells with vertices defined by the coordinates $(r_j, z_j),\ j = 1, 2 \ldots J$, (see Fig.1a). The cells are tracked anti-clockwise. Their centers are denoted by i, vertices by j, and edges by $j_{s+1/2}$, where s is the number of the edge in the cell.

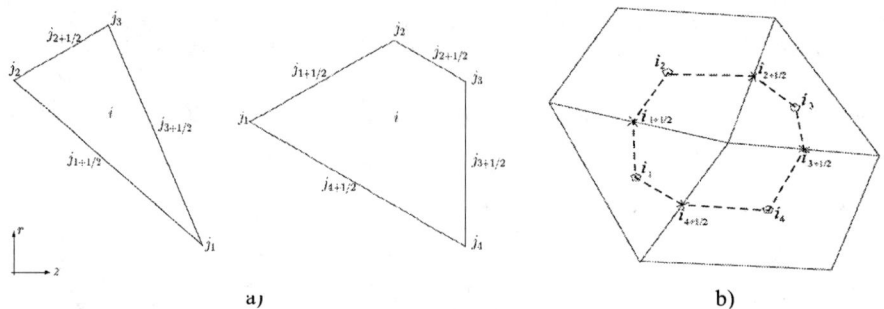

FIGURE 1. a) Mesh cells; b) The contour of integration to determine gradients.

Density and temperature are stored in cell centers and velocity is stored in cell vertices.

The finite-difference approximation to the equation of motion in (1) takes the form

$$\frac{\vec{u}_j^{n+1} - \vec{u}_j^n}{\tau} + \upsilon_j^n \left(grad_h \left(P \right) \right)_j^n = 0. \tag{3}$$

The artificial viscosity is introduced in (3) using the viscous pressure P_i in the center written as

$$P_i^n = p_i^n - \delta h_i^n a_i^n \left(div_h \left(\vec{u} \right) \right)_i^n, \tag{4}$$

where

$$\delta = \delta_1 + \delta_2 \frac{\rho_i^n h_i^n}{2 a_i^n} \left(\left| div_h \left(\vec{u} \right) \right|_i^n - \left(div_h \left(\vec{u} \right) \right)_i^n \right). \tag{5}$$

Here a_i^n is mass sound speed in the cell, $h_i = \sqrt{S_i}$ is the linear dimension of the cell, and δ_1 and δ_2 are, respectively, the coefficients of linear and quadratic viscosity.

The energy equation is solved in two steps. At first we determine the intermediate internal energy

$$\tilde{\varepsilon}_i = \varepsilon_i^n - P_i^n \left(\upsilon_i^{n+1} - \upsilon_i^n \right). \tag{6}$$

With $\tilde{\varepsilon}_i$, we determine the corresponding temperature $\tilde{T}_i$. After that we determine a new viscous pressure $\tilde{P}_i$ for the temperature $\tilde{T}_i$, the density ρ_i^{n+1} and the new velocity field $\vec{u}_i^{n+1}$.

Now we return to the equation for internal energy and define ε_i^{n+1} as

$$\varepsilon_i^{n+1} = \varepsilon_i^n - \frac{\tilde{P}_i + P_i^n}{2}\left(\upsilon_i^{n+1} - \upsilon_i^n\right) + \tau H_i. \tag{7}$$

Noh viscosity [3] is introduced in (7). It is defined as

$$H_i = h_N a_i^n \cdot div_h\left(S \cdot grad\left(\varepsilon\right)\right)_j^n. \tag{8}$$

Here h_N is Noh viscosity coefficient.

The temperature T_i^{n+1} in the cell is determined through linearization of the equation of state for internal energy:

$$T_i^{n+1} = T_i^n + \frac{1}{\varepsilon_T^n}\left(\varepsilon^{n+1} - \varepsilon^n - \varepsilon_\upsilon^n\left(\upsilon^{n+1} - \upsilon^n\right)\right)_i. \tag{9}$$

To improve invariance, gradients in vertices are calculated with the non-conservative formula

$$\left(grad_h\left(P\right)\right)_j = \frac{1}{S_j}\sum_{s=1}^{\tilde{s}_i}\left(\vec{n}\Delta\ell\right)_{i_{s-1/2},i_s} P_{i_{s-1/2},i_s} + \left(\vec{n}\Delta\ell\right)_{i_s,i_{s+1/2}} P_{i_s,i_{s+1/2}}. \tag{10}$$

Here S_j is the area of a polygon created by the integration contour, and $\tilde{s}_i$ is the number of cells around the vertex. The pressures $P_{i_{s-1/2},i_s}$ and $P_{i_s,i_{s+1/2}}$ are calculated as a half-sum of pressures in the cell center and in the center of the corresponding edge. The contour passes through the cell centers and the centers of the edges contiguous to the vertex (see Fig.1b). The formula is used to calculate the gradient of internal energy.

We tried several contours including a contour created by straight line segments between cell centers. However, the contour shown in Fig. 1b was found to work best because in this case the node of a strongly deformed mesh cannot appear beyond the integration contour. The contour also matches triangular and quadrilateral cells well.

Where necessary, we use divergence written (in the plain case) as

$$\left(div_h\left(\vec{u}\right)\right)_i = \frac{1}{S_i}\sum_{s=1}^{\tilde{s}_i}\left(\vec{n}\Delta\ell\vec{u}\right)_{j_{s+1/2}}. \tag{11}$$

This formula is also used to determine divergence of the internal energy gradient in (8).

NUMERICAL RESULTS

We first look into the linear phase in the evolution of Rayleigh-Taylor instability. Consider a plain system in which a heavy fluid is above a light one. The fluids are incompressible; their interface is sinusoidal. A constant acceleration pointed from the heavy fluid to the light one is applied.

Figure 2 illustrates instability evolution and a mesh fragment.

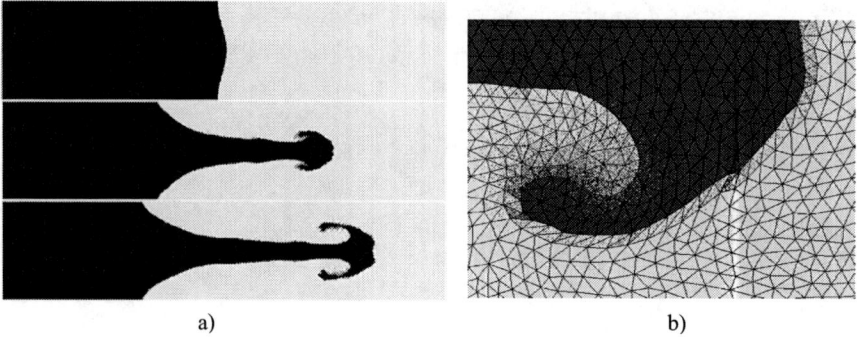

a)   b)

FIGURE 2. Rayleigh-Taylor instability: a) Instability evolution; b) A mesh fragment.

Obtained results satisfactorily reproduce the evolution of the instability how it is observed in experiment.

Initial conditions for Noh problem were taken from [3]. The problem was solved in a spherical setup on unstructured quadrilateral meshes. Results are shown in Fig. 3.

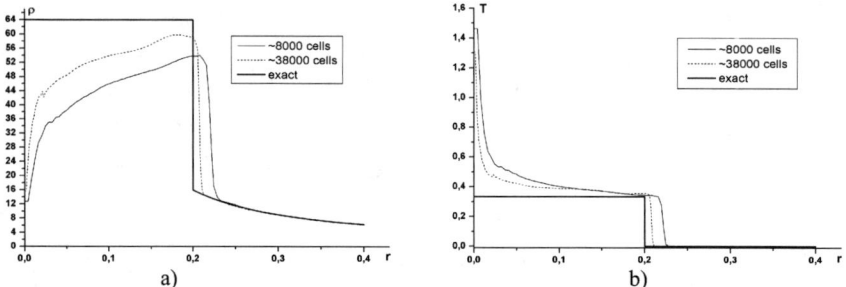

a)   b)

FIGURE 3. Noh problem: a) density profile; b) temperature profile.

Numerical results are seen to satisfactorily agree with the exact solution, and it is also seen that accuracy gets better as the number of cells increases. As for the entropy effect, it is a characteristic feature of this problem exhibited on structured meshes as well.

CONCLUSION

A simple explicit hydro scheme on unstructured meshes has been proposed and tested. Calculations show that unstructured meshes work effectively in the simulation of such problems as Rayleigh-Taylor instability which is difficult to model with structured meshes. With unstructured meshes, calculations of this type are easy and quite accurate. However, even 1D cylindrical and spherical problems cause difficulty. Calculations with triangular cells produce strong oscillations and accuracy degrades. This seems to result from the staggered mesh which is very sensitive to mesh uniformity. In triangular meshes, mass non-uniformity may be rather high in case of curvilinear coordinates. Finite-volume schemes which are less sensitive to uniformity and more monotone may work better in these cases.

REFERENCES

1. Anuchina, N.N., Babenko, K.I., Godunov, S.K., Dmitriyev, N.A. et al., A lagrangian hydro method to simulate 2D unsteady flows. Book "Theory and design of numerical algorithms for computational physics", Moscow, NAUKA Publishers, 1979, p. 174 – 200.
2. Skovpen, A.V., An improved algorithm for unstructured quadrilateral meshing. Snezhinsk, RFNC-VNIITF, 2004, 62 p.
3. Noh W.F., Errors for Calculations of Strong Shocks Using an Artificial Viscosity and an Artificial Heat Flux, J. Comput. Phys. 72 (1987), p.78

About Gas Motion Behind Diverging Detonation Wave From A Point On Free Surface

V.A. Suchkov, A.S. Shnitko, L.R. Islamova

RFNC-VNII Technical Physics
named after acad. E.I. Zababakhin, Snezhinsk, Russia

456770, box. 245, Snezhinsk, Chelyabinsk region , Vasilyev Street, 13
telephone: (351-46) 3-26-25, 5-43-67, fax: (351-46) 5-22-33, 5-55-66, 3-26-25
teletype: series "P", 124137 SNOW RU, telex: 124137 SNOW RU
E-mail: otdeldou@vniitf.ru

Abstract. The paper considers the modeling of detonation wave propagation. Verification of GRAD simulation methods on test problems which have exact solutions is carried out. The solutions are given as formulas and simulation results obtained with guaranteed degree of accuracy. Statements and results of the test problems simulations with the GRAD code are presented.

INTRODUCTION

The solution of the problem of explosive detonation initiation in a point on free surface of halfspace is constructed in the paper. The problem is automodel, required functions u,v, c depend on two variables ξ =x/t, η=y/t, where u, v are components of velocity vector, c is speed of sound, x, y are spatial coordinates, t is time. The equation of state of explosion products is p=$\gamma\rho^{\gamma}$ [1], where p is pressure, c is density, r is the adiabatic parameter. Similar problems of detonation wave propagation along free surface were considered in papers [2,3,4]. The basic asymptotics in vicinity of special points are received there.

The structure of the problem solution consists of the main detonation wave, the boundary characteristic separating the one-dimension solution behind the wave from two-dimension underpressure flow such as Prandtle-Mayer one, ending with free boundary P=0. The solution consists of simple wave in the field of one-dimension propagation to double underpressure wave . The problem was studied in two statements: plane and axially symmetric. In the first case detonation wave of constant intensity is cylindrical, in the second one it is spherical.

Numerical calculations of the appropriate problems with the GRAD program code [5] were carried out, graphic results are presented, analytical and numerical solutions are discussed. Problems have independent interest and can be used for improvement of detonation waves calculations in vicinity of free boundary.

CP849, *Zababakhin Scientific Talks - 2005*,
edited by E. N. Avrorin and V. A. Simonenko
© 2006 American Institute of Physics 0-7354-0345-7/06/$23.00

GRAD-METHODS

The methods and set of program codes GRAD are intended for the solution of non-stationary problems about motion of the non-uniform continuous medium on the electronic computer. In the GRAD methods for the description of the medium hydrodynamic and elastic - plastic models are used, there is a wide set of the equations of state , the calculations of "spreaded" shock and detonation waves are carried out. For the calculation of detonation waves the kinetics of conversion of explosive into explosion products is used. At calculation of elastic - plastic flows the computation of material destruction along ultimate main stresses with formation of cracks (one, two, three - complete destruction) is possible. The method of concentrations is applied to modeling of large deformations of the non-uniform medium.

The GRAD methods has passed verification on various test problems. The basic attention at improvement of difference schemes of GRAD methods is given to improvement of the detonation waves computation and their influences on elements of designs.

STATEMENTS OF TEST PROBLEMS

In the space x> 0, y > 0, occupied by the explosive the following parameters are given: $u^0=v^0=0$, $c^0=0.75$, $c^0=0$, $p^0=0$. The initiation of the explosive is made in a point on border $x=0$, $y=0$ at the moment of time $t=0$. The propagation of the detonation wave is examined at $t > 0$. The problem is automodel, all parameters u, v, c depend on spatial variables ξ =x/t, η=y/t. Here u, v - components of velocity vector , c – speed of sound, satisfy to the differential equations in conical variables:

$$(u-\xi)\frac{\partial u}{\partial \xi} + (v-\eta)\frac{\partial u}{\partial \eta} + \frac{\partial \varphi}{\partial \xi} = 0$$

$$(u-\xi)\frac{\partial v}{\partial \xi} + (v-\eta)\frac{\partial v}{\partial \eta} + \frac{\partial \varphi}{\partial \eta} = 0$$

$$(u-\xi)\frac{\partial \varphi}{\partial \xi} + (v-\eta)\frac{\partial \varphi}{\partial \eta} + c^2\left(\frac{\partial u}{\partial \xi} + \frac{\partial v}{\partial \eta}\right) + \frac{v \cdot c^2 \cdot v}{\eta} = 0 \qquad (1)$$

$$\varphi = c^2/(\varepsilon-1)$$

The problem can be studied in two statements: plane v=0 and axially symmetric v=1.

Test 1. The Motion Of The Cylindrical Detonation Wave

The problem is considered in plane statement. In this case the detonation wave of constant intensity is cylindrical.

Test 2. The Motion Of The Spherical Detonation Wave

The problem is considered in axially symmetric statement. In this case the detonation wave of constant intensity is spherical.

These problems were considered with different boundary conditions on the left border of the explosive x=0: u=0 - one-dimensional problem and p=0 – two-dimensional problem.

To investigate the symmetric motion the equations (1) are written down in polar coordinates :

$$x=rcosu, \ y=rsinu, \ u=ur \cdot cosu-uu \cdot sinu, \ v= ur \cdot sinu+ uu \cdot cosu \qquad (2)$$

$$ur=u \cdot cos\theta+v \cdot sin\theta, \ u\theta=-u \cdot sin\theta+vcos\theta, \ 0 \leq \theta < \pi/2$$

One-Dimension Problem

For one-dimension symmetric motion:

$$u\theta=0, \ ur=U, \ u= \xi \cdot U/\zeta, \ v=\eta \cdot U/\zeta, \ \zeta = \sqrt{\xi^2 + \eta^2}$$

The system of basic equations, describing one-dimension spherical and cylindrical waves is as following:

$$\frac{dU}{d\zeta} = \frac{v}{a^2} \frac{Uc^2}{\zeta}$$

$$\frac{dc}{d\zeta} = \frac{v(\gamma - 1)}{2} \frac{Uc}{\zeta a^2} (\zeta - U) \qquad (3)$$

$U=U(\textit{ж})$- symmetric velocity, c – speed of sound, $\textit{ж}$ – symmetric coordinate, $н=1,2$ –condition of symmetry: $н=1$ – cylindrical problem, $н=2$ – spherical problem, $a^2=(\textit{ж}-U)^2-c^2, \ 0 \leq \zeta \leq D$.

The system of differential equations (3) is solved with initial conditions: $\textit{ж}=D= в+1, \ U=1, \ c=\gamma$.

The result of integration of system (3) is the solutions of one-dimension problems $U=U(\textit{ж}), \ c=c(\textit{ж})$.

For the cylindrical detonation wave in domain $0 \leq \zeta \leq \zeta^0, \ U=0, \ c=\zeta^0=1.8876$.

For the spherical detonation wave in domain $0 \leq \zeta \leq \zeta^0, \ U=0, \ c=\zeta^0=1.8202$.

Two-Dimension Problem

On figure 1 the structure of the problem solution and density isolines at the moment of time t = 1.0 are presented.

472

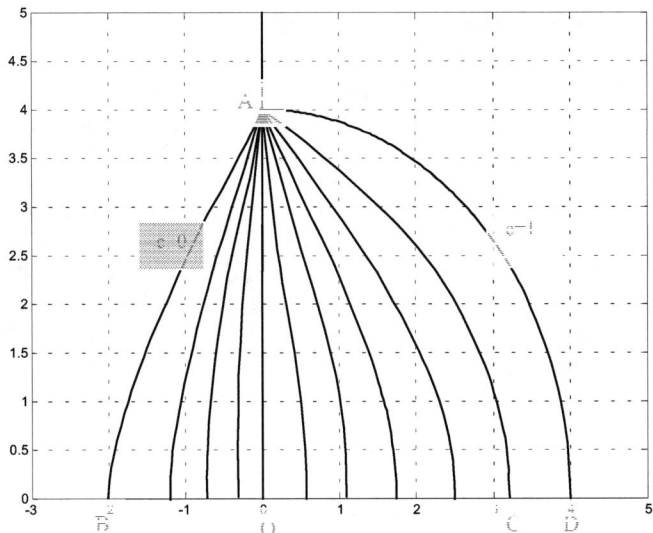

FIGURE 1. Density isolines at the moment t=1.0

The boundary characteristic AC is described by the following differential equation:

$$\frac{d\zeta}{d\delta} = \frac{\zeta a(D\sin\delta - y)}{cy - ax} \qquad (4)$$

$$\zeta = \sqrt{\xi^2 + \eta^2} \,,$$

$$a = \sqrt{(\zeta - U)^2 - C^2} \,,$$

$$x = D\cos\delta \,,$$

$$y = \begin{cases} \sqrt{\zeta^2 - x^2}, 0 \le \delta \le \delta_0 \\ -\sqrt{\zeta^2 - x^2}, \delta > \delta_0 \end{cases} \,,$$

with initial condition $\zeta = D$ at $\delta=0$, where $z=3$, $c=c$, $D=z+1$.

In equation (4) $\bar{\delta} = \sqrt{h}\delta$, $h = \frac{\gamma+1}{\gamma-1}$, the bar is omitted.

In the vicinity of special point $\delta=0$, $x=D$ the following decomposition is used:

$$x=D-64/81z\delta^4$$

Functions U and c depend on variable x:

$U=U(x)$, $c=c(x)$ – solutions of symmetric problems.

By integration of equation (4) we can obtain the boundary characteristic AC .

473

Test 3. The Problem About Explosive Sphere Detonation At Initiation In A Point On Free Surface

In a sphere of radius $r_0=4$ sm with the centre at the beginning of coordinates $x=y=z=0$ the spherical grid is constructed and parameters of explosive are set $u^0=v^0=0$, $c^0=0.75$, $c^0=0$, $p^0=0$.

On the axis of symmetry $y=0$ zero normal velocity $v=0$ is given. Boundary pressure $Pg=0$ is given on external border of the sphere. The detonation begins at the moment of time $t=0$ in the point $x=4$, $y=0$, $z=0$. The motion of the detonation wave in the sphere is examined at $t > 0$. In all cells it is possible to calculate the time of arrival of normal detonation wave. On figure 2 the initial state of the explosive surface and the position of detonation wave at the consecutive moments of time are shown.

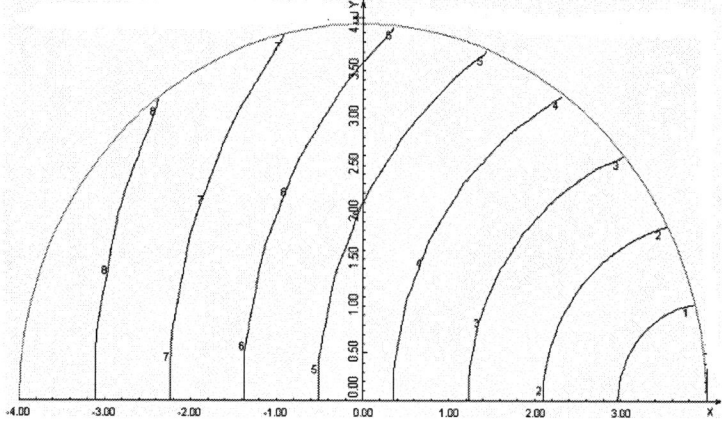

FIGURE 2. The initial state of the explosive surface.
Lines (2-9) - the detonation wave position at consecutive moments of time.

At the moment of time $t=0$ the wave is in the point $x= 4$, $y=0$, $z=0$, at the moment $t=2.0$ – in the point $x= - 4$, $y=0$, $z=0$.

CALCULATIONS WITH GRAD-CODE

Numerical calculations of test problems were carried out with the GRAD program code and graphic results were presented with the special visualization program code .

CONCLUSION

In the paper test problems about detonation waves motion from a point on free surface are presented. Exact solutions of test problems can serve for verification of numerical methods with the programs of computation of detonation problems . Statements and results of the computations of test problems with GRAD program code are stated, discussion of problems and comparison of results of the computation with exact solutions are carried out. The satisfactory agreement of results of numerical solutions with exact solutions is received. Test problems can be recommended for improvement of numerical methods of the detonation computation, especially for graphic presentation of calculations.

REFERENCES

1. Zababakhin E.I. Some problems of explosion gasdynamics, pp 163-166. Snezhinsk. 1997.
2. Kazhdan J.M. About motion of gas behind deverging detonation wave in space with the cut out cone. AMM. v. 31,
   issue 5, 1967.
3. Kazhdan J.M. Research of the vicinity of free border at motion of gas behind deverging detonation wave in space with the cut out cone. AMM, v. 32, issue. 2, 1968.
4. Suchkov V.A. About motion of gas behind plane detonation wave of orthogonal free surface. AMM, v 39, issue 6, 1971.
5. Suchkov V.A., Shnitko A.S., Zhilina R.A., Islamova L.R. Gasdynamic modeling of detonation wave propagation. Preprint № 210. RFNC – VNIITF, Snezhinsk, 2004.

Ovsyannikov Vortex:
Theory and Applications to Model of Hurricane

A.A. Cherevko, A.P. Chupakhin

Laboratory of Differential Equations, Lavrentyev Institute of Hydrodynamics,
Novosibirsk, 630090, Russia

Abstract. The new exact solution of gas dynamics equation, called Ovsyannikov vortex, is investigated. It is the generalization of radial symmetrical solutions. These solutions describe gas source with nonzero curl. The applications for simulation of hurricanes are discussed.

Keywords: exact solution, gas dynamics, vortex, hurricanes.
PACS: 47.32.Ef, 92.60.Bh.

INTRODUCTION

The classical spherically symmetric solutions and their applications to the various problems of gas dynamics, astrophysics, physics, of atmosphere are well known and described in many papers [1–3].

In 1995 Ovsyannikov found a new class of solutions generated by the rotation group [4] and called it "singular vortex". Later this remarkable object was called by *"Ovsyannikov vortex"* (OV). These solutions are regular partially invariant solutions of the gas dynamics equations. In contrast to the spherically symmetric solutions, the velocity $\vec{u}$ has a nonzero tangential to the sphere component $\vec{u}_\tau \neq 0$ and curl $\Omega = \operatorname{rot} \vec{u} \neq 0$. The angle of the velocity deflection from the meridian — the quantity $\omega = \omega(t, r, \theta, \varphi)$ is a function of all independent variables: $r = \sqrt{x^2 + y^2 + z^2}$, latitude $0 < \theta < \pi$ and longitude $0 \leq \varphi < 2\pi$ on the sphere $r = r_0$. Part of the unknown functions: the radial velocity component U and the modules $H = |\vec{u}_\tau|$ of the tangential velocity component and all thermodynamic parameters (pressure p, density ρ and entropy S) are spherically symmetric; i. e. they depend only on the time t and r (Figure. 1).

Thus, the OV is partially invariant solution of defect one and rank two [5]. There are one noninvariant function ω and two invariant independent variables r and t. In [4] Ovsyannikov proved the existence of solutions of this form, obtained some properties and gave a number of examples. The investigation of OV in gas dynamics and hydrodynamics was continued in [6–9]. The applications of OV in magnetodyrodynamics are in [10].

CP849, *Zababakhin Scientific Talks - 2005,*
edited by E. N. Avrorin and V. A. Simonenko

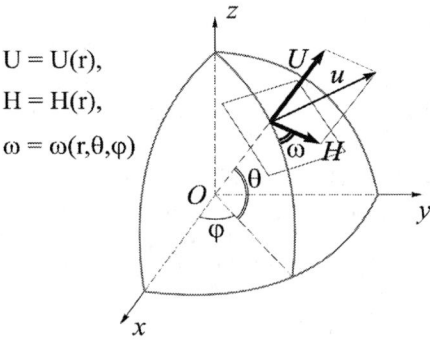

$U = U(r),$

$H = H(r),$

$\omega = \omega(r,\theta,\varphi)$

FIGURE 1. Ovsyannikov vortex velocity.

The main property of OV is the decomposition of equations to noninvariant over-determinated subsystem for function and invariant subsystem for function U, H, ρ, S, p. This decomposition defines the nonlinear superposition radial and spherical motions of gas particles. The first subsystem was integrated by Ovsyannikov in [5]. The principal problem is the investigation of invariant subsystem. This system reduced to ordinary differential equations for invariant models of OV.

STATIONARY OVSYANNIKOV VORTEX

The OV model admits a certain symmetry group. It is possible to construct its invariant solutions under this group called invariant submodels of Ovsyannikov vortex. We consider the stationary OV generated by the algebra Lie $L_4 = \langle so(3),T \rangle$, where $T = \partial_t$ [9]. This algebra L_4 specifies a (1, 1) type regular partially invariant solutions, where the invariant independent variable is r.

In this case the invariant system reduced to the following *key equation* for auxiliary functions $h = h(R)$, $R = r_*^{-1}r$, where r_* is scale factor

$$\left|h_R\right|^{\gamma+1} - \beta_0^2 \left(\frac{R^2-1}{R^2}\right)\left(1+h^2\right)^{(\gamma-1)/2} h_R^2 + \beta_0^2 \frac{(1+h^2)^{(\gamma+3)/2}}{R^4} = 0 \qquad (1)$$

Where γ is the isentropic exponent, $\beta_0 = \text{const}$. We have the following representation of invariant function

$$U = U_0\left(h^2+1\right)/\left(R^2 h_R\right), \quad H = H_0/R, \quad c^2 = c_0^2\left|h_R\right|^{\gamma-1}/\left(1+h^2\right)^{(\gamma-1)/2}, \qquad (2)$$

Where c is the speed of sound and U_0, $H_0 \neq 0$, c_0 are constants. There are two principal obstacles in analysis of key equation (1). It is implicit equation with respect derivative h_R. The basic theories of such equations are described in [11]. The equation

(1) contains modulus of derivative therefore we have some different equations in different domain on plane R and h.

The complete global analysis of all solutions of equation (1) was fulfilled in [8]. Now we mention some basic mathematical properties and physical consequences.

1) All nontrivial solutions of equation (1) are strictly monotone functions. They are defined in domain $\Omega \subset R^2(R,h)$ bounded by *discriminant curve* $\partial\Omega$. This curve is the singular curve and it is the source of all integral curves of equation (1).

2) There are four integral curves of key equation in every point $P \in \Omega$. Two of these curves for which $h_R > 0$ generate gas source and other two, for which $h_R < 0$, correspond gas drain. The gas source (drain) has finite minimal size R_{min} (Figure. 2).

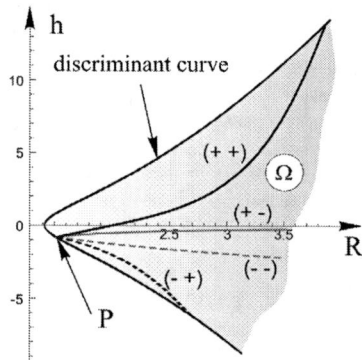

FIGURE 2. The four integral curves.

3) One of the integral curves of (1) with $h_R > 0$ describes supersonic gas flow from the spherical source. Solution and gas flow exist for arbitrary $R > R_{min}$. Second curve with $h_R > 0$ corresponds to gas flow wit continuous transition from supersonic to sub sonic motion. These solution and gas flows exist on finite interval $R \in [R_{min}, R_{max}]$ and ended the drain (Figure.3). There exists gas flow with spherical shock wave.

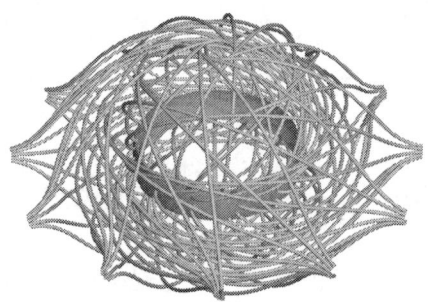

FIGURE 3. Lines of flow.

OVSYANNIKOV VORTEX AS THE MODEL OF HURRICANE

Natural phenomenon such that hurricanes, tornado attracts attention of many researchers. We prove that steady Ovsyannikov vortex simulates the developed stage of the tropical hurricanes. This model explains and describes following important attributes of this object.

1) The existence of the typhoon's eye. It is interior domain $R < R_{min}$ for OV. There are states of rest in this domain (Figure. 4).

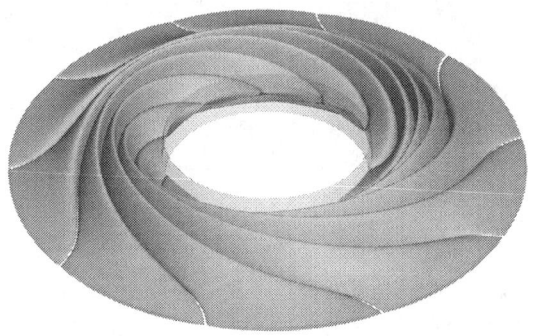

FIGURE 4. The typhoon's eye.

2) The existence of spiral branches in hurricane. They are described by elliptical drains of gas from its boundary $R = R_{min}$ (Figure. 5).

FIGURE 5. The spiral branch.

3) The coincidence of qualitative behavior of basic parameters in hurricane: pressure and velocity (Figure. 6).

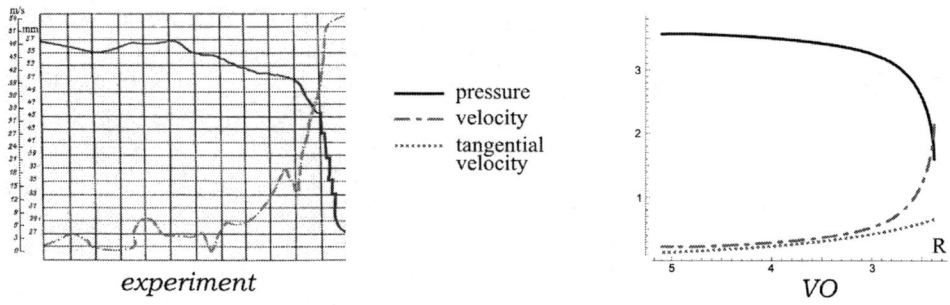

experiment   VO

— pressure
– – – velocity
· · · · · · · tangential velocity

FIGURE 6. The qualitative behavior of basic parameters.

ACKNOWLEDGMENTS

The work was supported by RFBR grant 05–01–00080 and by Council of Support of the Leading Schools, grants Sc. Sch.–440.2003.1.

REFERENCES

1. N. E. Kochin, I. A. Kibel', N. V. Roze, *Theoretical Hydromechanics*, Translated from the fifth Russian edition, New York–London–Sydney, Interscience Publishers, 1964.
2. L. I. Sedov, *Similarity and Dimensional Methods in Mechanics*, Transl. from the Russian, Moscow, Mir Publishers, 1982.
3. K. P. Stanyukovich, *Unstedy Motions of a Continuous Medium*, Nauka, Moscow, 1971.
4. L. V. Ovsyannikov, *Group Analysis of Differential Equations*, Academic Press, New York, 1982.
5. L. V. Ovsyannikov, Singular Vortex, Prikl. Mekh. Tekh. Fiz., 1995, 36, No. 3, p. 45–52; translation in J. Appl. Mech. Tech. Phys, 1995, 36, No. 3, p. 360–366.
6. A. P. Chupakhin, *Invariant Submodels of Singular Vortex*, J. Appl. Math. Mech., 2003, 67, p. 390–405.
7. A. A. Cherevko, A. P. Chupakhin, *Homogeneous Singular Vortex*, J. Appl. Mech. Tech. Phys., 2004, 45, p. 209–221.
8. A. P. Chupakhin, *Singular Vortex in Hydro- and Gas Dynamics*, Analytical Approaches to Multidimensional Balance Laws, Nowa Publishers, 2005, p. 81–109.
9. A. A. Cherevko, A. P. Chupakhin, *Stationary Ovsyannikov's Vortex*, Prep. Lavrentyev Institute of Hydrodynamics, 2005, 1–2005.
10. S. V. Golovin, *Singular Vortex in Magnetogydrodynamics*, J. Phys. A: Math. Gen., 2005, 38, p. 4501-4516.
11. V. I. Arnold, *Geometrical Methods in the Theory of Ordinary Differential Equations*, Udmurt State University, 2000.

The Specialized Systems of Scientific Visualization

Vladimir L. Averbukh

Institute of Mathematics and Mechanics, Ekaterinburg

Abstract. In the paper the necessity of development of the scientific visualization specialized systems is shown on examples and the tendencies in methods of specialization are described. The methodology of development of the computer visualization specialized systems is given. The project of interactive visualization system for parallel computations is described.

INTRODUCTION

The scientific visualization was one of the main objectives of using of computer graphics systems since the 60-s' years. The first computer graphics packages contained a significant set of means directed just for representation of computer modeling results.

It is necessary to note two tendencies of visualization systems design and development. On the one hand - developments of universal visualization systems, and on the another - specialization on all directions. The important feature of universal visualization systems is the presence of typical set of views for typical mathematical objects. Thus, it may support the tool for development of specialized visualization. The user task lies in describing of connections between standard views and modeling entities that is necessary to visualize. Universal systems contain some standard set of ("universal") views and techniques of visualization.

The specialized systems facilitate the user's work, and, in case of principally new model objects researching only through these systems using the evident representation about model's nature and particularities features is possible to get. Our experience gives a number of examples of specialized visualization systems using for the analysis and interpretation of the data that is the results of the difficult problems decisions.

A SPECIALIZED SYSTEM FOR VISUALIZATION OF TUBE-LIKE OBJECTS

The tube-like objects are the results of numerical solution of control theory and differential games, where the main element is the value function. Researchers are interested in finding such function characteristics degeneration of sections of these tubes, the presence of non-smoothness, so called narrow throats (see Fig. 1) and other tube details. It is impossible to represent these subtle details using universal visualization systems. Only specific information from mathematicians about the nature of mathematical objects had allowed showing them. Features of this system allow investigating singular surfaces in linear differential games and maximal stable bridges [1].

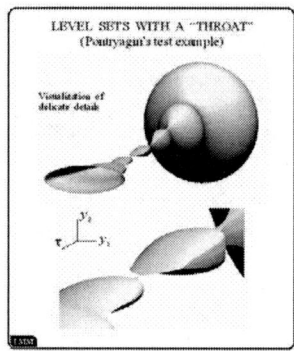

FIGURE 1. **"Narrow throats" in differential games**

VISUALIZATION OF FOUR-DIMENSIONAL INFORMATION SETS

The parallel program for search of reactions speeds at mathematical modelling of chemical processes was developed. Parameters of model are speeds of elementary reactions. After transformations the problem was reduced to finding the quaternaries of speeds of these elementary reactions. The set of all allowable series of speeds is information set, which in this case consists of about one million quaternaries of numbers. Thus, a problem of a finding of these numbers was reduced to a problem of a information set finding.

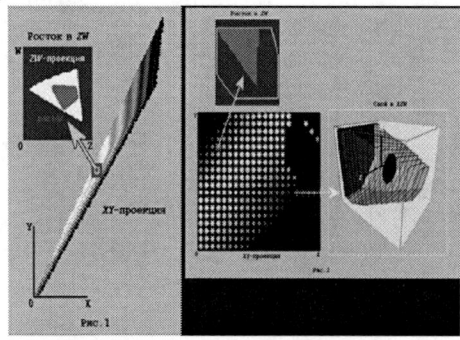

FIGURE 2. View of four-

Studying of a geometrical structure of this set has allowed investigating interrelation of the elementary reactions participating in process of splitting. Particularly, it is necessary to develop visualization methods to investigate local and global characteristics of sets arrangements in four-dimensional space. For analyses of the information about these four-dimensional set it is insufficiently to use only a one known view of multidimensional sets. A battery of visualization methods has been offered be the knowledge of aprioristic sets structure and the visualization goals. After consultations with mathematicians, (potential) users, priority views have been chosen among the suggested views. The complex approach to visualization has been developed. The idea of interaction with the user during the visualization and using of some complex views is a basis of this approach. Thus each view is connected with others. It allows to realize the "navigation" on the set and to analyze all its necessary properties. Thus, during the realization of the system the new approach for visualization of concrete information four-dimensional sets have been offered. The problem of their representation was basically solved [2]. See Fig. 2.

THE SPECIALIZED SYSTEM OF VISUALIZATION FOR ONE PROBLEM OF OPTIMUM CONTROL

The search of reachability sets in one of optimum control problems is a source of the statement of the following problem. A number of reasons, connected with the realization of the algorithm, has led to the fact that the body of computed data make about tens millions, and even billions points in a bitmap-format. The mathematicians would study a general view and an internal structure of the specific problem reachability set. A program complex allowing to do a data filtration not deforming neither external and nor internal object structures has been created. The system allows to display as process of the reachability set constructions and as object in different foreshortenings. The program complex consists of a set of functions to deal with a huge cloud of points and its subsequent visualization. The data are processed on the specialized "visualization pipeline", consisting of such stages as processing initial bitmap-files; lightness computing converting into a voxel format and the creation of structures for scene storage; smoothing of voxel objects; converting into a polygonal format. In the example so-called Lorentz's sphere (See Fig. 3) are considered. After computing of this sphere the volume of an initial file was approximately 6,5 millions points. In a result the graphical object consisting from about 40000 polygons was created. It is obviously, that now computer hardware does not allow to manipulate interactively by millions points objects in contrast to tens thousand polygons. Thus as a result of the specialized system development the volume of the data has considerably decreased with lost-free qualities. It has allowed to study results of numerical research without excessive requirements to hardware resources [3].

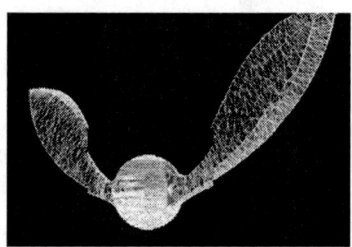

FIGURE 3. Lorentz's sphere.

METHODOLOGY OF SPECIALIZED COMPUTER VISUALIZATION SYSTEMS DEVELOPMENT

Our researches and the realization of concrete systems have enucleated the methodology of the development of the specialized computer visualization systems. First of all techniques of views construction were offered. Among these techniques are visual exaggeration of features due to dependent from ad hoc knowledge about physical (biological information, etc.) and/or mathematical entities of the data; choice of necessary perspective for graphical displays; construction of the special visual objects which are not having accordance to modelling objects, but providing their the analysis and interpretation; application of multiviews, showing different aspects of the investigated phenomenon, creation of views systems, including graphics, animation, spreadsheets and text representations, and also navigation by direct manipulations of visual objects; using in views the natural and familiar figurativeness, and also analogues and "drifting" views; application in visualization systems techniques adopted from a cinema and animated cartoon, and also some virtual reality techniques.

Our procedure of design and development of specialized visualization systems it is possible to describe as some plan to order systems realization. First of all, plan consists of positions describing visualization system design. The plan includes also the sequence questions connected to these positions, and also the analysis of problems arising during the designing process. The basic positions in turn determine a set of roles of design and development process participants.

Among specialists who may participate in the process of design and development of visualization systems there are a potential user, a designer of visualization, a expert in methods of computer graphics and human-computer interaction, and a system programmer. We have to understand thus that in the real project the same people can play different roles.

Among the basic positions in the plan of the specialized visualization systems design and development there are the problem under consideration; the user for whom it is supposed to construct visualization means; the program solving the given problem; figurativeness of visualization and methods of interaction; techniques of graphics generation and rendering; system questions. For each position there are the set of questions which answer have to describe the plan of system realization.

SYSTEM OF INTERACTIVE VISUALIZATION
OF PARALLEL COMPUTATION

The developed technique of designing specialized visualization systems is used now on the creation of the specialized visualization systems problems connected with "large" parallel computation (time of the execution - from several hours to several days and more). As concrete examples the problems of mathematical physics are considered. These problems frequently use for solving grid methods.

To begin the system realization the design of visualization was carried out together with the future users. Visualization and the interface metaphors, and also views were developed. The system should evidently map the grid data of large volumes with the opportunity of searching of anomalies in grid structures. A complex view with elements of interaction was developed. In this view the ability to change a lot of parameters is incorporated. Dimension, methods of rendering, set of the interface functions may change. Using of several rendering methods in one view raises appreciable its universality. Opportunities of polygonal graphics allow depicting in details the block structure of a grid, the hierarchy of components and features of its internal construction. The voxel graphics emphasizes applied value of grid problems, mapping not discrete set of points but three-dimensional model as a continuous array of the data.

The system will be capable to visualize rectangular, hexahedral, pyramidal (see Fig. 4) and other structures of a large volume grid - from tens thousand up to hundred millions points. In the last case parallel processing is supposed [4].

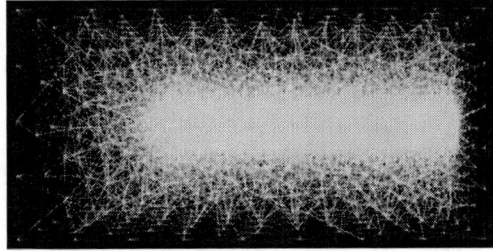

FIGURE 4. The display of the hole grid of 70300 points.

The work is supported by Russian Foundation for Basic Researches, the project N 04-07090120.

REFERENCES

1. Averbukh V.L., Kumkov S.S., Shilov E.A., Yurtaev D.A., Zenkov A.I. Specialized scientific visualization systems for optimal control application // Nonsmooth and Discontinuous Problems of Control & Optimization: Proc. IFAC Workshop Chelyabinsk, Russia, 17-20 June, 1998. New York etc.: Pergamon, 1999. P.71-76.

2. Vasev P.A., Pervalov D.S. One method of visualization of four-dimension sets // Proceedings of conference "Supercomputing technologies in physical and chemical researches", Chernogolovka, Russia, 2001, pp. 32-36.
3. Moshkov A.V., Pakhotinskikh V.Yu., Reshetniak V.O. Specialized visualization system for one of the problem of optimal conrol // THE 14th International Conference on Computer Graphics and Vision GraphiCon'2004, 2004 Moscow, Russia. Conference Proceedings. Pp.299-301.
4. Gorbashevskiy D.Yu., Kazantcev A/Yu. Visualization of grid data of large volume // THE 15th International Conference on Computer Graphics and Vision GraphiCon'2005, 2005, Novosibirsk, Russia. Conference Proceedings. Pp.366-367.

ASTRAL Code for Problems of Astrophysics and High Energy Density Physics

N.E. Chizhkova, G.V. Ionov, N.G. Karlykhanov, V.A. Simonenko

Russian Federal Nuclear Center - All-Russia Research Institute of Technical Physics
PO BOX 245, Snezhinsk, 456770, Russia

Abstract. The paper gives a brief description of ASTRAL code package for astrophysics simulations, including features in the implementation of basic physical processes and two tests. A sketch of the object code structure is provided.

Keywords: code, 1D, C++, OOP, astrophysics, star, high-energy-density.
PACS: 26.20.+f; 44.05.+e; 47.11.Bc

INTRODUCTION

There are a lot of 1D codes which simulate hydrodynamics, heat conduction and other phenomena independently. However, the simulation of physical phenomena in stars often requires that all these processes be modeled in a coupled fashion. Open source software we are developing will cover all phenomena, making it possible to model processes on the surface of neutron stars, the internal structure and evolution of stars and explosive processes in them.

Currently the phenomena are treated in 1D plain, cylindrical and spherical approximations. The entire problem is split into physical processes which can be combined arbitrarily. Now these processes are hydrodynamics, gravitation, heat conduction and thermonuclear reactions. Two numerical techniques are implemented for both hydrodynamics and heat conduction. Equations of state can be defined in analytical and tabulated forms.

Phases of stellar evolution sharply differ in the time scale. Therefore, we give special attention to the automated adjustment of time steps and even combine quasi-static and slow processes, e.g., slow thermonuclear reactions occur in a main sequence star being in quasi-static hydrodynamic equilibrium.

Great attention is also given to matching the dimensions of physical quantities. When different techniques are used simultaneously, it is important to match dimensions and properly select the system of units. This problem is fully resolved in our code because all physical quantities are automatically checked and appropriately converted.

We have chosen C++ because object codes written in this language exhibit the highest degree of code structure optimization and run most operating systems available. An object code is the most natural way of representing the physical picture

CP849, *Zababakhin Scientific Talks - 2005,*
edited by E. N. Avrorin and V. A. Simonenko

of the world; it prevents undesirable interactions between code fragments, confining the growth of chaos in the long development of large codes.

From a practical point of view, the code is broken into two components, namely a task solver and a task manager. The components can be improved and extended independently. The task is an object which contains all information on a system to be modeled, physical processes, on the source code, input data and data collected by gauges arranged in the system.

PHYSICAL PROCESSES

Consider heat conduction, hydrodynamics and thermonuclear reactions in more detail.

Heat conduction

Heat conduction is described by two differential equations for heat sources and heat flow:

$$
\begin{cases}
\dfrac{\partial}{\partial t}E + \Omega_v \dfrac{\partial}{\partial q}\left(r^{\nu-1}S\right) = Q \\[2ex]
-\kappa \cdot \Omega_v \rho r^{\nu-1} \dfrac{\partial}{\partial q}T(E) = S
\end{cases}
\tag{1}
$$

The difference scheme is implicit and first-order accurate with respect to time and second-order accurate with respect to mass coordinate on a uniform mesh. A solution is sought through successive corrections to the current solution from the tangents found.

The code was tested on linear and nonlinear heat conduction problems which have analytical solutions.

Hydrodynamic

The hydrodynamic set of equations includes three differential equations of motion, continuity and energy, and an equation of state.

$$
\begin{cases}
\dfrac{\partial u}{\partial t} = f - \Omega_v r^{\nu-1}\dfrac{\partial P}{\partial q} \\[2ex]
\dfrac{\partial V}{\partial t} = \Omega_v \dfrac{\partial}{\partial q}\left(r^{\nu-1}u\right) \\[2ex]
\dfrac{\partial E}{\partial t} = fU - \Omega_v \dfrac{\partial}{\partial q}\left(r^{\nu-1}Pu\right) \\[2ex]
E = \varepsilon + \dfrac{u^2}{2}; \quad P = P(V,T); \quad \varepsilon = \varepsilon(V,T)
\end{cases}
\tag{2}
$$

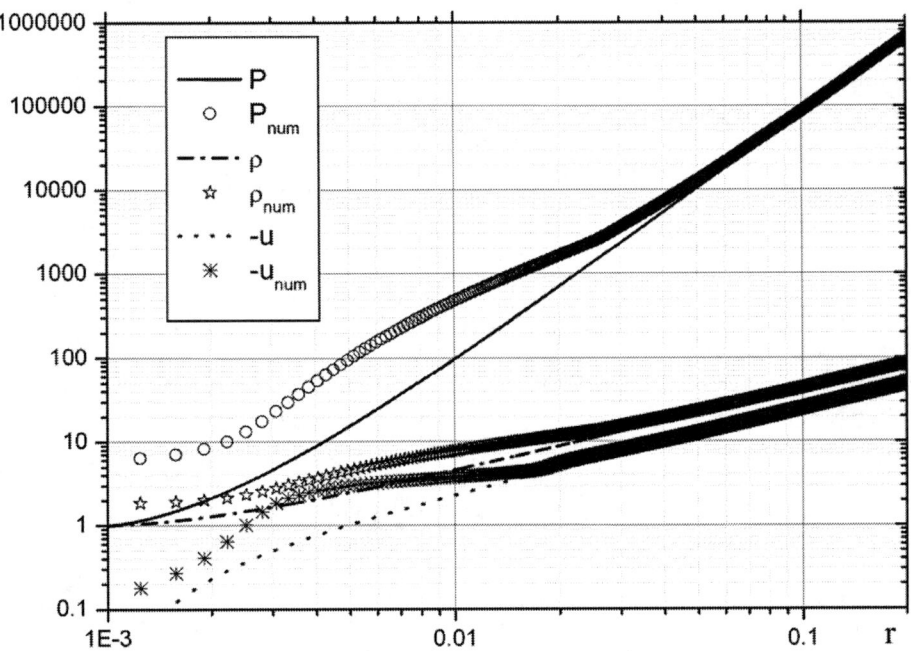

FIGURE 1. Symbols show the numerical solution and lines show the analytical solution. Density is shown by five-point stars, pressure is circle and absolute velocity is asterisks.

The difference equations are also implicit and first-order accurate with respect to time and second-order accurate with respect to mass coordinate. A solution is sought through successive corrections to the current solution from the tangents found.

Tests included various hydrodynamic problems in different symmetries. The final test involved comparison between the numerical solution to a problem and its analytical solution [1].

We simulated unlimited gas compression in a spherically symmetric configuration. The initial density is unity and the initial velocity is zero. We used the equation of state for ideal gas with and a heat capacity of one sixth, and the law of piston motion obtained in the analytical solution.

At time -0.001, differences between the analytical and numerical solutions are clearly seen (Figure 1). They are most likely the result of poor spatial resolution at the initial phases of motion where information from the piston is transferred through mesh periphery where spacing is large.

Thermonuclear reactions

The main source of energy in stars is thermonuclear reactions in which nuclides turn into heavier ones through chains of transformations. These reactions need to be modeled. For this end we must have temperature and density dependent reaction rates. We imported the most complete library of reactions proceeding in stars [2]; it was compiled from various sources by T. Rauscher and F. Thielemann.

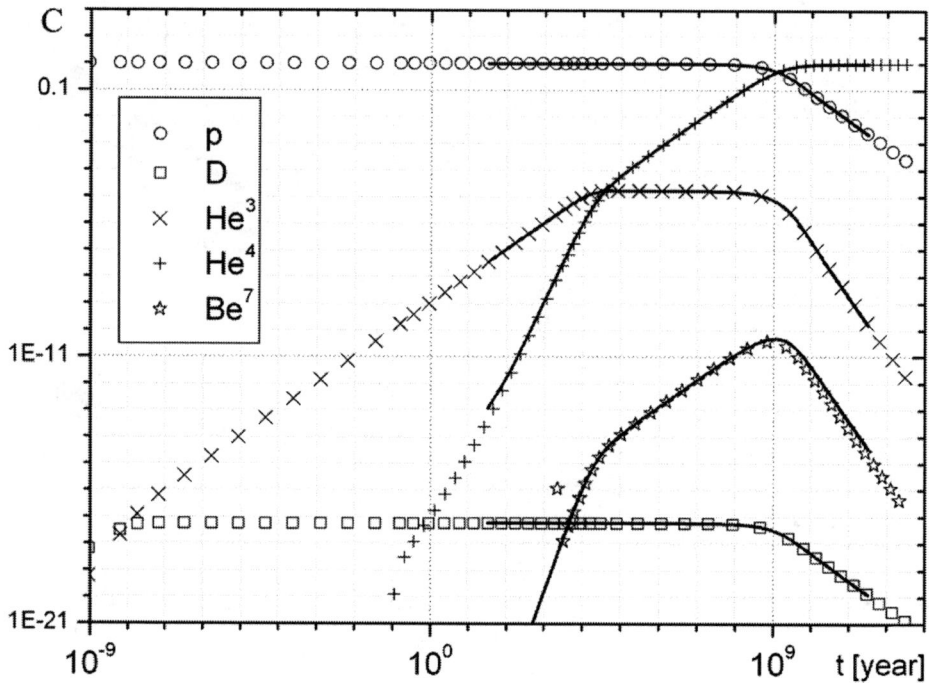

FIGURE 2. Molar concentrations versus time for Sun core conditions. Symbols show the numerical solution of ASTRAL and lines show the solution of ERA code.

The library contains 61336 reactions including beta decay and electron capture. It is clear that all these reactions cannot be modeled continuously. There are two levels where active reactions for specific conditions are selected. The sets of active reactions are dynamically changed as conditions in different parts of a system change with time.

For testing purposes we made a calculation on the evolution of a system at density and temperature corresponding to the conditions in the Sun core. The system was defined to be initially composed of pure hydrogen. The calculation was done with ASTRAL and ERA for comparison (Figure 2). Their results agreed excellently for all isotopes but $_4Be^7$ whose concentration is controlled by beta decay. However, results for beta decay were also in satisfactorily agreement.

CODE STRUCTURE

We should say a few words on the code structure. The code consists of a task manager and a solver. Information exchange between them is implemented via a file system as the most universal method of information exchange between different platforms.

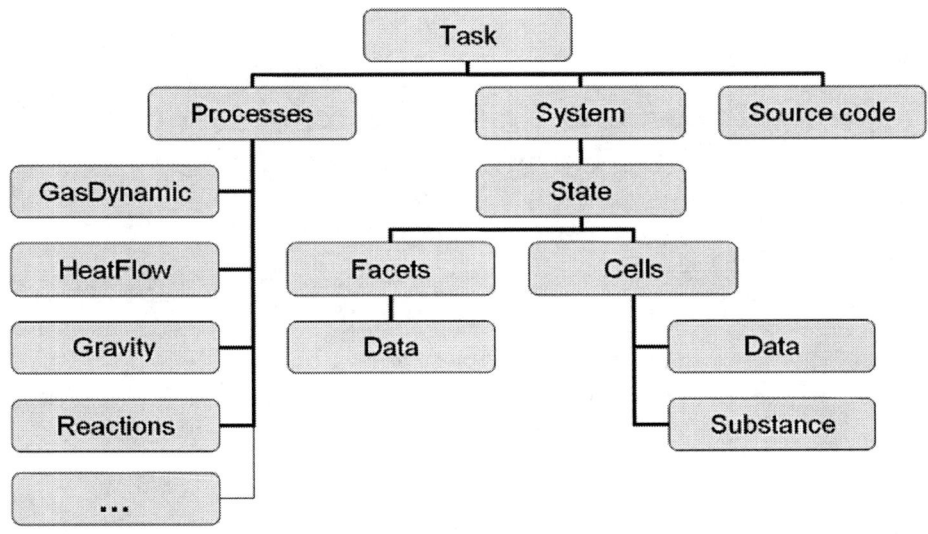

FIGURE 3. The general structure of classes.

All rests on the task that contains the system to be modeled, the vector of physical processes and the source code of the solver (Figure 3). The system knows its states at some time steps. The states are represented by cells and edges which contain physical data and substance compositions.

Further developments include
- Simulation of turbulent mixing.
- Automated multi-tasking control.
- Development of an input edit and task initiation system and a database of tasks.

ACKNOWLEDGMENTS

We are very much thankful to S. A. Baban for the development, programming and testing of the model of thermonuclear reactions. The authors are grateful to SFTI students who contributed to the development of ASTRAL code. We want to express gratitude to S. V. Peshekhonova for the development of the hydrodynamics model, to A. A. Alexandrova for the development of the heat conduction model, to L. A. Alexandrova for equations of state, and to S. F. Sergin for the his thorough investigation and elaboration of the k-ε model.

REFERENCES

1. V.A. Simonenko, N.E. Chizhkova, "Isentropic Self-similar Gas Compression", *This AIP publication*.
2. Tables of reaction rates based on F. Thielemann, et al., *Adv. Nucl. Astro.*, **525** (1987). 1991 updated version.

Three-dimensional Grid Generation Algorithms for the Volumes of Revolution.

T.N. Bronina, O.V. Ushakova

Institute of Mathematics and Mechanics, Ural Branch of the Russian Academy of Sciences
S. Kovalevskaya st. 16,Ekatherinburg, Russia, 620219
e-mail: btn@imm.uran.ru
e-mail: uov@imm.uran.ru

Abstract. We describe three-dimensional algorithms for computing optimal structured grids in the volumes of revolution. The specificity of computations of grids in considered cases is discussed. Grid cell classification and computational criteria for such a classification are given.

INTRODUCTION

The algorithms for generation of structured three-dimensional grids in the domains of complex geometries are suggested in [1-3]. These are the algorithms for the volume of revolution, for pipelines and algorithms for the global reconstruction of the grid. Algorithms [1-3] are elaborated within the approach [4] and are designed for multimaterial hydrodynamic simulation and for solving other physical and engineering problems.

The algorithms for the volume of revolution are developed for calculating grids in the domains obtained by the rotation of the generating curve about the axis. Algorithms in which generation of grids is carried out by the rotation of two-dimensional grids about the axis produce O-type grids containing degenerate cells on the axis of rotation, moreover these degenerate cells become too small for a small angle of rotation. Algorithms [1-3] are not reduced to the rotation of a two-dimensional grid about the axis and such singularities do not arise.

Utilization of algorithms [1-3] for computation of grids in the volumes of revolution, and also taking into account the computational algorithms requirements to grids resulted in emerging specificity of configurations for the considered three-dimensional domains and computations of grids there. This led us to the development of new algorithms.

Section 1 is devoted to the formulation of the problem. In this section, we define what we understand under the volume of revolution and its configuration. We enumerate the considered configurations of the volumes of revolution and analyze their specificities. In section 2, we characterize grid generation algorithms. Presented algorithms generate three-dimensional nondegenerate structured grids composed of hexahedral cells which are the images of unit cubes for trilinear mapping. Degeneration of cells is allowed only in the form of prisms with triangular base. In section 3, we present a hexahedral cells classification.

CP849, *Zababakhin Scientific Talks - 2005,*
edited by E. N. Avrorin and V. A. Simonenko

1. FORMULATION OF THE PROBLEM

The problem consists in constructing a structured three-dimensional grid $(x^1_{ijk}, x^2_{ijk}, x^3_{ijk})$, $i=0,..,N$, $j=0,..,M$, $k=0,..,L$ in a three-dimensional domain G of the variables x^1, x^2, x^3 obtained by the rotation of a generating curve given in the plane x^1, x^3 about the axis x^3 through the angle φ ($\varphi_0 < \varphi \leq \pi$). The generating curve can consist of straight line segments, arcs of circles and ellipses. Constructing a three-dimensional structured grid in a simply-connected three-dimensional domain means that we establish a one-to-one correspondence between an auxiliary parametric domain (parallelepiped or cube) (see [5]) and the physical domain which is a simply-connected domain where we want to construct a three-dimensional structured grid. Doing this we define the volume of revolution as a hexahedron with 8 vertices, 6 faces and 12 edges. Representing the volume of revolution as such a hexahedron defines the configuration of the domain. So, under configuration of the domain, we understand not only the geometry of the domain, but also the way of its representation in the form of curvilinear hexahedron by arranging its vertices and indicating edges and faces. Necessity of such representation is connected with the way of structured grid generation in simply-connected domains of complex form. The specificity of considered cases consists in that some faces of hexahedron lie in one plane. Such configurations are exotic and give rise to singularities in a grid (degenerate cells), but allow us to compute grids which provide efficient modeling the physical field phenomena and considering block constructions composed of some volumes of revolution put on each other.

Different types of generating curves and corresponding configurations of domains, one can find in [1,3]. Such domains very often arise in multimaterial hydrodynamics simulations. Three different configurations that are a ``body'', a ``shell'', and a ``cut'' or ``cut shell'' can be found also in [6]. The specificity of these configurations is the following: three faces of a curvilinear hexahedron (a volume of revolution) or more lie in one plane and all 8 vertices of a hexahedron lie in this plane too. Since we use for constructing a grid the mapping approach, in a general situation we are looking for a homeomorphic mapping of an auxiliary domain (a parallelepiped) into a physical domain (see [5]). We are looking for this mapping only in some points of an auxiliary domain. In other points we fill up this mapping by trilinear mappings of the unit cubes. By this, we construct hexahedral cells of a grid.

In [7-8], the sufficient conditions for mappings to be homeomorphic are obtained under different suppositions about their smoothness. These conditions ensure nondegeneracy of grids and can be applied to justify properties of algorithms for constructing them in cases of structured and unstructured grids in complicated domains. To check conditions of the obtained theorems, information on properties of the mapping on the domain boundary and information of local character is used. Because of above specificity, hexahedral cells along common edges of plane faces are characterized by zero Jacobians of the mapping used for generating cells, therefore the general mapping along these edges will not satisfy the conditions of theorems [8] providing the homeomorphism of a mapping. Along these edges hexahedral cells degenerate into prims and the general mapping is not a homeomorphism. This caused the development of new grid generation algorithms.

2. GRID GENERATION ALGORITHMS

The main idea of the algorithms is presented in [4]. The algorithms are variational and a construction of a grid is based on the minimization of the discrete functionals characterizing a grid quality. One of the used in [1-3] functionals is the functional of the form

$$D = A_U D_U + A_O D_O + A_A D_A. \qquad (1)$$

Functionals D_U, D_O, D_A formalize criteria of grid closeness to uniform (U), orthogonal (O) and adaptive grids (A), respectively. Constants A_U, A_O, A_A are non-negative weights in the general functional D, minimization of which provides an optimal grid generation. Here we construct only grids close to uniform and orthogonal grids and put $A_A = 0$.

For a three-dimensional case, this approach is for the first time described in [1] and then in [3]. It is described in details in [6]. Here we described the present state of its development on the examples of grid generation for the volume of revolution. The numerical method for constructing grids in the volumes of revolution as for a general case [4] consists of two stages. On the first stage, an initial grid satisfying some required properties is obtained. For generating an initial grid, algorithms based on the combination of geometric and variational approaches are suggested. These algorithms construct structured nondegenerate grids close in the sense of [4] to a uniform grid. The initial grid also satisfies to the requirement of closeness of linear sizes of cells [1,3]. For some configuration of the domains [1,3], an initial grid can contain hexahedral cells degenerating into prisms with triangular bases or into a union of such prisms. Suggested algorithms for constructing an initial grid admit such types of degenerate cells and permit the joining of different volumes of revolutions (from node to node) that gives the opportunity to consider multi-block constructions of such configurations of domains. The algorithms provide the correspondence of the symmetries of a grid to the symmetries of a domain.

The approach used for generating the initial grid is the same as in [1,3]. It is based on geometrical principals. The algorithm suggested for considered configurations is the following. The required three-dimensional grid in the domain of revolution is composed of L two-dimensional grids computed on surfaces of revolutions generating curves l_k of which are arranged in the plane of a generating curve for the considered domain. Actually in the plane of a generating curve, a two-dimensional grid is constructed and corresponding generating curves for constructing surfaces of revolutions are chosen. We construct two non-self-intersecting families of coordinate lines on the surfaces of revolutions. For constructing grids on surfaces of revolution, we use the fact that the surfaces are the surfaces of revolutions. Then for constructing an inner grid nodes, a linear interpolation is applied. More detailed description is given in [6].

The algorithms generating an initial grid only on the basis of a geometric approach have restricted possibilities in regulating the grid quality, therefore even on the stage

494

of constructing an initial grid some optimization procedures are used. Since we construct non-self-intersecting two-dimensional grids on the surfaces of revolutions and then use a linear interpolation for constructing a three-dimensional grid, a resulting grid as a three-dimensional grid can be degenerate. For example, an initial grid (see [9]) besides nondegenerate hexahedral cells, can contain degenerate hexahedral cells. Among nondegenerate hexahedral cells we also single out cells of two different structures: usual cells with the structure of a cube and ``unscrewed'' cells (for detail, see next section). Optimization of a two-dimensional grid on the surfaces allows to get rid of degenerate cells and ``unscrewed'' nondegenerate cells.

Full grid optimization based on the minimization of the functional D is applied on the second stage. The algorithms for optimization are developed on the basis of the global reconstruction algorithms [1-3]. The optimization algorithms reconstruct the initial grid into a smooth nondegenerate curvilinear grid close to a uniform orthogonal grid (optimal in the sense of [4]). An optimal grid satisfies also to the criterion of closeness of linear sizes of cells [3]. The optimization of a grid can take place both inside the volume of revolution and on its boundary. In the case of grid reconstruction on the boundary, the nodes of a grid are moving on the surface composed of ruled surfaces of cell faces of an initial grid. In the initial grid, nodes are arranged on the surface given analytically. Main features of the grid reconstruction algorithm are listed in [1]. In detail, this step of grid optimization is described in [6]. Now the following six algorithms have been developed.

Algorithm 1. Fixed boundary nodes. The reconstruction of a grid is carried out on the basis of the above criteria of optimality inside the physical domain or in its subdomain. In this case, boundary nodes are considered to be fixed (both for the whole domain or its subdomain).

Algorithm 2. Free boundary nodes. The grid generation is carried out in the whole domain. This algorithm assumes grid node reconstruction from the above criteria of optimality as on the boundary of the domain (on edges and faces) and inside the domain. Only vertices of a hexahedron are fixed.

Algorithm 3. Conditions of orthogonality to faces. The reconstruction of nodes is carried out also for the whole domain. The reconstruction of nodes on the faces is carried out from the condition of grid lines orthogonality to faces of the domain, and on edges and inside the domain from the optimality criteria.

Algorithm 4. Conditions of orthogonality to faces and edges. The reconstruction of nodes is carried out for the whole domain, on faces and edges from the conditions of grid line orthogonality to faces and edges, inside the domain from the optimality criteria. In this case, because of concordance conditions for nodes on edges, adjoining faces have to be orthogonal to each other.

Algorithm 5. Reconstruction of selected nodes. The reconstruction of the grid is carried out in the nodes selected by the values of indices. This algorithm allows to fix the node arrangement on the single faces, to fix the location of single nodes on the boundary for the preservation of such constructive features of the form of the domain as sharp turn of the boundary, and to define the location of single grid nodes from the condition of grid line orthogonality to the boundary.

Algorithm 6. Smooth adjoining of coordinate lines on common edges of one plane faces. In this algorithm nodes on common edges of one plane faces are found from the condition of smooth adjoining of coordinate lines.

A grids presented in Fig. 1 is an example of algorithm 5. Nodes on the plane faces are fixed, on other faces and inside the domain are free.

The main feature of the suggested algorithms is their barrier properties. It means that a construction of the functional and minimizing procedure have a barrier against degenerate cells i.e., if an initial grid has not got degenerate cells, the optimal grid also has not got degenerate cells. Besides, the algorithm allows us to start from an initial grid containing single degenerate cells. It is again possible because of barrier features of the algorithm (see [6]). It allows to get rid of degenerate cells which are not caused by the configuration of the domain. Degeneration of cells is permitted for an optimal grid only in the form of ruled prisms with a triangular base. For example, an optimal grid with 111x111x11 grid nodes shown in Fig.1 has 20 cells degenerating into prisms arranged along two edges of adjoining faces lying in one plane.

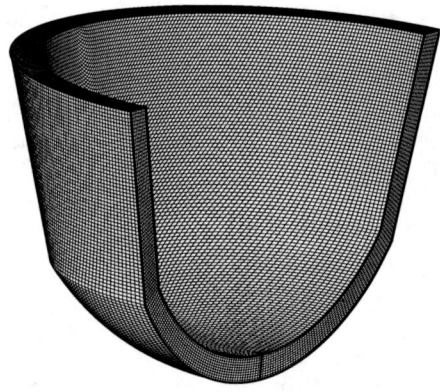

FIGURE 1. An optimal grid for the volume of revolution.

3. ANALYSIS OF CELLS

To carry out such an analysis of cells we need to have special criteria for them.We construct hexahedral cells by means of a trilinear mapping of a unit cube. Constructed in such a way cells are called ruled hexahedral cells. Nondegeneracy conditions for them are obtained in [10,11]. What is understood under the nondegeneracy of computational grids and cells along with the special theorems formulating nondegeneracy conditions, can be found in [6,12]. Among special theorems obtained in [7,8], in [6,12] three main theorems widely used in the theory and practice of grid generation are formulated. These theorems are formulated for two different cases: 1) when a grid is given analytically by smooth functions and 2) when a grid is given by discrete set of nodes (in this case the grid can be given analytically by a piecewise smooth mapping). One of the important conditions of these theorems for establishing grid nondegeneracy is the positivity of the Jacobian of a mapping utilized for generation of a grid or a cell. On the basis of such theorems conditions of nondegeneracy have been obtained. Besides ruled hexahedral cells nondegeneracy conditions are obtained for other types ruled cells: prisms and pyramids. Nondegeneracy conditions for prisms with triangular base and ruled faces, and for pyramids with ruled base and triangular face are obtained in [13,14]. In [6,12] another nondegeneracy conditions have been found in the form which is more convenient for the suggested algorithms. For efficient computation of volume of cells, we use the formulas obtained in [10,11] for hexahedral cells and obtained in [12] for pyramids and prisms. The distinctive feature of obtained nondegeneracy conditions, hexahedral cell classification and formulas of volume of cell is that for complicated curvilinear cells having in general non-planar faces, both the conditions, classification and the formulas of volumes are obtained in terms of volumes of polyhedrons and tetrahedrons with planar faces. This gives the opportunity for an essential simplification of numerical analysis and constructing the numerical algorithms on suggested grids (see, for example, [15]). Nondegeneracy conditions [6] for prisms and pyramids are also used for estimating type of cells into which hexahedral cells can degenerate. The question about admissibility of these types of degenerate cells is discussed in [6,12]. Prismatic cells can arise for some configurations of domains, for example, for those which are given in this paper. Degeneration of hexahedral cells into prisms takes place due to a specificity of the chosen configuration of the domain since two adjoint faces of a hexahedron lie in one plane. The two-dimensional analogs of such cells are triangles which can arise in configurations for some two-dimensional domains, an example of such domain is a semicircle when one of the vertices of the curvilinear quadrangle in the form of which the domain is represented lies on the straight line segment of the boundary of a semicircle. Some initial grid (for example an initial grid for an optimal grid in Fig.1), besides degenerate cells in the form of prisms with triangular base can contain degenerate cells which can be represented as the unions of such prisms. These cells are characterized by one or two non-positive values of the Jacobian of a trilinear mapping in vertices of a cell. The two-dimensional analog of the unions of two prisms is non-convex quadrangles. Pyramidal and tetrahedral cells can arise in some curvilinear coordinate systems with degeneracy, for example, inherent in spherical coordinate system.

Among nondegenerate hexhedral cells we can single out an ``unscrewed" hexahedral cell. The emergence of such cells is can be undesirable, for example in [15], because one of the two dodecahedrons which correspond to a given hexahedral cell is degenerate. This kind of nondegenerate hexahedral cells is characterized by one negative volume of a diagonal tetrahedra (constructed on the diagonals of faces of a hexahedron) or negative coefficient κ_{000} or κ_{111} (see [11]). Diagonals of a hexahedral cell can be chosen by two different ways, so dodecahedrons can be constructed by two different ways too. Each of these dodecahedrons can be represented as a union of images of five tetrahedrons of a unit cube (corresponding to four coefficients from eight α_{i1i2i3}, see also [11], and one of coefficients κ_{000}, κ_{111}) for a piecewise linear mapping of the unit cube. Since one of κ_{000}, κ_{111} is negative, the Jacobian of a piecewise linear mapping for this element is negative and for rest four others the Jacobians are positive, therefore a dodecahedron with negative volume of a diagonal tetrahedra is degenerate according to [5]. If such cells arise in optimal grids, one can try to enlarge the number of grid nodes or to optimize the form of cells. Presented algorithms allow to get rid of such cells (see Fig.1).

ACKNOWLEDGMENT

This work was supported by the Russian Foundation for Basic Research, project 06-01-00306-a.

REFERENCES

1. 1.   T. N. Bronina, I. A. Gasilova, O. V. Ushakova, "Application of the Sidorov's approach to generation of three-dimensional structured grids", in *Proceedings of 8th International Conference on Numerical Grid Generation in Computational Field Simulations,* edited by B. K. Soni, J. Hauser, J. F. Thompson, P. R. Eiseman, J. R. Chawner, N. P. Weatherill, and K. Nakahashi, Waikiki, Hawaii: ISGG, 2002, pp. 445-454.
2. T. N. Bronina, I. A. Gasilova, O. V. Ushakova, *International Conference "Zababakhin Scientific Talks", abstracts, Snezhinsk:RFNC-VNIITF,* 2003, pp. 229-230.
3. T. N. Bronina, I. A. Gasilova , O. V. Ushakova, *Zh. Vychisl. Mat. Mat. Fiz.,* 6, 2003, pp. 875-883.
4. O. B. Khairullina, A. F. Sidorov, and O. V. Ushakova, Variational methods of construction of optimal grids. In *Handbook of Grid Generation*, edited by J. F. Thompson, B. K. Soni, and N. P. Weatherill, CRC Press, Boca Raton, FL,1999, pp. 36-1-36-25.
5. O. V. Ushakova. *Proceedings of the Steklov Institute of Mathematics,* Suppl. 1, 2004, pp.S78--S100.
6. T. N. Bronina, O. V. Ushakova, "Application of Optimal Grid Generation Algorithms to the Volumes of Revolution." in *Advances in Grid Generation.* ed. By O. V. Ushakova. Novascience Publishers, to be published.
7. N. A .Bobylev, S. A. Ivanenko, I. G. Ismailov, *Mat. Zametki,* 60, 4 (1996), pp.593-596.
8. N. A. Bobylev, S. A. Ivanenko, A. V. Kazunin, *Zh. Vychisl. Mat. Mat. Fiz.,* 6, 2003, pp. 808-817.
9. T. N. Bronina, O. V. Ushakova. "Generation of Optimal Grids for the Volume of Revolution", in *Proceedings of 9th International Conference on Numerical Grid Generation,* edited by P. Papadopolous, B. Soni, J. Hauser, P. Eiseman and J. Thompson, Birmingham, Alabama: ISGG, 2005, pp. 270-279.
10. O. V. Ushakova, *"Conditions of nondegeneracy of three-dimensional cells: A formula of a volume of cells"* in Numerical Grid Generation in Computational Field Simulations, edited by B. K. Soni, J. Haeuser, J. F. Thompson, and P. Eiseman, Mississippi State, MS,: International Society of Grid Generation (ISGG), 2000, pp. 659-668.
11. O. V. Ushakova, *SIAM J. Sci. Comp*, 23, 4, 2001, pp. 1273-1289.
12. O. V. Ushakova "Nondegeneracy Conditions for Different Types of Grids " *in Advances in Grid Generation,* edited by O. V. Ushakova, NY: Novascience Publishes, to be published.
13. P. Knabner, G. Summ, *Numerical mathematics,* 88, 2001, 661-681.
14. P. Knabner, S. Korotov, G. Summ, *Finite elements in analysis and design,* 2, 2003, pp.159--172.
15. B. N. Azarenok, Conservative Remapping on Hexahedral Meshes. *in Advances in Grid Generation,* edited by O. V. Ushakova, NY: Novascience Publishes, to be published.

2-D Mathematical Models
OF THE Unlimited Cumulation Phenomena

A.D. Zubov

Russian Federal Nuclear Center - Zababakhin Institute of Technical Physics FNC-VNIITF,P.O. Box 245, Snezhinsk, Chelyabinsk Region, Russia 456770

Abstract. With the help of the simple differential equations describing evolution by a closed curve, it is shown, that, basically, there may be a self-focusing of a curve in a point or a segment. The offered models show an opportunity of unlimited cumulation even at the asymmetrical initial data; small perturbations may be suppressed.

We present simple enough mathematical models showing an opportunity of the unlimited cumulation phenomena even at the asymmetrical initial data. Small indignations thus are suppressed. The simple partial differential equations describing the evolution of the closed or cylindrical surface moving along a normal with speed or acceleration, dependent on curvature of this surface are used. Well-known examples of such models are the canal approximation CCW (Chester-Chisnell-Whitham) [1] for the description of evolution of the shock wave front, and also crystal growth and flame propagation [2]. Similar asymmetrical focusings are theoretically possible in gas dynamics, for example, so-called many-dimensional motions with the homogeneous deformation, resulting to focusing not only in a point or on an axis, but also in a segment or a figure limited to an ellipse [3—4].

The case in point was studied before E.I.Zababakhin [5] - [6]. With the help of set-theory analysis he shown, that ""arbitrary unlimited cumulation is unstable or its probability is equal to zero and no supposed sometimes property of self-focusing" [5]. In the same place it was specified, that the energy dissipation (viscosity and thermal conductivity) usually do not eliminate the cumulation. At the same time, for a case of shock waves in [7] it was shown, that "dissipative effects chosen in physically proved manner, eliminate the unlimited growth of cumulation parameters.

In our work some is entered some model dissipative mechanism reminding a surface tension, and the question on allocation of a class of possible initial conditions, and also classes of curvature-dependent functions which may result to unlimited, probably, asymmetrical, cumulation is discussed.

Let's notice thus, that described below mathematical models and results of numerical calculations on them at all should not be counted mathematical proofs

We shall be limited here to a bidimentional case of motions with flat symmetry.

CP849, *Zababakhin Scientific Talks - 2005,*
edited by E. N. Avrorin and V. A. Simonenko
© 2006 American Institute of Physics 0-7354-0345-7/06/$23.00

Let at the initial moment of time $t=0$ in $\mathbf{R}^2$ it is given closed simple regular (twice differentiable) a curve $\Gamma:f(x,y)=0$ or, otherwise, $\gamma(0)$ with natural parametrization - length of a curve s.

Let $\gamma(t)$ — 1-parametrical set of curves, where $t\in[0,\infty)$ - time. This set is formed by movement of an initial curve $\gamma(0)$ along a normal vector field with speed $V=V(\kappa)$ where $\kappa=\kappa(x,y)$ — curvature of a line $\gamma(t)$ in a point $M(x,y)\in\Gamma$:

$$\kappa=(y_{ss}x_s-x_{ss}y_s)/(x_s^2+y_s^2)^{3/2}. \tag{1}$$

The motion equations we shall write down as:

$$x_t=V(\kappa)\frac{y_s}{(x_s^2+y_s^2)^{1/2}}, \quad y_t=-V(\kappa)\frac{x_s}{(x_s^2+y_s^2)^{1/2}}. \tag{2}$$

Closed curve Γ represents in this case front of a wave, contact border or a line of a level of any parameter of hydrodynamical flow (density, pressure, temperature etc.). Dependence on curvature allows to smooth border Γ and represents some hypothetical mechanism of a surface tension in hydrodynamics.

Despite of a complex view of the equations (1—2), there is a theory (see for example [2]), allowing to consider this system from the point of view of the equation of Hamilton-Jackobi known in analytical dynamics.

At a choice $V(\kappa)=a\cdot\kappa^2$, where $a>0$ — the constant, occurs self-focusing of a curve in a point, and before it the form of a curve becomes round (fig.1 — the first quadrant is represented). Relative difference from a circle in calculation makes less then 10^{-4}.

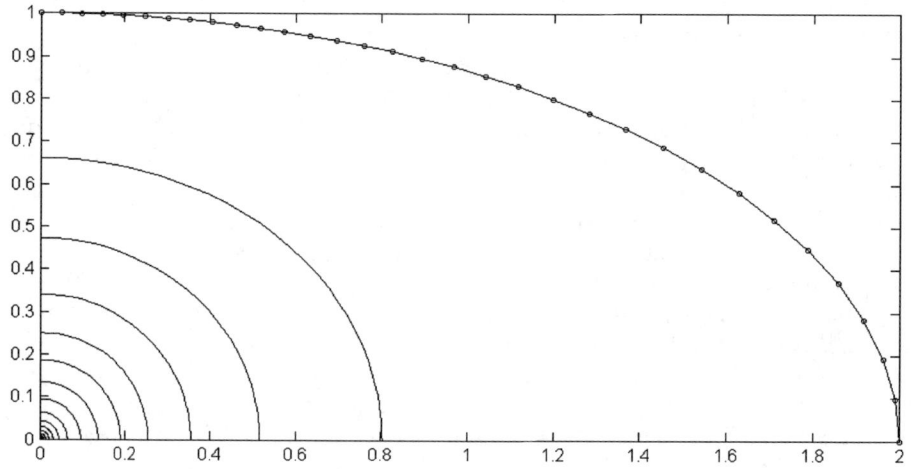

FIGURE 1. Focusing of an ellipse in a point.

Choice of non-convex initial function $f(x,y)=0$ also results in symmetric self-focusing in a point (fig.2).

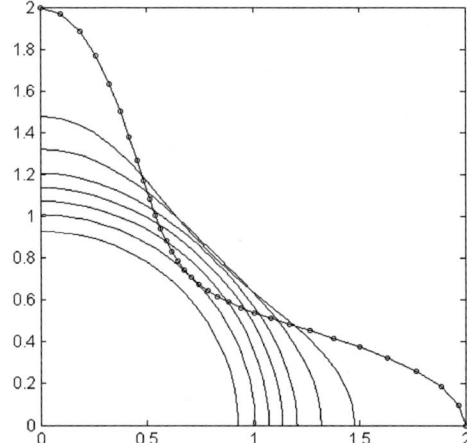

FIGURE 2. Focusing of a non-convex curve in a point.

Instead of the equation of the first order it is possible to apply the equation **of the second** order (it also can be written down in the form similar (2) to the description of evolution of border Γ):

$$\frac{\partial^2 q}{\partial t^2}=W\big(\kappa(x,y)\big).\tag{3}$$

Here $q=q(t,x,y)$ — a vector field, normal to a curve Γ, $W=W\big(\kappa(x,y)\big)$ — the given function determining acceleration of a point $M(x,y)\in\Gamma$ along it. The equation (3) allows taking into account "prehistory" of process and, in the certain sense, is related known to a **canal approximation CCW** (Chester-Chisnell-Whitham) [1] for the description of evolution of front of a shock wave (so-called "the geometric theory of a shock wave").

The set of focusing states thus extends. On fig.3 the example of self-focusing (is given at zero initial velocities $\partial q/\partial t=0$ for all points $M(x,y)\in\Gamma$) in a piece of axis Oy is shown.

Similar asymmetrical focusings are theoretically possible in gas dynamics, for example, so-called many-dimensional motions with the homogeneous deformation [3]-[4], resulting to focusing not only in a point, but also in a segment or a figure limited to an ellipse.

It would be interesting to allocate a class of possible initial conditions within the framework of model (3) $f(x,y)=0$, $\partial q/\partial t|_{t=0}=\psi(x,y)$, $M(x,y)\in\Gamma$, and also a class of functions $W=W\big(\kappa(x,y)\big)$ which result in similar focusings.

501

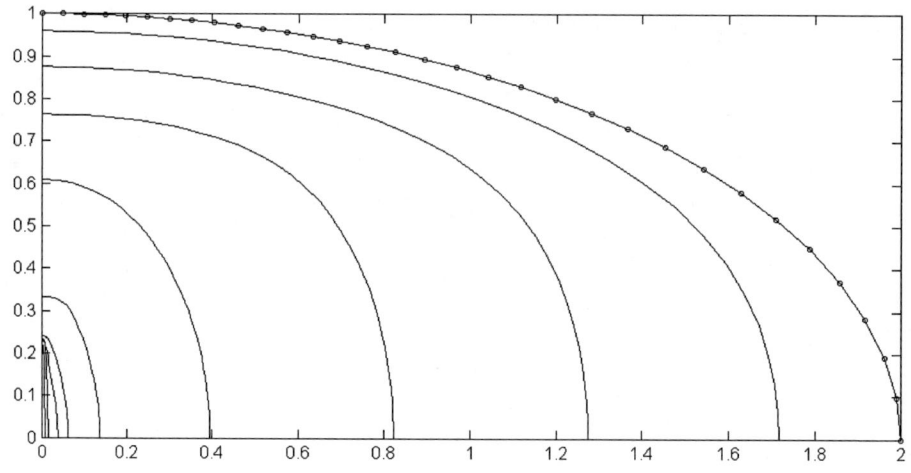

FIGURE 3. Focusing of an ellipse in a segment

On fig. 4 the example of calculation accordance with the model (3) of a cylindrical wave of square section is given. Evolution of border reminds periodic cumulation of a shock wave of the same kind (see for example [6]), but there are also essential differences. We shall notice, that speed of the perturbation propagation along the sides of a square can be considered as certain "sound speed" of the given mathematical model.

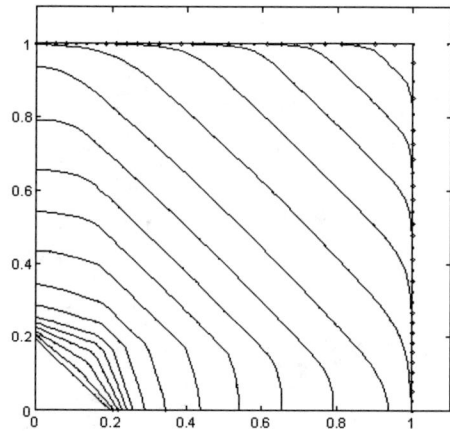

FIGURE 4. Periodic focusing of a square.

REFERENCES

1. G.B.Whitham, *Linear and Nonlinear Waves,* Wiley Interscience, New York, 1974.
2. S.Osher and J.A.Sethian, J. of Comput. Phys., **79**, N1, 12-49 (1988).
3. O.I.Bogojavlenski, *Methods of the qualitative theory of dynamic systems in astrophysics and gas dynamics,* "Science", Moscow, 1980 (in Russian).
4. A.D.Zubov and V.A.Simonenko, "Voprosi Atomnoy Nauki i Tekhniki", Series: Theor. and applied physics, N1 (3), 3-12 (1986) (in Russian).
5. E.I.Zababakhin, Letters in JETPh, **30**, N2, 97-99 (1979) (in Russian).
6. E.I.Zababakhin and I.E.Zababakhin, *The phenomena of unlimited cumulation,* "Science", Moscow, 1988 (in Russian).
7. V.S.Imshennik, Prikl. Matem. i Tekhn. Fizika, N6, 10-19 (1980) (in Russian).

Analytical Statement of a Problem of Axis-symmetrical Reflection of Weak Shock Wave

K.V. Kurmaeva

The Ural State University of Railway Transport, 620034, Ekaterinburg, Russia

Abstract. In this article the possible applied application of the analytical solution of the equation of the velocity potential of a steady axis-symmetrical ideal gas flow- problem of reflection of a weak shock wave from a symmetry axis is considered in case of shallow water.

The majority of physical appearances and processes investigated by gas dynamics are featured with the help of nonlinear differential equations in partial derivatives with singularity [1]. Singularities are understood as the points, stipulated by a physical model, of a non-analyticity of a solution of these equations. At a determination of solutions of the equations of a similar type there are complicated, linked with impossibility of application of Kovalevskaya's theorems. In such cases the problem can be reduced to an analytical solution of a characteristic Cauchy problem as a series on degrees of a characteristic variable. The systematic application of characteristic series to a solution of nonlinear problems in gas dynamics was promoted by A.A. Dorodnitsyn [2] and L.V. Ovsyannikov [3], describe stationary hypersonic and transonic rotationally symmetric flows of gas. The theory of representation of solutions as characteristic series is introduced by A.F. Sidorov [4]. The specificity of statement of a characteristic Cauchy problem has allowed finding the answers to many problems of gas dynamics, which are by M.Yu. Kozmanov [5], V.M. Teshukov [6], S.P. Bautin [7-9] etc. One of possible applications of characteristic series is linked to a determination of solutions of quasi-linear partial differential equations of the second order. For the equation of potential of velocities $\Phi(x_1, x_2, x_3, t)$ in common spatial case

$$\Phi_{tt} + 2\sum_i \Phi_{x_i}\Phi_{x_it} + \sum_{ik}(1-\delta_{ik})\Phi_{x_i}\Phi_{x_k}\Phi_{x_ix_k} - \sum_i\left(\Theta - \Phi_{x_i}^2\right)\Phi_{x_ix_i} = 0, \qquad (1)$$

here $\Phi = \Phi(x_1, x_2, x_3, t)$, ($x_i$ are the Cartesian coordinates), δ_{ik} – Kronecker delta,

$$\Theta = (\gamma - 1)\left(K - \frac{1}{2}\sum_i \Phi_{x_i}^2\right), \qquad (2)$$

where Θ – velocity of sound squared, $K > 0$, $K = const$, γ – ratio of specific heats.

L.V. Ovsyannikov [3] proved an analog of Kovalevskaya's theorem, which justifies the use of series in powers of the distance to the axis of symmetry in the inverse problem of gas flows in axis-symmetric nozzles. The arbitrariness of solution of this problem is one analytical function of the stream-wise variable (gas velocity at the axis of symmetry). The analytical solution with arbitrariness in two functions for the equation of a velocity potential of stationary motion of rotationally

CP849, *Zababakhin Scientific Talks - 2005*,
edited by E. N. Avrorin and V. A. Simonenko

symmetric ideal gas flow was constructed in a neighborhoods of the given point on the symmetry axis as a double series on degrees of distance up to the symmetry axis and its logarithm [11]. Model of shallow water is the equation of a velocity potential of non-stationary motion ideal gas of with the ratio of specific heats $\gamma = 2$. Let's write out the equation (1), (2) at lack of a longitudinal variable in axis-symmetrical a case for shallow water:

$$x_3 = z = 0, \; r = \sqrt{x_1^2 + x_2^2}, \; \Phi = \Phi(r,t), \; \Phi_z = 0,$$

$$\Phi_{tt} + 2\Phi_r\Phi_{rt} + \Phi_r^2\Phi_{rr} - \Theta(\Phi_{rr} + \Phi_r / r) = 0, \; \Theta = K - \Phi_t - \Phi_r^2 / 2. \tag{3}$$

We shall give the analysis of a possibility of reflection of weak shock wave from a symmetry axis as weak shock wave.

Let in some initial instant t there occurs a spreading of shallow water on a horizontal rigid surface. An abscissa axis we designate r – radial variable, an axis of ordinates we shall designate t – time. In the origin of coordinates O we locate an instant t^* – time of arrival of weak shock wave to centre either symmetry axis, or (in a flat case) – collision of this wave with a wall. In this point directly there is an adjoin of two analytical solutions presenting an incident and reflex wave. The domain I features a rest of shallow water, in which the process of a spreading does not happen, i.e. the velocity of motion is equal to zero, and the potential Φ, we assume $\Phi = C_1$.

The domain II characterizes shallow water being in a state of a spreading on the surface. We guess after [12], that this state is analytically featured as a series on fractional degrees of the distance to the axis of symmetry (with a singularity)

$$\Phi(r,t) = \sum_n \Phi_n(t)r^n, \tag{4}$$

where fractional degrees r^n are determined by the set of values $n \in \{0, 2/3, 4/3, 2, 8/3, ...\}$, therefore we shall rewritten series (7) as

$$\sum_n \Phi_n(t)r^n = \eta(r,t) = \sum_{n=0}^{\infty} \eta_{n'}(t)r^{2n'/3}. \tag{5}$$

The domain III characterizes a state of shallow water after reflection of weak shock wave from the symmetry axis and it is supposed rest, as well as in domain I, i.e. the velocity of motion is equal to zero, and for the potential Φ we assume $\Phi = C_2$. Line OA - incident wave has the equation in a case of rest (in the elementary case) in domain I: $R_1(t) = -k_1 t$. Line OB - reflex wave, which has the equation in a case rest of water in domain III in form: $R_2(t) = -k_2 t$.

The constants k_1, k_2 characterize a velocity of motion of weak shock wave. More precisely, k_1 characterizes a velocity of a sound in gas, which adjoins to the symmetry axis before arrival of weak shock wave, k_2 – after its departure.

We assume the following: 1) condition of a rotational symmetry, valid on a symmetry axis $r = 0$, except the point t^*: $\Phi_r = 0$, 2) condition of the adjoin on incident wave OA (adjoin to domain of rest): $\eta(-k_1 t) = C_1$. 3) condition of the adjoin on reflex wave OB: $\eta(-k_2 t) = C_2$. Our purpose is the construction even one flow implementing the given configuration, namely, construction in domains I, II, III, these

505

domains, continuously docked on boundaries $(OA,\ OB)$, excluding, probably, point O. We construct a solution of the equation (3) as a series (5). For this purpose originally we construct a solution of the equation (3) as a series by the whole degrees of distance to the symmetry axis r (without singularities)

$$\Phi(r,t)=\psi(r,t)=\sum_{n=0}^{\infty}\psi_n(t)r^n.\qquad(6)$$

The equation (3) is conversed by multiplying by r^4: $\left(r\psi_r\right)^2\left[\dfrac{3}{2}r^2\psi_{rr}+\dfrac{1}{2}r\psi_r\right]=$

$$=K\left[r^4\psi_{rr}+r^3\psi_r\right]-r^2\psi_t\left[r^2\psi_{rr}+r\psi_r\right]-r^4\psi_{tt}-2\left(r^2\psi_r\right)\left(r^2\psi_{rt}\right).\qquad(7)$$

Collecting formally terms at r^n with $n\geq 0$, after substituting series (6) in to (7), we obtain the n-th the equation of the system for the series for coefficients

$$\sum_{k+l+m=n}k\psi_k l\psi_l\left[\dfrac{3}{2}m-1\right]m\psi_m=K\left(n-2\right)^2\psi_{n-2}-\sum_{k+l=n-2}\left(2k+1\right)\dot\psi_k l\psi_l-\ddot\psi_{n-4}.\qquad(8)$$

Let's remark, that at a deduction (8) was not used of any suppositions about numbers k,l,m,n (being whole etc.). Therefore, in expression (8) for coefficients of series on the whole degrees r, we have the right to put, according to [12], $k=2k'/3,\quad l=2l'/3,\quad m=2m'/3,\quad n=2n'/3$. Then we receive a relation on coefficients of series (5) on fractional degrees for $\psi_n=\psi_{2n'/3}=\eta_{n'}$, and deleting (for brevity) primes

$$\dfrac{2}{3}\sum_{k+l+m=n}k\eta_k l\eta_l\left[m-1\right]m\eta_m=K\left(n-3\right)^2\eta_{n-3}-\sum_{k+l=n-3}\left(2k+1\right)\dot\eta_k l\eta_l-\dfrac{9}{4}\ddot\eta_{n-6}.\qquad(9)$$

From equality (9) we express higher coefficient of series (5)

$$\eta_{n-2}=\dfrac{3}{2(n-2)(n-3)\eta_1^2}\left\{-\dfrac{2}{3}\sum_{\substack{k+l+m=n\\0\leq k,l,m<n-2}}k\eta_k l\eta_l\left[m-1\right]m\eta_m+\right.$$

$$\left.+(n-3)^2\left(K-\dot\eta_0\right)\eta_{n-3}-\sum_{\substack{k+l=n-3\\k\neq0}}\left(2k+1\right)\dot\eta_k l\eta_l-\dfrac{9}{4}\ddot\eta_{n-6}\right\}.\qquad(10)$$

Let's analyses coefficients of series (5), using this expression for higher coefficient.

For $n=4$ $\qquad\eta_2=\dfrac{3}{4\eta_1^2}\left(K\eta_1-\dot\eta_0\eta_1\right)=\dfrac{3}{4\eta_1}\left(K-\dot\eta_0\right)$, for $\eta_1\neq0.$ (11)

Thus, we obtain, that the zero and first coefficients of series (5) are arbitrary functions of a variable t:

$$\eta_0=\eta_0(t),\qquad \eta_1=\eta_1(t).\qquad(12)$$

For $\quad n=5\qquad \eta_3=\dfrac{3}{12\eta_1^2}\left(-\dfrac{4}{3}\cdot2\eta_2\eta_1 2\eta_2+4\left(K-\dot\eta_0\right)\eta_2-3\dot\eta_1\eta_1\right)=-\dfrac{3\dot\eta_1}{4\eta_1}.\qquad(13)$

We see that the denominator of each coefficient of series (5), start from the second, contains some degree of coefficient η_1. For its definition it is necessary to find out the greatest degree η_1, included in the denominator of one of items in the right part of

506

equation (10) for higher coefficients of series (5). For simplicity of the further reasoning we designate through s_n this degree. In order to the sequence $\{s_n\}$, composed from degrees of coefficient η_1, organized an arithmetical progression, it was necessary $s_3 = 3$. Further, in the obtained arithmetical progression we have formula

$$s_n = -3 + 2n. \tag{14}$$

We find the greatest degree η_1 in a denominator of higher coefficient η_{n-2} of series (5) by a system, which each row corresponds to a degree of a denominator of each item from a right part of equation (10), with the usage (14)

$$s_{n-2} = \max \left\{ \begin{array}{l} s_{n-2}^{(1)} = 2 + s_k + s_l + s_m \\ s_{n-2}^{(2)} = 2 + s_{n-3} \\ s_{n-2}^{(3)} = 2 + 1 + s_k + s_l \\ s_{n-2}^{(4)} = 2 + 2 + s_{n-6} \end{array} \right\} = \max \left\{ \begin{array}{l} s_{n-2}^{(1)} = -7 + 2n \\ s_{n-2}^{(2)} = -7 + 2n \\ s_{n-2}^{(3)} = -9 + 2n \\ s_{n-2}^{(4)} = -11 + 2n \end{array} \right\}, \tag{15}$$

Here, and further, the superscript designates number of an item from a right part of expression (10). In each row we add number two because each item contains number η_1^2 in the denominator. From the system (15) it is seen, that the maximum degree of η_1 is contained in the first item of equation (10) and is determined by the formula

$$s_{n-2} = -7 + 2n. \tag{16}$$

For the further analysis of structure describing the coefficients of series (5) we enter weight $w(\eta_n)$ of coefficient η_n and the weight we define by the following rule:

$$w(\eta_n) = n, \quad w(\dot{\eta}_n) = n + 3, \quad w(\eta_k \cdot \eta_l) = w(\eta_k) + w(\eta_l). \tag{17}$$

In particular, $w(\eta_1) = 1$, $w(\dot{\eta}_0) = 3$, $w(\ddot{\eta}_0) = 6$, $w(K - \dot{\eta}_0) = 3$. Besides, we remark, that the difference of weights of numerator and denominator in equation (14) for higher coefficients of series (5) coincides with the number of computed coefficient. Really, the weight of the first item in the bracket of the right parts of equation (10) is equal: $w^{(1)} = w(\eta_k \eta_l \eta_m) = k + l + m = n$.

Weight of the second item: $w^{(2)} = w((K - \dot{\eta}_0)\eta_{n-3}) = 3 + n - 3 = n$.

Weight of the third item: $w^{(3)} = w(\dot{\eta}_k \eta_l) = 3 + k + l = 3 + n - 3 = n$.

Weight of the fourth item: $w^{(4)} = w(\ddot{\eta}_{n-6}) = 2 \cdot 3 + n - 6 = n$. Further, taking into account the weight of the denominator of the right part of equation (10): $w(\eta_1^2) = 2$. We receive the weight of the right part of equation (10): $w = n - 2$. This weight, by the rule entered above, coincides with weight of the left-hand part of equation (10): $w(\eta_{n-2}) = n - 2$. Thus, on the base of the reasoning above, we obtain, that the structure of coefficients of series (5) can be find as follows

$$\eta_n = \frac{B}{A\eta_1^{-3+2n}}, \tag{18}$$

507

Here $A = const$, B - homogeneous polynomial from $(K - \dot{\eta}_0), \ddot{\eta}_0, \dddot{\eta}_0, \ldots, \eta_1^2, \eta_1^3,$ $\ldots$, which weight is equal $w(B) = 3(n-1)$. The found structure of coefficients of series (5) allows us to pose the problem of the adjoin analytical solutions on incident and reflex waves. The posed problem will be solved approximately under condition of a determination of correspondence between arbitrary functions $\eta_0 = \eta_0(t), \eta_1 = \eta_1(t)$. For this purpose, using conditions of the adjoin on incident and reflex waves, we would decide a following system, throwing out terms of series (5) from the fifth

$$\begin{cases} \eta_0 + \eta_1 R_1^{2/3}(t) + \eta_2 R_1^{4/3}(t) + \eta_3 R_1^2(t) + \eta_4 R_1^{8/3}(t) = C_1, \\ \eta_0 + \eta_1 R_2^{2/3}(t) + \eta_2 R_2^{4/3}(t) + \eta_3 R_2^2(t) + \eta_4 R_2^{8/3}(t) = C_2. \end{cases} \tag{19}$$

In case of the third approximation, the system (19) is

$$\begin{cases} \eta_0 + \eta_1 \sqrt[3]{k_1^2} \sqrt[3]{t^2} + \dfrac{3(K - \dot{\eta}_0)}{4\eta_1} \sqrt[3]{k_1^4} \sqrt[3]{t^4} = C_1, \\ \\ \eta_0 + \eta_1 \sqrt[3]{k_2^2} \sqrt[3]{t^2} + \dfrac{3(K - \dot{\eta}_0)}{4\eta_1} \sqrt[3]{k_2^4} \sqrt[3]{t^4} = C_2. \end{cases} \tag{20}$$

From the system (20) we obtain a relation between functions $\eta_0 = \eta_0(t), \eta_1 = \eta_1(t)$

$$\eta_0 \left(\sqrt[3]{k_1^2} + \sqrt[3]{k_2^2} \right) + \eta_1 \sqrt[3]{t^2} \sqrt[3]{k_1^2 k_2^2} = \dfrac{C_2 \sqrt[3]{k_1^4} - C_1 \sqrt[3]{k_2^4}}{\sqrt[3]{k_1^2} - \sqrt[3]{k_2^2}}.$$

We can write: $\eta_0 = \mu \eta_1 \sqrt[3]{t^2} + \varepsilon$, where $\mu = -\dfrac{\sqrt[3]{k_1^2 k_2^2}}{\sqrt[3]{k_1^2} + \sqrt[3]{k_2^2}}$, $\varepsilon = \dfrac{C_2 \sqrt[3]{k_1^4} - C_1 \sqrt[3]{k_2^4}}{\sqrt[3]{k_1^4} - \sqrt[3]{k_2^4}}$

Obtained expression does not determine the functions $\eta_0 = \eta_0(t), \eta_1 = \eta_1(t)$ in an explicit form. But, nevertheless, it determines some relation between them. In particular, the relation between functions $\eta_0 = \eta_0(t), \eta_1 = \eta_1(t)$ contains (for coefficient μ) an estimate, defined by inequality between medial arithmetical and medial geometrical.

Thus, we made the analysis of structure of coefficients of a characteristic series describing flow of gas in a neighborhood of the symmetry axis. The knowledge of this structure of coefficients allows posing the problem about reflection of weak shock wave from the symmetry axis in model of shallow water. Namely, the conditions of adjoining by using series with a singularity (5) on fractional degrees by formulas (10) - (17) allow reducing this problem to a mathematical problem of the analysis of properties of the relevant series. The approximate solution of a problem is reduced to a system of two differential equations with two unknown functions. General problem of choice of these two arbitrary functions is reduced to complicated but only mathematical problem of a determination of individual solution of a system of two nonlinear ordinary differential equations of by infinite order. In contrast from systems of linear differential equations of infinite order, which theory was by designed Gel'fond [13], the theory of such nonlinear systems misses. The determination of individual solution of this nonlinear system would have both only mathematical and applied value: namely solution of the gas-dynamics problem about a possibility of reflection of weak shock wave from the symmetry axis also as weak shock wave .

This work was supported by the RFBR (Grant No. 04-01-00205, 05-01-00217).

REFERENCES

1. L.V. Ovsyannikov, *Lectures on Fundamentals of Gas Dynamics*, Inst. Computer Research, Moscow-Izhevsk, 2003.
2. A.A. Dorodnitsyn, *"Some Cases of Rotationally Symmetric Supersonic Flows of Gas"* in *Collection of Theoretical Works on aerodynamics*, Moscow, 1957, pp. 77-88.
3. L.V. Ovsyannikov, *"Convergence of the Meyer Series for an Axisymmetrical Nozzle. Supplement 1,"* in *Calculation of the Transonic Part of Plane and Axissymmetrical Nozzles with Curved Transition Lines,* edited by E. Martesen, R. von Sengbush, Sib. Otd. Akad. Nauk SSSR, Novosibirsk, 1962, pp. 41- 43.
4. A.F. Sidorov, *Selected Works. Mathematics. Mechanics,* Fizmatlit, Moskow, 2001.
5. M.Yu. Kozmanov, *"A Method of a Solution of Some Boundary Value Problems for Systems of the Quasilinear Equations of the First Orde"* in *Numerical Methods of a Mechanics of a Continuous Media*, 1976, Vol. 7, No. 2, pp. 44-53.
6. V.M. Teshukov, *"Centered Waves in Spatial Gas Flows"* in *Dynamics of Continuous Media*, 1979, No. 39, pp. 102-118.
7. S.P. Bautin, *" Analytical solutions of a problem about motion the piston"* in *Numerical Methods of a Mechanics of a Continuous media*, 1973, Vol. 4, No. 1, pp. 3-15.
8. S.P. Bautin, *"The Characteristic Cauchy Problem for a Quasilinear Analytical System"* in *Differential equations*, 1976, Vol. 12, No. 11, pp. 2052-2063.
9. S.P. Bautin, *"Reduction of Some Problems of Gas Dynamics to Characteristic Cauchy Problem of standard form"* in *Analytical and Numerical Research Techniques of Problems of Mechanics of a Continuous Media,* IMM UNC AN SSSR, Sverdlovsk, 1987, pp. 4-22.
10. S.S. Titov, *Solving of Equations with Singularities in Analytical Scales of Banach Space,* Preprint/UralGAHA, Ekaterinburg, 1999.
11. K.V. Kurmaeva, *"Quasianalytic Transonic Flows of Gas"* in *Theses of the Reports XX of the All-Russian School – Seminar "Analytical Methods and Optimization of Processes in Mechanics of Fluid and Gas "*, 2004, pp. 48-49.
12. K.V. Kurmaeva, *"A Cauchy Problem for flows of gas with the data on a symmetry axis"* in *Problems of theoretical and applied mathematics: proceedings of XXXVI Regional Youth Conference,* URO RAN, Ekaterinburg, 2005, pp.151-157.
13. A.O. Gel'fond, *Calculus of Finite Differences*, Nauka, Moscow, 1967.

Vortex-in-Cell Method for Plane Flows of Compressible, Nonuniform and Relaxing Gases

Yu.N. Grigoryev

Institute of Computational Technologies Siberian Branch of RAS , 630090, Novosibirsk, Russia

Abstract. A new vortex-in-cell method for a plane compressible flows is presented. The method is easily spreaded for calculation of vortical flows of nonuniform fluids and relaxing gases.

Keywords: Vortex-in-cell method, compressible gas flow, nonuniform fluid, relaxing gas.
PACS: 47.11.-j + 47.15.ki + 47.32.C-

1. INTRODUCTION

The pure Lagrangian method of discrete (point) vortices was originally realized in [1] for numerical modeling of vortex sheet instability in an ideal incompressible fluid. The first combined Eulerian – Lagrangian vortex-in-cell (VIC) algorithm for the plane flows of incompressible fluid was introduced at the 70-s in [2, 3]. But despite of successful calculations of fine vortex structure [2], VIC-methods were not widely adopted in the next years. In present time some new Lagrangian methods such as SPH (Smoothed Particle Hydrodynamics) [4, 5] have appeared in computational fluid mechanics. They stimulated a progress of known Lagrangian vortex methods. In recent works [6, 7] such an algorithm was developed for plane compressible flows. At the same time the Lagrangian vortex methods are not economical ones. In this case corresponding volume of computations is estimated as $O(N^2)$ (N is a number of Lagrangian particles). From the point of view VIC-method [2, 3, 8] is economical .algorithm and an estimation of its computational cost is $O(N \ln N)$. Therefore a development of VIC-method for compressible flows is of interest both for computational mathematics and for applications to problems like considered in [6, 7].

2. METHOD DESCRIPTION

For constructing of particle-in-cell algorithms it is crucial to choose a splitting scheme of original problem in Eulerian and Lagrangian subsystems and a set of features carried by model particles [9]. We start with the system of gas dynamics equations in the form [10]

$$\frac{\partial \rho}{\partial t} + \nabla(\rho \mathbf{u}) = 0, \tag{1}$$

CP849, *Zababakhin Scientific Talks - 2005*,
edited by E. N. Avrorin and V. A. Simonenko
© 2006 American Institute of Physics 0-7354-0345-7/06/$23.00

$$\frac{\partial \mathbf{u}}{\partial t} + \nabla(\frac{\mathbf{u}^2}{2}) - \mathbf{u} \times \nabla \times \mathbf{u} = -\frac{1}{\rho}\nabla p, \tag{2}$$

$$\frac{\partial \varepsilon}{\partial t} + \nabla(\varepsilon \mathbf{u}) = -(\gamma - 1)\varepsilon \nabla \mathbf{u}. \tag{3}$$

$$p = (\gamma - 1)\varepsilon. \tag{4}$$

Here ρ is gas density a, $\mathbf{u} = (u, v)$ is a velocity vector, p is a pressure, ε is a specific internal energy, γ is a ratio of specific heats, ∇ is a gradient operator on the plane $\mathbf{r} = (x, y)$. Since 2D-flows are considered, vorticity has only a single component

$$\omega = \nabla \times \mathbf{u} = \text{rot}\mathbf{u} = \frac{\partial v}{\partial x} - \frac{\partial u}{\partial y} \equiv v_x - u_y, \tag{5}$$

that is perpendicular to flow plane. Momentum equation (2) in the Lamb's form it is convenient to transform into two equations for scalars ω and $\theta = \nabla \mathbf{u} = \text{div}\mathbf{u}$. The function θ we define as a dilatation. Applying to Eq. (2) curl and divergence operations we have

$$\frac{\partial \omega}{\partial t} + (\mathbf{u}\nabla)\omega = -\omega\theta + \frac{1}{\rho^2}\nabla\rho \times \nabla p. \tag{6}$$

$$\frac{\partial \theta}{\partial t} + (\mathbf{u}\nabla)\theta + (\nabla\mathbf{u}):(\nabla)^T = -\frac{1}{\rho}\Delta p + \frac{1}{\rho^2}\nabla\rho\nabla p, \tag{7}$$

where $(\nabla\mathbf{u}):(\nabla)^T = u_x^2 + v_y^2 + 2u_y v_x$. Term $\frac{1}{\rho^2}\nabla\rho \times \nabla p$ is a producer of vorticity by baroclinic mechanism [10] and can be essential behind the front of curved shock waves or under development of the Richtmayer – Meshkov instability.

In compressible flows the velocity field is decomposed in a sum of solenoidal and potential components [10]

$$\mathbf{u}(x, y) = \mathbf{u}_s(x, y) + \mathbf{u}_g(x, y). \tag{8}$$

Solenoidal component $\mathbf{u}_s$ is calculated through vector-potential $\mathbf{A} = \{0, 0, \psi(x, y)\}$, the single z-coordinate of which is a stream function $\psi(x, y)$. Potential component $\mathbf{u}_g$ is evaluated through scalar potential $\varphi(x, y)$. We have

$$\mathbf{u}_s = \nabla \times \mathbf{A} = \{\psi_y, -\psi_x\}, \qquad \mathbf{u}_g = \nabla\varphi.$$

Taking into account these expressions and applying to Eq. (8) div -and rot -operations one can obtain the next boundary problems for functions $\varphi(x, y)$ and $\psi(x, y)$

$$\Delta\varphi = \theta, \qquad \frac{\partial\varphi}{\partial n} = \mathbf{un}\,|_{\partial\Omega}; \qquad \Delta\psi = -\omega, \qquad \frac{\partial\psi}{\partial n} = \mathbf{us}\,|_{\partial\Omega}\,. \tag{9}$$

Here Δ is the Laplacian, $\mathbf{n}, \mathbf{s}$ are unit vectors of outer normal and tangent to the boundary $\partial\Omega$ of current domain. We transform the l.h.s in Eqs. (6), (7) to divergent form and in right side exclude a pressure p with using of state equation (4). As a result we bring out the next system of equations

$$\frac{\partial\rho}{\partial t} + \nabla(\rho\mathbf{u}) = 0, \tag{10}$$

$$\frac{\partial\omega}{\partial t} + \nabla(\omega\mathbf{u}) = \frac{\gamma-1}{\rho^2}(\rho_x\varepsilon_y - \rho_y\varepsilon_x),$$

$$\frac{\partial\theta}{\partial t} + \nabla(\theta\mathbf{u}) = -2(u_y v_x - u_x v_y) + \frac{\gamma-1}{\rho^2}(\rho_x\varepsilon_x + \rho_y\varepsilon_y) - \frac{\gamma-1}{\rho}\Delta\varepsilon,$$

$$\frac{\partial\varepsilon}{\partial t} + \nabla(\varepsilon\mathbf{u}) = -(\gamma-1)\varepsilon\theta.$$

To the system [10] a general approach for constructing of particle-in-cell methods [9] is applied in apparent manner. On the time step τ splitting is introduced. The system of equations of the Lagrangian stage takes the form

$$\frac{\partial\tilde\rho_1}{\partial t} + \nabla(\tilde\rho_1\tilde{\mathbf{u}}_1) = 0, \qquad \tilde\rho_1(t) = \rho(t); \frac{\partial\tilde\omega_1}{\partial t} + \nabla(\tilde\omega_1\tilde{\mathbf{u}}_1) = 0, \qquad \tilde\omega_1(t) = \omega(t); \tag{11}$$

$$\frac{\partial\tilde\theta_1}{\partial t} + \nabla(\tilde\theta_1\tilde{\mathbf{u}}_1) = 0, \qquad \tilde\theta_1(t) = \theta(t); \qquad \frac{\partial\tilde\varepsilon_1}{\partial t} + \nabla(\tilde\varepsilon_1\tilde{\mathbf{u}}_1) = 0, \qquad \tilde\varepsilon_1(t) = \varepsilon(t);$$

Correspondingly, for Eulerian stage we obtain

$$\frac{\partial\tilde\rho_2}{\partial t} = 0, \qquad \tilde\rho_2(t) = \tilde\rho_1(t+\tau); \frac{\partial\tilde\omega_2}{\partial t} = \frac{\gamma-1}{\tilde\rho_1^2}\tilde\rho_{1x}\tilde\varepsilon_{1y} - \tilde\rho_{1y}\tilde\varepsilon_{1x}, \qquad \tilde\omega_2(t) = \tilde\omega_1(t+\tau); \tag{12}$$

$$\frac{\partial\tilde\theta_2}{\partial t} = -2(\tilde u_{1y}\tilde v_{1x} - \tilde u_{1x}\tilde v_{1y}) + \frac{\gamma-1}{\tilde\rho_1^2}(\tilde\rho_{1x}\tilde\varepsilon_{1x} + \tilde\rho_{1y}\tilde\varepsilon_{1y}) - \frac{\gamma-1}{\tilde\rho_1}\Delta\tilde\varepsilon_1, \qquad \tilde\theta_2(t) = \tilde\theta_1(t+\tau);$$

$$\frac{\partial\tilde\varepsilon_2}{\partial t} = -(\gamma-1)\tilde\varepsilon_1\tilde\theta_1, \qquad \tilde\varepsilon_2(t) = \tilde\varepsilon_1(t+\tau).$$

The equations for Eulerian stage are comleted by equations for vortical and potential components of velocity field

$$\Delta \tilde{\varphi}_2 = \tilde{\theta}_2, \quad \frac{\partial \tilde{\varphi}_2}{\partial n} = \mathbf{un} \,|_{\partial\Omega}, \quad u_{g2} = \tilde{\varphi}_{2x}, v_{g2} = \tilde{\varphi}_{2y};$$

$$\Delta \tilde{\psi}_2 = -\tilde{\omega}_2, \quad \frac{\partial \tilde{\psi}_2}{\partial n} = \mathbf{us} \,|_{\partial\Omega}, \quad u_{s2} = \tilde{\psi}_{2y}, v_{s2} = -\tilde{\psi}_{2x}.$$

The splitting pick out two groups of physical processes. Eqs. (11) describe a transport of the features characterized a local state of compressible gas — density, vorticity, dilatation and internal energy by velocity field (8). The system (12) reflects the processes of a local generation of the carried features.

Now we present briefly a realization of the Lagrangian stage with using of standart scheme of particle-in-cell method [9]. The system (11) in vector form is written as follows

$$\frac{\partial \mathbf{q}}{\partial t} + \nabla(\mathbf{u}\mathbf{q}) = 0. \tag{14}$$

Solution of (14) represents by the next interpolation formula

$$\mathbf{q}(\mathbf{r}) = \sum_{j=1}^{N} \mathbf{Q}_j R(\mathbf{r}, \mathbf{r}_j(t)). \tag{15}$$

on the Lagrangian mesh of particles which in this case we denote as *vortices*. In (15) vectors $\mathbf{Q}_j$ of the full (summary) features carried by particles are defined as sets

$$\mathbf{Q}_j = \{m_j, \Gamma_j, \Theta_j, E_j\}, \qquad j = 1, ..., N. \tag{16}$$

Here m_j is a mass, Γ_j is a circulation, Θ_j is a dilatation, E_j is an internal energy of j-particle. The function $R(\mathbf{r}, \mathbf{r}_j(t))$ is a kernel of model particle with the next universal properties

$$R(\mathbf{r}, \mathbf{p}) = R(\mathbf{p}, \mathbf{r}), \quad \frac{\partial R(\mathbf{r}, \mathbf{p})}{\partial \mathbf{r}} = -\frac{\partial R(\mathbf{r}, \mathbf{p})}{\partial \mathbf{p}}, \quad \int_{\Omega} R(\mathbf{r}, \mathbf{p}) d\mathbf{r} = 1.$$

Let $\varphi(\mathbf{r})$ is an arbitrary smooth finite function in Ω. Substitution of (15) in (14) and integration with a weight of $\varphi(\mathbf{r})$ over Ω domain give us

$$\sum_{j=1}^{N} \mathbf{Q}_j \int_{\Omega} R(\mathbf{r}, \mathbf{r}_j) \nabla \varphi(\mathbf{r}_j) [\frac{d\mathbf{r}_j}{dt} - \mathbf{u}(\mathbf{r}_j)] d\mathbf{r}_j = 0.$$

Because of an arbitrarinessy of functions $\varphi(\mathbf{r})$ и $R(\mathbf{r}, \mathbf{r}_j)$ integrals in this equality have to vanish identically. For this it is necessary to carry out the conditions

$$\frac{d\mathbf{r}_j}{dt} - \mathbf{u}(\mathbf{r}_j) = 0, \qquad j = 1, ..., N. \tag{17}$$

513

This a dynamical system describes a motion of model vortices. Integrating Eqs. (17) along time step τ we by virtue of Eqs. (14), (15) reproduce approximately space-time evolution of a solution of system (11).

After ending of the Lagrangian stage it is necessary to project obtained solution (15) from the points $\{\mathbf{r}_j^{p+1}\}_{j=1}^{N}$ into nodes of Eulerian mesh. Let a set $\{\delta_\alpha\}_{j=1}^{M}$ is a regular Cartesian partition of a flow domain Ω into cells of which centers are the Eulerian mesh nodes. Under interpolation it is needed to fulfill the integral conservation laws of carried features that are written in the form

$$\sum_{j=1}^{N} \mathbf{Q}_j \int_\Omega R(\mathbf{r}, \mathbf{r}_j) q d\mathbf{r} = \sum_{j=1}^{N} \mathbf{Q}_j = \sum_{\alpha=1}^{M} \tilde{\mathbf{q}}_\alpha |\delta_\alpha|.$$

Here $|\delta_\alpha|$ is a square of the Cartesian mesh with index α. The quantities $\tilde{q}_\alpha$ are piecewise-constant densities of carried features per unit square. It is clear that conservation laws will be take place if values of mesh densities are calculated according to formulas

$$\tilde{\mathbf{q}}_\alpha = \frac{1}{|\delta_\alpha|} \int_{\delta_\alpha} \mathbf{q} d\mathbf{r} = \frac{1}{|\delta_\alpha|} \sum_{j=1}^{N} \mathbf{Q}_j \int_{\delta_\alpha} R(\mathbf{r}, \mathbf{r}_j) d\mathbf{r} = \sum_{j=1}^{N} \mathbf{Q}_j \overline{R}(\mathbf{r}, \mathbf{r}_j), \quad \alpha = 1, ..., M.$$

In a result we obtain a set of mesh functions on ($p+1$)-time step

$$\mathbf{q}_\alpha = \{\tilde{\rho}_{1\alpha}^{p+1}, \tilde{\omega}_{1\alpha}^{p+1}, \tilde{\theta}_{1\alpha}^{p+1} \tilde{\varepsilon}_{1\alpha}^{p+1}\}, \quad \alpha = 1, ..., M,$$

that are the initial data for Eulerian stage calculated on the base of some finite-difference approximation of Eqs. (12). As it is seen from (12) the mesh gas density $\{\tilde{\rho}_{2\alpha}\}$ stays constant along Eulerian stage. When calculating of dilatation θ one need to know values of mesh velocity function. In particular, we can use velocity values calculated at ending of Eulerian stage on previous time step. By obtained voricity and dilatation mesh functions through difference approximations of the boundary problems (9)

$$\Delta_h \tilde{\varphi}_{2\alpha}^{p+1} = \tilde{\theta}_{2\alpha}^{p+1}, \quad l_h \tilde{\varphi}_{2\alpha}^{p+1} = \mathbf{u}_{nh}; \quad \Delta_h \tilde{\psi}_{2\alpha}^{p+1} = -\tilde{\omega}_{2\alpha}^{p+1}, \quad l_h \tilde{\psi}_{2\alpha}^{p+1} = \mathbf{u}_{sh},$$

mesh potential and stream functions $\{\tilde{\varphi}_{2\alpha}^{p+1}, \tilde{\psi}_{2\alpha}^{p+1}\}$ are found. Then mesh velocity function $\mathbf{u}_\alpha^{p+1}$ is calculated. Before the beginning of the next Lagrangian step it is necessary to find new values of vectors $\mathbf{Q}_j$ and to interpolate velocity function $\mathbf{u}_\alpha^{p+1}$ into locations of particles. Description of these procedures are given in [9].

The developed VIC-algorithm is easily applied for calculations of flow problems with small density variations. Such flows arise, for example, under free convection in a gravity field. In general case similar flows are describe by the Oberbeck-Boussinesk system [11]. Furthermore presented method is also generalized for solving of vortrical compressible flow problems accompanied by some relaxation process. This can be

chemical reaction, dissociation-recombination, thermalization of inner modes in molecular gases [12].

REFERENCES

1. L. Rosenhead, *Proc. Roy. Soc. London A*. 134, 170-192 (1931).
2. J.P. Christiansen, *J. Comp. Phys*. 13, 363 – 379 (1973).
3. G.R. Baker, *J. Comp. Phys.*.31, 76 – 95 (1979).
4. R.A. Gingold and J.J. Monoghan, *Mon. Not. R. Astron*. Soc. 181, 375 – 387(1970).
5. J.J. Monoghan, *Comp. Phys. Rep*. 3, 71 – 85 (1985).
6. J.D. Eldredge, T Colonius and A.Leonard, *J. Comp. Phys*.179, 371 – 399 (2002).
7. J.D. Eldredge, T Colonius and A.Leonard, *J. Turb*. 3, 1 -10 (2002).
8. A.Leonard, *J. Comp. Phys.*.37, 289-335(1980).
9. Yu.N. Grigoryev, V.A. Vshivkov and M.P. Fedoruk, Numerical "Partical - in – Cell" methods. Theory and Applications. Utreht-Boston: VSP, 2002.
10. G.K. Batchelor, An Introduction in Fluid Mechanics. Cambridge: Univ. Press, 1967.
11. D.D. Joseph, Stability of Fluid Motions .Berlin-Heidelberg-New York: Springer-Verlag, 1976.
12. .S.A.Losev. Gasdynamical Lasers. Moscow: FM, 1977. (in Russian).

Author Index

A

Abakumov, A. I., 51
Aduev, B. P., 129, 196
Akimova, I. V., 242
Akunets, A. A., 363
Alekseev, A. V., 133, 174
Alexeev, N. S., 57
Aluker, E. D., 129, 196
Andrianov, V. P., 237
Andriyash, A. V., 35
Andryushin, V. V., 35
Antsiferova, E. V., 121
Artamonov, M. V., 453
Avdeyev, P. A., 453
Averbukh, V. L., 481
Avramenko, M. I., 329

B

Bachurina, I. V., 213
Bagavetdinov, N. G., 191
Bakhrakh, S. M., 317, 453
Baksht, R. B., 262
Bakulina, E. A., 453
Beilis, I., 262
Belenovskii, Yu. A., 206
Belokurov, G. M., 129, 196
Belyaev, V. S., 237
Bezrukova, I. Yu., 317, 453
Bliznetsov, M. V., 323, 341
Bodrenko, S. I., 179
Bogdanov, V. V., 121
Borisenko, N. G., 237, 242, 363
Boriskin, A. S., 273, 278, 290, 297
Borlyaev, V. V., 453
Borshchevsky, A. O., 349
Brodsky, A. Ya., 273
Bronina, T. N., 492
Burtsev, V. V., 179
Bushmelev, P. S., 179

C

Chefonov, O. V., 35
Chekmarev, A. M., 237
Chemagina, I. V., 174

C

Cherevko, A. A., 476
Chirin, N. A., 363
Chizhkov, M. N., 35, 251
Chizhkova, N. E., 82, 487
Chupakhin, A. P., 476

D

Demchenko, N. N., 237
Demidov, V. A., 273, 278, 290, 297
Derebenko, E. V., 121
Devyatkin, I. V., 51
Dimant, E. M., 290
Dimant, Ye. M., 273
Dmitrov, D. A., 35
Domnichev, V. V., 179
Dorogotovtsev, V. M., 363
Dremov, V. V., 368, 380
Duday, P. V., 278
Dzyobark, S. V., 257

F

Fedorenko, A. G., 51
Fedorov, A. V., 311
Fel'dman, V. I., 444
Filin, V. P., 133, 174, 196
Filippov, A. V., 273, 278
Filonenko, A. D., 257
Fomin, V. M., 311
Fridlyander, I. N., 416
Frolov, Yu., 164
Frolova, N. Yu., 62

G

Gadzhiev, A. D., 465
Gagarin, A. L., 133
Garanin, S. G., 237
Garmasheva, N. V., 133, 174
Gerasimov, A. V., 57
Glazyrin, I. V., 262
Glazyrin, V. P., 421
Golosov, S. N., 273, 278, 290, 297

517

Shukailo, V. P., 133
Shushlebin, A. N., 329
Shuvalova, E. V., 453
Sil'vestrov, V., 439
Simonenko, V. A., 3, 82, 487
Sin'ko, G. V., 375
Sirenko, A. V., 154
Skovpen, A. V., 465
Slobodenjukov, V. M., 191, 349
Smirnov, N. A., 375
Smirnov, S. P., 150
Sokolovskaya, V. L., 89
Solovyev, V. P., 51
Sotskov, E. A., 323, 341
Spiridonov, V. F., 453
Stadnik, A .L., 179
Starodubov, S. V., 453
Stavrietsky, V. I., 393
Storozhenko, P. A., 363
Stryakhnin, V. L., 133
Suchkov, V. A., 470
Svitsin, R. A., 363
Syrunin, M. A., 51

T

Taibinov, N. P., 174
Talala, K. A., 268
Tanakov, Z. V., 179, 185
Taradai, I. Yu., 453
Taranik, M. V., 140
Tarzhanov, V. I., 393, 406, 433
Tatsenko, O. M., 273, 278
Taybinov, N. P., 133
Telichko, I. V., 393, 433
Tikhonova, A. P., 453
Timakova, M. S., 251
Tkachev, O. V., 191
Tochilina, L. V., 323, 341
Treshalin, S. M., 51
Tsarenkova, C. K., 140
Tselinsky, I. V., 213
Tsiberev, K. V., 453

U

Ugodenko, A. A., 35
Ushakova, O. V., 492

Ustinenko, V. A., 323, 341

V

Valiev, R. Zh., 251
Vasil'ev, M. L., 179
Vaziev, E. M., 465
Velichko, S. V., 453
Velichutin, N. V., 110
Veremeenko, I. L., 94
Vershinin, A. V., 206
Vihklyaev, D. A., 35
Vildanov, V. G., 349, 380
Vinogradov, V. I., 237
Vlasov, Yu. V., 273, 278, 290
Vlasova, M. A., 179
Volkov, N. B., 268
Volodina, N. A., 453
Vorobyov, A. V., 393
Vorontsov, A. S., 363

Y

Yakovlev, I. V., 305
Yakunin, A. K., 399, 406
Yalovets, A. P., 268
Yanenko, V. A., 290
Yanilkin, Yu. V., 179
Yelkin, V. M., 406
Yunoshev, A., 439

Z

Zabrodin, A. V., 363
Zahikin, V. T., 380
Zaikin, V. T., 191
Zapysov, A. L., 35
Zarko, V. E., 159
Zel'dovich, V. I., 62
Zhabitskii, S. K., 179
Zherebtsov, A. L., 140, 201
Zhugin, Yu. N., 380
Zhukov, V. T., 221
Zubarev, N. M., 99
Zubov, A. D., 89, 499
Zuev, Yu. N., 35